AF411300

Novel Approaches for Structural Health Monitoring

Novel Approaches for Structural Health Monitoring

Editor

Cecilia Surace

MDPI • Basel • Beijing • Wuhan • Barcelona • Belgrade • Manchester • Tokyo • Cluj • Tianjin

Editor
Cecilia Surace
Department of Structural,
Building and Geotechnical
Engineering
Italy

Editorial Office
MDPI
St. Alban-Anlage 66
4052 Basel, Switzerland

This is a reprint of articles from the Special Issue published online in the open access journal *Applied Sciences* (ISSN 2076-3417) (available at: http://www.mdpi.com).

For citation purposes, cite each article independently as indicated on the article page online and as indicated below:

LastName, A.A.; LastName, B.B.; LastName, C.C. Article Title. *Journal Name* **Year**, *Volume Number*, Page Range.

ISBN 978-3-0365-2404-7 (Hbk)
ISBN 978-3-0365-2405-4 (PDF)

Cover image courtesy of Cecilia Surace

Contents

About the Editor

Prof. Dr. Cecilia Surace is an associate professor of structural mechanics at Politecnico di Torino, Turin, Italy, since 2014. She is also a member of the committee for doctoral students in aerospace engineering and head of the laboratory of bio-inspired nanomechanics in the Department of Structural, Building, and Geotechnical Engineering. Furthermore, she has been active in the fields of structural dynamics and structural health monitoring since the early 1990s, with more than 100 publications on these topics.

Preface to "Novel Approaches for Structural Health Monitoring"

This Special Issue on novel approaches for structural health monitoring aims at reporting the latest and more interesting improvements in the field of vibration-based and machine learning-based structural health monitoring. It collects 15 published papers, each of them making a relevant contribution on a specific related fundamental aspect. All the involved authors are gratefully acknowledged for their original contributions. Special thanks go to Dr. Marco Civera and Mr. Davide Martucci for their support in the preparation of this book.

Cecilia Surace
Editor

applied sciences

Editorial

Special Issue on Novel Approaches for Structural Health Monitoring

Cecilia Surace

Department of Structural, Geotechnical and Building Engineering, Politecnico di Torino, 10129 Turin, Italy; cecilia.surace@polito.it

Citation: Surace, C. Special Issue on Novel Approaches for Structural Health Monitoring. *Appl. Sci.* **2021**, *11*, 7210. https://doi.org/10.3390/app11167210

Received: 27 July 2021
Accepted: 1 August 2021
Published: 5 August 2021

Publisher's Note: MDPI stays neutral with regard to jurisdictional claims in published maps and institutional affiliations.

1. Introduction

Crucial mechanical systems and civil structures or infrastructures, such as bridges, railways, buildings, wind turbines, aeroplanes and more are subjected during their lifetime to natural deterioration of their structural integrity. This is due to several concomitant factors (i.e., environmental conditions, operating loads, etc.). Worsening of this degradation or accidental events lead to an impairment of the functionality and a severe decrease of the safety level, while extremely critical interruptions of service and catastrophic collapses with possible loss of life may occur. Moreover, taking into consideration also the economic relevance of these systems, sustainable management requires the implementation of specific maintenance strategies to reduce the related repair costs.

Due to all these reasons, the last decades have seen a growing interest in the field of structural health monitoring (SHM), involving multidisciplinary academics and practitioners aimed to develop effective technologies, procedures and algorithms for damage diagnosis. Continuous scientific advancement and technological evolution provide the means to face old and new challenges. These are overcome by designing new approaches, defining new damage-sensitive features, and enhancing the already-existing devices for this scope. However, although in recent years numerous interesting findings and applications have been made, further efforts are still needed.

In this context, the "Special Issue on Novel Approaches for Structural Health Monitoring" in *Applied Sciences* collects 15 published papers, each of them a relevant contribution on a specific related fundamental aspect. These are briefly reviewed here for the reader's convenience.

2. The Extreme Function Theory for Damage Detection: An Application to Civil and Aerospace Structures

The research reported in [1] describes an approach, based on the extreme function theory (EFT), for mode shape-based damage detection. Applications to both aerospace and civil structures are proposed as (numerical and experimental) case studies. The EFT can be considered as the extension of the classic extreme value theory (EVT) to whole functions—in this specific case, to mode shapes, extracted from the target system and benchmarked against the ones identified from the pristine baseline. More into detail, the continuous mode shapes are defined by means of Gaussian process regression (GPR), applied over a limited set of data points. These correspond to the output channels distributed over the structure under investigation. To compensate for the shortage of training data, a data augmentation strategy is included as well. This is intended to deal with the practical issue of data scarcity, which often hampers the applicability of machine learning approaches for SHM. The performances of the procedure are addressed in terms of true positives, true negatives, and type I and II errors. The rationale is to investigate not only the damage detection capabilities of the algorithm but also its robustness to false alarms. The robustness to artificially added measurement noise is tested as well on several finite element models—a simple beam with several boundary conditions, the spar of a prototype high aspect ratio wing, and a shear-type 3-stories frame structure. Finally, a direct comparison between EFT-based damage detection and the EVT-based alternative is reported, considering in the latter case the same algorithm applied pointwise only at the output channels. This last study shows a statistically significant reduction of false alarms.

3. Full-Field Strain Reconstruction Using Uniaxial Strain Measurements: Application to Damage Detection

The accurate full-field reconstruction of the strain and displacement fields of a structure using a set of uniaxial strain measurements is the prime focus of the work presented in [2]. The use of uniaxial strain measurements, as obtained from a fibre-optic sensor, for two-dimensional displacement field reconstruction leads to difficulties due to insufficient strain information at a point and can potentially lead to a breakdown of the reconstruction procedure. This work proposes a solution to the problem based on the inverse finite element method (iFEM), combined with a pre-processing step for strain smoothing using the smoothing element analysis (SEA) approach. The iFEM is a variationally-based approach, where the structural domain is discretized using finite elements, and the displacement field is reconstructed by minimizing an error functional representing the least-squares error between analytic and experimental strain measures. The effect of sensor position and orientation on reconstruction results is investigated and used to identify effective strain-sensor patterns ensuring reconstruction accuracy. The iFEM performance is evaluated numerically using the problem of a thin plate, subject to several internal damage scenarios. Damage detection capability depends on an accurate reconstruction of the local internal strain perturbations, and the iFEM reconstructed strain fields successfully revealed the damage locations as regions of strain concentration containing information regarding damage size, position, and orientation. Additionally, a sensitivity analysis demonstrates the proposed methodology's robustness to measurement noise, although it hinted at difficulties in detecting small-sized damages. The main achievement of the paper is in showing the potential of strain measurements based on fibre optic sensors for practical SHM applications.

4. Rail Diagnostics Based on Ultrasonic Guided Waves: An Overview

The diagnostic of rail tracks damage conditions is the topic investigated in [3]. The authors give an extensive overview from the general context to the latest innovations, focusing on different non-destructive testing (NDT) methodologies. The authors describe the implementation and analysis of the performance of inspection strategies based on ultrasonic guided waves (UGW). Firstly, a detailed introduction about various types of rail track defects and different rail diagnostic techniques developed in the last decades is reported. Then, with a proper distinction between the main classes of diagnostic systems (on-board, land-based) and types of approaches (active, passive), the specific architecture and data processing approaches related to UGW methodologies are presented. Next, comprehensive sections are explicitly dedicated to the latest land-based systems, dealing with core systems of ultrasonic broken rail detector, early rail defect detection capability, mixed evolved techniques and other commercial projects. Finally, the performance analyses of all the aforementioned rail diagnostics are shown, followed by a discussion about their advantages and disadvantages. The authors additionally outline several potential future developments based on the limits or open issues of current, state-of-the-art ultrasonic systems. For its completeness and detailed considerations, the works presented represents an effective baseline for new researches and further improvements.

5. Mooring-Failure Monitoring of Submerged Floating Tunnel Using Deep Neural Network

A study of structural health monitoring of submerged floating tunnels (SFT) is presented in [4]. This kind of infrastructure presents many advantages in comparison with suspended or floating bridges since these latter solutions are very responsive to environmental solicitations (wave, seismic actions, etc...). Moreover, SFTs naturally allow sea crossing without any physical impediment. Nevertheless, due to their submerged condition, any failure occurrence of the balancing mooring can be catastrophic if not promptly repaired. Therefore, the monitoring and maintenance of these infrastructures result challenging and expensive. Thus, the authors propose a deep learning-based algorithm, able to overcome all the limitations of traditional maintenance strategies such as the need for visual inspection, the requirement of numerous sensors, or the prior knowledge of structural parameters. In particular, a deep neural network (DNN) has been implemented to analyse lateral and vertical displacements measured through different accelerometers settings under different wave conditions. Numerical simulations have been initially used to select the best features of the neural network, as its architecture (i.e., the number of hidden layers and neurons), the activation function, and the loss function. The different investigations show that a very high rate of accuracy for correct detection can be reached with a reasonably low number of sensors (from three to seven if opportunely located). These results have been confirmed under different wave conditions and by comparing them with experimental results. The work presented here, thus, enhances the feasibility perspective in SFT spread, since it deploys an effective and attainable method for a valid monitoring strategy.

6. Damping of Beam Vibrations Using Tuned Particles Impact Damper

The work presented in [5] introduces an innovative tuned particle impact damper (TPID). The device is proposed as an improvement of the well-known particle impact damper vibration adsorbers, with the main feature of rapid tuning its damping parameters. The dissipation of the kinetic energy occurs through friction and viscous and inelastic impacts of the particles among themselves and the container. The governing mechanism of damping depends on several factors, such as geometry, material, mass and stiffness of the grains, coefficient of restitution, the composition of the container, filling ratio, particles degradation, and temperature increase. These are investigated by experimental verifications. The conclusions show that the most relevant parameters are the volume ratio between the grains and the container. The authors propose a simple yet highly efficient design: the grains are encapsulated into an inflatable balloon provided by an external valve, allowing the easy and quick tuning of the available volume by varying the internal pressure. Numerous tests and experiments assess the characteristics of the novel device. Results show that damping performances increase when dealing with a higher mass of grains and higher balloon volume. Regarding the composition of the grains, a plastic material is found as the best solution among the options investigated by the authors. Indeed, the selected material proved to be very responsive to volume variations while ensuring high vibration attenuation. In conclusion, this study lays the basis for a truly innovative, low-cost and ecologically sustainable device, enhancing the opportunities of semi-active vibration attenuation strategy thanks to its tunable damping feature.

7. Health and Structural Integrity of Monitoring Systems: The Case Study of Pressurized Pipelines

The authors of [6] present the physical principles of structural health monitoring related to the specific topic of pressurised pipelines, giving a strong overview supported by multiannual experience in the field and real results coming from operating systems. The main issues causing damages in pressurised pipelines (which often result in catastrophic failures) are traced back to unpredictable variations in operating conditions, such as changes in internal pressures, landslides, subsidence of foundation ground soil, or steel corrosion. Moreover, these root causes involve secondary effects like induced vibrations and additional stresses, often interacting simultaneously. Since the unpredictability of these phenomena implies the impossibility to provide adequate design solutions, the authors individuate in long-term real-time monitoring systems the only feasible tool to overcome these issues. In particular, stress and strain measurements are envisaged as fundamental to determine high-risk conditions while taking into consideration some crucial guidance. For instance, fatigue cycles should be analysed individually in order to properly assess the anelasticity limit for fatigue crack. Sensing devices applied for long-term monitoring must appropriately compensate for the thermal effects to be reliable. In the case of additional stress generated by soil movements, the sensors location setting should cover the whole circumference for a complete observation of the stress field. Finally, corrosion insurgence has to be rightly avoided with opportune protection and continuously monitored, since it can cause wall weakening and sudden collapses. The possible choices for its monitoring include analytical solutions of kinematic growth equations or prescriptions from current normative. Ultimately, the guidelines provided by the authors in this paper give useful considerations resulting from direct field experience.

8. Strain Response Characteristics of RC Beams Strengthened with CFRP Sheet Using BOTDR

The research reported in [7] focuses on the crucial aspect of monitoring reinforced concrete (RC) structures, in particular those strengthened with carbon-fibre-reinforced polymer (CFRP) sheets. The rationale is that the performance of such structural elements can be seriously affected by the de-bonding phenomenon. This is particularly relevant in the initial cracking stage, with the significant risk of a relevant reduction of their bearing capacity. In comparison with classical strain gauge sensors, the authors present numerically and experimental results obtained by using a Brillouin optical time-domain reflectometer (BOTDR) fibre sensor. The use of this type of sensor shows several advantages: although it might be less accurate than standard strain gauges, the BOTDR is not affected by the surface conditions, assuring, thus, steady measurements. Numerical finite elements simulations conducted with the commercial software LS-DYNA take into account all the aspects involved in the deflectional process, such as the mechanical properties of reinforced concrete and carbon fibres, the anisotropy of composite concrete, the orthotropic behaviour of polymeric sheets, the interface interactions, and the failure tiebreak contact model. Results coming from the four-point bending experiments conducted in displacement control over several specimens in different sheets bonding conditions highlight the prominent influence of de-bonding of CFRP in the initial cracking stage, while the ultimate failure state results uncorrelated. Moreover, experimental measurements and numerical analysis show a satisfactory match that grants the reliability of the latter as a predictive method. Thus, for all the aforementioned aspects, the methodology proposed by the authors constitutes a valid and robust technique for RC beams monitoring.

9. Bayesian Calibration of Hysteretic Parameters with Consideration of the Model Discrepancy for Use in Seismic Structural Health Monitoring

In the study reported in [8], the authors investigate model-driven seismic structural health monitoring procedures, based on a Bayesian uncertainty quantification framework. The variety of schemes and uncertainties that are typical of civil structures make the prediction of their actual mechanical behaviour and structural performance a difficult task. In this regard, computer simulations are useful engineering tools to design complex systems and assess their performance. These simulations aim at reproducing the underlying physical phenomena under investigation, providing a solution for the governing equations. However, accurate modelling of the structural systems requires the numerical models to be calibrated and validated with direct observations and measured experimental data. For this aim, the authors applied a Bayesian inference strategy to calibrate a nonlinear hysteretic Bouc–Wen model, derived from real data acquired on a monitored masonry building, in terms of both most probable values (MPV) and discrepancy posterior distribution. This pointed to the importance of correlating the choice of the discrepancy model function to the possible degradation amount and the characteristics of the external seismic input. The findings of the study define a non-arbitrarily of the choice of the discrepancy model. According to their findings, the selection of this model should be subordinated to the statistical nature of the external force (e.g., amplitude and frequency) and the statistical nature of the modal characteristics (e.g., natural frequencies) of a system, evaluated in operational conditions. For instance, for external forces with frequency content close to the natural frequencies of the system, there is a high chance of high degradation to occur, and thus a discrepancy model distribution close to a Gaussian distribution could bring trivial results. In conclusion, the authors make relevant considerations for the use of model-driven solutions for seismic structural health monitoring, especially for applications to masonry structures.

10. Piezoelectric Electro-Mechanical Impedance (EMI) Based Structural Crack Monitoring

In [9], the application of the piezoelectric electro-mechanical impedance (EMI) method is proposed as an effective active sensing approach to localise small cracks in beam- and plate-like structures. The integrity of the structure is investigated by analysing the imaginary peak frequency of the piezoelectric admittance spectrum. The rationale is that, in accordance with the coupled behaviour of the electro-mechanical systems, the occurrence of damage in the structure causes changes in both its piezoelectric and dynamic properties, with a close correlation between the modal resonance frequencies and the piezoelectric ones. This allows assuming the latter as a valid and sensitive indicator of the deviations from the pristine mechanical state. The proposed method is validated through numerical finite element (FE) and experimental simulations, where the crack size has fixed depth and width and variable length. The numerical analysis shows off the feasibility of the method and establishes it as a useful tool for the selection of the scanning bands in the piezoelectric admittance frequency spectrum due to the relationship with the harmonic response. The experimental results, conducted on aluminium specimens, are consistent with the numerical model and confirm the theoretical expectations of effectiveness in crack detection and growth monitoring. They also highlight a high accuracy level and stability when dealing with real-life, noisy observations. The decrement of local stiffness due to damage presence is related to lower admittance peak frequency, while the shift can be assumed as a feature for evaluating the severity of the crack and modulated through an increase in the detection frequency band. In conclusion, the proposed method has great potential as a compelling crack detection and monitoring strategy.

11. Robust Structural Damage Detection Using Analysis of the CMSE Residual's Sensitivity to Damage

A consistent improvement in damage identification is presented in [10]. The well-known vibration-based method of cross-modal strain energy (CMSE) presents several intrinsic advantages in modal analysis with respect to traditional modal strain energy (MSE). For instance, it does not strictly require the same number of intact and damaged mode shapes. However, the resulting linear inverse problem is ill-conditioned, raising issues related to excessive perturbations propagation and incorporating some ineffective, otherwise counterproductive, redundant equations. To face this issue, the authors introduce a sensitivity analysis to identify and remove those inessential equations and enhance the effectiveness and the robustness of detection. Ancillary, two improvements of the iterative Tikhonov regularisation method have been proposed in the selection of the regularisation parameter of the adaptive strategy and in the formulation of the regularisation operator. These two aspected aimed at increasing both the rate of convergence and the accuracy. The effects of different damage levels and locations are investigated through noisy numerical simulations and confirmed by experimental validation. In all the investigated cases, the so-called robust cross-modal strain energy (RCMSE) method showed better performances than conventional CMSE. In particular, RMCSE outputs a considerably reduced number of false positives, which is an aspect of primary importance in the field of SHM. The robustness of results has been also confirmed by investigating the influence of different noise levels, with minor discrepancies at lower damage levels. Convincingly, the work presented in this paper progresses the effectiveness of MSE-based techniques.

12. A Novel Dense Full-Field Displacement Monitoring Method Based on Image Sequences and Optical Flow Algorithm

An innovative methodology for deformation monitoring using image acquisition techniques is proposed in [11]. Specifically, a deep learning algorithm is applied, conjunctly to several vision technologies. This is intended to achieve a full-field displacement measurement, globally for the whole large-scale structure. This mainly overcome the typical issues of discrete target points observation, poor in the characterisation of the overall structure, while maintaining the advantages of long-range accuracy and cost-effectiveness. The proposed approach allows as well big data acquisition, which is essential for feeding machine learning algorithms. The designed noncontact remote sensing (NSR) device is able to acquire from multiple perspectives the time-space labelled static images sequences, according to the overlapped camera fields of view. Afterwards, edge detection, pixel virtual marker methods, and the scale-invariant feature transform (SIFT) algorithm are applied to holographically reconstruct the dense full-field displacement. The proposed device and method are applied to test a reduced-scale model of a self-anchored suspension bridge under several load and damage conditions. The bridge was equipped with dial gauges for displacement measurements and a numerical Finite Element model was developed for additional comparison. The experimental results show a level of accuracy high enough for engineering applications, reaching a maximum error of 12% with respect to the observation coming from conventional measurement devices and numerical predictions. Although this study represents only the first step towards a dense optical monitoring strategy and still requires further studies, the proposed methodology has great potentiality for real-time and long-term monitoring applications.

13. Railway Wheel Flat Recognition and Precise Positioning Method Based on Multisensor Arrays

The study published in [12] concerns the major problem of wheel defects detection and their long-term monitoring, to enhance a more sustainable maintenance planning of trains and rails while ensuring high standards of serviceability and safety. In particular, this work focuses on wheel flats, well-known defects responsible for anomalous impacts on the track that accelerate the degradation of both the track and the wheel itself. The proposed land-based measurement method consists of multisensor arrays, able to assess the condition of the wheels by evaluating the dynamic vertical strain response of the track during the train passage. A multibody dynamic system is numerically modelled using the finite element method; the rail web compression method is chosen to measure the wheel impact, due to its sensitivity to vertical strain, its stability to bending and torque moments caused by interfering lateral forces, and the low number of sensors required. The transverse, longitudinal and plane sensors layouts are also investigated to enhance accurate and unbiased measurements. Finally, the designed algorithm exploits multiple sensors data fusion to establish when the anomalous impacts occurred (by analysing the outliers space-time distribution), where those impacts occurred (through the sensors spatial correlation), and which wheel causes these abnormal fluctuations (by associating average speed with time and position to individuate the impact processes). The subsequent offline experimental validation confirmed the capability of the algorithm to effectively recognise and locate the presence of wheel flats. The main contribution of this paper is the proposal of a real-time monitoring solution that, for its effectiveness and its feasibility, is able to easily detect wheel flat; this can concretely improve the maintenance operations for railroad owners and operators.

14. Monitoring and Analysis of Dynamic Characteristics of Super High-Rise Buildings Using GB-RAR: A Case Study of the WGC under Construction, China

The importance of monitoring the dynamic characteristics of skyscrapers and tall buildings is well outlined in [13], as extreme displacements can cause severe damage to the structures or compromise their operational safety, especially during the construction phase. This research describes a field application of displacement measurements with the technique of interferometric ground-based real aperture radar (GB-RAR), able to overcome the limitations of classical methodologies, such as the need to place sensors, resorting to expensive devices. It also compares favourably, from a cost-efficient point of view, with satellite-based imaging technologies, which are affected by low resolution, potential misrepresentation due to atmospheric effects, and delay in delivery of the results. The main application of this study consists of monitoring a high-rise building under construction in mainland China. The proposed strategy reached a sub-millimetric level of accuracy in measurements. In the paper, the influence of temperature is investigated through a meteorological station. The procedure adopted by the authors include also several correction methods to correct the intrinsic and extrinsic influence of the acquired signal data. In particular, a windowing procedure is used to eliminate sidelobe effects. Gross error due to external vibrations and environmental factors are detected and removed and a wavelet denoising procedure is applied to the observations. Although working in a construction site context, characterised by the presence of several interfering structures (such as service platforms, cranes, etc...) and disturbing environmental elements like wind, sunlight and temperature, the methodology proposed by the authors is effectively capable to identify trajectories and displacements of the buildings, confirming to be a highly accurate, low-cost and non-invasive valid solution, even in presence of construction vibrations and unfavourable conditions.

15. State-of-the-Art Review on Determining Prestress Losses in Prestressed Concrete Girders

The monitoring of prestress losses in prestressed concrete (PC) girders is fundamental to preserve the structural integrity of bridges. Indeed, the deleterious effects of prestress losses require the elapse of a long time to emerge and, thus, their identification and predictive maintenance result to be very challenging. A state-of-art review on this topic is presented in [14], with a particular focus on the existing non-destructive testing methods and related strategies. The authors firstly review the context of the application for PC elements. Here, they propose an overview of previous works aimed to measure the prestressing force and to predict related losses, considering over 30 articles concerning different experimental and numerical methodologies. The following sections extensively collect more than 60 papers of experimental and numerical research works, ranging over different metrics (i.e., mechanical parameters, vibrational features, etc.) and approaches (considering destructive, semi-destructive, and non-destructive testing options). Finally, the study focuses on static NDT methods, remarking the higher reliability and sensitiveness to prestressing losses in comparison with dynamic techniques. The overview offered for such an important topic will be essential to both academic researchers and practitioners.

16. Application of the Subspace-Based Methods in Health Monitoring of Civil Structures: A Systematic Review and Meta-Analysis

Subspace system identification (SSI) methods have been widely studied and applied in the last two decades, investigating both mechanical and civil structures with numerical and experimental analyses. Since its large spread and abundance of academic research papers, the need for a systematic literature review is faced in [15]. The authors follow the Preferred Reporting Items for Systematic Reviews and Meta-Analyses (PRISMA) approach to conduct rigorous selection, screening, classification and examination of reviewed works, summarising at the end of the process a total of 69 articles from 31 international journals published in the period 2008–2019. Criteria of classification include the typology of test structures, nature of processing algorithms, a-priori knowledge of the input and/or the output, and influence of operational and environmental conditions. Moreover, the authors propose several comparisons and considerations on different methods, discussing the advantages and disadvantages of the different techniques. Conclusively, the extensive literature review presented by the authors poses the basis for further studies in the field of subspace-based methods, comprehensively outlining research gaps and future perspectives to enhance future developments.

Funding: This research received no external funding.

Acknowledgments: This Special Issues collected the efforts of all the authors, reviewers, and members of the Editorial Office of Applied Sciences. We would like to thank all the professional contributions to this publication. The Special Issue benefited from the coordination efforts and the support from Marco Civera and Davide Martucci.

Conflicts of Interest: The author declares no conflict of interest.

References

1. Martucci, D.; Civera, M.; Surace, C. The Extreme Function Theory for Damage Detection: An Application to Civil and Aerospace Structures. *Appl. Sci.* **2021**, *11*, 1716. [CrossRef]
2. Roy, R.; Gherlone, M.; Surace, C.; Tessler, A. Full-Field Strain Reconstruction Using Uniaxial Strain Measurements: Application to Damage Detection. *Appl. Sci.* **2021**, *11*, 1681. [CrossRef]
3. Bombarda, D.; Vitetta, G.M.; Ferrante, G. Rail Diagnostics Based on Ultrasonic Guided Waves: An Overview. *Appl. Sci.* **2021**, *11*, 1071. [CrossRef]
4. Kwon, D.-S.; Jin, C.; Kim, M.; Koo, W. Mooring-Failure Monitoring of Submerged Floating Tunnel Using Deep Neural Network. *Appl. Sci.* **2020**, *10*, 6591. [CrossRef]
5. Żurawski, M.; Zalewski, R. Damping of Beam Vibrations Using Tuned Particles Impact Damper. *Appl. Sci.* **2020**, *10*, 6334. [CrossRef]
6. Chmelko, V.; Garan, M.; Šulko, M.; Gašparík, M. Health and Structural Integrity of Monitoring Systems: The Case Study of Pressurized Pipelines. *Appl. Sci.* **2020**, *10*, 6023. [CrossRef]
7. Hong, K.-N.; Shim, W.-B.; Yeon, Y.-M.; Jeong, K.-S. Strain Response Characteristics of RC Beams Strengthened with CFRP Sheet Using BOTDR. *Appl. Sci.* **2020**, *10*, 6005. [CrossRef]
8. Ceravolo, R.; Faraci, A.; Miraglia, G. Bayesian Calibration of Hysteretic Parameters with Consideration of the Model Discrepancy for Use in Seismic Structural Health Monitoring. *Appl. Sci.* **2020**, *10*, 5813. [CrossRef]
9. Wang, T.; Tan, B.; Lu, M.; Zhang, Z.; Lu, G. Piezoelectric Electro-Mechanical Impedance (EMI) Based Structural Crack Monitoring. *Appl. Sci.* **2020**, *10*, 4648. [CrossRef]
10. Xu, M.; Wang, S.; Guo, J.; Li, Y. Robust Structural Damage Detection Using Analysis of the CMSE Residual's Sensitivity to Damage. *Appl. Sci.* **2020**, *10*, 2826. [CrossRef]
11. Deng, G.; Zhou, Z.; Shao, S.; Chu, X.; Jian, C. A Novel Dense Full-Field Displacement Monitoring Method Based on Image Sequences and Optical Flow Algorithm. *Appl. Sci.* **2020**, *10*, 2118. [CrossRef]
12. Zhou, C.; Gao, L.; Xiao, H.; Hou, B. Railway Wheel Flat Recognition and Precise Positioning Method Based on Multisensor Arrays. *Appl. Sci.* **2020**, *10*, 1297. [CrossRef]
13. Zhou, L.; Guo, J.; Wen, X.; Ma, J.; Yang, F.; Wang, C.; Zhang, D. Monitoring and Analysis of Dynamic Characteristics of Super High-rise Buildings using GB-RAR: A Case Study of the WGC under Construction, China. *Appl. Sci.* **2020**, *10*, 808. [CrossRef]
14. Bonopera, M.; Chang, K.-C.; Lee, Z.-K. State-of-the-Art Review on Determining Prestress Losses in Prestressed Concrete Girders. *Appl. Sci.* **2020**, *10*, 7257. [CrossRef]
15. Shokravi, H.; Shokravi, H.; Bakhary, N.; Heidarrezaei, M.; Rahimian Koloor, S.S.; Petrů, M. Application of the Subspace-Based Methods in Health Monitoring of Civil Structures: A Systematic Review and Meta-Analysis. *Appl. Sci.* **2020**, *10*, 3607. [CrossRef]

 *applied sciences*

Article

The Extreme Function Theory for Damage Detection: An Application to Civil and Aerospace Structures

Davide Martucci [1], Marco Civera [2,*] and Cecilia Surace [1]

[1] Department of Structural, Geotechnical and Building Engineering, Politecnico di Torino, 10129 Turin, Italy; s220332@studenti.polito.it (D.M.); cecilia.surace@polito.it (C.S.)

[2] Department of Mechanical and Aerospace Engineering, Politecnico di Torino, 10129 Turin, Italy

* Correspondence: marco.civera@polito.it

Featured Application: This paper presents an application of Extreme Function Theory (EFT) and Gaussian Processes (GPs) to perform mode shape-based damage detection and Structural Health Monitoring (SHM).

Abstract: Any damaged condition is a rare occurrence for mechanical systems, as it is very unlikely to be observed. Thus, it represents an extreme deviation from the median of its probability distribution. It is, therefore, necessary to apply proper statistical solutions, i.e., Rare Event Modelling (REM). The classic tool for this aim is the Extreme Value Theory (EVT), which deals with uni- or multivariate scalar values. The Extreme Function Theory (EFT), on the other hand, is defined by enlarging the fundamental EVT concepts to whole functions. When combined with Gaussian Process Regression (GPR), the EFT is perfectly suited for mode shape-based outlier detection. In fact, it is possible to investigate the structure's normal modes as a whole rather than focusing on their constituent data points, with quantifiable advantages. This provides a useful tool for Structural Health Monitoring, especially to reduce false alarms. This recently proposed methodology is here tested and validated both numerically and experimentally for different examples coming from Civil and Aerospace Engineering applications. One-dimensional beamlike elements with several boundary conditions are considered, as well as a two-dimensional plate-like spar and a frame structure.

Keywords: Structural Health Monitoring; machine learning; damage detection; extreme function theory; non-destructive testing; extreme value theory; generalised extreme distribution

 check for updates

Citation: Martucci, D.; Civera, M.; Surace, C. The Extreme Function Theory for Damage Detection: An Application to Civil and Aerospace Structures. *Appl. Sci.* **2021**, *11*, 1716. https://doi.org/10.3390/app11041716

Academic Editor: César M. A. Vasques

Received: 28 December 2020
Accepted: 11 February 2021
Published: 15 February 2021

Publisher's Note: MDPI stays neutral with regard to jurisdictional claims in published maps and institutional affiliations.

1. Introduction

The identification and localisation of damage in one-, two-, or three-dimensional structures through their mode shapes are widespread in the Structural Health Monitoring (SHM) community. It is well known that by inserting a localised discontinuity, the mode shapes diverge from their usual deflection path [1]. This principle has been extensively applied to investigate damage-induced variations in the mode shapes slope or curvature for both 1-dimensional beam-like [2] and 2-dimensional plate-like [3,4] structures. In this regard, a review of classic approaches can be found in Reference [5]. These approaches consider modal curvatures [6], mode shape rotations [7], and/or several other Damage-Sensitive Features (DSFs) based on the structure's eigenvectors. In particular, mode shapes-based methods are preferred over other modal parameters for damage localisation, since they inherently have the spatial resolution needed for this specific task [8,9]. However, even having established that the mode shapes and derived quantities can be exploited as a reliable and spatially dense DSFs, it remains to define how these features can be used for anomaly detection.

The changes in the mode shapes can be detected, e.g., through a Machine Learning (ML) process, trained exclusively on the mode shapes extracted from the current state of

9

the structure. This can be applied both to a known pristine condition or to an already-damaged structure, since the basis of outlier detection is to identify variations from the configuration "as it is" [10]. Indeed, no method can actually detect "damage", but rather its effects on the structural properties [11]. That is to say, the damage detection algorithms are inherently subject to false positives as they could mislabel noisy data for actual deviations. This problem is very commonly encountered with measurements from real-life applications.

The economic, social, and safety concerns of structural false alarms must not be neglected, as they are one of the major reasons which still hampers the perceived usefulness of SHM and continuous monitoring, limiting their spreading. For example, for any fixed- or rotary-wing aircraft, an onboard Health and Usage Monitoring System (HUMS) would be useless, if not even dangerous, if constantly sending false alarms. These are all major practical issues; the latter one, for instance, can be arguably considered the main factor behind the industry reluctance to apply SHM systems at large scale in the last two decades [12], and the reduction of false alarms is still a matter of research as of 2020 [13].

The main concept of this work departs from the consideration that structural damages are rare events. Therefore, assuming a normal distribution for damage occurrences is both statistically unprincipled and potentially inaccurate with all the resulting risks. Thus, Rare Event Modelling (REM) should be preferably applied. For scalar values, this can be done following the well-known Extreme Value Theory (EVT), also known as Extreme Value Statistics (EVS) [14]. As the name suggests, the EV framework—a part of the more general theory of the order statistics—deals with the statistics of extremes. This represents an ideal framework for data-driven thresholding [13], not only for damage assessment but generally for several applications of novelty detection (as reviewed in Reference [15]). The EVT framework for novelty detection has been firstly proposed by Roberts [16], especially for applications with biomedical signals [17]. For SHM purposes, the EVT was firstly introduced by Sohn et al. [18] and applied on time series. Park et al. [19] further deepened these studies considering ARX models in the frequency domain. Sohn and colleagues subsequently used the EVS framework for the analysis of delaminated composite panels, with the use of wavelets [20] and time-reversal acoustics [21]. Sundaram et al. [22] used multivariate EV Statistics and Gaussian Mixture Models on 8 performance and 4 vibrational parameters extracted from data collected from aerospace gas-turbine engines.

However, the basic concepts of EVT can be easily extended to whole functions to define the so-called Extreme Function Theory (EFT, [23]). The rationale is that the same statistical tools applied commonly on scalar values can be applied on 1- or multi-dimensional functions (like the mode shapes), not differently from what, e.g., is done for Gaussian Processes (GPs) in comparison to Gaussian distributions. This was, for instance, utilised by Papatheou et al. [24] to perform damage detection in offshore wind turbines based on their recorded power curves.

Another important aspect is that, according to the Fisher–Tippett–Gnedenko theorem [25,26], any extreme distribution on the lower or upper tail converges to only three possible extreme distributions—the Gumbel, Fréchet, and Weibull distribution families, also known as type I, II and III EV distributions—which can be unified in the Generalised Extreme Value (GEV) formulation. This happens independently from the parent distribution; that is to say, the proposed approach is not limited by the assumption of a Gaussian distribution, which—while being commonly used in structural mechanics and dynamics on the basis of the well-known Central Limit Theorem (CLT) [27]—could be not always consistent with the experimental data (e.g., in case of skewed distributions).

In this work, GP Regression and EFT are combined to define a "normality" data-driven model, fitted over the mode shapes collected from the structure "as it is", and then to check for damage-related deviations from this model [28]. The use of a statistical framework purposely crafted on rare events greatly reduced the number of false positives. The rest of this paper is organised as follows: The theoretical framework of the procedure is described in Section 2. The results of the investigations performed on the numerically simulated case studies are reported in Section 3. The experimental validation is described in Section 4. Section 5 (Conclusions) ends this paper.

2. Gaussian Processes and Extreme Function Theory

In the specific case of interest, the problem can be stated as follows. Let $i = 1, \ldots, n$ be the index of n examples taken from a population of functional data. Each i-th example is an identified mode shape $\boldsymbol{\varphi}_i$, i.e., a vector of $j = 1, \ldots, n_i$ modal coordinates $\varphi_{i,j}$ observed at a finite number of corresponding spatial coordinates $x_{i,j}$. From a theoretical standpoint, there is no requirement that the n mode shapes have the same number of coordinates nor the same location. This is the first strong point of EFT in comparison to EVT, as the functions can be collected from different sensor arrangements—even if, for practical concerns, it is more reliable to consider a fixed number and position of the output channels. Importantly, $x_{i,j}$ can be a scalar for a 1-dimensional beam element, or a pair $x_{i,j} = \{\xi_{i,j}, \eta_{i,j}\}$ for a 2-dimensional plate-like structure (and so on). For simplicity's sake, all the rest of this discussion will focus on the 1-dimensional case but the formulation for higher dimensionalities can be straightly derived.

There are two subsequent aims. The first one is to build the "normality" model $\mathcal{M}$ from the pairs of training data $\{x_i, \boldsymbol{\varphi}_i\}$, collected from a given structure excited in the current conditions (assumed as a baseline for outlier detection). Once the model is defined, the second step is to discern if a test set $\{x_i^*, \boldsymbol{\varphi}_i^*\}$ (in this case, $i = 1, \ldots, n^*$) belongs or not to the tails of the distribution $\mathcal{M}$. The first part can be achieved with the classic GP Regression, which is here briefly recalled for completeness, while the EFT comes in the second part.

2.1. The Gaussian Process (GP) Regression

Being the mode shapes φ a function of the space variable x, it is possible to define a GP prior over this latent variable, in the general form

$$f(x) \sim GP(m(x), k(x, x')),\tag{1}$$

not dissimilarly from how a normal distribution is uniquely defined by its mean and variance, the GP of the general process $f(x)$ can be defined uniquely by its mean function

$$m(x) = E[f(x)]\tag{2}$$

and on the covariance matrix

$$k(x, x') = E[(f(x) - m(x))(f(x') - m(x'))]\tag{3}$$

where $E[\cdot]$ denotes the expectation. Throughout this whole discussion, a zero-mean and squared-exponential (SE) covariance function will be applied (the exact formulation will be discussed later; other less frequent options exist and a discussion on the topic can be found in Reference [29]).

If $f = f(x)$ indicates the set of the function values at the training points X (note that the training vectors have been assembled into a matrix form) and f^* a set of test outputs at a new set of points X^*, according to the zero-mean prior it is possible to define

$$\begin{bmatrix} f \\ f^* \end{bmatrix} \sim \mathcal{N}\left(0, \begin{bmatrix} K(X,X) & K(X,X^*) \\ K(X^*,X) & K(X^*,X^*) \end{bmatrix}\right),\tag{4}$$

where $K(X^*, X) \in \mathbb{R}^{n^* \times n}$ is a matrix with the terms defined by the covariances evaluated at all pairs of test and training points (the same applies for $K(X,X)$, $K(X,X^*)$, and $K(X^*,X^*)$).

The posterior distribution is defined by conditioning the joint prior on the observations, resulting (for an ideally noise-free scenario) in

$$p(f^*|X^*,X,f) \sim \mathcal{N}\Big(K(X^*,X)K(X,X)^{-1}f, K(X^*,X^*) \\ -K(X^*,X)K(X,X)^{-1}K(X,X^*)\Big). \tag{5}$$

By considering the measurement noise as i.i.d. Gaussian noise ϵ with variance σ_n^2, one can define

$$\varphi = f(x) + \epsilon \tag{6}$$

thus, Equation (5) becomes

$$p(f^*|X^*,X,\varphi) \sim \mathcal{N}(m(f^*), cov(f^*)) \tag{7}$$

with the predictive mean

$$m(f^*) = K(X^*,X)[K(X,X) + \sigma_n^2 I]^{-1}\varphi \tag{8}$$

and the predictive covariance matrix

$$cov(f^*) = K(X^*,X^*) - K(X^*,X)\Big[K(X,X) + \sigma_n^2 I]^{-1}K(X,X^*)\Big). \tag{9}$$

At this point, it is evident that the only step remaining before obtaining the normality model is to define its covariance and the added noise. It was already assumed that this latter depends solely on σ_n^2. In the simpler (i.e., 1-dimensional) case, the aforementioned SE covariance function can be formulated as

$$k_\varphi(x_p, x_q) = \sigma_f^2 \exp\left[-\frac{1}{2l}(x_p - x_q)^2\right] + \sigma_n^2 \delta_{p,q} \tag{10}$$

where k_φ is the covariance of the noisy target set φ, σ_f^2 is the variance of the function, l is the length-scale parameter, $\delta_{p,q}$ is the Kronecker delta, and x_p, x_q are the locations of the training data sets corresponding to the respective indices. σ_f^2, l, and σ_n^2 are the free parameters who need to be set from the data and are collectively known as the GP Regression hyperparameters θ. Their setting requires an optimisation, which is generally performed by maximising the log marginal likelihood with respect to the hyperparameters. However, in general, the minimisation, rather than the maximisation, is performed; therefore, it is more practically convenient (without any theoretical drawback) to define directly the negative log marginal likelihood (NLML) as

$$logp(\varphi\Big|X, \theta = l, \sigma_f, \sigma_n) = \frac{1}{2}\varphi^T K_\varphi^{-1}\varphi + \frac{1}{2}log|K_\varphi| - \frac{n}{2}log2\pi \tag{11}$$

where K_φ indicates the covariance matrix of the noisy test set. The iterative minimisation is performed through a Conjugated Gradient Optimisation (CGO) algorithm; the process stops when the difference in NLML values are smaller than a set tolerance. At that point, the hyperparameters can be considered as the best fitting for the SE covariance function on the training data considered. The GP Regression over the training data can be therefore easily achieved and the metamodel $\mathcal{M}$ defined.

2.2. The Extreme Function Theory and the Complete Procedure

Consider the posterior distribution as defined in Equation (7), set accordingly to the hyperparameters optimised as described in the previous Section. By definition, a GP distribution over functions is a multivariate normal (Gaussian) distribution, thus the probability distribution $z = p(\mathbf{f}^*|\mathbf{X}^*, \mathbf{X}, \boldsymbol{\varphi})$ will follow the form

$$z = \frac{1}{\sqrt[2]{2\pi^D|K^*|}} \exp\left[-\frac{1}{2}(f^* - m^*)^T K^{*-1}(f^* - m^*)\right] \tag{12}$$

where $m^* = m(\mathbf{f}^*)$, $K^* = cov(\mathbf{f}^*)$, and D is the dimensionality of $\mathbf{x}^*$ (here onwards it is assumed that a single mode shape is tested at a time; even if it is theoretically possible to test jointly a set of more mode shapes at once, this will have little practical sense). Therefore, the whole test set $\{\mathbf{x}_i^*, \boldsymbol{\varphi}_i^*\}$ is reduced to a single scalar value corresponding to its likelihood. In turn, this can be used to fit an EV distribution over its left tail (i.e., for a very low probability of occurrence, as expected for damaged conditions). The lower 10% of the validation dataset was used to this aim. Specifically, the Cumulative Distribution Function (CDF) is here fitted on low values of the Gaussian log probability $\ln(z)$. For the GEV minima distribution, the exact formulation of the CDF is

$$L(\ln(z), \lambda, \delta, \gamma) = 1 - \exp\left(-\left(1 + \gamma\left(\frac{\lambda - \ln(z)}{\delta}\right)\right)^{\frac{1}{\gamma}}\right) \tag{13}$$

where $\lambda, \delta,$ and γ represent, in the same order, the location, scale, and shape parameter of the GEV distribution, which needs to be inferred from the data. Importantly, the shape parameter γ controls the specific form of the limit minima distribution, which simplifies in a Gumbel distribution for $\gamma \to 0$, in a Weibull distribution if $\gamma < 0$, or in a Fréchet distribution if $\gamma > 0$ [30].

The estimation of the GEV parameters has been achieved utilising a Differential Evolution (DE) algorithm, as suggested in Reference [31]; specifically, the Self-Adaptive Differential Evolution (described in Reference [32]) was applied, considering 10 runs, 100 generations, and population size equal to 30. The Normalised Mean Square Error (NMSE) was set as the target cost function, minimised against the empirical CDF calculated according to the Hazen plotting position; an example is depicted in Figure 1. Having constructed the CDF it is then possible to define a threshold value $\ln z_{\text{lim}}$ in correspondence of an arbitrarily chosen quantile α. Here in this study, $\alpha = 1\%$ was imposed. This completes the description of the whole procedure.

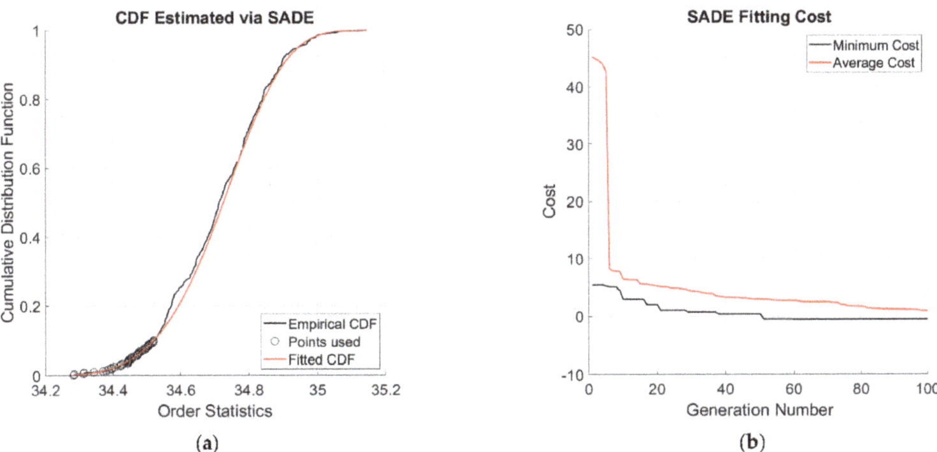

Figure 1. Example of SADE optimisation. (**a**) Cumulative Distribution Function (CDF) fitting; (**b**) evolution of cost with number of generations (costs reported in logarithmic scale).

2.3. Test of the Performances of the Procedure

To quantitatively estimate the performances of the described approach, this was tested on numerically simulated and experimental datasets with known damage conditions. For this purpose, three datasets—training, validation, and test—are defined. The training and the validation datasets, needed to preliminary set and validate the normality model offline, are made up of functions drawn from the undamaged conditions, while the test set includes both damaged and undamaged states. For clarity's sake, the several steps described in the previous Subsection are graphically depicted in Figure 2.

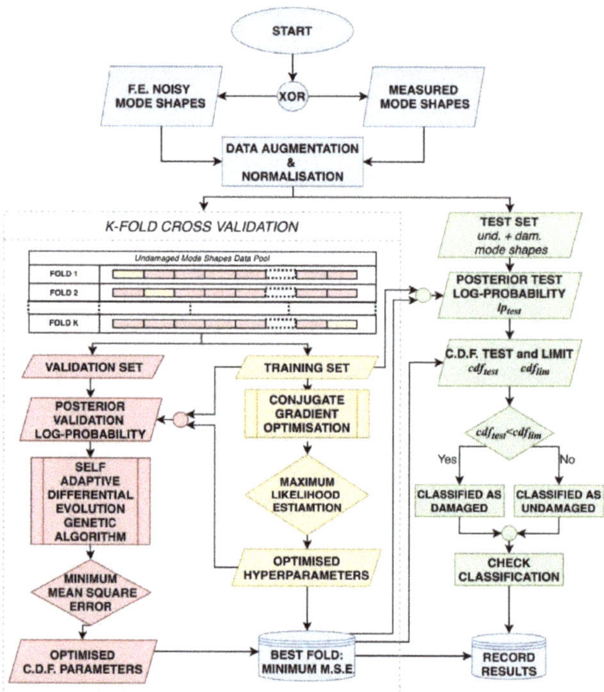

Figure 2. The flow chart of the complete procedure.

In comparison to a preliminary version of the algorithm described in Reference [28] and tested for simple cantilever beams, the following enhancements have been included.

Firstly, a data augmentation strategy was applied to overcome the limited amount of available measured data (numerical data were generated with the same scarcity as for the real data from the experimental case study). Dealing with the feature of transverse displacements, the latent function $f(x)$ of Equation (6) is the resulting mean of all the original mode shapes and the resulting standard deviation indicates how data are spread out from the average. Both the mean and the standard deviation were evaluated for each channel, to consider the distribution for the specific measurement point and not of the whole dataset. After generating the desired number of noise-free copies, white Gaussian noise was added by iteratively reducing the Signal-to-Noise-Ratio (SNR) until the artificially simulated data fell completely within the set limits. In this case, the range was arbitrarily chosen to be $\pm 2\sigma$ and not to cover the whole $\pm 3\sigma$ range, since this restriction strengthens the "normality" condition (i.e., undamaged conditions, not far from the median value). It should be noted that data coming from the augmentation process are involved in the training and validation steps, while only original data are tested to assess the efficiency of the algorithm.

As a second improvement, a step of k-fold cross-validation was added, to avoid overfitting of model parameters estimations. In contrast to the usual validation procedure, the k-th holdout fold subset was used to train the Gaussian Process and estimate its hyperparameters, while keeping the remaining $k - 1$ folds for the CDF estimation and model testing. This inversion in the common paradigm of using the larger part of data for training step is due to the need to take into consideration only the lower 10% of the available data to estimate the minima form of CDF as defined in Equation (13). Subsequently, the training (TR), validation (VA), and test (TS) subsets were assembled from the given d structural configurations considered as the "normality" model (generally $d = 1$, the integer measurements data set) and d^* structural configurations considered as damaged, each with n actual observations and n^* artificial copies. Table 1 describes the procedure settings, as applied for all the numerical and the experimental case studies for consistency.

Table 1. The procedure settings of the damage detection.

Number of Observations (n)	50 for the Experimental Case Study 100 for the Numerical Case Studies
Data Augmentation (n*)	500
Range Augmentation	$\pm 2\sigma$
K-Fold (k)	5
Training Set (TR)	$\frac{1}{k}\left(\frac{n}{2} + n^*\right)$
Validation Set (VA)	$\left(1 - \frac{1}{k}\right)\left(\frac{n}{2} + n^*\right)$
Test Set (TS)	$\frac{n}{4}$

As said, the test set contains both undamaged (φ_{und}) and damaged (φ_{dam}) mode shapes. When a φ_{und} is classified as damaged, it is recorded as a false positive; on the other hand, misclassified φ_{dam} are considered as false negatives. Otherwise, as pictorially described in Figure 3, the results are labelled as true positives and true negatives.

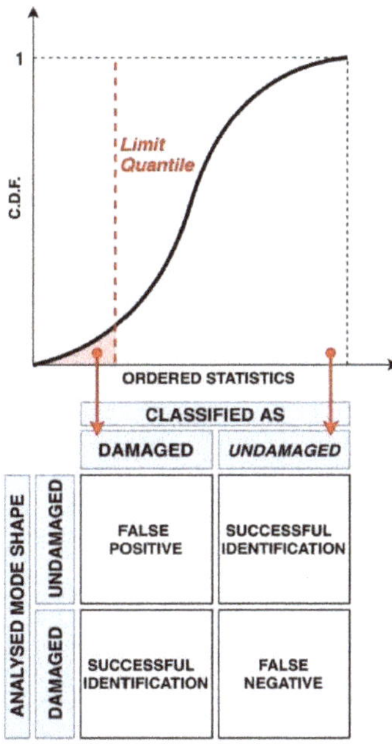

Figure 3. False positives and negatives in damage detection, according to the CDF limit quantile.

3. Numerical Simulations

The procedure is firstly validated on a numerically-simulated beam with different boundary conditions. Then, for more complex structures, the proposed approach is firstly validated on two other numerical datasets.

The first case study comes from an aerospace application and involves the High Aspect Ratio (HAR) aluminium spar of the XB-1 wing prototype [33]. The spar—portrayed in Figure 4a—can be seen as a thin, plate-like structural element with a peculiar shape. Due to its large flexibility, the spar may undergo large flap-wise oscillations [34,35] and it is therefore highly subject to the insurgence of fatigue damage, especially close to its clamped end and in its mid-length at the section where the width taper angle change. The simple Finite Elements (FE) model utilised here was realised in Ansys® Mechanical APDL; 400 8-noded, 6-degrees-of-freedom-per-node quadratic shell elements were utilised. The main mechanical and geometrical properties are reported in Table 2 (the values of the parameters derive from preovius studies described in [36]). The model parameters were updated with vibrational data coming from laser Doppler vibrometer (LDV) and video acquisitions [36], extracted by means of the Fast Relaxed Vector Fitting (FRVF) approach [37,38]. Five output channels (located at the positions of the 4 Inertial Measurement Units and of the LDV target used during the experimental acquisitions, as represented in Figure 4a) were utilised to define the mode shapes. Several levels of reduced Young's modulus were applied to the portions highlighted in Figure 4b to simulate 20 scenarios [10]. In particular, the first two states (01 and 02) are intended to represent small perturbations from the undamaged baseline, where the variations are not sufficiently marked to be defined certainly as actual damage. This simulates the possible statistical variability of the identified mechanical properties; therefore, these states should preferably be labelled as false positives by a reliable (i.e., not hypersensitive) damage detection procedure. Indeed, it was found that, when trained on noiseless mode shapes only, the procedure does not discern these small perturbations from the normal (undamaged) model.

The second numerical case study comes from a civil engineering application and is intended to model a simple multi-storey frame structure, which will serve as the experimental benchmark in the next Section. The FE (Figure 5) model, developed utilising the StaBil 2.0 MatLab Toolbox [39], was set as follows. 8 beam elements per column and per beam were used, totalling 72. The Timoshenko model was applied for all elements.

To mimic the experimental setup which will be described in the next section, the Young's modulus of the beams E_{beam} was set as two orders of magnitude larger than its counterpart for the columns (E_{column}). This allows to approximate the structural system as a shear-type frame structure; therefore, a single channel per floor is enough to capture its main vibrational modes. Three output channels were considered as virtual sensors, located at nodes #4, #12, and #20 (indicated by the purple dots in Figure 5). The main technical details are reported in Table 3. The column tracts highlighted in green are the ones where the damage was inserted (as a stiffness reduction equal to 25% or 50%).

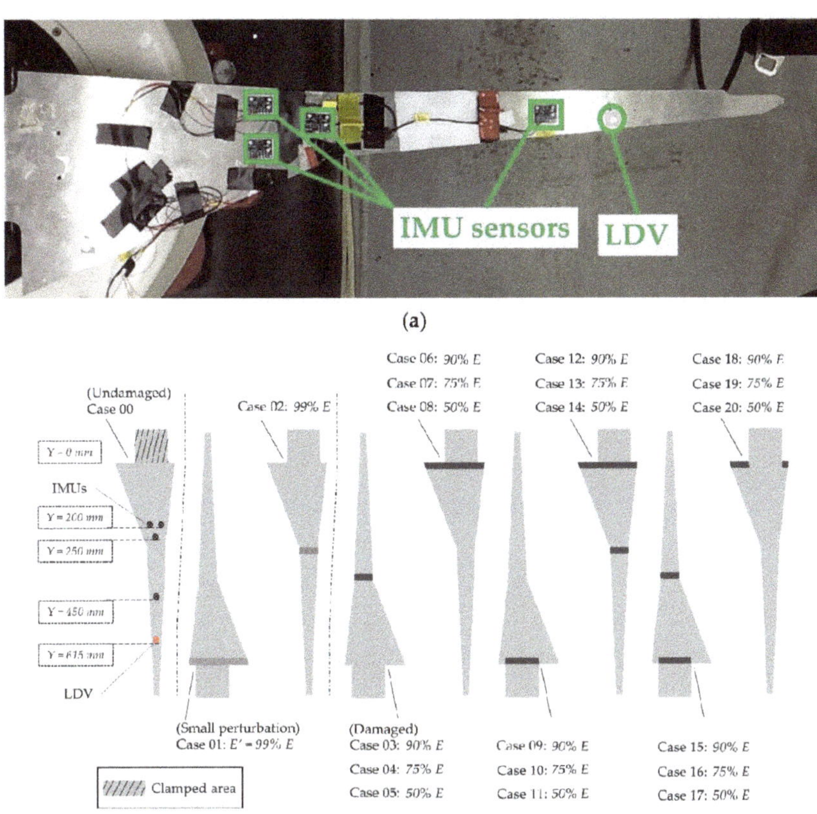

Figure 4. The High Aspect Ratio (HAR) wing spar. (**a**) The experimental prototype. (**b**) The damage scenarios considered in the FE analysis.

Table 2. Mechanical and geometrical properties of the spar.

Density ρ	2893	kg/m^3
Young's modulus E	5.90×10^{10}	Pa
Poisson's ratio v	0.26	-
Damping ratio	0.8634	%
Free length (clamp to tip) l_{tip}	706	mm
Thickness t	2	mm
Max width at clamped section b_{max}	180.00	mm
Width at the taper angle change (l = 258 mm) b_{258}	56.10	mm
Min width at the tip section b_{min}	17.04	mm

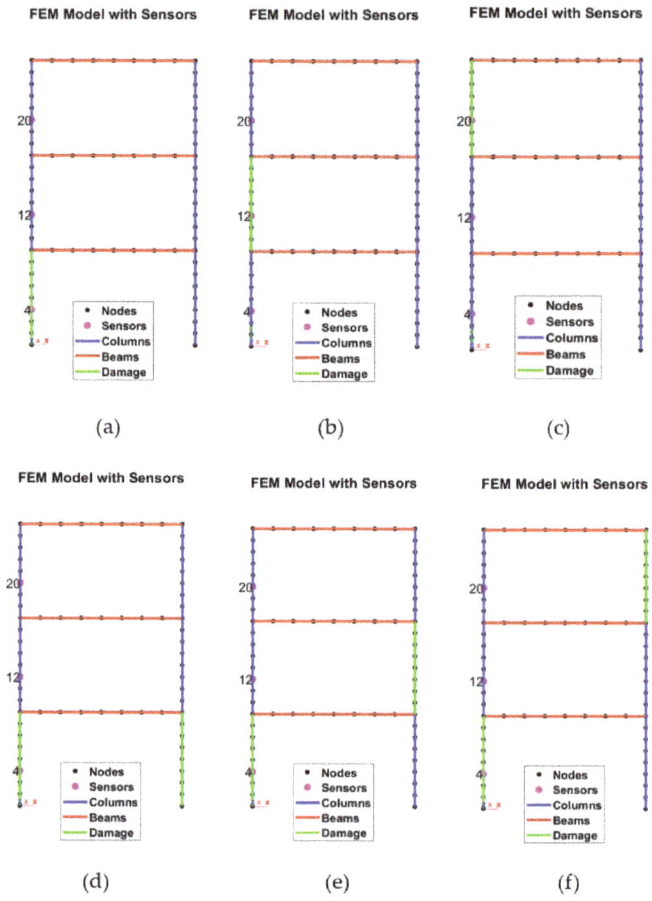

(a)   (b)   (c)

(d)   (e)   (f)

Figure 5. The multi-storey FE Model. (**a**) with one column damaged at the 1st inter-storey (-25% or -50%); (**b**) with one column damaged at the 2nd inter-storey; (**c**) with one column damaged at the 3rd inter-storey; (**d**) with both columns damaged at the 1st inter-storey; (**e**) with two columns damaged, one at the 1st inter-storey and one at the 2nd inter-storey; and (**f**) with two columns damaged, one at the 1st inter-storey and one at the 3rd inter-storey. The purple dots indicate the position of the output channels.

Table 3. Details of the frame FE model.

Density ρ	2700	kg/m^3
Young's modulus (columns) E_{column}	6.90×10^{10}	Pa
Young's modulus (beams) E_{beam}	6.90×10^{12}	Pa
Poisson's ratio v	0.3	-
Column length (per storey)	177	mm
Beam length	305	mm
Width of the column/beam cross-section	6	mm
Height of the column/beam cross-section	25	mm

3.1. Results for the Beam (1-Dimensional) FE Model

Figure 6 shows the results for two examples, the symmetrical pinned-pinned structural configuration and the asymmetrical clamped-pinned configuration. The method can be nevertheless applied to any statically determined or undetermined scenario. As it will be shown in the next Section, the same is valid for frame structures as well as 2-dimensional or more complex structures. Two levels of artificially added noise have been considered—SNR = 60 dB and 40 dB, respectively. The first state depicted in Figure 6 corresponds to the undamaged structure. Consider that in this state, no damage was to be detected, so the successful damage identifications and false alarms are both zeroed. The remaining states are defined for nine crack locations, equally spaced of 10% *l* between 0 and *l*, and for a crack depth equal to 10% of the thickness (states 2 to 10) or 20% (states 11 to 19). It can be noticed how these states, being affected by larger damage, are more easily detected by the algorithm as foreseeable. Only the first mode shape is reported for conciseness, calculated at the nodes #10, #30, and #49 (numbered sequentially, left to right, with the beam discretised in 50 elements).

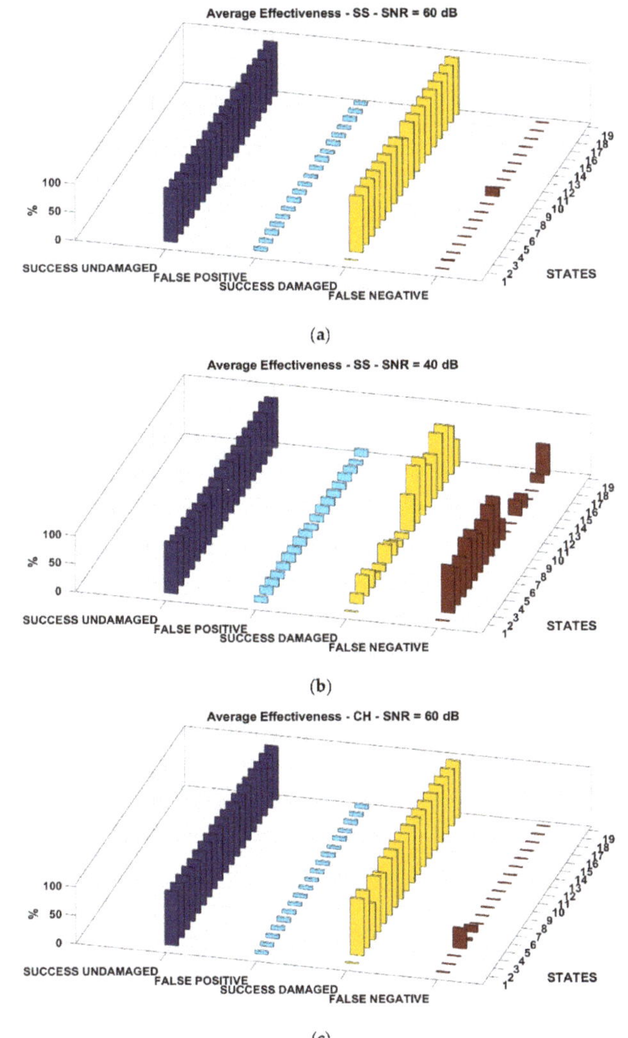

Figure 6. *Cont.*

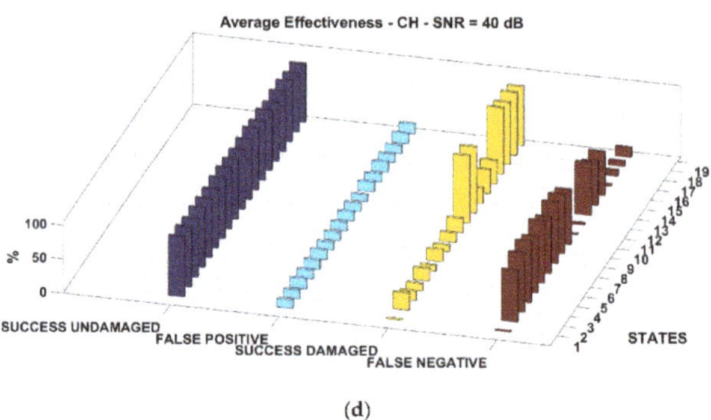

(**d**)

Figure 6. The beam-like structures with different boundary conditions. (**a**) Simply supported (SS) with SNR = 60 dB, (**b**) SS with SNR = 40 dB; (**c**) clamped-hinged (CH) with SNR = 60 dB; (**d**) CH with SNR = 40 dB.

It can be appreciated that the rate of success strictly depends on the position of the damage. This issue is well-known for any mode shape-based approach. In general, this depends on the position of the mode shapes nodes and antinodes and can be solved by considering, e.g., different mode shapes at once [40].

For a baseline-based approach as the one proposed here, the normalisation to a unit maximum displacement also causes some regions of the damaged mode shape to be very close to the ones of the undamaged scenario, reducing the detectability of the damage. This arises from the eigenproblem being underdetermined and therefore the eigenvectors being defined only up to scale.

This difference between the normalised damaged and baseline mode shapes is here considered in terms of cumulative Euclidean distance (CED) and it is exemplified in Figure 7 (for better interpretability, the complemental similarity 1-CED is shown). The numerical example represents the 1st mode shape of a cantilevered aluminium box beam, modelled from the experimental case study analysed in Reference [9] (the concept can be extended to any structural configuration and set of boundary conditions and expanded to 2- and 3-dimensional structures). To make the damage effects more visible, an unrealistically large damage (a crack with deep equal to the 75% of the beam height) has been inserted at 20 equally spaced locations. The mode shapes are defined at any centimetre over the whole beam length (totalling 715 nodes). Figure 8 a,b represent the same concept adapted for the two structural configurations discussed previously. For indicative purposes only, damage equal to a 10% reduction in the beam stiffness was considered, roved between 0 and l at steps of 1%. Only for this graphic, the mode shapes have been defined using all the available nodes as output channels and with noise-free data, i.e., considering ideal conditions.

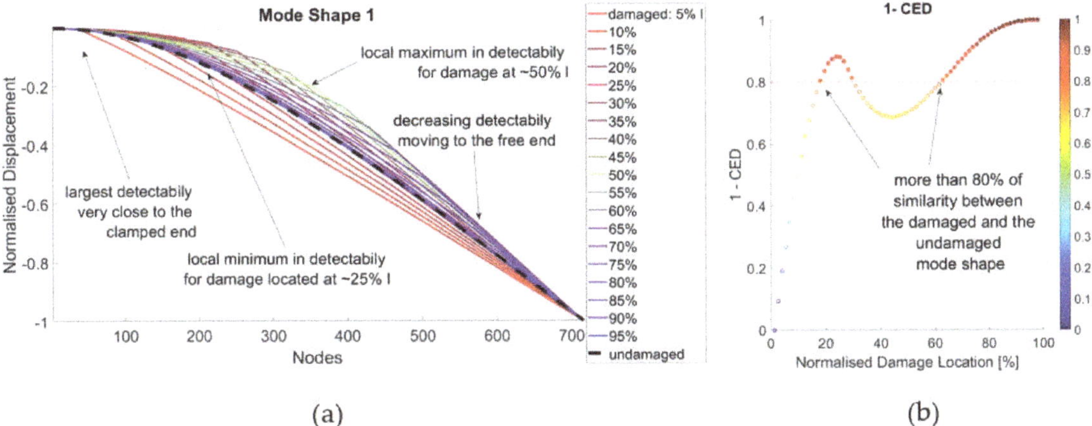

(a)
(b)

Figure 7. The first mode shape of a cantilevered box beam, without damage and with roving damage. (**a**) Damaged and undamaged mode shapes superimposed; (**b**) similarity between the damaged and undamaged mode shapes in terms of 1-CED (points with similarity equal or larger than 80% are highlighted in shades of red, above the dash-dotted line).

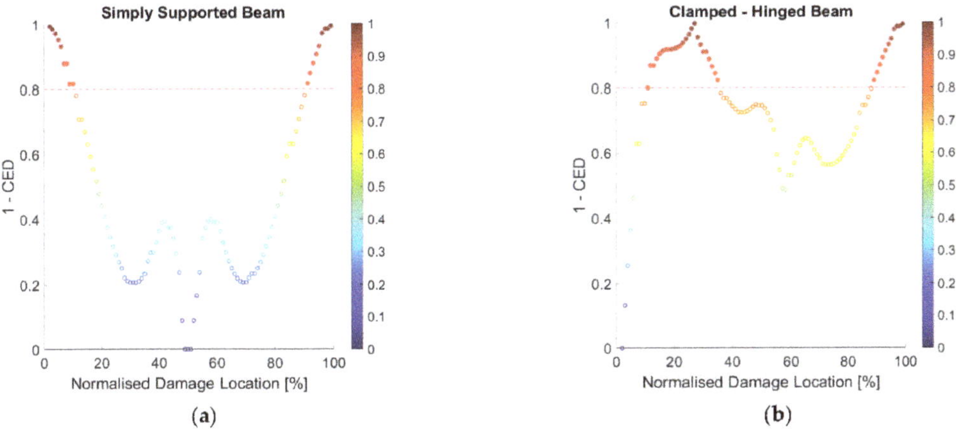

(**a**)
(**b**)

Figure 8. The similarity between the damaged and undamaged mode shapes in terms of cumulative Euclidean distance (**a**) simply supported beam (**b**) clamped-hinged beam. Points with similarity equal or larger than 80% are highlighted in shades of red, above the dash-dotted line.

3.2. Results for the HAR Wing Spar (2-Dimensional) FE Model

The results for the damage scenarios depicted in Figure 4b are portrayed in Figure 9. Only the 1st mode shape is showed for brevity; the 2nd and 3rd mode shape returned similar results in terms of accuracy (with similar rates of occurrence for both false positives and false negatives). Artificial white Gaussian was added considering SNR = 60 dB and 40 dB.

One can see that the method always successfully labels the undamaged and damaged states correctly for SNR = 60 dB, which is already a relatively large amount of noise.

Note that no damage was inserted in state 00; therefore, no damage was to be identified there, similarly to what expressed for the results of the beam structure (in Figure 6).

The states 01 and 02, which have too small variations of Young's modulus to be effectively considered as damaged, are rightly identified as false negatives. This proves the reliability of the method in discerning small perturbations, which may happen in real-life acquisitions due to the statistical variance of the acquisitions, from the actual damage. On the other hand, with even more white Gaussian noise artificially added to the input data (SNR = 40), the approach still correctly identifies the undamaged states, while it loses accuracy on the damaged mode shapes of the test set. Specifically, it remains able to recognise the most damaged situations (with a 25% or 50% reduction of Young's modulus) whenever the damage is inserted at the clamped base. As expected, also by comparison with previous works [10], the scenarios with the damage inserted solely at the mid-length cross-section (states 3 to 5) or on the sides of the base cross-section not fixed to the centre wing box (states 18 to 20).

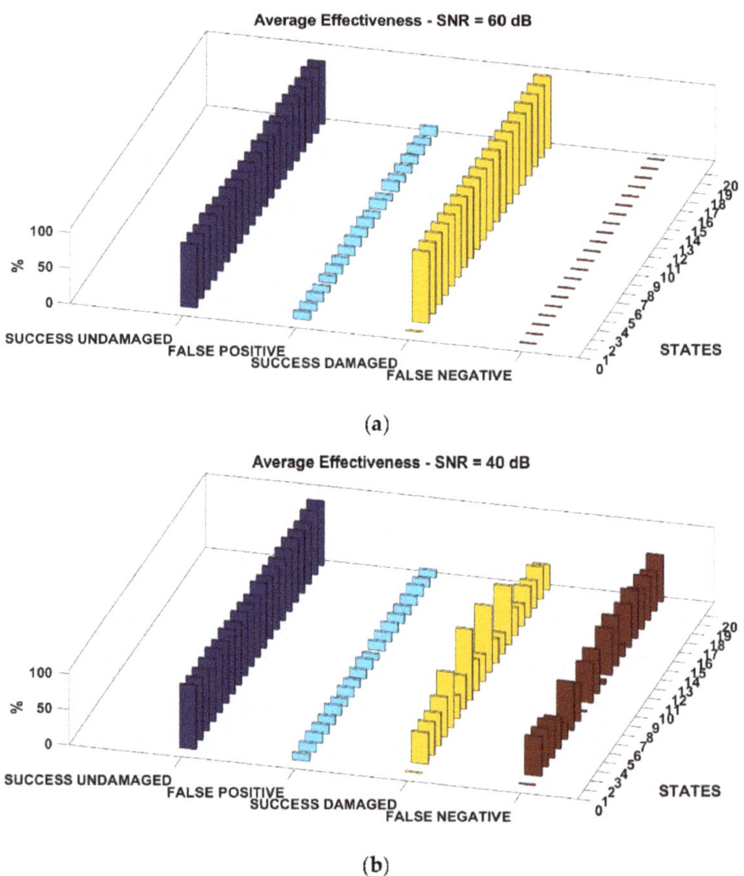

(**a**)

(**b**)

Figure 9. Extreme Function Theory (EFT) results for the HAR wing FEM, 1st mode shape with artificially added White Gaussian Noise (WGN) (**a**): SNR = 60 dB; (**b**): SNR = 40 dB. Effectiveness expressed in percentage.

Regarding the computational efficiency of the algorithm, the whole simulation of the complete HAR wing dataset was performed in around 31 s. The non-optimised MatLab script elapsed on average 2.2 s to perform a single training fold, 3.9 s for CDF validated estimation and 0.02 s for the damage detection task on a single damage state, with no major differences among the 20 states, on an Intel® Core™ i7-10750H CPU with 2.6 GHz base frequency and 15.8 GB of available RAM. Similar elapsed times were found for the other numerical and experimental cases.

3.3. Results for the Frame FE Model

The settings for the second numerical investigation are the same as described in Table 1. Figure 9 reports the results of the frame structure FE model for an increasing level of artificially added noise. Again, as for the previous case studies, consider that no damage was inserted in the baseline (state 1). For what concerns the noise level, the results in Figure 10b show how, for a realistic noisy scenario consistent with current measurement technology, a good performance is reached in the states with high damage severity (stiffness reduction ~75%) or when the damage location is highly influent (states 8 and 9, i.e., with both base columns damaged). Otherwise, Figure 10a shows a high performance level obtained in correspondence of a small noise reduction, achievable with improvement in measurement accuracy or by applying data cleansing and noise reduction techniques. All these findings are consistent with what preliminary assessed in the previous works [28] for the simpler cantilevered beam model.

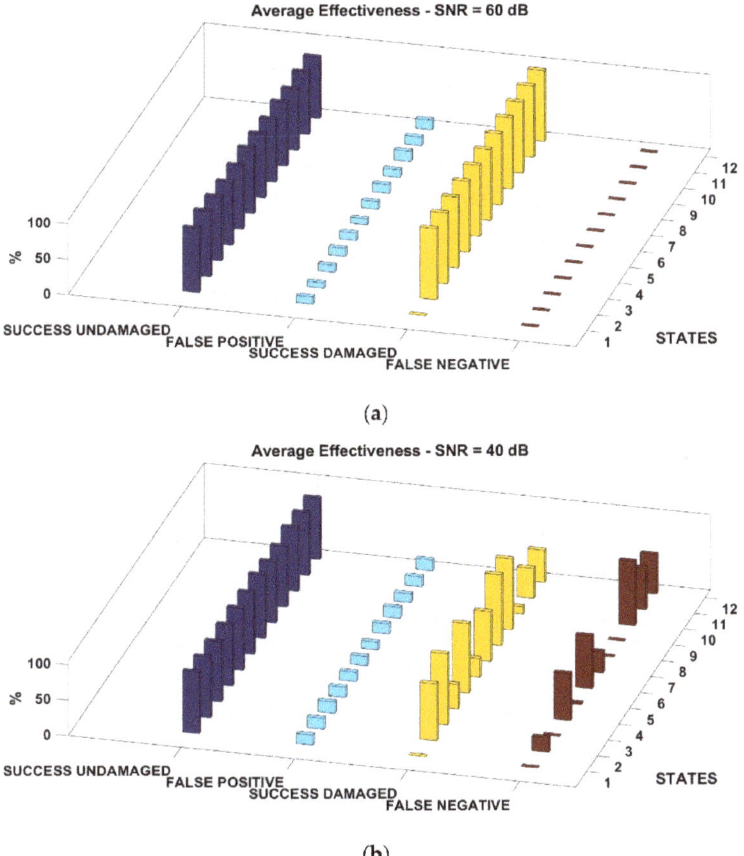

Figure 10. Results for the Frame FEM. (**a**) SNR = 60 dB (**b**) SNR = 40 Db Sensors channels deployed as indicated by the purple dots in Figure 5.

4. Experimental Validation

The experimental validation was performed on a well-known database realised at the Los Alamos National Laboratory (LANL). As for the frame numerical model described in the previous Section, the structure (represented in Figure 11) is a three-storey shear-type frame. Seventeen scenarios were considered, as described in Table 4. They include one or more stiffness reductions applied to the beams (to simulate damage) and/or one or more added masses (to simulate changing operational conditions). The dataset also included some scenarios with a bumper-column apparatus attached at the top floor. The original intent of this mechanism was to emulate the nonlinear effects induced by a breathing crack [41], which is known to act as a pointwise source of bi-linearity (a deeper discussion can be found in Reference [42]).

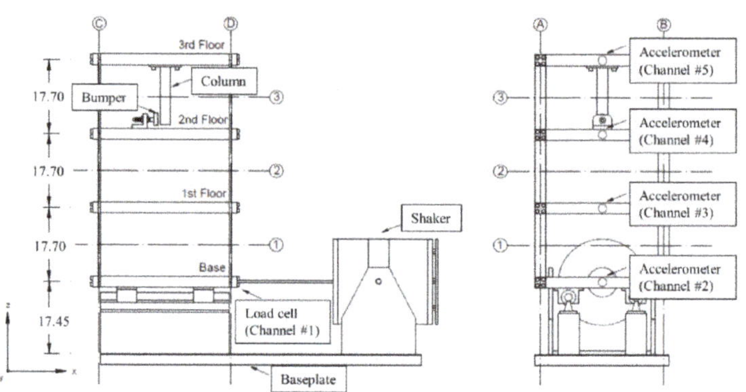

Figure 11. Scheme of the Los Alamos National Laboratory (LANL) frame. Left: Side view. Right: Front view.

Table 4. Damage States for the Los Alamos National Laboratory (LANL) frame.

	Case	Description
	1	Undamaged Baseline
	2	Added mass of 1.2 kg at the base
	3	Added mass of 1.2 kg at the 1st floor
	4	87.5% stiffness reduction in one column of the 1st inter-storey
Linear States	5	87.5% stiffness reduction in two columns of the 1st inter-storey
	6	87.5% stiffness reduction in one column of the 2nd inter-storey
	7	87.5% stiffness reduction in two columns of the 2nd inter-storey
	8	87.5% stiffness reduction in one column of the 3rd inter-storey
	9	87.5% stiffness reduction in two columns of the 3rd inter-storey
	10	Distance between bumper and column tip 0.20 mm
	11	Distance between bumper and column tip 0.15 mm
	12	Distance between bumper and column tip 0.13 mm
Nonlinear States	13	Distance between bumper and column tip 0.10 mm
	14	Distance between bumper and column tip 0.05 mm
	15	Bumper 0.20 mm from column tip, 1.2 kg added at the base
	16	Bumper 0.20 mm from column tip, 1.2 kg added on the 1st floor
	17	Bumper 0.10 mm from column tip, 1.2 kg added on the 1st floor

Generally, the presence of such nonlinearities is used to detect and localise surface cracks in otherwise linear systems [43]. However, these data are here used with another intent. As it can be noticed from Figure 12c,d, the insertion of the mechanism causes an increase in the system stiffness. Therefore, the natural frequencies grow as well, while a breathing crack would decrease them. However, the bumper-column mechanism affects the mode shapes making them detectable as a deviation from the undamaged baseline. Moreover, since the structural nonlinearities generate noise-like distortions in the frequency response, the damage scenarios 10–17 can be utilised as well to prove the robustness of the procedure when dealing with distorted signals.

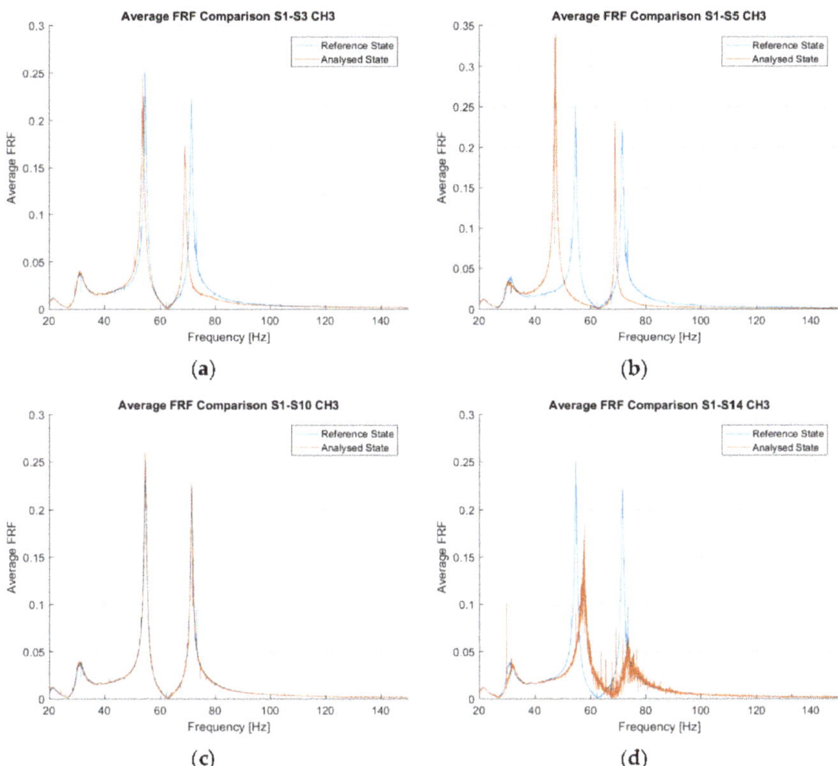

Figure 12. The frequency shifts induced in the averaged frequency response function (FRF). (**a**) Due to the mass added at the 1st floor (State 3); (**b**) due to the stiffness reduction inserted in two columns of the 1st inter-storey (State 5) (**c**) due to the column-bumper with 0.20 mm gap (State 10, weakly nonlinear system); (**d**) due to the column-bumper with 0.05 mm gap (State 13, strongly nonlinear system).

The acquisition procedure can be summed up as follows (more details can be found in the original source [44]). Fifty band-limited (20–150 Hz) white Gaussian noise realisations were applied as input for 25.6 s at the structure base. The system response was recorded at four points corresponding to the three levels plus the base, with a sampling frequency $f_s = 320$ Hz. The mode shapes were then extracted from the frequency response functions (FRFs) defined between the acceleration output time histories and the force input utilising the FRVF procedure described previously.

Importantly, for states 10–17, the concept of "mode shapes" in a nonlinear context could be not totally accurate. It is necessary to remark that the specific experimental setup (single-input multi-output acquisitions with a random noise as input) only allows linear system identification [45]. The experimental identification of nonlinear normal modes would require more complex procedures which are beyond the objective of this study. Therefore, the extracted mode shapes should be considered as the ones of the underlying linear system for these eight nonlinear states [45].

The Data Augmentation procedure described in Section 2 was applied to increase the number of training and validation data. For illustrative purposes, Figure 13 shows the resulting model obtained by applying the GP Regression over the three mode shapes identified from the experimental data.

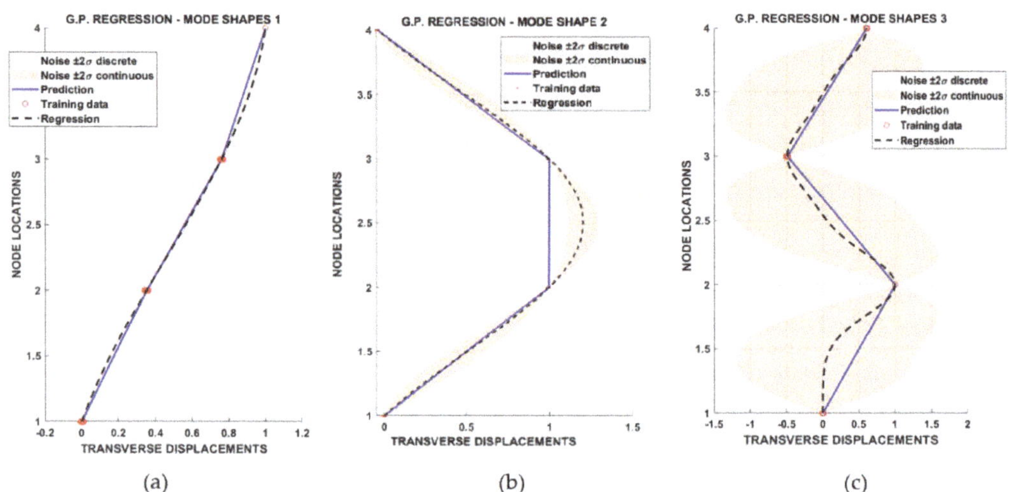

Figure 13. An example of Gaussian Process (GP) Regression over the experimental data for the first (**a**), second (**b**), and third (**c**) mode shapes.

The results of the experimental validation are shown in Figure 14. As for all the previous examples, no damage was to be identified in the first state, therefore there are no false alarms nor successful damage identifications. The EFT-based procedure correctly identifies all the structural changes except for the configurations with the bumper-column gap larger than 10 mm and no added mass (i.e., states 10 to 12). However, as it can be seen from Figure 12c, the nonlinear distortions are minimal for this larger gap and the response of the structure is almost indistinguishable from the pristine baseline. The algorithm struggled to recognise the last two states (16 and 17) as well when fed with the first mode shape; this issue did not arise either with the second or the third eigenmode. On the other hand, due to the larger confidence intervals (Figure 13c), the fitting of the third mode shape returned a relatively larger number of false positives (almost constantly 17% for any damage scenario).

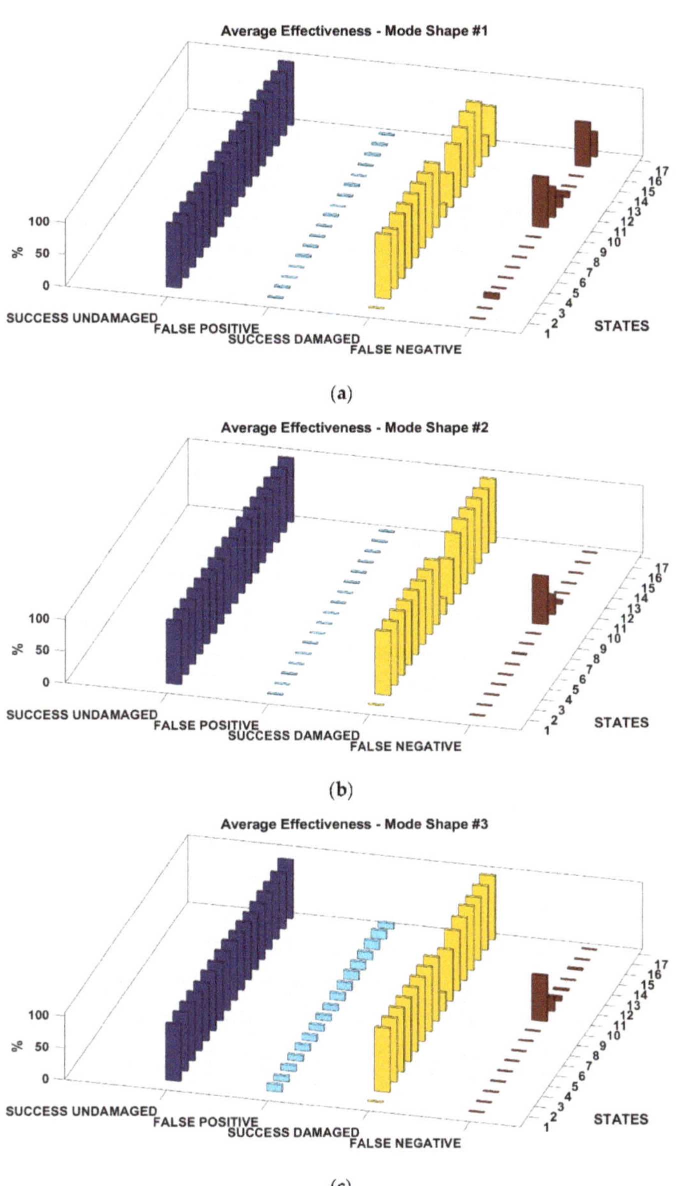

Figure 14. The results for the experimental dataset. (**a**) mode #1; (**b**) mode #2; (**c**) mode #3.

5. EFT vs. EVT

To conclude this work, a comparison between the extreme value and the extreme function theory has been performed on all the numerical and experimental data. This comparison can be done directly since the same spatial coordinates were utilised to define all the damaged and undamaged mode shapes. That means that any i-th mode shape is defined on the same n_i points. For EFT, this results in n vectors, which return a single distribution over functions, i.e., 1-dimensional data; while in EVT, there are n_i distributions over scalar values, each one with n samples.

The results are reported in Appendix A (Tables A1–A5) for all the numerical and the experimental case studies presented in the previous Sections (in the same order). Everywhere, the increment is calculated as FALSE POSITIVE EVT—FALSE POSITIVE EFT, expressed in percentage.

As expected, the incidence of false positives decreased significantly, with an improvement everywhere larger than 6% for the HAR spar Finite Element Model (even with a double-digit increment in the false alarm reduction for the case with SNR = 40 dB) and still very relevant for the second numerical case study. The EVT outperformed the EFT in terms of fewer false positives only for the simply supported beam and for the third mode shape with the lowest SNR. This is most probably due to the combination of the larger confidence intervals of the GP regression over the 3rd mode shape, the larger variability induced by the noise, and the structural symmetry. The experimental data confirmed the key findings encountered on the simulated dataset. Specifically, the percentage of false positives decreased for all the three mode shapes.

6. Conclusions

This study investigated the validity of EFT as a framework for mode shape-based SHM in 2-dimensional plate-like and frame structures, which are representative of common applications in Aerospace and Civil Engineering, respectively. The rationale is that damage is a rare occurrence and therefore the data-driven models defined over the undamaged conditions must take this consideration into account. Experimental and numerical data were utilised to this aim for different damage locations and severity levels. The robustness of the procedure to artificially added white Gaussian noise has been methodically studied and the results have been benchmarked against the results of the same algorithm trained with pointwise EVT values, showing a statistically relevant decrease of the number of false positives. Moreover, the use of EFT allows to compare points where the data were not directly collected; furthermore, it assigns only one possible outcome ("normal" or "abnormal") to the whole function. On the other hand, the single components of the mode shape, if taken one by one as in the EVT, could be under the threshold at some modal coordinates and over it at other locations for the same identification.

This study provides a strong foundation for future works in the field of EFT and EVT for structural health monitoring purposes. It will be important to validate the proposed approach in situ on real-life civil structures. Another related field of research, which the Authors are intended to investigate deeper in the next future, regards the scarcity of data from damaged structures and the need to compensate with numerically simulated data.

Author Contributions: Conceptualization, M.C. and C.S.; Data curation, D.M. and M.C.; Methodology, D.M., M.C., and C.S.; Project administration, C.S.; Resources, C.S.; Software, D.M. and M.C.; Supervision, M.C. and C.S.; Validation, D.M. and M.C.; Visualization, D.M. and M.C.; Writing—original draft, D.M. and M.C.; Writing—review and editing, C.S. All authors have read and agreed to the published version of the manuscript.

Funding: This research received no external funding.

Institutional Review Board Statement: Not Applicable.

Informed Consent Statement: Not Applicable.

Data Availability Statement: The data presented in Section 3 are available on request from the corresponding author.

Acknowledgments: The Authors would like to thank Keith Worden from the University of Sheffield for his precious help and advice and LANL for the experimental dataset.

Conflicts of Interest: The Authors declare no conflict of interest.

Appendix A

Table A1. The comparison between Extreme Value Theory (EVT) and EFT for the simply supported beam numerical case study.

	BEAM FEM (Hinged-Hinged, SNR = 60 dB)		
	FALSE POSITIVE EFT	FALSE POSITIVE EVT	IMPROVEMENT
	[%]	[%]	[%]
Mode Shape #1	6.44	15.53	9.09
Mode Shape #2	9.22	16.87	7.64
Mode Shape #3	4.64	10.76	6.11
	BEAM FEM (hinged-hinged, SNR = 40 dB)		
	FALSE POSITIVE EFT	FALSE POSITIVE EVT	IMPROVEMENT
	[%]	[%]	[%]
Mode Shape #1	12.04	15.60	3.56
Mode Shape #2	8.40	13.56	5.16
Mode Shape #3	16.07	14.87	−1.20

Table A2. The comparison between EVT and EFT for the clamped-hinged beam numerical case study.

	BEAM FEM (Clamped-Hinged, SNR = 60 dB)		
	FALSE POSITIVE EFT	FALSE POSITIVE EVT	IMPROVEMENT
	[%]	[%]	[%]
Mode Shape #1	6.00	16.62	10.62
Mode Shape #2	8.11	14.40	6.29
Mode Shape #3	1.64	8.82	7.18
	BEAM FEM (clamped-hinged, SNR = 40 dB)		
	FALSE POSITIVE EFT	FALSE POSITIVE EVT	IMPROVEMENT
	[%]	[%]	[%]
Mode Shape #1	12.91	20.71	7.80
Mode Shape #2	10.60	15.51	4.91
Mode Shape #3	7.40	12.87	5.47

Table A3. The comparison between EVT and EFT for the HAR wing spar numerical case study.

	HAR FEM (SNR = 60 dB)		
	FALSE POSITIVE EFT	FALSE POSITIVE EVT	IMPROVEMENT
	[%]	[%]	[%]
Mode Shape #1	11.26	19.28	8.02
Mode Shape #2	7.6	14.96	7.36
Mode Shape #3	9.56	16.16	6.6
	HAR FEM (SNR = 40 dB)		
	FALSE POSITIVE EFT	FALSE POSITIVE EVT	IMPROVEMENT
	[%]	[%]	[%]
Mode Shape #1	12.32	28.04	15.72
Mode Shape #2	10.42	23.02	12.6
Mode Shape #3	11.52	26.28	14.76

Table A4. The comparison between EVT and EFT for the shear-type frame numerical case study.

	FRAME FEM (SNR = 60 dB)		
	FALSE POSITIVE EFT	FALSE POSITIVE EVT	IMPROVEMENT
	[%]	[%]	[%]
Mode Shape #1	11.19	20.49	9.30
Mode Shape #2	8.81	15.81	7.00
Mode Shape #3	12.24	13.78	1.56
	FRAME FEM (SNR = 40 dB)		
	FALSE POSITIVE EFT	FALSE POSITIVE EVT	IMPROVEMENT
	[%]	[%]	[%]
Mode Shape #1	14.06	20.28	6.22
Mode Shape #2	7.06	10.70	3.64
Mode Shape #3	14.97	22.66	7.69

Table A5. The comparison between EVT and EFT for the experimental case study.

	FRAME EXPERIMENTAL		
	FALSE POSITIVE EFT	FALSE POSITIVE EVT	IMPROVEMENT
	[%]	[%]	[%]
Mode Shape #1	1.25	3.08	1.83
Mode Shape #2	0.43	2.50	2.07
Mode Shape #3	10.38	17.55	7.16

References

1. Ho, Y.K.; Ewins, D.J. On the structural damage identification with mode shapes. In Proceedings of the European COST F3 Conference on System Identification & Structural Health Monitoring, Madrid, Spain, June 2000.
2. Surace, C.; Archibald, R.; Saxena, R. On the use of the polynomial annihilation edge detection for locating cracks in beam-like structures. *Comput. Struct.* **2013**, *114–115*, 72–83. [CrossRef]
3. Gherlone, M.; Mattone Massimiliano Surace, C.; Tassotti, A.; Tessler, A. Novel Vibration-Based Methods for Detecting Delamination Damage in Composite Plate and Shell Laminates. *Key Eng. Mater.* **2005**, *293*, 289–296. [CrossRef]
4. Surace, C.; Saxena, R.; Gherlone, M.; Darwich, H. Damage localisation in plate like-structures using the two-dimensional polynomial annihilation edge detection method. *J. Sound Vib.* **2014**, *333*, 5412–5426. [CrossRef]
5. Maia, L.M.M.; Silva, J.M.M.; Almas, E.A.M.; Sampaio, R.P.C. Damage detection in structures: From mode shape to frequency response function methods. *Mech. Syst. Signal Process.* **2003**, *17*, 489–498. [CrossRef]
6. Pandey, A.K.; Biswas, M.; Samman, M.M. Damage detection from changes in curvature mode shapes. *J. Sound Vib.* **1991**, *145*, 321–332. [CrossRef]
7. Abdo, M.A.B.; Hori, M. A numerical study of structural damage detection using changes in the rotation of mode shapes | Request PDF. *J. Sound Vib.* **2002**, *251*, 227–239. [CrossRef]
8. Kim, J.T.; Ryu, Y.S.; Cho, H.M.; Stubbs, N. Damage identification in beam-type structures: Frequency-based method vs mode-shape-based method. *Eng. Struct.* **2003**, *25*, 57–67. [CrossRef]
9. Civera, M.; Zanotti Fragonara, L.; Surace, C. An experimental study of the feasibility of phase-based video magnification for damage detection and localisation in operational deflection shapes. *Strain* **2020**, *56*, e12336. [CrossRef]
10. Civera, M.; Ferraris, M.; Ceravolo, R.; Surace, C.; Betti, R. The Teager-Kaiser Energy Cepstral Coefficients as an Effective Structural Health Monitoring Tool. *Appl. Sci.* **2019**, *9*, 5064. [CrossRef]
11. Farrar, C.; Manson, G.; Worden, K.; Farrar, C.R.; Park, G. The Fundamental Axioms of Structural Health Monitoring. *Proc. R. Soc.* **2007**, *463*, 1639–1664. [CrossRef]
12. Tumer, I.; Bajwa, A. A survey of aircraft engine health monitoring systems. In Proceedings of the 35th Joint Propulsion Conference and Exhibit, Los Angeles, CA, USA, 20–24 June 1999. AIAA Meeting Paper.
13. Toshkova, D.; Asher, M.; Hutchinson, P.; Lieven, N. Automatic alarm setup using extreme value theory. *Mech. Syst. Signal Process.* **2020**, *139*, 106417. [CrossRef]
14. Pickands, J. Statistical Inference Using Extreme Order Statistics. *Ann. Stat.* **1975**, *3*, 119–131.
15. Clifton, D.A.; Hugueny, S.; Tarassenko, L.; Clifton, D.A.; Hugueny, S.; Tarassenko, L. Novelty Detection with Multivariate Extreme Value Statistics. *J. Sign. Process. Syst.* **2011**, *65*, 371–389. [CrossRef]
16. Roberts, S.J. Novelty detection using extreme value statistics. *IEEE Proc. Vis. Image Signal Process.* **1999**, *146*, 124–129. [CrossRef]

17. Roberts, S.J. Extreme Value Statistics for Novelty Detection in Biomedical Signal Processing. In Proceedings of the 2000 First International Conference Advances in Medical Signal and Information Processing, Bristol, UK, 4–6 September 2000; pp. 168–172. [CrossRef]

18. Sohn, H.; Allen, D.W.; Worden, K.; Farrar, C.R. Structural damage classification using extreme value statistics. *J. Dyn. Syst. Meas. ControlTrans. Asme* **2005**, *127*, 125–132. [CrossRef]

19. Park, G.; Rutherford, A.C.; Sohn, H.; Farrar, C.R. An outlier analysis framework for impedance-based structural health monitoring. *J. Sound Vib.* **2005**, *286*, 229–250. [CrossRef]

20. Sohn, H.; Park, G.; Wait, J.R.; Limback, N.P.; Farrar, C.R. Wavelet-based active sensing for delamination detection in composite structures. *Smart Mater. Struct.* **2004**, *13*, 153–160. [CrossRef]

21. Sohn, H.; Hyun Woo Park Law, K.H.; Farrar, C.R. Combination of a Time Reversal Process and a Consecutiv Outlier Analysis for Baseline-free Damage Diagnosis. *J. Intell. Mater. Syst. Struct.* **2007**, *18*, 335–346. [CrossRef]

22. Sundaram, S.; Strachan, I.G.D.; Clifton, D.A.; Tarassenko, L.; King, S. Aircraft Engine Health Monitoring using Density Modelling and Extreme Value Statistics. In Proceedings of the Sixth International Conference on Condition Monitoring and Machinery Failure Prevention Technologies, British Institute of Non-Destructive Testing. Northampton, UK, 23–25 June 2009.

23. Clifton, D.A.; Clifton, L.; Hugueny, S.; Wong, D.; Tarassenko, L. An extreme function theory for novelty detection. *IEEE J. Sel. Top. Signal Process.* **2013**, *7*, 28–37. [CrossRef]

24. Papatheou, E.; Dervilis, N.; Maguire, A.E.; Campos, C.; Antoniadou, I.; Worden, K. Performance monitoring of a wind turbine using extreme function theory. *Renew. Energy* **2017**, *113*, 1490–1502. [CrossRef]

25. Fisher, R.A.; Tippett, L.H.C. Limiting forms of the frequency distribution of the largest or smallest member of a sample. *Math. Proc. Camb. Philos. Soc.* **1928**, *24*, 180–190. [CrossRef]

26. Gnedenko, B. Sur La Distribution Limite Du Terme Maximum D'Une Serie Aleatoire. *Ann. Math.* **1943**, *44*, 453. [CrossRef]

27. Rosenblatt, M. A Central Limit Theorem and a Strong Mixing Condition. *Proc. Natl. Acad. Sci. USA* **1956**, *42*, 43–47. [CrossRef] [PubMed]

28. Martucci, D.; Civera, M.; Surace, C.; Worden, K. Novelty Detection in a Cantilever Beam using Extreme Function Theory. *J. Phys. Conf. Ser.* **2018**, *1106*, 012027. [CrossRef]

29. Rasmussen, C.E. Gaussian Processes in machine learning. In *Lecture Notes in Computer Science (Including Subseries Lecture Notes in Artificial Intelligence and Lecture Notes in Bioinformatics)*; Springer: Berlin/Heidelberg, Germany, 2004; Volume 3176, pp. 63–71. [CrossRef]

30. Park, H.W.; Sohn, H. Parameter estimation of the generalized extreme value distribution for structural health monitoring. *Probabilistic Eng. Mech.* **2006**, *21*, 366–376. [CrossRef]

31. Worden, K.; Manson, G.; Sohn, H.; Farrar, C.R. Extreme Value Statistics from Differential Evolution for Damage Detection. In Proceedings of the 23rd International Modal Analysis Conference, Orlando, FL, USA, 3 February 2005.

32. Qin, A.K.; Suganthan, P.N. Self-adaptive Differential Evolution Algrithm for Numerical Optimization. *Proc. IEEE Congr. Evol. Comput.* **2005**, *2*, 1785–1791.

33. Pontillo, A.; Hayes, D.; Dussart, G.X.; Lopez Matos, G.E.; Carrizales, M.A.; Yusuf, S.Y.; Lone, M.M. Flexible High Aspect Ratio Wing: Low Cost Experimental Model and Computational Framework. In *2018 AIAA Atmospheric Flight Mechanics Conference*; American Institute of Aeronautics and Astronautics: Reston, VA, USA, 2018. [CrossRef]

34. Civera, M.; Fragonara, Z.; Surace, C. Video Processing Techniques for the Contactless Investigation of Large Oscillations. *J. Phys. Conf. Ser.* **2019**, *1249*, 12004. [CrossRef]

35. Civera, M.; Zanotti Fragonara, L.; Surace, C. Using Video Processing for the Full-Field Identification of Backbone Curves in Case of Large Vibrations. *Sensors* **2019**, *19*, 2345. [CrossRef]

36. Civera, M.; Zanotti Fragonara, L.; Surace, C. A Computer Vision-Based Approach for Non-Contact Modal Analysis and Finite Element Model Updating. *Lect. Notes Civ. Eng.* **2020**. [CrossRef]

37. Civera, M.; Calamai, G.; Zanotti Fragonara, L. Experimental Modal Analysis of Structural Systems by Using the Fast Relaxed Vector Fitting Method. *Struct. Control Health Monit.* **2021**, e2695. [CrossRef]

38. Civera, M.; Calamai, G.; Zanotti Fragonara, L. System Identification and Structural Health Monitoring of Masonry Bridges by Means of Fast Relaxed Vector Fitting. *Structures* **2021**, *30*, 277–293. [CrossRef]

39. Dooms, D.; Jansen, M.; De Roeck, G.; Degrande, G.; Lombaert, G.; Schevenels, M.; François, S. *StaBIL: A Finite Element Toolbox for Matlab*, 2nd ed.; Structural Mechanics Section of the Department of Civil Engineering, KU Leuven: Leuven, Belgium, 2010.

40. Civera, M.; Boscato, G.; Zanotti Fragonara, L. Treed gaussian process for manufacturing imperfection identification of pultruded GFRP thin-walled profile. *Compos. Struct.* **2020**, *254*, 112882. [CrossRef]

41. Figueiredo, E.; Park, G.; Farrar, C.R.; Worden, K.; Figueiras, J. Machine learning algorithms for damage detection under operational and environmental variability. *Struct. Health Monit. Int. J.* **2011**, *10*, 559–572. [CrossRef]

42. Bovsunovsky, A.; Surace, C. Non-linearities in the vibrations of elastic structures with a closing crack: A state of the art review. *Mech. Syst. Signal Process.* **2015**, *62–63*, 129–148. [CrossRef]

43. Civera, M.; Zanotti Fragonara, L.; Surace, C. A novel approach to damage localisation based on bispectral analysis and neural network. *Smart Struct. Syst.* **2017**, *20*, 669–682.

44. Figueiredo, E.; Park, G.; Figueiras, J. *Structural Health Monitoring Algorithm Comparisons Using Standard Data Sets*; Los Alamos National Lab. (LANL): Los Alamos, NM, USA, 2009.

45. Schoukens, J.; Pintelon, R.; Dobrowiecki, T.; Rolain, Y. Identification of linear systems with nonlinear distortions. *Automatica* **2005**, *41*, 491–504. [CrossRef]

 *applied sciences*

Article

Full-Field Strain Reconstruction Using Uniaxial Strain Measurements: Application to Damage Detection

Rinto Roy [1], Marco Gherlone [1,*], Cecilia Surace [2] and Alexander Tessler [3]

1   Politecnico di Torino, Department of Mechanical and Aerospace Engineering, Corso Duca degli Abruzzi 24, 10129 Torino, Italy; rinto.roy@polito.it
2   Politecnico di Torino, Department of Structural, Geotechnical and Building Engineering, Corso Duca degli Abruzzi 24, 10129 Torino, Italy; cecilia.surace@polito.it
3   Structural Mechanics and Concepts Branch, NASA Langley Research Center, Mail Stop 190, Hampton, VA 23681-2199, USA; alexander.tessler-1@nasa.gov
*   Correspondence: marco.gherlone@polito.it

Abstract: This work investigates the inverse problem of reconstructing the continuous displacement field of a structure using a spatially distributed set of discrete uniaxial strain data. The proposed technique is based on the inverse Finite Element Method (iFEM), which has been demonstrated to be suitable for full-field displacement, and subsequently strain, reconstruction in beam and plate structures using discrete or continuous surface strain measurements. The iFEM uses a variationally based approach to displacement reconstruction, where an error functional is discretized using a set of finite elements. The effects of position and orientation of uniaxial strain measurements on the iFEM results are investigated, and the use of certain strain smoothing strategies for improving reconstruction accuracy is discussed. Reconstruction performance using uniaxial strain data is examined numerically using the problem of a thin plate with an internal crack. The results obtained highlight that strain field reconstruction using the proposed strategy can provide useful information regarding the presence, position, and orientation of damage on the plate.

Keywords: shape sensing; inverse Finite Element Method; structural health monitoring; inverse problem; fiber optics; full-field reconstruction

Citation: Roy, R.; Gherlone, M.; Surace, C.; Tessler, A. Full-Field Strain Reconstruction Using Uniaxial Strain Measurements: Application to Damage Detection. *Appl. Sci.* **2021**, *11*, 1681. https://doi.org/10.3390/app11041681

Academic Editor: Motoharu Fujigaki

Received: 26 November 2020
Accepted: 10 February 2021
Published: 13 February 2021

Publisher's Note: MDPI stays neutral with regard to jurisdictional claims in published maps and institutional affiliations.

1. Introduction

Structural health monitoring (SHM) has been identified as a key technology for the operation and maintenance of future civil, naval, and aerospace structures. An ideal SHM system uses sensors embedded on the structure to provide a real-time assessment of structural integrity. This leads to a reduction in maintenance cost, time, and an overall improvement in structural safety. A variety of SHM methodologies are currently available in the open literature. The primary approach for damage detection is a comparison between the damaged and healthy state of the structure, using certain damage sensitive mechanical features. Some of the most popular SHM methods are based on modal parameters of the structure, where changes in the natural frequencies or mode shapes are used as damage indicators [1]. Similarly, techniques that investigate slope or curvature discontinuities (caused by damage) in the mode shapes have been applied to beam [2,3] and plate [4–6] structures. Data-normalization procedures based on machine learning have also been developed to improve SHM performance under the influence of different operational and environmental conditions [1]. Aside from modal parameters, methods that analyze the strain or displacement field are also used. Here, damage can be detected using inverse modeling approaches [7] or directly by examining the strain or displacement field for any violations of the governing differential equations of the structure [8,9]. The use of fiber optic strain sensors has become increasingly common [10], due to their small size, resistance to electromagnetic interference, reliability, and resistance to weathering and

Appl. Sci. **2021**, *11*, 1681. https://doi.org/10.3390/app11041681   https://www.mdpi.com/journal/applsci

corrosion, making them ideal candidates for long-term health monitoring applications. These advantages have seen them being widely used for SHM of civil structures such as bridges and tunnels [11,12], structures subjected to seismic loads [13], and for monitoring offshore wind turbine structures [14]. The possibility of embedding the fiber within a structure has seen its growing use in aerospace applications. Some of these applications include monitoring of future inflatable space habitats [15] and composite structures, like an aircraft wing box [16,17].

In this context, the use of shape sensing methods for developing strain or displacement-based SHM systems is hugely appealing. Shape sensing refers to the inverse problem of reconstructing the displacement field of a structure using discrete surface strains. The reconstructed strain or displacement field can be analyzed to reveal the presence of damage on the structure. Current shape-sensing methods vary depending on their theoretical approaches to displacement reconstruction. Methods based on integrating experimental strains [18] and using basis functions to approximate the displacement field [19] have been used widely, while the use of neural networks (NN) has also been explored [20].

Another approach for shape sensing is based on a variational principle, such as the inverse Finite Element Method (iFEM). The iFEM is based on discretizing the structural domain into a set of finite elements. The displacement field is obtained by minimizing an error functional defined as the least-square error between the analytic and experimental strain measures [21,22]. The iFEM has been developed for 1D beams and frames [23], 2D plates and shells [24,25], and multi-layered composite and sandwich plates [26,27]. The iFEM can analyze both the static and dynamic response of a structure [23,28], in the linear and non-linear displacement regimes [29], without any prior knowledge of the structure's material properties or loading conditions. The use of iFEM for SHM has been demonstrated on simple beams using fiber optic strain measurements [30] and on thin plates using strain measurements from a grid of strain rosettes [31,32].

The majority of aforementioned iFEM applications used tri-axial strain rosette measurements. In the few cases, where uniaxial strain data were considered, the primary focus was on reconstructing simple membrane or bending deformations of the structure [27,30]. Due to the high measurement density and operational convenience of a fiber optic system, there is enormous potential in using fiber optic strain measurements for the iFEM reconstruction. A possible application is for the SHM of plates or shells, where potential damage can cause local perturbations in the strain field. An accurate reconstruction of these 2D strain perturbations can provide useful information to identify the size, position, and orientation of the damage. However, in comparison with a strain rosette, strain measurements from a fiber optic sensor are uniaxial, i.e., only the component of strain along the local fiber direction is measured. iFEM reconstruction using only uniaxial strain data could lead to errors due to insufficient information regarding the strain field of a structure.

This work addresses the problem of damage detection in a thin cracked plate under the action of in-plane loading. Strain reconstruction is performed using the iFEM methodology in the presence of the discrete uniaxial strain data resulting from several fiber-optic strain sensor patterns. The approach also examines the use of a one-dimensional smoothing algorithm for generating additional input strain data along the paths of the fiber optic sensors. The paper is organized as follows. In Section 2, an iFEM formulation for plate/shell structures is briefly described. In Section 3, the use of iFEM for strain field reconstruction using uniaxial strain data is demonstrated using an example problem of a biaxially loaded thin plate under various internal damage scenarios. The effect of position and orientation of uniaxial strain data are investigated, and the damage detection performance of the reconstructed strain field is also discussed. Finally, Section 4 presents the major conclusions and directions for future work.

2. The Inverse Finite Element Method

The iFEM formulation for plate/shell structures is based on the kinematic assumptions of Mindlin plate theory [21,22]. For a plate of thickness, $2t$, lying in a cartesian coordinate system, the components of the displacement vector are expressed as

$$
\begin{aligned}
u_x(x,y,z) &= u + z\theta_y\,, \\
u_y(x,y,z) &= v - z\theta_x\,, \\
u_z(x,y,z) &= w\,,
\end{aligned}
\tag{1}
$$

where u, v, w, θ_x, and θ_y are the kinematic variables associated with the mid-plane of the plate and are used to describe the displacement vector at any point of the structure. Variables u, v, and w are average displacements along the x, y, and z axis, respectively; θ_x, θ_y are the bending rotations about the x and y axis, respectively (see Figure 1).

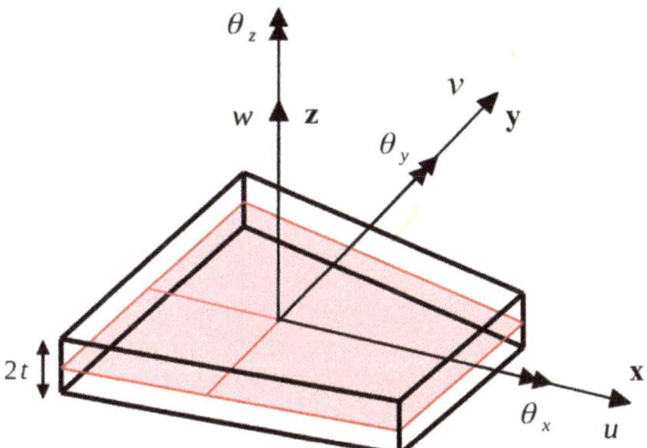

Figure 1. Sign conventions used for the kinematic variables of a 4-node quadrilateral inverse element.

The linear strain displacement relations are used to obtain the strain field from the displacement assumptions of Equation (1)

$$
\left\{
\begin{array}{c}
\varepsilon_{xx} \\
\varepsilon_{yy} \\
\gamma_{xy}
\end{array}
\right\}
=
\left\{
\begin{array}{c}
\varepsilon_{x0} \\
\varepsilon_{y0} \\
\gamma_{xy0}
\end{array}
\right\}
+ z
\left\{
\begin{array}{c}
\kappa_{x0} \\
\kappa_{y0} \\
\kappa_{xy0}
\end{array}
\right\}
\equiv \mathbf{e}(\mathbf{u}) + z\mathbf{k}(\mathbf{u})\,.
\tag{2}
$$

The strain field of Equation (2) is represented by six strain measures; three membrane strain measures, $\mathbf{e}(\mathbf{u})$, representing the in-plane stretching of the mid-plane, and three bending strain measures, $\mathbf{k}(\mathbf{u})$, representing the bending and twisting curvatures. Additionally, Mindlin theory gives rise to two transverse shear strain measures, $\mathbf{g}(\mathbf{u})$. The eight strain measures are given as

$$
\begin{aligned}
\mathbf{e}(\mathbf{u}) &= \left\{\varepsilon_{x0},\ \varepsilon_{y0},\ \gamma_{xy0}\right\}^T = \left\{u_{,x},\ v_{,y},\ u_{,y} + v_{,x}\right\}^T, \\
\mathbf{k}(\mathbf{u}) &= \left\{\kappa_{x0},\ \kappa_{y0},\ \kappa_{xy0}\right\}^T = \left\{\theta_{y,x},\ -\theta_{x,y},\ -\theta_{x,x} + \theta_{y,y}\right\}^T, \\
\mathbf{g}(\mathbf{u}) &= \left\{\gamma_{xz0},\ \gamma_{yz0}\right\}^T = \left\{w_{,x} + \theta_y,\ w_{,y} - \theta_x\right\}^T.
\end{aligned}
\tag{3}
$$

Throughout this study, a four-node plate/shell element, iQS4 [25], is used. The element is formulated using a set of anisoparametric C^0-continuous shape functions that enabled improved kinematic interdependency between the bending and shear deformations. Using the element shape functions, the strain measures can be expressed in terms of the element nodal degrees-of-freedom (DOF)

$$\left\{ \begin{array}{c} \varepsilon_{xx} \\ \varepsilon_{yy} \\ \gamma_{xy} \end{array} \right\} \equiv \mathbf{e}(\mathbf{u}^e) + z\mathbf{k}(\mathbf{u}^e) = \mathbf{B}^m \mathbf{u}^e + z\mathbf{B}^b \mathbf{u}^e, \left\{ \begin{array}{c} \gamma_{xz} \\ \gamma_{yz} \end{array} \right\} \equiv \mathbf{g}(\mathbf{u}^e) = \mathbf{B}^s \mathbf{u}^e, \tag{4}$$

where $\mathbf{B}^m$, $\mathbf{B}^b$ and $\mathbf{B}^s$ are matrices of shape function derivatives corresponding to the membrane, bending, and transverse shear strains, respectively. The vector, $\mathbf{u}^e$, representing the element DOF of each element, e, can be expressed as

$$\mathbf{u}^e = [\mathbf{u}_1^e \ \mathbf{u}_2^e \ldots \mathbf{u}_n^e]^T, \tag{5}$$

where $\mathbf{u}_i^e$ denotes the vector of nodal DOF for each node i, and n denotes the total number of nodes of the element. The membrane and bending strain measures, $\mathbf{e}_j^\varepsilon$ and $\mathbf{k}_j^\varepsilon$, corresponding to experimental surface strain measurements at any location (x_j, y_j), are evaluated as

$$\mathbf{e}_j^\varepsilon = \left\{ \begin{array}{c} \varepsilon_{x0}^\varepsilon \\ \varepsilon_{y0}^\varepsilon \\ \gamma_{xy0}^\varepsilon \end{array} \right\}_j = \frac{1}{2} \left(\left\{ \begin{array}{c} \varepsilon_{xx}^+ \\ \varepsilon_{yy}^+ \\ \gamma_{xy}^+ \end{array} \right\}_j + \left\{ \begin{array}{c} \varepsilon_{xx}^- \\ \varepsilon_{yy}^- \\ \gamma_{xy}^- \end{array} \right\}_j \right), j = 1, \ldots, N, \tag{6}$$

$$\mathbf{k}_j^\varepsilon = \left\{ \begin{array}{c} \kappa_{x0}^\varepsilon \\ \kappa_{y0}^\varepsilon \\ \kappa_{xy0}^\varepsilon \end{array} \right\}_j = \frac{1}{2t} \left(\left\{ \begin{array}{c} \varepsilon_{xx}^+ \\ \varepsilon_{yy}^+ \\ \gamma_{xy}^+ \end{array} \right\}_j - \left\{ \begin{array}{c} \varepsilon_{xx}^- \\ \varepsilon_{yy}^- \\ \gamma_{xy}^- \end{array} \right\}_j \right), j = 1, \ldots, N, \tag{7}$$

where $\left\{ \varepsilon_{xx}^+, \varepsilon_{yy}^+, \gamma_{xy}^+ \right\}$ and $\left\{ \varepsilon_{xx}^-, \varepsilon_{yy}^-, \gamma_{xy}^- \right\}$, are the in-plane normal and shear strains measured on the top and bottom surface of the structure, respectively, and N refers to the total number of strain-sensor locations corresponding to the mid-plane coordinates (x_j, y_j).

The iFEM variational formulation is based on a weighted-least squares error functional that minimizes the least-square errors between the analytic and experimental strain measures. The structural domain is discretized using the customary finite element framework, and the individual inverse finite elements are formulated on the basis of the error functional, given as

$$\Phi_e(\mathbf{u}^e) = \mathbf{w}_e \big\| \mathbf{e}(\mathbf{u}^e) - \mathbf{e}^\varepsilon \big\|_2 + \mathbf{w}_k \big\| \mathbf{k}(\mathbf{u}^e) - \mathbf{k}^\varepsilon \big\|_2 + \mathbf{w}_g \big\| \mathbf{g}(\mathbf{u}^e) - \mathbf{g}^\varepsilon \big\|_2, \tag{8}$$

where $\mathbf{w}_e$, $\mathbf{w}_k$ and $\mathbf{w}_g$ are vectors of weighting coefficients associated with the squared norms. The weighting coefficients are used to enforce a stronger or weaker correlation between the experimental strain measures and those described analytically. In elements where experimental strain measures are known, the coefficient vectors were assigned a value of unity ($\mathbf{w}_e = \mathbf{w}_k = \{1, 1, 1\}$, $\mathbf{w}_g = \{1, 1\}$), and the squared norms are expressed as

$$\big\| \mathbf{e}(\mathbf{u}^e) - \mathbf{e}^\varepsilon \big\|_2 = \tfrac{1}{A_e} \int_{A_e} [\mathbf{e}(\mathbf{u}^e) - \mathbf{e}^\varepsilon]^2 dA,$$

$$\big\| \mathbf{k}(\mathbf{u}^e) - \mathbf{k}^\varepsilon \big\|_2 = \tfrac{(2t)^2}{A_e} \int_{A_e} [\mathbf{k}(\mathbf{u}^e) - \mathbf{k}^\varepsilon]^2 dA, \tag{9}$$

$$\big\| \mathbf{g}(\mathbf{u}^e) - \mathbf{g}^\varepsilon \big\|_2 = \tfrac{1}{A_e} \int_{A_e} [\mathbf{g}(\mathbf{u}^e) - \mathbf{g}^\varepsilon]^2 dA,$$

where A_e is the element area. In specific situations of a pure membrane or bending behavior, Equations (6) and (7) can be simplified further. When the plate is only under in-plane loads, the in-plane normal and shear strains on the top and bottom surfaces of the plate are equal. Strain measurements on only one surface of the plate are required to calculate $\mathbf{e}^\varepsilon$ using Equation (6). In this case, the bending curvatures $\mathbf{k}^\varepsilon$ in Equation (7) vanish identically. Similarly, when the plate is under pure bending, the strains on the top and bottom surfaces have the same magnitude with opposing signs (tension and compression). Measurements on only one surface of the plate are required to calculate $\mathbf{k}^{\varepsilon}$, whereas the membrane strain measures $\mathbf{e}^{\varepsilon}$ vanish identically. In elements, where the experimental strain measures are unknown (due to the absence of experimental data), the coefficient vectors are assigned a relatively small value ($\mathbf{w}_e = \mathbf{w}_k = \{10^{-4},\, 10^{-4},\, 10^{-4}\}$, $\mathbf{w}_g = \{10^{-4},\, 10^{-4}\}$), and the corresponding squared norms reduce to

$$
\left\| \mathbf{e}(\mathbf{u}^e) - \mathbf{e}^{\varepsilon} \right\|_2 = \tfrac{1}{A_e} \int_{A_e} \mathbf{e}(\mathbf{u}^e)^2 dA \,,
$$

$$
\left\| \mathbf{k}(\mathbf{u}^e) - \mathbf{k}^{\varepsilon} \right\|_2 = \tfrac{(2t)^2}{A_e} \int_{A_e} \mathbf{k}(\mathbf{u}^e)^2 dA \,, \tag{10}
$$

$$
\left\| \mathbf{g}(\mathbf{u}^e) - \mathbf{g}^{\varepsilon} \right\|_2 = \tfrac{1}{A_e} \int_{A_e} \mathbf{g}(\mathbf{u}^e)^2 dA.
$$

In cases where only specific components of membrane, bending, or transverse shear strains are experimentally measured at a point, two different strategies can be used for calculating the error functional of Equation (8). The first approach uses a suitable definition of the corresponding weighting coefficient vector. When using uniaxial strain sensors, where only specific components of the in-plane strains are measured at a point, the weighting coefficient vector for the membrane squared norms, $\mathbf{w}_e$, is defined accordingly. For example, if the in-plane strain along the x-axis ($\varepsilon_{x0}^{\varepsilon}$) is the only strain component measured at a point. Then the corresponding weighting coefficient vector could be set as, $\mathbf{w}_e = \{1,\, 10^{-4},\, 10^{-4}\}$. The use of a small value for the weighting coefficient reduces the contribution of those unknown strain components to the element error functional. The second approach employs a smoothing technique. The smoothing technique uses the existing $\varepsilon_{x0}^{\varepsilon}$ strain measurements to obtain a smoothed value of $\varepsilon_{x0}^{\varepsilon}$ at points with no strain data. Using this approach, all unknown components of the membrane, bending or transverse shear strain measures, not experimentally measured at a point, can be obtained by smoothing strain data from other measurement locations of the plate. The latter approach is used in the current work; a detailed explanation of the steps involved is discussed in Section 3.

The transverse shear strain measure, $\mathbf{g}^{\varepsilon}$, cannot be obtained directly using experimental strain measurements. Hence, the transverse shear squared norm (defined in Equation (10)) is associated with a small value of the corresponding weighting coefficient vector ($\mathbf{w}_g = \{10^{-4},\, 10^{-4}\}$) for all elements. The error functional, Φ_e, is solved by minimizing with respect to the nodal DOF, yielding a set of linear algebraic equations

$$
\frac{\partial \Phi_e(\mathbf{u}^e)}{\partial \mathbf{u}^e} = \mathbf{k}^e \mathbf{u}^e - \mathbf{f}^e = 0 \,, \tag{11}
$$

where the matrix, $\mathbf{k}^e$, and vector, $\mathbf{f}^e$, are functions of the strain sensor positions and measured surface strain data, respectively. Both $\mathbf{k}^e$ and $\mathbf{f}^e$ can be expanded and given as

$$
\begin{aligned}
\mathbf{k}^e &= \tfrac{1}{A_e} \int_{A_e} \left[\mathbf{w}_e (\mathbf{B}^m)^T \mathbf{B}^m + \mathbf{w}_k (2t)^2 (\mathbf{B}^b)^T \mathbf{B}^b + \mathbf{w}_g (\mathbf{B}^s)^T \mathbf{B}^s \right] dA, \\
\mathbf{f}^e &= \tfrac{1}{A_e} \int_{A_e} \left[\mathbf{w}_e (\mathbf{B}^m)^T \mathbf{e}^{\varepsilon} + \mathbf{w}_k (2t)^2 (\mathbf{B}^b)^T \mathbf{k}^{\varepsilon} + \mathbf{w}_g (\mathbf{B}^s)^T \mathbf{g}^{\varepsilon} \right] dA.
\end{aligned} \tag{12}
$$

Since the above terms involve area integrals, a suitable numerical integration scheme was used. The global matrices and vectors can be assembled by summing up the contributions from all the inverse elements, N_e,

$$
\mathbf{K} = \sum_{}^{N_e} (\mathbf{T}^e)^T \mathbf{k}^e \mathbf{T}^e \,, \quad \mathbf{F} = \sum_{}^{N_e} (\mathbf{T}^e)^T \mathbf{f}^e \,, \quad \mathbf{U} = \sum_{}^{N_e} (\mathbf{T}^e)^T \mathbf{u}^e, \tag{13}
$$

where matrix, $\mathbf{T}^e$, represents the element coordinate transformation matrix. The finite element assembly results in the global system of algebraic equations, given as

$$\mathbf{KU} = \mathbf{F}. \tag{14}$$

The solution of Equation (14) involves the application of the requisite displacement boundary conditions to restrain the structure against a rigid-body motion. Subsequently, provided $\mathbf{K}$ is non-singular, the DOF vector, $\mathbf{U}$, can be uniquely determined.

3. Numerical Studies

The application of the iFEM method for strain field reconstruction using uniaxial strain data is presented using an example problem of a biaxially loaded square plate. The reconstructed strain field is further assessed to detect the presence of damage on the plate. The plate had a length $L = 3.8$ m, thickness $2t = 3.8$ mm, and is made of an Aluminium alloy (Young's modulus, $E = 73$ GPa, and Poisson's Ratio, $v = 0.3$). The plate is subjected to a uniform biaxial load of magnitude 10^5 N/m. The damage on the plate is modeled as a crack, and three different damage scenarios of the plate (varying size, position, and orientation of the crack) are investigated (see Figure 2a),

- Damage Case-1: a 25 cm long crack is embedded at the center of the plate (crack position coordinates, $\{x_c, y_c\} = \{1.9, 1.9\}$), with the crack front parallel to the vertical axis.
- Damage Case-2: a 10 cm long crack is embedded near the corner of the plate ($\{x_c, y_c\} = \{3.1, 3.1\}$), with the crack front parallel to the vertical axis.
- Damage Case-3: a 25 cm long crack is embedded at the center of the plate ($\{x_c, y_c\} = \{1.9, 1.9\}$), with the crack front oriented at 45^0 with the vertical axis.

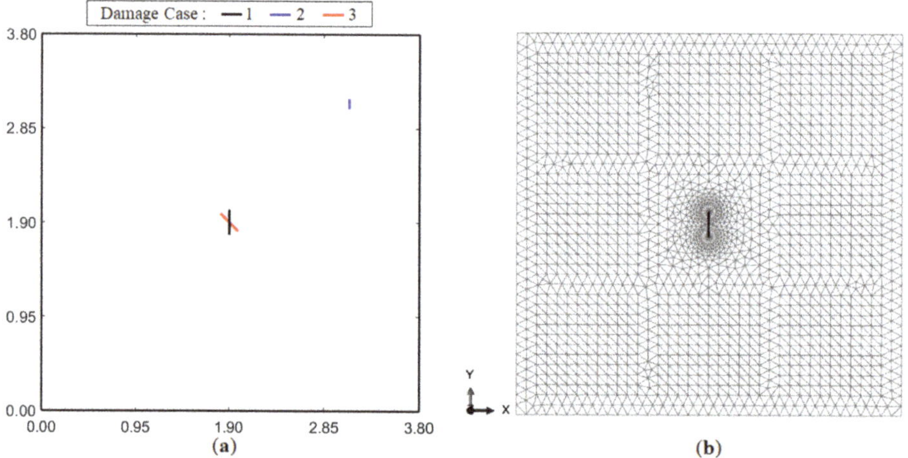

Figure 2. The damaged plate model showing: (**a**) crack size and position for all three damage cases (each crack represents a separate damage scenario), and (**b**) FE mesh used for Damage Case-1.

Because of the pure membrane loading involved in this example problem, only the first term in Equation (8) is relevant for this iFEM application [33]. This is because the membrane response is totally decoupled from the bending and transverse-shear deformations.

A high-fidelity FE model of the plate is developed in ABAQUS using the S3R element, a 3-node constant-strain shell element with reduced integration. A mesh convergence study is performed to ensure the convergence of the FE results. The FE model provided the simulated experimental strain measurements required for the iFEM analysis. The embedded crack is modeled in ABAQUS using the seam feature [34], which created a set of overlapping duplicate nodes along the crack front. The FE model for Damage Case-1 used a total of 3368 elements, with a dense mesh near the crack tip (see Figure 2b). The uniform biaxial loading condition is implemented in the FE model by prescribing a uniform distributed load on the top and right edge of the plate and imposing symmetric displacement boundary conditions on the left and bottom edge of the plate. The iFEM analysis is carried out using the 4-node iQS4 element [25]. The iFEM mesh used has 16 elements along the plate length, leading to a total of 256 inverse elements (mesh is shown in Figure 3). The absence of suitable displacement boundary conditions in the iFEM model can lead to a singular system matrix, $\mathbf{K}$ (see Equation (14)), resulting in no available iFEM solution. Hence, the present iFEM model used symmetric displacement boundary conditions on the left ($u = w = \theta_y = \theta_z = 0$), and bottom ($v = w = \theta_x = \theta_z = 0$) edge of the plate (similar to the FE model).

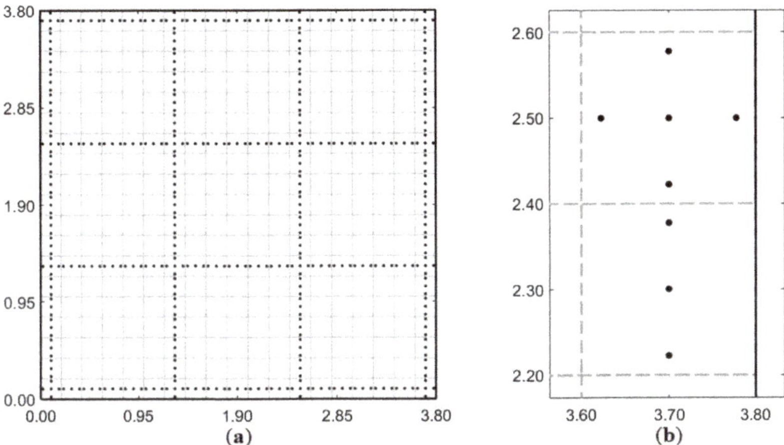

Figure 3. Sensor configuration of the benchmark inverse Finite Element Method (iFEM) model: (**a**) strain rosette grid (each point represents a strain rosette and the sensors divide the plate into nine square cells), and (**b**) magnified view of boundary elements with a maximum of five, and a minimum of three rosettes per element.

The results of the iFEM reconstruction are presented as contour plots of the maximum principal strain

$$\varepsilon_p = \frac{\varepsilon_{xx} + \varepsilon_{yy}}{2} + \sqrt{\left(\frac{\varepsilon_{xx} - \varepsilon_{yy}}{2}\right)^2 + \gamma_{xy}^2}. \tag{15}$$

A damage index, I_D, which provides a normalized value of ε_p within the range $[0, 1]$ is also proposed

$$I_D = \frac{\varepsilon_p - \varepsilon_p\big|_{min}}{\varepsilon_p\big|_{max} - \varepsilon_p\big|_{min}}, \tag{16}$$

where $\varepsilon_p\big|_{min}$ and $\varepsilon_p\big|_{max}$ define the minimum and maximum values of ε_p for a reconstructed strain field. The damage index is used in setting thresholds for damage localization.

3.1. Benchmark iFEM Results

Before investigating the effects of uniaxial strain data, iFEM reconstruction using strain data measured by a dense grid of strain rosettes is used to establish a set of benchmark iFEM results. These results corresponded to a high accuracy iFEM reconstruction and are used as reference results for forthcoming comparisons. As the plate is under biaxial loading, in-plane strains are uniform across the plate thickness. Hence, only strain measurements made either on the top or bottom surface of the plate are required for calculating the experimental strain measures. In the present work, strain measurements made on the top surface of the plate are used. Membrane strain measures, $\mathbf{e}^\varepsilon$ could be calculated using

Equation (6)), while the bending strain measures, $\mathbf{k}^{\varepsilon}$ are equal to zero (Equation (7)). The sensor configuration used for the benchmark results is shown in Figure 3a, where each point designates a strain rosette, and a total of 440 strain rosettes are used. Such a sensor grid divided the plate into nine square cells, where strain measurements are performed only along the cell boundaries. At each sensor position, the ε_x, ε_y and γ_{xy} components of strain are measured. Figure 3b shows a magnified view of the sensor positions within some boundary elements. Depending on element position, strain measurements are available in at least three and at most five points within each element. For this analysis, the integrals of Equation (12) are integrated numerically using the 3×3 Gauss scheme.

Initially, the benchmark iFEM model is used to reconstruct the strain field of an undamaged plate under biaxial loading. The FE and benchmark iFEM results for an undamaged plate are shown in Figure 4. As the plate is under uniform biaxial loading, the in-plane normal strains are uniform, and the shear strain is negligible. Hence, ε_p is a constant over the plate, as evident from the FE results (see Figure 4a). The contour plot of iFEM results also showed a constant value of ε_p, with minor variations of the order of 10^{-8} (which are negligible). Accurate reconstruction of the undamaged far-field strains is essential for the damage detection strategy presented, as it acted as the baseline strain field of the structure. The present results clearly demonstrated the accuracy of the benchmark iFEM model in reconstructing the undamaged strain field. Next, the reconstruction accuracy of a damaged strain field is investigated.

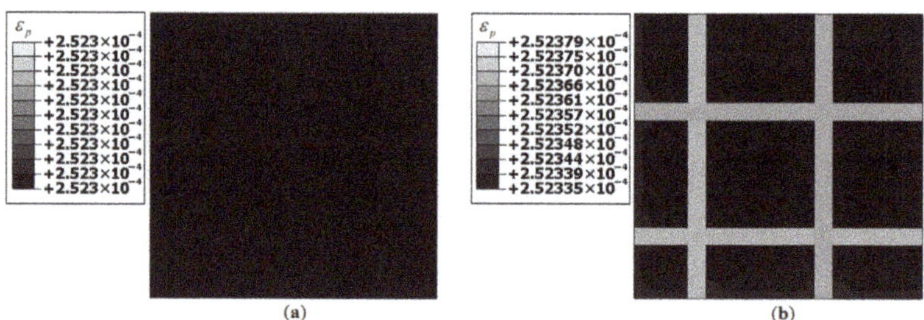

Figure 4. Contour plots of ε_p for an undamaged plate: (**a**) FE results, and (**b**) benchmark iFEM results.

The FE and benchmark iFEM results for Damage Case-1 are shown in Figure 5. Both plots showed a strain concentration in the vicinity of the crack, with the iFEM results having a more diffuse distribution and a lower magnitude of ε_p. This is due to the absence of strain measurements close to the crack tip. The iFEM reconstruction utilized strains measured by the grid of sensors around the crack (see Figure 3), and the sensors measured a lower strain magnitude than at the crack tip. This resulted in a lower magnitude and a more diffuse strain distribution in the iFEM results. Nevertheless, for damage detection, the iFEM results of Figure 5b are promising as they successfully reconstructed a strain concentration at the damage site.

The FE and benchmark iFEM results for Damage Case-2 are shown in Figure 6. In this case, the magnitude of ε_p is smaller due to the smaller damage size. The FE results are highly localized, and the iFEM results showed a more diffuse ε_p distribution at the damage site. The minimum value of ε_p (corresponding to far-field strains) is similar in both plots of Figure 6. This further corroborated the conclusions derived from Figure 4, that the iFEM is accurate in reconstructing the undamaged far-field strains of the plate.

The results for Damage Case-3 are shown in Figure 7. The inferences here are similar to those of Damage Case-1, except for a key feature, i.e., the strain field orientation. Both FE and iFEM results showed a greater ε_p distribution perpendicular to the crack front, i.e., oriented at 45° with respect to the horizontal axis. This indicated that information regarding damage orientation could also be obtained by analyzing the reconstructed strain field.

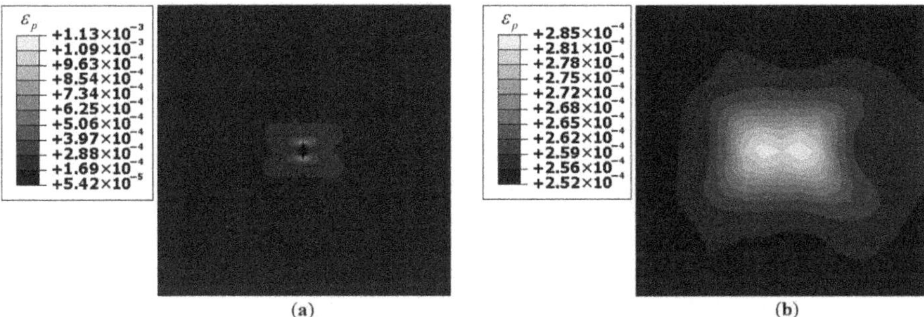

Figure 5. Contour plots of ε_p for Damage Case-1: (**a**) FE results, and (**b**) benchmark iFEM results.

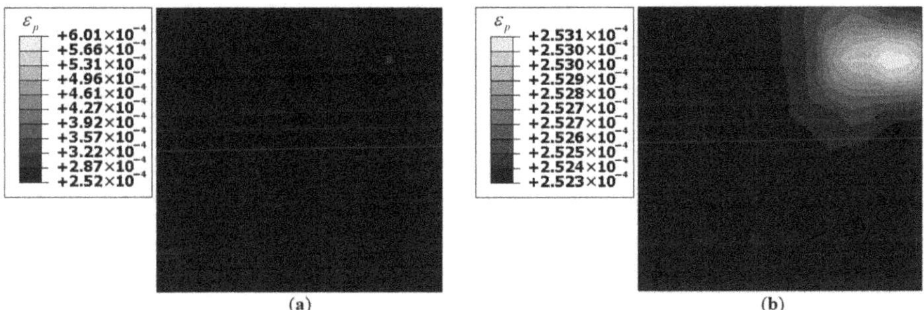

Figure 6. Contour plots of ε_p for Damage Case-2: (**a**) FE results, and (**b**) benchmark iFEM results.

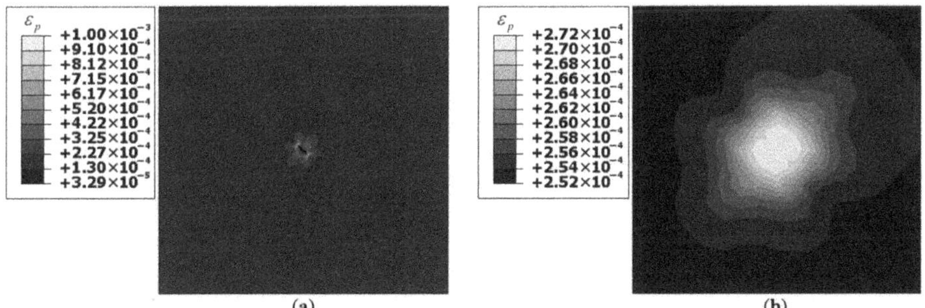

Figure 7. Contour plots of ε_p for Damage Case-3: (**a**) FE results, and (**b**) benchmark iFEM results.

The following conclusions could be drawn from the benchmark iFEM results. The iFEM procedure successfully reconstructed the far-field strains since the strain sensors are located far from the vicinity of the crack. Nevertheless, the presence and location of damage could be inferred from the iFEM results by contrasting the local strain peaks with the far-field strains (that acted as a healthy baseline state of the structure). It is noted that although a highly accurate reconstruction of the damaged strain field across the entire plate domain is preferred, it is not a prerequisite for the presented damage detection methodology. The current study aimed to maximize reconstruction accuracy and minimize the number of sensors used so that meaningful conclusions could be drawn from the reconstructed strain field for the purpose of damage detection. For the damage cases investigated, damage detection is successful regardless of the damage size and orientation. Although these benchmark results are promising, using such a large number of strain rosettes (440) may be generally impractical. This limitation could be overcome by using fiber optic strain sensors as discussed in the next section.

3.2. iFEM Studies Involving Uniaxial Strain Data

Fiber optic sensors are ideal candidates for strain field reconstruction using iFEM as they can provide a dense set of strain measurements over the fiber length. A typical strain rosette measurement system comprises sensors attached via a cable network to a Data Acquisition (DAQ) system. The measurements are processed into viable information and are subsequently stored and analyzed for decision making. In comparison, the cable network of a fiber optic system is composed entirely of the fiber with strain measurements all along the fiber length. The instrumentation systems used for fiber optic sensors depend on the specific sensing technology used [10]. A distributed fiber optic system uses an interrogator to transmit optical pulses into the fiber and capture the backscattered light. The backscattered light is subsequently analyzed to obtain strain measurements along the fiber. In contrast to strain rosettes, fiber optic strain measurements are uniaxial, i.e., along the local fiber direction. Uniaxial strain measurements cannot provide all the components of in-plane strain (ε_x, ε_y, γ_{xy}) at a point, leading to inaccuracies in the iFEM reconstruction. The number and position of strain sensors used also affects the iFEM solution. Sensor configurations with an insufficient number of boundary elements with available strain data can lead to a singular system matrix and a breakdown of the iFEM solution. Similar is the case for configurations with discontinuous sensor patterns. Any proposed sensor configuration should avoid these conditions to ensure accurate iFEM results. In what follows, we describe efforts to reconstruct the strain field in a damaged plate using spatially distributed uniaxial fiber-optic strain sensors. The results are compared with the benchmark iFEM results of Section 3.1.

As the location and orientation of the strain measurements depend on the fiber arrangement, different fiber patterns are investigated. The plate is assumed to be instrumented with one or two continuous fibers to recreate the sensor pattern of the benchmark model (Figure 3). The fiber arrangement should also maximize the quality and quantity of strain measurements, minimize fiber length, and avoid any self-intersections. Based on these requirements, two fiber arrangements are proposed (see Figure 8),

- Configuration-1: a continuous non-intersecting fiber arranged as a wave on the plate. The fiber arrangement within an element is referred to as 'Unitcell-1' (Figure 9a), having the same pattern repeated across multiple elements. Strain measurements corresponding to Unitcell-1 are along the directions: $\{0, \pm90\}$, which is a mixture of ε_x and ε_y strain data at different 3×3 Gauss integration points within an element. At most, three strain measurements are made per element.
- Configuration-2: two continuous non-intersecting fibers are used to create a strain rosette within each element [35]. The fiber arrangement within an element is referred to as 'Unitcell-2' (Figure 9b), and the pattern is repeated across multiple elements. Strain measurements corresponding to Unitcell-2 are along directions: $\{0, \pm60\}$. These three measurements formed a strain rosette and are used to calculate the ε_x, ε_y and γ_{xy} components of strain at the centroid of each element (Figure 9b).

Strain measurements corresponding to Unicells-1 and 2 suffered from certain limitations. The fiber arrangements of Figure 8 are unable to replicate the benchmark pattern of Figure 3, resulting in certain boundary elements of the plate without any strain data (see Figure 10), potentially leading to a singular system matrix, **K**. This issue could be readily resolved by a pre-processing procedure that would enable these elements to be populated with strain data. For this purpose, a one-dimensional (1D) version of the Smoothing Element Analysis (SEA) [36] is used herein.

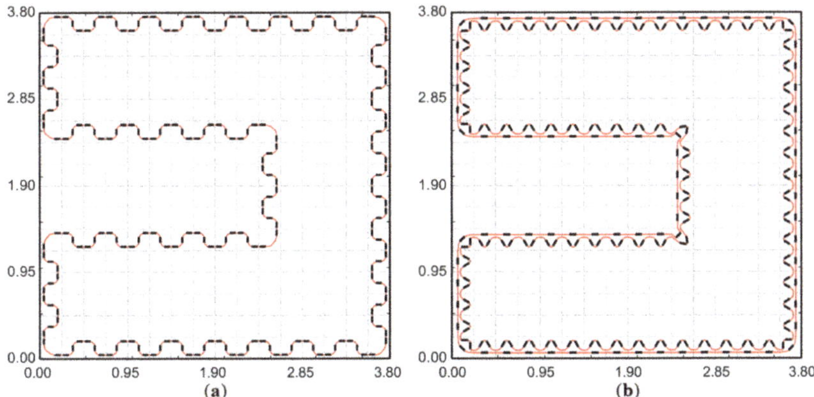

Figure 8. Fiber arrangements investigated: (**a**) Configuration-1 (with a wave-like fiber pattern, referred to as 'Unitcell-1', repeated across multiple elements), and (**b**) Configuration-2 (with a strain rosette fiber pattern, referred to as 'Unitcell-2', repeated across multiple elements).

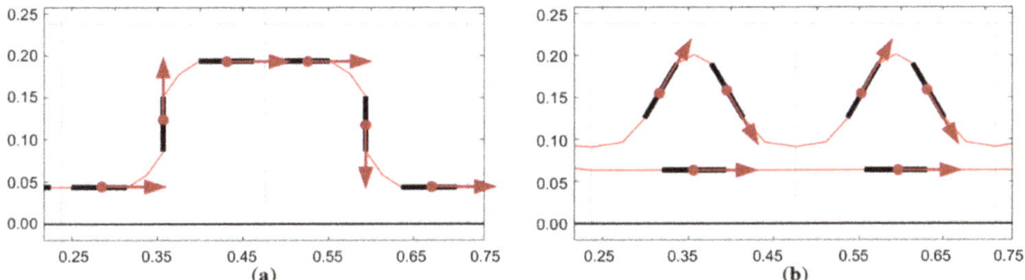

Figure 9. Unitcells corresponding to the fiber arrangements: (**a**) Unitcell-1 (single fiber arrangement measuring a mixture of ε_x and ε_y strains within each element), and (**b**) Unitcell-2 (dual fiber arrangement measuring ε_x, ε_y and γ_{xy} strains at the centroid of each element).

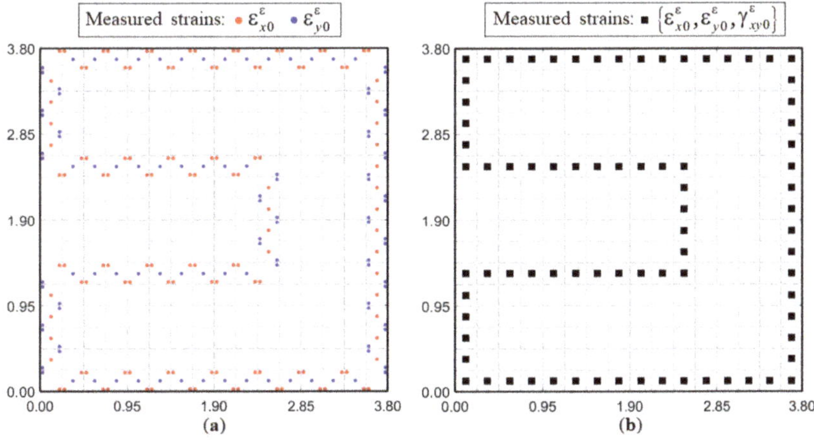

Figure 10. In-situ strain measurements and their locations corresponding to: (**a**) Unitcell-1, and (**b**) Unitcell-2.

The SEA is a finite element (FE) based method that has been used as an FE post-processing tool for stress or strain 'recovery' (improved prediction) and error estimation. It is a variational based approach where the structural domain is discretized using smoothing elements. The problem is solved by minimizing a discrete least-squares error functional enforcing the continuity of the strains and their derivatives. Thus, the SEA recovers C^1-continuous strains with C^0-continuous derivatives. As the present work focused on a 1D formulation of the SEA, a two-node linear smoothing element with quadratic interpolation of strains and linear interpolation of the strain derivatives is used. The accuracy of the smoothed, SEA-generated strain data depended on the interpolation order of the smoothing element used and the complexity of the plate's strain field. The SEA is expected to provide high accuracy results for an undamaged plate under uniform biaxial loading as the in-plane normal strains had uniform distributions, and the contribution of in-plane shear strain is negligible. For a damaged plate, recovering the strain field due to the crack is more challenging. In the presented context, however, the role of the SEA-based strain smoothing is to provide relatively accurate strain fields along the boundary of the plate. In addition to the in-situ measured strains, these smoothed strains would become input strains in the iFEM analysis.

Figure 11 shows the use of the SEA on the strain data set of Unitcell-2. Along the grid lines, the SEA is used to interpolate existing FE strain data. Each grid line is discretized into a series of 1D smoothing elements, with at least one FE strain data point per element. The 1D SEA is used to smooth each in-plane strain component (ε_x, ε_y and γ_{xy}) individually. At the end of the smoothing procedure, each element of the grid would have tri-axial strain data at the element centroid. A similar SEA approach is used for Unitcell-1 as well. The use of fiber optics enabled strain measurements along long one-dimensional sensor paths, making it suitable for the 1D SEA methodology. However, different smoothing strategies, such as using 2D SEA elements or an alternate smoothing strategy, may also be used depending on the problem.

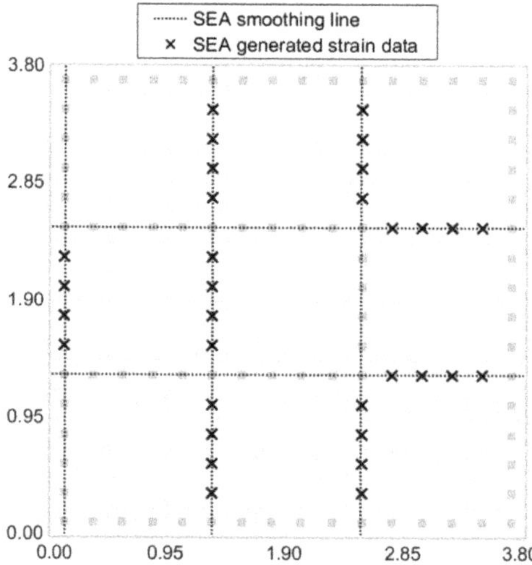

Figure 11. Use of Smoothing Element Analysis (SEA) for the strain data set of Unitcell-2; the position of the smoothing lines and the smoothed strain data generated have been shown.

Note that Unitcell-1 lacked shear strain measurements, potentially leading to an inaccurate iFEM reconstruction. This issue is resolved by assuming a constant value of in-plane shear strain for all elements along the grid lines. A few additional strain rosettes are introduced in Configuration-1 to calculate shear strains at some discrete locations, and their average is used to represent the constant shear strain value for all elements along the grid lines. Although theoretically incorrect, this assumption is used to improve the quality of strain data available for the iFEM reconstruction. As the presented problem explored the plate cases under uniform biaxial loading, this assumption is expected to be moderately successful. However, in alternative problems involving more complex loading scenarios, the effect of in-plane shear strain would be more prominent, and this assumption would lead to erroneous results. The validity of this assumption will be discussed further when discussing the results in the following sections.

Different numerical integration strategies are employed for the iFEM analysis using strain data corresponding to the two Unitcells. As Unitcell-1 provided at most three strain measurement points within each element, the 3 × 3 Gauss scheme is used. In contrast, Unitcell-2 had only one measurement point within each element. Hence, the 2 × 2 Gauss scheme is used for Unitcell-2, with the same centroidal strain data used at all 4 Gauss points. This also led to an interesting investigation into using a constant set of strain data for integrating within an iQS4 element.

3.2.1. Results for Damage Case-1

The iFEM results for Damage Case-1, using strain data corresponding to Unitcells-1 and 2 are shown in Figure 12.

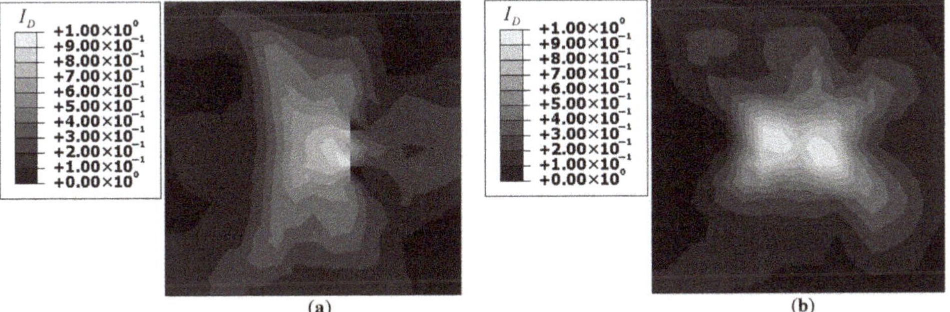

Figure 12. iFEM reconstructed I_D distribution for Damage Case-1 using: (**a**) Unitcell-1, and (**b**) Unitcell-2.

Results for Unitcell-2 are similar to the benchmark results, while Unitcell-1 results are deficient. Both plots showed a I_D concentration at the center of the plate; however, the results of Unitcell-1 are asymmetric. A greater I_D concentration is observed in the vicinity of those elements along which actual strain measurements are made using the optical fiber (see Figure 12a). This asymmetry is not observed in the results of Unitcell-2. A comparison with the benchmark results revealed differences in the strain field distribution around the damage site. Compared to the benchmark results, the results of Unitcell-2 showed a greater dispersion of the damaged strain field to far-field locations. As the benchmark model used a greater number of strain-sensors along with a symmetric sensor grid, the results are symmetric with a well-defined strain peak at the center (see Figure 5b). Despite these shortcomings, the plots of Figure 12 offered sufficient information for successful damage detection.

Next, the effect of measurement noise on the results is investigated. The strain data is contaminated using random noise, added as a percentage of the strain magnitude. The noise distribution is based on a Gaussian curve with zero mean and the value of three standard deviations equal to 5%. The use of random noise in the measured strain data allowed for the introduction of various external factors that affected the method's practical implementation. One key issue is the sensor bonding to the host structure [37]. As the strain is transmitted from the host structure through the bonding surface (typically an epoxy adhesive) to the sensor, a strong and complete bond must be ensured for ideal strain transfer to the sensor and avoid any measurement errors. A similar issue is faced when fiber optic sensors are embedded within structures. Incorrect inclusion could lead to stress concentration between the structure and the sensor leading to erroneous measurements and potential damages to the host structure. Strain transfer analysis between the fiber cable, protective layer, bonding material, and the host structure [38,39] offers a way of identifying the parameters influencing the actual and measured strains and could be used for improving the strain measurements using fiber optic sensors. The current study adopted a more simplistic approach by using random noise to introduce these experimental effects in the numerical examples. The contour plots of I_D using contaminated strain data are shown in Figure 13.

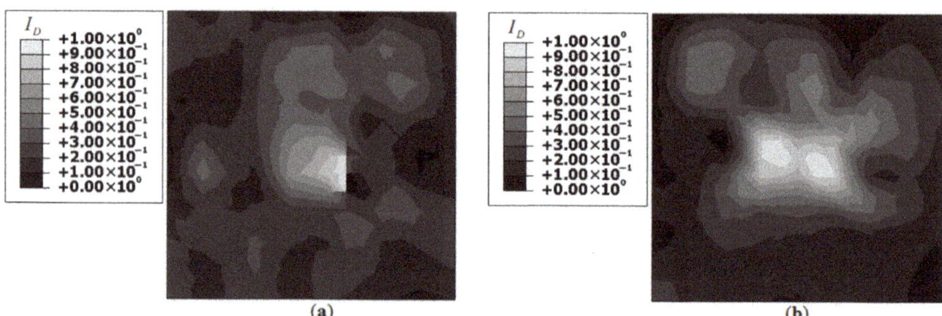

Figure 13. iFEM reconstructed I_D distribution for Damage Case-1 (5% noise) using: (**a**) Unitcell-1, and (**b**) Unitcell-2.

Figure 13 shows that the introduction of noise lead to further diffusion of the strain field, with a prominent I_D peak at the center and smaller I_D peaks at far-field locations. This could be better illustrated using a I_D threshold of 0.5, i.e., the lower limit of the I_D plot is restricted to 0.5. The contour plots with this threshold level are shown in Figure 14.

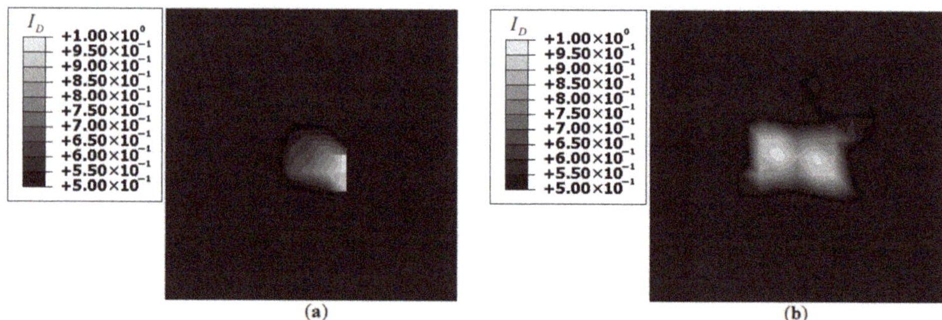

Figure 14. iFEM reconstructed I_D distribution for Damage Case-1 (5% noise and threshold = 0.5) using: (**a**) Unitcell-1, and (**b**) Unitcell-2.

The use of a threshold helped to discriminate between regions with and without damage. Figure 14 shows that the damage is located at the center of the plate. As discussed previously, asymmetry in Figure 14a may lead to a false conclusion that the damage is not at the center but slightly off centric. Similarly, the dual peaks observed in Figure 14b may also lead to a false interpretation of multiple damages present. Instead, these I_D distributions could be considered a feature of the method. These results are compared with benchmark iFEM results for Damage Case-1 using strain data contaminated with 5% noise and enforcing a threshold of 0.5 for isolating the damage location. These benchmark results are shown in Figure 15. The threshold enforced I_D plot (Figure 15b) depicted similar results to those of Unitcell-2 in terms of the distribution and location of the I_D peaks. Both cases showed the dual I_D peak at the center of the plate and no peaks at far-field locations. As demonstrated here, the use of a threshold to define a damage region rather than an exact point constituted an improved damage detection strategy for the present problem.

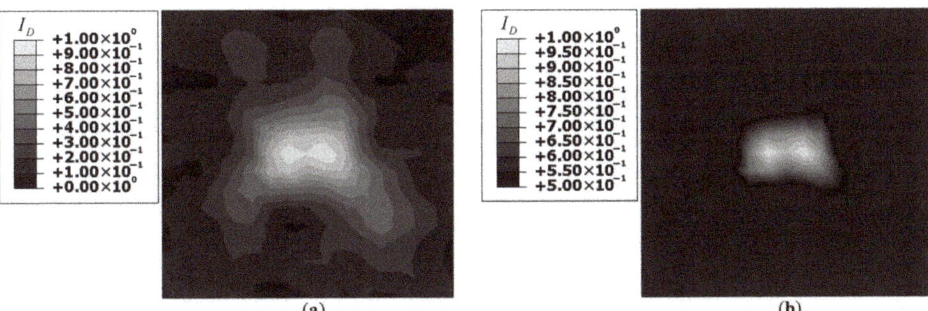

Figure 15. Benchmark iFEM results of I_D for Damage Case-1 (5% noise) using: (**a**) no threshold, and (**b**) threshold = 0.5.

3.2.2. Results for Damage Case-2

The iFEM results for Damage Case-2 are shown in Figure 16. Despite the smaller damage size, the plots clearly showed a I_D peak near the actual damage site. Compared to the benchmark results (Figure 6), the strain concentration is more diffuse; nevertheless, the plots illustrated the presence of damage near the corner of the plate. An interesting point to be noted is that even for Unitcell-2, the strain distribution far from the damage site is more prominent. This is not the case in the benchmark results where the far-field strains are virtually unaffected by the damage. The leakage of the strain field and subsequent contamination of the far-field strains are limitations of the Unitcell strategy. This illustrated potential difficulties in using the present strategy for detecting small-sized damages.

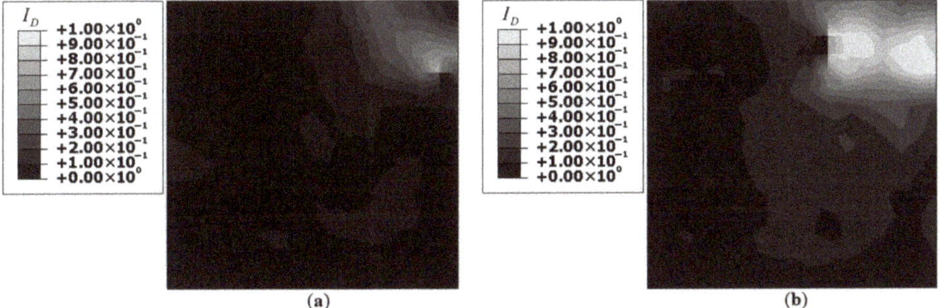

Figure 16. iFEM reconstructed I_D distribution for Damage Case-2 using: (**a**) Unitcell-1, and (**b**) Unitcell-2.

The iFEM results obtained using strain data contaminated with 5% noise are shown in Figure 17.

The plots of Figure 17 showed numerous I_D peaks at various locations of the plate. None of the peaks coincided with the actual damage site, leading to erroneous conclusions when used for damage detection. These results are compared with benchmark iFEM results for Damage Case-2 using contaminated strain data (5% noise). These benchmark results are shown in Figure 18, and they also indicated a drop in damage detection performance due to the addition of noise. Compared to Unitcell-2, the benchmark results showed a prominent I_D peak near the damage site (Figure 18b). The benchmark results also had the same limitations seen for Unitcell-2, where minor I_D peaks are seen at far-field locations. However, these results provided one significant inference: there is a lower limit to damage size that could be successfully detected using the proposed methodology. Successful damage detection is not possible as the strain perturbations due to the damage, measured by the strain-sensors, are smaller than the strain perturbations due to the added noise. It also pointed to possible alternative scenarios where similar-sized damages could be detected, e.g., when the damage is closer to one of the sensors and measures a higher strain perturbation.

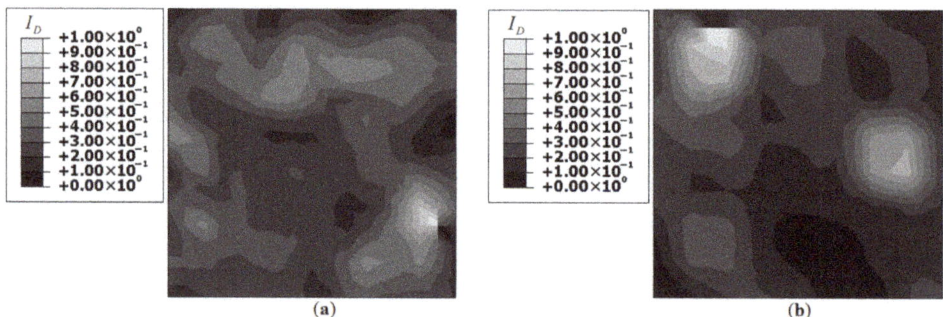

Figure 17. iFEM reconstructed I_D distribution for Damage Case-2 (5% noise) using: (**a**) Unitcell-1, and (**b**) Unitcell-2.

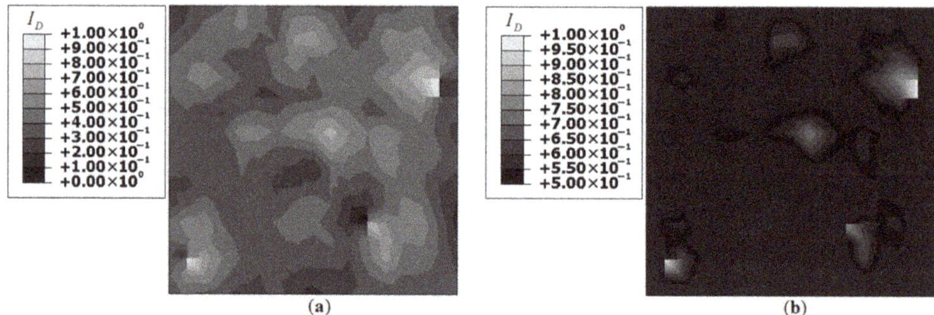

Figure 18. Benchmark iFEM results of I_D for Damage Case-2 (5% noise) using: (**a**) no threshold, and (**b**) threshold = 0.5.

3.2.3. Results for Damage Case-3

The iFEM results for Damage Case-3 are shown in Figure 19. As seen in the benchmark results, Figure 19 showed a I_D peak at the center of the plate and the damage orientation is reflected in the I_D distribution. Compared to the previous two cases, the results of Unitcell-2 are quite similar to the benchmark results, particularly in the regions around the damage and the strain distribution far away from the damage location.

Results of Unitcell-2 (Figure 19b) showed a greater distribution of I_D perpendicular to the crack front. However, Unitcell-1 results are relatively symmetric, and no significant inference could be made regarding the damage orientation. The iFEM results obtained using strain data contaminated with 5% noise are shown in Figure 20.

The contour plots of Figure 20 showed notable changes due to the addition of noise. Unitcell-1 results no longer presented an obvious damage location; however, improved predictions are obtained using Unitcell-2. These results are further refined using a threshold of 0.5; the corresponding plots are shown in Figure 21. The use of a threshold showed multiple I_D peaks in the results of Unitcell-1, making accurate damage detection difficult. In contrast, the results of Unitcell-2 are of improved quality, with a well-defined I_D peak at the center. These results are again compared with the benchmark iFEM results for Damage Case-3 using strain data contaminated with 5% noise. The benchmark results are shown in Figure 22. Compared to Unitcell-2, the benchmark results offered better directional information, with larger I_D distribution perpendicular to the crack front and a less diffuse strain field near the damage site. These results offer improved predictions of the crack position and orientation.

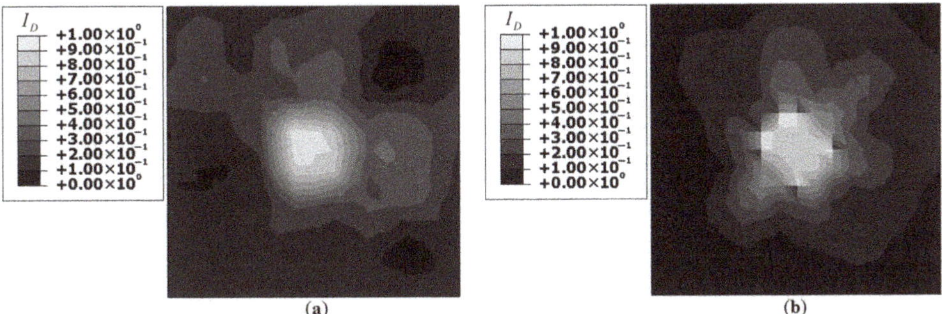

Figure 19. iFEM reconstructed I_D distribution for Damage Case-3 using: (**a**) Unitcell-1, and (**b**) Unitcell-2.

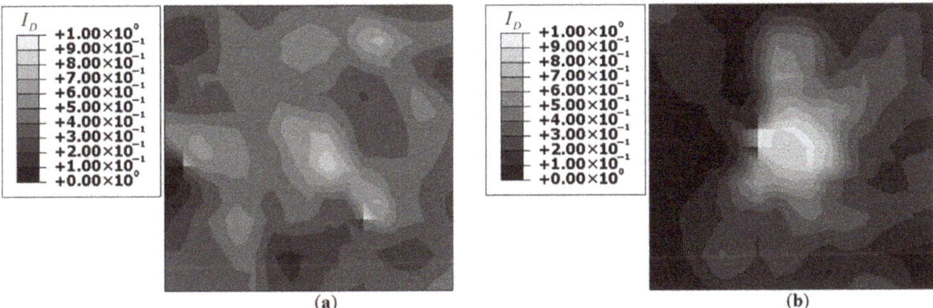

Figure 20. iFEM reconstructed I_D distribution for Damage Case-3 (5% noise) using: (**a**) Unitcell-1, and (**b**) Unitcell-2.

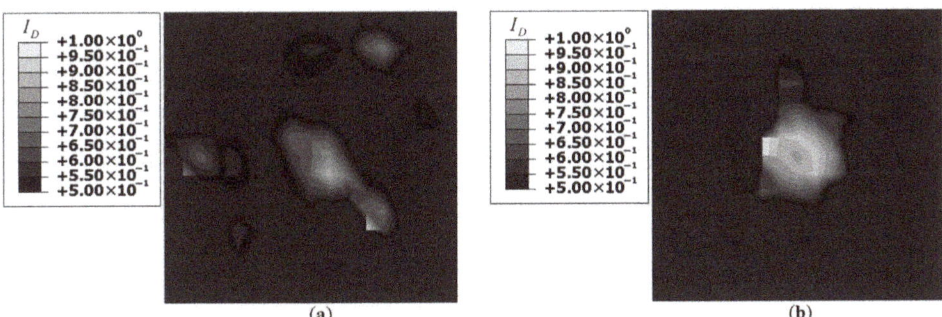

Figure 21. iFEM reconstructed I_D distribution for Damage Case-3 (5% noise, threshold = 0.5) using: (**a**) Unitcell-1, and (**b**) Unitcell-2.

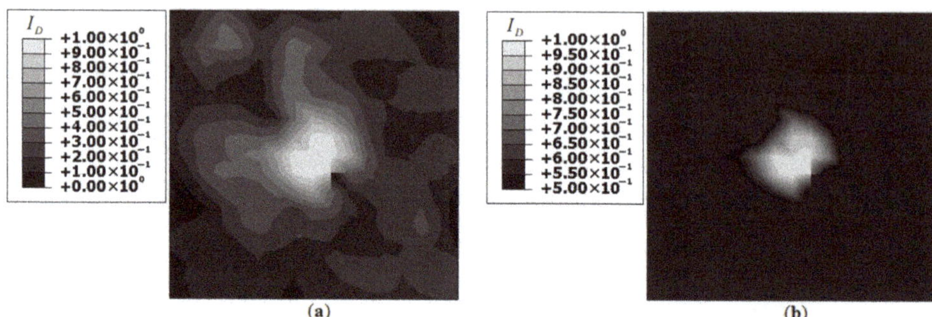

Figure 22. Benchmark iFEM results of I_D for Damage Case-3 (5% noise) using: (**a**) no threshold, and (**b**) threshold = 0.5.

The superior predictions achieved with Unitcell-2 are attributed to its ability to provide the complete three strain component information within an element. Although the presented results focused on a relatively simple load case of a plate under biaxial loading, the real challenge is the strain field reconstruction near the damage site. Near the damage, the strain field distribution is complex, and the in-plane shear strain effect is significant. In this context, it is understandable that those sensor configurations with more in-plane shear strain measurements, i.e., the iFEM benchmark and Unitcell-2, yielded more accurate results. Although Unitcell-1 results are of theoretical interest, in most practical applications, the assumption of a constant in-plane shear strain over the plate is expected to produce somewhat erroneous results. In such situations, it would be of interest to investigate alternate configurations inspired by Unitcell-2.

Aside from the strain-sensor arrangement, the numerical integration scheme used and the smoothing strategy employed also influenced the iFEM accuracy. Both the 2×2 and 3×3 Gauss integration schemes are employed for the current set of results. Although an explicit claim regarding the superiority of one scheme over the other cannot be made, the accuracy of Unitcell-2 when using a constant set of centroidal strain data at all integration points of the 2×2 Gauss scheme is promising. This strategy reduced the number of strain measurement points required within an element, is more computationally efficient, and proved to be robust in the face of measurement noise. The choice of smoothing strategy also affected the iFEM results. Although the 1D smoothing strategy adopted in the present work produced good results, alternate strategies can also be explored and are expected to produce varying degrees of success. Regardless of the specific method considered, the relevant problem is the interpolation order of the smoothing strategy used and the complexity of the strain field investigated. A possible alternative is a 2D SEA scheme, where the plate is discretized using triangular smoothing elements. The refinement of the mesh could also be a contributing factor. The effect of the numerical integration scheme and the smoothing strategy employed are factors worthy of future investigation.

3.3. Damage Detection as a Function of Noise Level in Measured Strain Data

This section presents the results of a numerical study that explored the damage detection quality as a function of the noise level. Section 3.2 reported iFEM results using strain data contaminated with 5% noise, where the two Unitcells reported a mixed level of success. The results of Unitcell-2 for Damage Cases-1 and 3 are successful in detecting damage and provided information regarding damage position and orientation. The effect of noise is further investigated here using Unitcell-2 strain data for reconstructing the strain field of Damage Case-1. The strain data is contaminated incrementally using eight different noise levels from 2.5% to 20%. The results of the study are presented as contour plots of I_D with a threshold of 0.5, and the plots are shown in Figure 23.

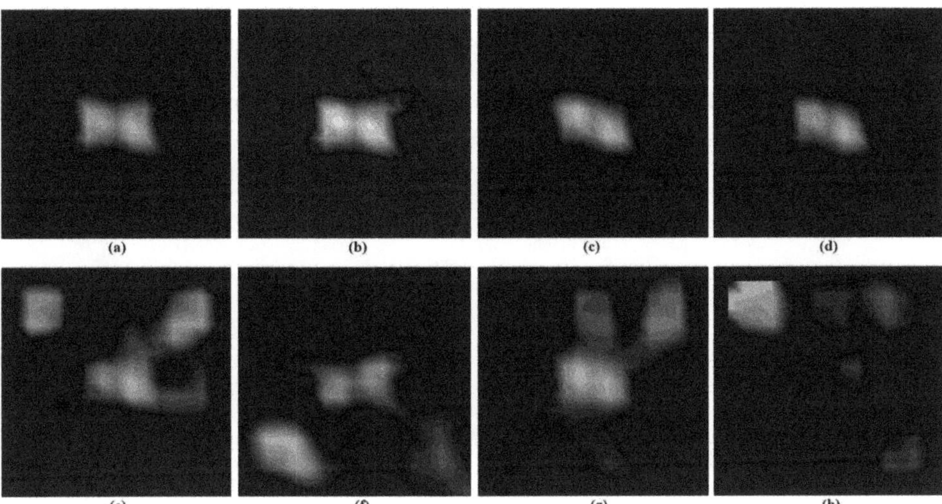

Figure 23. Noise study considering Unitcell-2 results for Damage Case-1; contour plots of I_D for different noise levels (threshold = 0.5): (**a**) 2.5%, (**b**) 5%, (**c**) 7.5%, (**d**) 10%, (**e**) 12.5%, (**f**) 15%, (**g**) 17.5%, and (**h**) 20%.

The study results showed that successful damage detection is possible up to a noise level of 10%. For noise levels below 10%, the prominent I_D peak is observed at the center of the plate, coinciding with the damage location. As the noise level increased further, additional peaks at far-field locations started to become more prominent. Even up to a noise level of 17.5%, it could be claimed that the prominent peak is at the center of the plate, and the use of a higher threshold could lead to successful damage detection by isolating the central peak. But that is no longer the case for noise levels of 20% or greater, as far-field I_D peaks are dominant, and successful damage detection was no longer possible. The results further demonstrate the robustness of Unitcell-2 results in the presence of random noise and are promising for potential practical applications.

4. Conclusions

This paper presented a numerical study using the inverse Finite Element Method (iFEM) to reconstruct the two-dimensional displacement and strain fields of a cracked plate undergoing membrane deformations. The study's main goal was damage detection. The plate was instrumented using fiber-optic strain sensors that measure only uniaxial strains. Several fiber arrangements were investigated, and strain-interpolation strategies were used to improve the quality of strain data used for the iFEM analysis. Various damage scenarios were investigated by varying the damage size, position, and orientation. The effects of measurement noise were also investigated. The location of the damage occurred as a region of strain concentration in the reconstructed strain field, and the iFEM results were successful in detecting and localizing the damage. It was also shown that information regarding the damage orientation is represented in the iFEM results. The addition of measurement noise led to difficulties in detecting small-size cracks because the noise overshadows the strain perturbations due to the actual damage. The use of a fiber pattern resembling a strain rosette produced superior results because it provided greater information regarding the strain field within an element. These results highlight the potential of strain measurements based on fiber optic sensors for practical SHM applications. This research has also revealed that additional post-processing procedures can be investigated to obtain requisite information regarding the position and orientation of damage.

Author Contributions: Methodology, R.R.; software, R.R.; review, A.T., M.G. and C.S.; editing, R.R.; supervision, A.T., M.G. and C.S. All authors have read and agreed to the published version of the manuscript.

Funding: This research received no external funding.

Institutional Review Board Statement: Not applicable.

Informed Consent Statement: Not applicable.

Data Availability Statement: No new data were created or analyzed in this study. Data sharing is not applicable to this article.

Conflicts of Interest: The authors declare no conflict of interest.

References

1. Farrar, C.R.; Worden, K. *Structural Health Monitoring: A Machine Learning Perspective*; John Wiley & Sons Ltd.: West Sussex, UK, 2013.
2. Surace, C.; Archibald, R.; Saxena, R. On the use of the polynomial annihilation edge detection for locating cracks in beam-like structures. *Comput. Struct.* **2013**, *114–115*, 72–83. [CrossRef]
3. Corrado, N.; Durrande, N.; Gherlone, M.; Hensman, J.; Mattone, M.; Surace, C. Single and multiple crack localization in beam-like structures using a gaussian process regression approach. *J. Vib. Control* **2018**, *24*, 4160–4175. [CrossRef]
4. Gherlone, M.; Mattone, M.; Surace, C.; Tassotti, A.; Tessler, A. Novel vibration-based methods for detecting delamination damage in composite plate and shell laminates. *Key Eng. Mater.* **2005**, *293–294*, 289–296. [CrossRef]
5. Surace, C.; Saxena, R.; Gherlone, M.; Darwich, H. Damage localisation in plate like structures using the two-dimensional polynomial annihilation edge detection method. *J. Sound Vib.* **2014**, *333*, 5412–5426. [CrossRef]
6. Corrado, N.; Gherlone, M.; Surace, C.; Hensman, J.; Durrande, N. Damage localisation in delaminated composite plates using a gaussian process approach. *Meccanica* **2015**, *50*, 2537–2546. [CrossRef]
7. Bezerra, L.M.; Saigal, S. A boundary element formulation for the inverse elastostatics problem (IESP) of flaw detection. *Int. J. Numer. Meth. Eng.* **1993**, *36*, 2189–2202. [CrossRef]
8. Wildy, S.J.; Kotousov, A.G.; Codrington, J.D. A new passive defect detection technique based on the principle of strain compatibility. *Smart Mater. Struct.* **2008**, *17*, 045004. [CrossRef]
9. Wildy, S.; Codrington, J. An algorithm for identifying a crack within a measured displacement field. *J. Nondestruct. Eval.* **2017**, *36*, 37. [CrossRef]
10. Glišić, B.; Inaudi, D. *Fiber Optic Methods for Structural Health Monitoring*; John Wiley & Sons Ltd.: West Sussex, UK, 2007.
11. Glisic, B. Long-term monitoring of civil structures and infrastructure using long-gauge fiber optic sensors. In Proceedings of the 2019 IEEE Sensors Conference, Montreal, QC, Canada, 27–30 October 2019; pp. 1–4.
12. Domaneschi, M.; Casciati, S.; Catbas, N.; Cimellaro, G.P.; Inaudi, D.; Marano, G.C. Structural health monitoring of in-service tunnels. *Int. J. Sustain. Mater. Struct. Syst.* **2020**, *4*, 268–291. [CrossRef]
13. Zhang, C.; Alam, Z.; Sun, L.; Su, Z.; Samali, B. Fibre bragg grating sensor-based damage response monitoring of an asymmetric reinforced concrete shear wall structure subjected to progressive seismic loads. *Struct. Control Health Monit.* **2019**, *26*, 2307. [CrossRef]
14. Mieloszyk, M.; Ostachowicz, W. An application of structural health monitoring system based on FBG sensors to offshore wind turbine support structure model. *Mar. Struct.* **2017**, *51*, 65–86. [CrossRef]
15. Ohanian, O.J.; Davis, M.A.; Valania, J.; Sorensen, B.; Dixon, M.; Morgan, M.; Litteken, D. Embedded fiber optic SHM sensors for inflatable space habitats. In *ASCEND 2020, Virtual Event*, 16–18 November 2020; AIAA: Reston, VA, USA, 2020; p. 4049. [CrossRef]
16. Güemes, A.; Fernández-López, A.; Díaz-Maroto, P.F.; Lozano, A.; Sierra-Perez, J. Structural health monitoring in composite structures by fiber-optic sensors. *Sensors* **2018**, *18*, 1094. [CrossRef] [PubMed]
17. Martins, B.L.; Kosmatka, J.B. Detecting damage in a UAV composite wing spar using distributed fiber optic strain sensors. In Proceedings of the 56th AIAA/ASCE/AHS/ASC Structures, Structural Dynamics, and Materials Conference, Kissimmee, FL, USA, 5–9 January 2015.
18. Ko, W.L.; Richards, W.L.; Fleischer, V.T. *Applications of Ko displacement Theory to the Deformed Shape Predictions of the Doubly-Tapered Ikhana Wing*; NASA/TP-2009-214652; NASA Dryden Flight Research Center: Edwards, CA, USA, 2009.
19. Glaser, R.; Caccese, V.; Shahinpoor, M. Shape monitoring of a beam structure from measured strain or curvature. *Exp. Mech.* **2012**, *52*, 591–606. [CrossRef]
20. Bruno, R.; Toomarian, N.; Salama, M. Shape estimation from incomplete measurements: A neural-net approach. *Smart Mater. Struct.* **1993**, *3*, 92–97. [CrossRef]
21. Tessler, A.; Spangler, J.L. *A Variational Principle for Reconstruction of Elastic Deformations in Shear Deformable Plates and Shells*; NASA/TM-2003-212445; NASA Langley Research Center: Hampton, VA, USA, 2003.
22. Tessler, A.; Spangler, J.L. A least-squares variational method for full-field reconstruction of elastic deformations in shear-deformable plates and shell. *Comp. Meth. Appl. Mech.* **2005**, *194*, 327–339. [CrossRef]

23. Gherlone, M.; Cerracchio, P.; Mattone, M.; Di Sciuva, M.; Tessler, A. Shape sensing of 3D frame structures using an inverse finite element method. *Int. J. Solids Struct.* **2012**, *49*, 100–112. [CrossRef]
24. Tessler, A.; Spangler, J.L. Inverse FEM for full-field reconstruction of elastic deformations in shear deformable plates and shells. In Proceedings of the 2nd European Workshop on Structural Health Monitoring, Munich, Germany, 7–9 July 2004.
25. Kefal, A.; Oterkus, E.; Tessler, A.; Spangler, J.L. A quadrilateral inverse-shell element with drilling degrees of freedom for shape sensing and structural health monitoring. *Eng. Sci. Tech. Int. J.* **2016**, *19*, 1299–1313. [CrossRef]
26. Cerracchio, P.; Gherlone, M.; Di Sciuva, M.; Tessler, A. A novel approach for displacement and stress monitoring of sandwich structures based on the inverse finite element method. *Comput. Struct.* **2015**, *127*, 69–76. [CrossRef]
27. Kefal, A.; Yildiz, M. Modeling of sensor placement strategy for shape sensing and structural health monitoring of a wing-shaped sandwich panel using inverse finite element method. *Sensors* **2017**, *17*, 2775. [CrossRef]
28. Roy, R.; Tessler, A.; Surace, C.; Gherlone, M. Shape sensing of plate structures using the inverse finite element method: Investigation of efficient strain–sensor patterns. *Sensors* **2020**, *20*, 7049. [CrossRef]
29. Tessler, A.; Roy, R.; Esposito, M.; Surace, C.; Gherlone, M. Shape sensing of plate and shell structures undergoing large displacement using the inverse finite element method. *Shock Vib.* **2018**, 8076085. [CrossRef]
30. Quach, C.; Vazquez, S.; Tessler, A.; Moore, J.; Cooper, E.; Spangler, J. Structural anomaly detection using fiber optic sensors and inverse finite element method. In Proceedings of the AIAA Guidance, Navigation, and Control Conference and Exhibit, San Francisco, CA, USA, 15–18 August 2005.
31. Roy, R.; Gherlone, M.; Surace, C. Damage localisation in thin plates using the inverse finite element method. In Proceedings of the 13th International Conference on Damage Assessment of Structures, Lecture Notes in Mechanical Engineering; Springer: Singapore, 2020. [CrossRef]
32. Colombo, L.; Sbarufatti, C.; Giglio, M. Definition of a load adaptive baseline by inverse finite element method for structural damage identification. *J. Mech. Syst. Signal Process.* **2019**, *120*, 584–607. [CrossRef]
33. Reddy, J.N. *Theory and Analysis of Elastic Plates and Shells*; CRC Press: Boca Raton, FL, USA, 2007.
34. Dassault Systemes Simulia Corp. *Abaqus Analysis User's Guide*; Dassault Systemes Simulia Corp: Providence, RI, USA, 2013.
35. Moore, J.P.; Przekop, A.; Juarez, P.D.; Roth, M.C. *Fiber Optic Rosette Strain Gauge Development and Application on a Large-Scale Composite structure*; NASA/TM–2015-218970; NASA Langley Research Center: Hampton, VA, USA, 2015.
36. Tessler, A.; Riggs, H.R.; Macy, S.C. A variational method for finite element stress recovery and error estimation. *Comp. Meth. Appl. Mech. Eng.* **1994**, *111*, 369–382. [CrossRef]
37. Measures, R.M. *Structural Monitoring with Fiber Optic Technology*; Academic Press: San Diego, CA, USA, 2001.
38. Sun, L.; Hao, H.; Zhang, B.; Ren, X.; Li, J. Strain transfer analysis of embedded fiber bragg grating strain sensor. *J. Test. Eval.* **2016**, *44*, 20140388. [CrossRef]
39. Sun, L.; Li, C.; Zhang, C.; Liang, T.; Zhao, Z. The strain transfer mechanism of fiber bragg grating sensor for extra large strain monitoring. *Sensors* **2019**, *19*, 1851. [CrossRef] [PubMed]

Article

Rail Diagnostics Based on Ultrasonic Guided Waves: An Overview

Davide Bombarda [1,*], Giorgio Matteo Vitetta [1] and Giovanni Ferrante [2]

[1] Department of Engineering "Enzo Ferrari", University of Modena and Reggio Emilia, 41125 Modena, Italy; giorgiomatteo.vitetta@unimore.it
[2] Alstom Ferroviaria S.p.A., 40128 Bologna, Italy; giovanni.ferrante@alstomgroup.com
* Correspondence: davide.bombarda@unimore.it

Abstract: Rail tracks undergo massive stresses that can affect their structural integrity and produce rail breakage. The last phenomenon represents a serious concern for railway management authorities, since it may cause derailments and, consequently, losses of rolling stock material and lives. Therefore, the activities of track maintenance and inspection are of paramount importance. In recent years, the use of various technologies for monitoring rails and the detection of their defects has been investigated; however, despite the important progresses in this field, substantial research efforts are still required to achieve higher scanning speeds and improve the reliability of diagnostic procedures. It is expected that, in the near future, an important role in track maintenance and inspection will be played by the ultrasonic guided wave technology. In this manuscript, its use in rail track monitoring is investigated in detail; moreover, both of the main strategies investigated in the technical literature are taken into consideration. The first strategy consists of the installation of the monitoring instrumentation on board a moving test vehicle that scans the track below while running. The second strategy, instead, is based on distributing the instrumentation throughout the entire rail network, so that continuous monitoring in quasi-real-time can be obtained. In our analysis of the proposed solutions, the prototypes and the employed methods are described.

Keywords: rail; guided wave ultrasound; broken rail detection; rail diagnostics; structural health monitoring; non destructive testing

Citation: Bombarda, D.; Vitetta, G.M.; Ferrante, G. Rail Diagnostics Based on Ultrasonic Guided Waves: An Overview. *Appl. Sci.* **2021**, *11*, 1071. https://doi.org/10.3390/app11031071

Received: 18 December 2020
Accepted: 19 January 2021
Published: 25 January 2021

Publisher's Note: MDPI stays neutral with regard to jurisdictional claims in published maps and institutional affiliations.

1. Introduction

The track is one of the basic elements of railroading. Since the weight of trains is discharged on a very small portion of rail surfaces, tracks require careful maintenance [1]. It is well known that: (a) rails are subject to intense bending and shear stresses, plastic deformation and wear, leading to progressive degradation of their structural integrity [2]; (b) they may contain internal fabrication defects undetected by quality control. All of this may result in rail breakage and, consequently, in train derailment with potential catastrophic consequences (traffic interruption, possible losses of rolling stock material and even lives, etc.).

According to the European Union agency for railways (ERA) safety overview for 2017 [3], in the years 2011–2015, broken rails have represented the main form of precursors to accidents, i.e., of inconveniences that, under other circumstances, could have led to an accident (other precursors to accidents include track buckles, signals passed at danger, wrong-side signaling failures, broken wheels, and axles). On the average, 4445 broken rails per year have been detected over a total of 11,222 precursors to accidents in the whole European Union (EU-28) in the same period. This explains why track inspection and maintenance play a fundamental role; moreover, these aspects are expected to become more and more crucial since operative loads, traffic, and speeds are progressively increasing [1].

Over the years, various systems have been developed to monitor the health of rails. Since all the currently available solutions do not fully match the requirements set by railway infrastructure management companies, substantial research efforts are still required in this

Appl. Sci. **2021**, *11*, 1071. https://doi.org/10.3390/app11031071

field. In particular, an important challenge is represented by the development of reliable methods for detecting potentially hazardous rail defects before they evolve in rail breakages. In fact, the availability of rail diagnostics methods allow for introducing predictive maintenance procedures instead of the preventive maintenance strategy commonly used nowadays. This will lead to more reliable railway networks, with a substantial increase of the efficiency and economic sustainability of maintenance procedures [4].

In the last two decades, few review articles about the monitoring of rail tracks have been published [4–6]. Moreover, in recent years, new technologies and approaches have been developed in this field, and the interest of authorities and companies has grown. This has motivated the writing of this manuscript that aims at providing a self-contained and comprehensive overview of the techniques and systems for rail defect detection based on *ultrasonic guided waves* (UGWs). This technology is relevant since it allows for accomplishing *non destructive testing* (NDT) and to sound a large area by means of a single transducer [7]. Two technical solutions are available for track monitoring based on UGWs. The first solution is based on a measuring equipment placed on board a moving diagnostic vehicle. It offers the important advantage of an increased inspection speed with respect to conventional bulk ultrasound or eddy currents methods; it is worth remembering that conventional methods suffer from slow inspection speed and, consequently, reduce the infrastructure availability for commercial services. The second solution, instead, relies on a monitoring equipment placed on the ground; this makes possible a constant monitoring of rail status if the employed sensors are distributed along the considered rail network.

The remaining part of this manuscript is organized as follows. In Section 2, a taxonomy of rail defects and the main techniques for their detection are described. A classification of the techniques based on guided ultrasound waves and a description of their architecture is provided in Sections 2 and 4, respectively. Various details about their implementation are illustrated in Sections 5–8, whereas their performance is analyzed in Section 9. Finally, some conclusions are offered in Section 10.

2. An Overview of Rail Defects and of the Techniques for Their Detection

The railroad track system is very complex, and involves many interactions [8]: damage can occur anytime and anywhere. The development of defects in rails is due to uncontrolled and random processes. If not detected in time, defects can lead to rail failures, which, in some cases, happen without any previous indication. The prediction of crack growth rates and of the size of defects at failure are both influenced by various parameters [4,9]. In the following two paragraphs, we focus on the different types of defects and on the techniques that can be employed for their detection.

2.1. Rail Defects

The transverse section of a rail is represented in Figure 1. Rails in the past were joined together using fishplates; this procedure has been replaced by welding that allows for developing long and continuous stretches of rails forming a *continuously welded rail*.

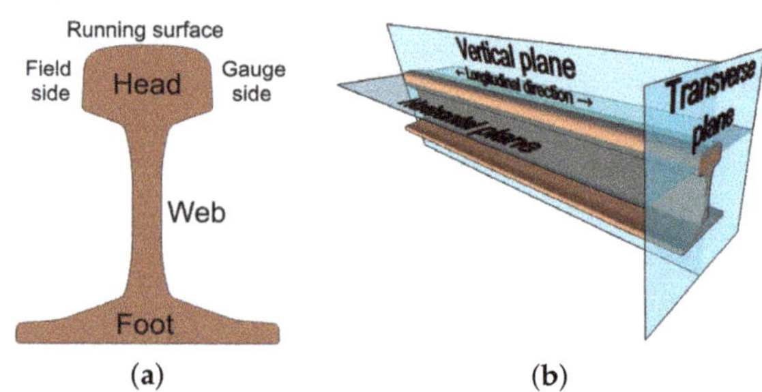

Figure 1. Representation of the terminology commonly adopted in the description of rails: (**a**) transverse section; (**b**) reference planes.

Various classifications of rail defects are adopted in the technical literature. Following Ref. [10], a rail track defect could be either of *geometry* or *structural* type. Track geometry defects are related to the geometric conditions of tracks; track structural defects, instead, refer to ill-conditioned structural parameters of tracks (including rails [8]). In the remaining part of this paragraph, we take into consideration track structural defects only; these are divided into the three broad categories described below.

2.1.1. Rail Manufacturing Defects

Rail manufacturing defects usually comprise inclusions or incorrect local mixings in rail steel; these generate localized stresses under operative load, which, in turn, can trigger a rail failure process [1,11]. Damages of this category include *transverse defects* (TD) and *longitudinal defects* (LD) [1,12,13]. The former type of defects consists of a progressive fracture developing in the railhead parallel to the transverse direction [1] (see Figure 2); the latter type, instead, is represented by an internal progressive fracture propagating longitudinally in rails. Longitudinal defects can be further subdivided into *vertical* and *horizontal* split heads (see Figure 3).

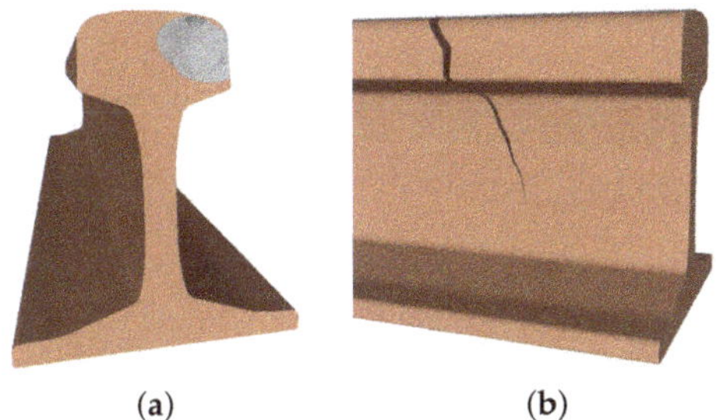

(a)   (b)

Figure 2. Three-dimensional representation of transverse defects: (**a**) section with shell; (**b**) lateral view.

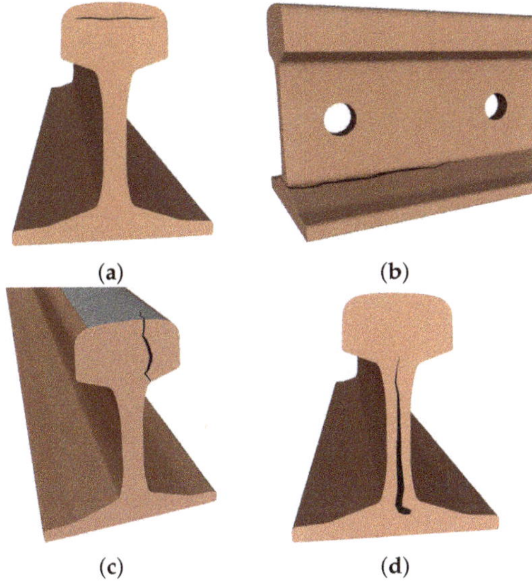

(a)   (b)

(c)   (d)

Figure 3. Three-dimensional representation of longitudinal defects: (**a**,**b**) horizontal fissures; (**c**,**d**) vertical fissures.

2.1.2. Defects Due to Improper Use or Handling of Rails

Defects related to improper use or handling of rails are usually caused by train wheels spinning on rails (this produces the so-called *wheelburn* defect, shown in Figure 4) or by sudden train brakes [1].

Figure 4. Three-dimensional representation of a wheelburn defect.

2.1.3. Defects Due to Rail Wear and Fatigue

Rail wear and fatigue defects are due to wearing mechanisms of the rolling surface and/or to fatigue. Well known examples of this category are: (a) *corrugation*; (b) *rolling contact fatigue* (RCF) damages; (c) *bolt-hole cracks* [1].

Corrugation is related to the wearing of railhead [14] and does not compromise rolling safety, but affects both track elements and rolling stock, since it increases noise emission, loading, and fatigue [15].

Rolling contact fatigue damages are much more severe from the point of view of structure integrity, as they could lead to complete rail failure [16–18]. Independently of any material defect, fatigue cracks initiate on (or very close to) the rail running surface [17]. The RCF damage can be further subdivided in: (a) *checking* and possible *spalling*; (b) *shelling*; (c) *squats* (see Figure 5) [19]. The occurrence rate of RCF is proportional to the speed of the train or to its weight [1]. To counteract RCF, the damaged rail can be prematurely removed in the most severe cases or ground to remove the surface-initiated crack. The running surface can also be lubricated; however, it is believed that fluid entrapment in metal can speed up the growth of a surface-initiated crack [16].

Bolt-hole cracks appear in joined tracks; as shown in Figure 6, they originate on the surface closer to bolt-holes and propagate with a $\pm45°$ angle from the vertical up to reaching web-railhead junctions. These defects may originate from the fretting fatigue caused by the bolt shank against the surface of bolt-holes [1].

In welded rails, critical points requiring careful inspections are represented by welds (see Figure 7); in fact, internal defects of welds can affect structural integrity and fatigue performance, bringing rapid failures [5].

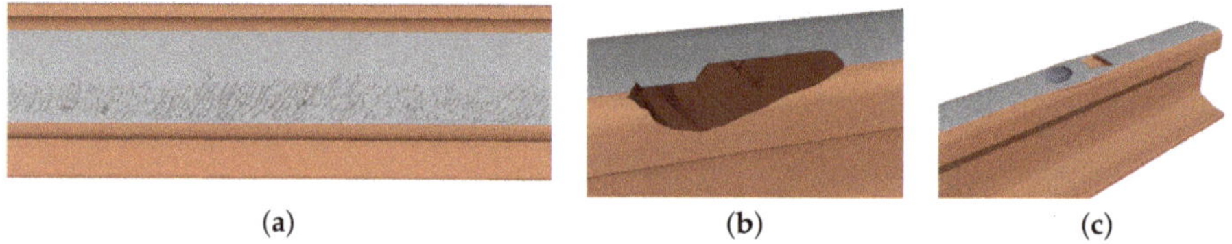

(a)   (b)   (c)

Figure 5. Three-dimensional representation of defects originating from RCF: (a) gauge corner checking and spalls can produce (b) shelling and (c) squats.

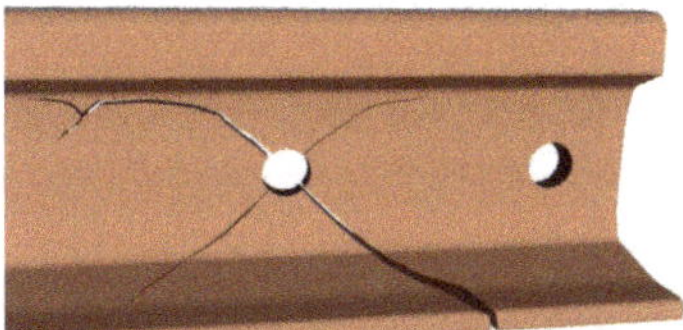

Figure 6. Three-dimensional representation of bolt-hole cracks.

Figure 7. Thermite weld joining two rails with different profiles (50E5, UNI-50, on the right and 60E1, UNI-60, on the left).

2.1.4. Defect Growth

The failure due to cracks of metallic structures goes through the following three phases: (a) *crack initiation*; (b) *crack growth*; (c) a subsequent *quick crack growth* culminating in a total fracture. Specific techniques can be used to detect cracks when they are in their earlier stages [8], so that the third phase is not reached [16]. Since defect development is due to uncontrolled and random processes or events, this process does not lend itself to a simple statistical description [4].

2.2. Rail Diagnostics Techniques

Different inspection techniques for *non destructive testing* (NDT) of rails are available and currently in use. These can be divided in the following classes [5]: (a) *eddy current (EC)-based methods*; (b) *ultrasonic techniques*; (c) *visual inspection techniques*; (d) *thermal techniques*; (e) *radiographic techniques*. These techniques require that some instrumentation is installed on a vehicle (i.e., a train or a special vehicle) to acquire the required measurements while moving on tracks.

2.2.1. EC-Based Methods

These methods include *eddy current* inspection and *magnetic flux leakage* (MFL). The method based on eddy current measures the material response to an induced electromagnetic field: the presence of a surface or near-surface defect produces a distributed electromagnetic field that can be measured. This method is contactless, but very sensitive to probe lift-off from the surface of test specimens [20]. The method based on MFL consists of magnetizing the object to be tested and in scanning its surface by means of a flux-sensitive sensor [21]. In the presence of a defect, the magnetic flux (usually contained into the magnetized test object) leaks into air [22]. This method is able to detect superficial or near-superficial transverse fissures (like RCF) but is not suited to longitudinal fissures, and its performance is influenced by the selected scanning speed [23].

The inspection speeds of the systems based on the two methods described above are limited (75 km/h for EC testing, 35 km/h for MFL) and smaller than those characterizing revenue service trains [5].

The limitations of EC inspection of rails are analyzed in Ref. [24]. Some of the most recent advances in EC-based testing of rails are illustrated in Refs. [25,26]. Finally, it is worth mentioning that a recent implementation of the MFL technology can be found in Ref. [27].

2.2.2. Ultrasonic Techniques

These comprise *conventional bulk wave ultrasonics*, *phased arrays*, and *ultrasonic guided waves* (A brief introduction to ultrasonic guided waves and to their propagation in rails is provided in Appendices A and B, respectively.) (UGW) techniques; the last class of techniques can be further divided into *on-board* and *land-based* systems. An introduction to ultrasonic techniques applied to rail diagnostics is given in Section 2.3.

2.2.3. Visual Inspection

Visual inspection can be exploited by a human expert walking on tracks and searching for defects, or by automated techniques. Automated visual inspection is based on video cameras, optical sensors, and custom-designed algorithms to analyze the rail surface as the diagnostic train rides over [28–31]. Test speeds can be as high as the velocity of high-speed trains [4], but only the visible surface of the rail is inspected and internal defects cannot be detected [5]. Furthermore, since detection performance is influenced by lighting conditions, proper countermeasures need to be adopted [32]. An example of a high-speed diagnostic train also making use of visual inspection techniques is shown in Figure 8 [33].

Figure 8. ETR.500 Y 2 "Dia.man.te", high speed (300 km/h) visual inspection EMU (courtesy of © Stefano Paolini).

2.2.4. Thermal Techniques

These techniques exploit the change in the thermal properties of rail material in the presence of a defect. The presence of an anomaly is inferred from a single frame or from a video sequence of the temperature distribution of the surface of the considered specimen [5,34]. Some recent approaches are illustrated in Refs. [35,36].

2.2.5. Radiographic Techniques

Radiography is the only technique able to analyze visually the interior of a rail or a weld. This method can detect cracks, flaws, and thickness reduction in detail, but has safety hazard concerns, is expensive, and time consuming. For these reasons, it is often used to analyze rails; only after that have other NDT techniques detected a specific problem [5].

2.2.6. Track Circuits

Track circuits are systems used to detect the presence of a train in a stretch (*section*) of track by means of an electrical current. A transmitter and a receiver placed at the two ends of the considered track section are exploited. In the absence of trains, an electrical signal flows through the rails from transmitter to receiver. On the contrary, when a train enters into the track section, its wheels short-circuit the pair of rails, so that the signal sent by the transmitter is unable to reach the receiver and the section status becomes busy. An occupied section is also detected if a complete rail failure occurs, since the electrical continuity is interrupted. The main problem of track circuits is represented by the fact that they are able to detect a rail failure only if the electrical continuity of the associated circuit is interrupted; consequently, early-stage cracks are not detected [37].

2.3. Defect Detection Techniques Based on Ultrasonic Waves

Detection techniques based on ultrasonic waves are divided into three main categories [5]: (a) *conventional ultrasound*, (b) *phased arrays*, and (c) *guided waves ultrasound*.

2.3.1. Conventional Ultrasonic Techniques

If these techniques are employed, the railhead is scanned by means of ultrasonic beams: defects are detected by measuring the reflected or scattered energy. The presence of a defect, along with its location and severity, is inferred from the amplitude of the resulting reflections and their arrival times [23]. The employed ultrasonic transducers are contained within a liquid-filled tire; this allows for improving signal quality by reducing the acoustic mismatch between rail steel and air. For the same reason, water is sprayed between the wheel containing the ultrasonic probes and rails [4]. Although ultrasonic testing is capable of inspecting the whole railhead [38], it has the following limitations: (a) limited car speed in ultrasonic inspection [16] (in practice, the achievable test speeds can be as low as 15 km/h [4] and, consequently, inspection has to be accomplished outside the operation period of commercial trains [1]); (b) *shallow crack shadowing* [13,16] (small shallow cracks can shade more severe and deeper cracks by reflecting ultrasonic beams [1]); (c) *false defect detection* [1] (this slows down inspection operations).

The ultrasonic testing vehicle "us1-Galileo" employed by the Italian railway infrastructure manager, *Rete Ferroviaria Italiana* (RFI), is shown in Figure 9.

Figure 9. Ultrasonic testing vehicle "us1-Galileo" (courtesy of © Benedetto Sabatini).

2.3.2. Phased Arrays

Phased arrays use multiple ultrasonic elements and electronic time delays to generate beams by exploiting constructive and destructive interference. Phased arrays provide ultrasonic beams that can be steered, scanned, swept, and focused electronically [39].

2.3.3. Guided Waves

In an effort to overcome some of the technical problems associated with ultrasonic wheel inspections, some research groups are investigating ultrasonic *guided waves* for rail inspections. Different types of guided waves can propagate in any bounded medium. Inspection tests can be accomplished in *pulse-echo mode*; this means that an ultrasonic transducer is employed to transmit a pulsed guided wave along the structure of interest and that the presence of defects or other structural features is inferred from analyzing the returning echoes sensed by the same transducer [7]. Operation in *pitch–catch mode* can be also employed; in this case, a transmitter generates a guided wave in the waveguide and a receiver (i.e., a different transducer, placed at the other end of the sample) is employed to detect the incoming wave. If a defect reflects a portion of the transmitted energy, the received signal is weaker than that observed in pristine waveguide conditions; this allows for detecting the presence of defects [40].

Ultrasonic guided waves offer the following advantages [41]: (a) *guided waves propagate along (rather than across) rails*, and are ideal for detecting critical TDs, even if they are relatively insensitive to parallel fissures [42]; (b) *guided waves allow for improving inspection coverage*, thereby relaxing limits on the maximum achievable speed (if the inspection equipment is installed on-board a moving vehicle); (c) *guided waves can travel underneath shelling and still interact with internal defects*, since they penetrate a finite depth of the surface of rails; iv) *guided waves can penetrate alumino-thermic welds*, hence potentially targeting weld cracks and discontinuities, since they travel in the mid-frequency range (say, in the 20 kHz–1 MHz range).

Unluckily, the complicated propagation behavior of guided waves cannot be easily managed [43]. In fact, ultrasonic guided waves, especially in complex waveguides such as rails, express a *multi-mode* character (many modes can propagate simultaneously) and a *dispersive* character (the propagation velocity depends on frequency) [7]. In practice, tens of different modes can be observed in the employed frequency range; moreover, the dispersion curves characterizing this propagation phenomena are very complicated [43], as it can be easily inferred from Figure 10.

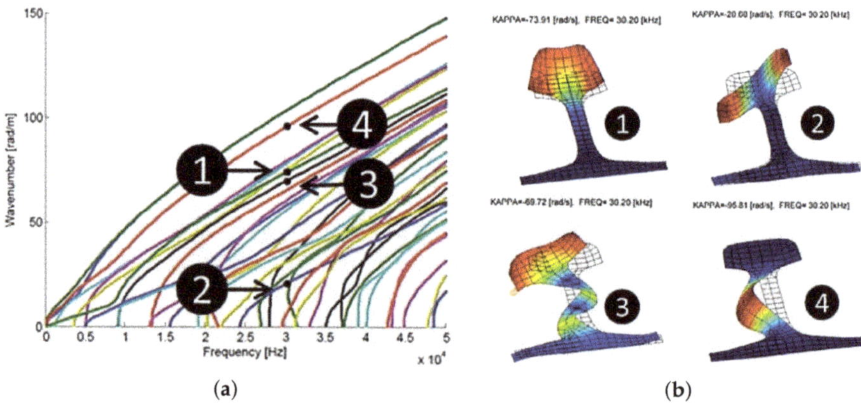

Figure 10. Dispersion curves (**a**) and mode shapes (**b**) observed in a rail waveguide. Some candidate modes for long range damage detection are highlighted. Modes 1 and 2 are symmetric and anti-symmetric, respectively, and their energy is concentrated in the crown of the considered rail; the energy of mode 3, instead, is more evenly distributed across the rail cross section (a small portion of it is observed in the foot), whereas that of mode 4 is concentrated in the web of the rail (pictures taken from Ref. [44]).

3. Systems for Rail Defect Detection Based on Ultrasonic Guided Waves: Classification

Substantial research efforts have been devoted to the study of systems exploiting ultrasonic guided waves for the detection of rail defects. More specifically, the following two broad categories of systems have been investigated:

1. *On-board systems* [40,45–49]-The measurement equipment is embarked on board a moving vehicle (e.g., an inspection cart), allowing for scanning the head of the rails on which it is riding. The detection of defects in the scanned rails is achieved through a pitch–catch mechanism. Initial implementations were based on an active approach, in which a source of ultrasounds was used to inject UGW into the rails, and sensing was accomplished through an array of transducers. More recent implementations adopt a passive approach, in which the ultrasonic excitation is generated by the rolling wheels of the moving vehicle. The approaches based on on-board systems have been promoted, among others, by the *Experimental Mechanics & NDE Laboratory of the University of California*, San Diego, USA.

2. *Land-based systems* [9,50–67]-The employed equipment is placed on the ground near to the railway infrastructure, and its actuators and sensors are attached to the rails under test. The detection of defects in the rails is achieved through a pulse-echo approach. A lot of contributions about this topic have been published by the *Sensor Science and Technology Department* of the *Council for Scientific and Industrial Research*, South Africa. The *ultrasonic broken rail detection* (UBRD) system has been developed by this institution and put in use on the Oreline in South Africa to detect the presence of a fully broken rail; the system is evolved to implement an early defect detection system. In recent years, important results in this research field have been achieved by two Chinese research groups. In fact, significant contributions about the best methods to excite and detect propagation modes in a rail, and about feasible methods for rail defect location have been published by various researchers working at the Beijing Jiaotong University (*School of Mechanical, Electronic, and Control Engineering, and Key Laboratory of Vehicle Advanced Manufacturing, Measuring and Control Technology*). Moreover, an electronic system able to efficiently excite ultrasonic guided waves in rails has been developed by various researchers working at the Xi'an University of Technology (*Department of Electronic Engineering*).

In the remaining part of this manuscript, both categories of systems are taken into consideration. In particular, the architecture of these systems is described in Section 4. Sections 5–7 are devoted to the implementation of on-board approaches, to that of the UBRD system, and to the evolution of the UBRD system towards early defect detection and monitoring. In Section 8, commercial projects like RailAcoustic (by Enekom), or university studies, like those accomplished by the Beijing Jiaotong University or the Xi'an University of Technology are summarized. Finally, various results about the performance provided by each of the considered systems are illustrated in Section 9.

4. Systems for Rail Defect Detection Based on Ultrasonic Guided Waves: Architecture

In the technical literature, various architectures have been proposed for on-board or land-based systems for rail defect detection based on UGWs. In this section, some essential features of the available technical options are illustrated.

4.1. On-Board Systems

Over the years, *active* and *passive* detection strategies have been proposed for on-board systems. The main characteristics of the systems employing such strategies can be summarized as follows:

*   An *active* rail inspection system is able to detect surface-breaking cracks and internal defects located in the railhead. It is based on UGWs and air-coupled contactless probing. Both the transmission of test signals and their reception are accomplished by means of air-coupled piezoelectric transducers arranged in a pitch–catch configu-

ration. The use of a laser transmitter has been also investigated [40]; however, this option has been judged too costly and is potentially hard to maintain. In an active system, receivers are arranged in pairs and placed at the two sides of the transmitter: this allows for implementing a differential detection scheme. The system output, after data analysis, is represented by the so-called *damage index* (DI), exhibiting peaks in correspondence of discontinuities (i.e., of potential defects) in rails. In recent years, the use of laser technology for both exciting UGWs in rails and sensing them has been investigated [68,69]. In this case, non-ablative laser sources are used to generate a guided tensional wave system in the rail; the response to such a system can be sensed by means of a rotational laser vibrometer.

- A *passive* rail inspection system aims at estimating the acoustic transfer function of rails; the excitation is represented by the rolling wheels of the moving instrumented train [48]. When a train is in motion, its rotating wheels generate a continuous dynamic excitation of the rail. Such an excitation is uncontrolled, non-stationary, and difficult to characterize [48]. Therefore, in this case, the challenge to be faced is to generate a stable rail transfer function that is not affected by the excitation [49]. The presence of a discontinuity in the rail can be inferred from the estimated transfer function, using again the DI indicator.

4.2. Land-Based Systems

The UBRD is a *broken rail detection* (BRD) system, designed to continuously detect full breaks in rails [50]. Its transducers are tied to the rail and allow for monitoring a span of up to 1 km of track in pitch–catch mode. This system has been already installed on 841 km of tracks in the high-haul Oreline network in South Africa [51,52]. The system architecture is represented in Figure 11; its operation is based on a simple transmit–receive confirmation protocol. As long as a reliable reception of the ultrasonic signal generated by the transmitter is possible, the integrity of the rail is verified; otherwise, an alarm is set [50,53].

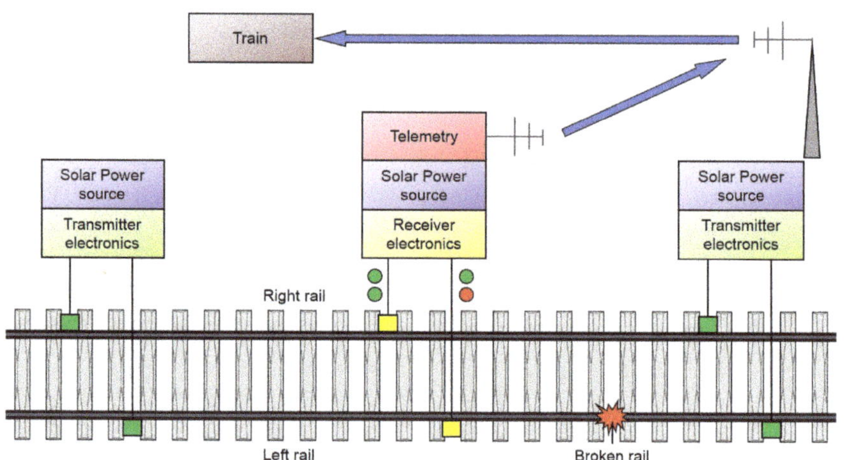

Figure 11. UBRD system concept [51].

The developers of the UBRD system are studying how UGWs can be exploited to detect and monitor a growing defect, and how specific guided modes can be excited in a rail. The use of techniques for the excitation of guided waves different from those based on piezoelectric transducers have been investigated. For example, the use of *electromagnetic acoustic transducers* (EMATs) has been taken into consideration [55]. In the meantime, the reflection characteristics of defects and their dependence on time and environmental conditions are under evaluation [9,56,57,65].

In principle, UGWs have the potential to monitor a long span of a waveguide using permanently installed transducers in pulse–echo mode. The same transducer array is used to generate the ultrasonic wave and then to measure the echo generated by the wave reflection on a defect. A continuously welded rail is a perfect example of such a waveguide and, as a matter of fact, defects (and welds) act as reflectors for guided waves. A system based on this approach is able to detect the *presence* of a defect, and to estimate its *magnitude* and *location*. These results are achieved by monitoring the echoes coming from defects (defect presence), measuring their amplitude (defect magnitude) and the time-of-flight required by the transmitted wave to travel back and forth from the transducer to the defect (defect location) [51].

An important advantage offered by land-based solutions is represented by the fact that monitoring can be accomplished in near real-time, so that defect growth can be analyzed over time and an alarm set only when a crack assumes a potentially critical size [9,65].

Further details about the architecture of the UBRD system are provided in Section 6. In Section 7, its evolution towards an early defect detection system is described. Other projects or studies about land-based systems are illustrated in Section 8.

5. Implementation of On-Board Systems

In this section, the implementation of on-board systems is described in detail; both active and passive approaches are considered.

5.1. Active Approach

5.1.1. Hardware Configuration

The configuration adopted in the active on-board system is shown in Figure 12: a single focused transmitter and four pairs of receivers are fixed to a frame which is in turn attached to a cart [40,45,47]. The lower active surface of the transducer is separated from the upper surface of the rail; their distance is known as *lift-off* and is on the order of a few centimeters for both the transmitter and the receivers. The transmitter feeds the top of the rail surface with a narrowband toneburst signal (a Hanning-windowed sinusoid); it has a tunable repetition rate (representing the frequency of the excitation and related to spatial resolution). This results in a (repeated) multimodal guided wave that propagates symmetrically with respect to the excitation line. The receivers detect the guided waves leaking into the surrounding air; for this reason, they are tilted according to Snell's law.

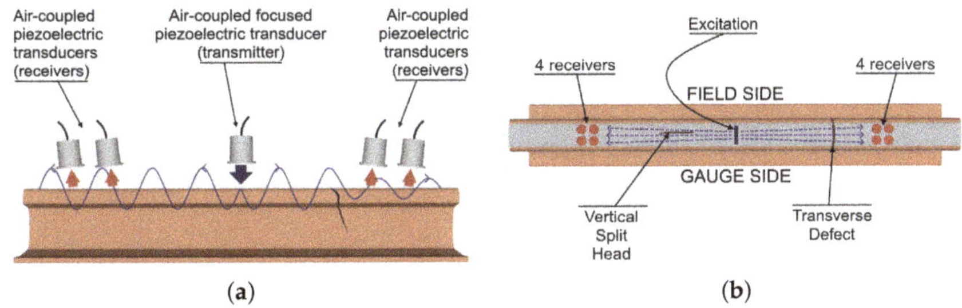

Figure 12. Schematic of the non-contact defect detection system based on ultrasonic guided waves: (**a**) side view and (**b**) top view. The two red lines indicate the possible locations of a *vertical split head* and a *transverse defect* [46].

5.1.2. Defect Detection Principles

The pair of receivers placed at the two sides of the transmitter are employed in a differential detection scheme; this compares the strength of the signal received at both sides. If an internal flaw is present on one side of the transmitter or on the other one, the transmitted wave is scattered and the outputs of the receiver pair are unbalanced. Some strength-related metrics are extracted from the received waveforms to generate a "feature supervector"; in practice, this is obtained by combining the "feature vectors" provided by the two receivers. These metrics are listed in Table 1, in which the symbols $PkPk(x)$ and $RMS(x)$ denote the *mean peak-to-peak* and *root-mean-square*, respectively, of the quantity x [46].

Table 1. Features extracted from the received time-domain waveforms x_1 and x_2; these are recorded by two receivers forming a symmetric pair [46].

Feature #	Feature Content
1	$\max\left(\dfrac{\max|x_1|}{\max|x_2|}, \dfrac{\max|x_2|}{\max|x_1|}\right)$
2	$\max\left(\dfrac{\mathrm{Pkpk}|x_1|}{\mathrm{Pkpk}|x_2|}, \dfrac{\mathrm{Pkpk}|x_2|}{\mathrm{Pkpk}|x_1|}\right)$
3	$\dfrac{RMS(x_2)}{RMS(x_1)}$

5.1.3. Signal Processing

The received signal needs to be properly processed (and, in particular, denoised) to overcome the inherently low *signal-to-noise ratio* (SNR) of air-coupled piezoelectric sensors [40]. A *multivariate outlier analysis* (MOA) is used to compensate for the natural signal variations expected during the test; the final result is represented by a *damage index* (DI) computed at each test position. Defects are detected by an high value of DI, i.e., by an unusually large discordancy index from the "normal" baseline of the rail. The DI is represented by the Mahalanobis squared distance discordancy metric D_ζ of a MOA [46], which is defined as

$$D_\zeta = \left(\mathbf{x}_\zeta - \bar{\mathbf{x}}\right)^T \cdot \mathbf{K}^{-1} \cdot \left(\mathbf{x}_\zeta - \bar{\mathbf{x}}\right), \tag{1}$$

where $\mathbf{x}_\zeta$ is the potential outlier vector, $\bar{\mathbf{x}}$ is the mean vector of the baseline, $\mathbf{K}$ is the covariance matrix of the baseline, and $(\cdot)^T$ represents the transpose operator. A new observation is classified as an outlier if the corresponding value of D_ζ is higher than a proper threshold, previously established [45].

To reduce the probability of false alarms due to the occurrence of isolated noise-related high peaks in the DI trace, system redundancy can be exploited [46]. When scanning a defect, the presence of a real crack should result in multiple peaks in the observed trace. Therefore, a peak is detected by the system only when a given number of DI values exceeds the above-mentioned threshold.

5.1.4. Reverberation of Airborne Signals Caused by an Acoustic Mismatch

The presence of reverberations of the airborne waves between the transmitter and the top of the excited rail has been found. This phenomenon is caused by the unavoidably high acoustic impedance mismatch between the air medium and the solid boundaries of the rail steel and of the piezoelectric transducer. The strong intensity of the reverberations affects the detection of acoustic waves traveling through the rail steel associated with the presence of rail defects. Although time gating can effectively separate the signal of interest, a lower SNR is observed when the repetition rate (i.e., the frequency of the excitation) is sufficiently high such that the airborne reverberations from a previous excitation overlap with the waves of a new excitation. This limits the test speed to ~1.6 km/h (~2.4 km/h when a slightly worse SNR is acceptable) because of the required spatial resolution. The most effective solution developed for this problem until now consists of inserting a sponge between the transmitter and the rail to attenuate the airborne reverberations. The main drawback of this solution is represented by the fact that the sponge has friction with the rail, and this raises the noise level affecting the received waves as the train speed increases; in practice, a reliable signal detection is possible up to 24 km/h [47]. Some test results are illustrated in Section 9.1.

5.2. Passive Approach

5.2.1. Hardware Configuration

In this case, as illustrated in Refs. [49,66], the sensing head is made of two receivers, denoted A and B, separated by a known distance; both are sensitive only to waves propagating unidirectionally (from left to right in Figure 13). The sensing head is mounted on a beam rigidly connected to the front-axle of the test car; similarly as in the active approach, contactless probing is employed. The sensors are tilted with respect to the rail surface (according to Snell's law) to best capture the leaky surface waves propagating in the railhead. The orientation of receivers ensures directional sensing of the waves excited by the wheels located on only one side of the arrays (front end), with virtually no sensitivity to waves propagating in the opposite direction (i.e., to reflections or waves excited from the wheels located to the other side of the arrays' back end).

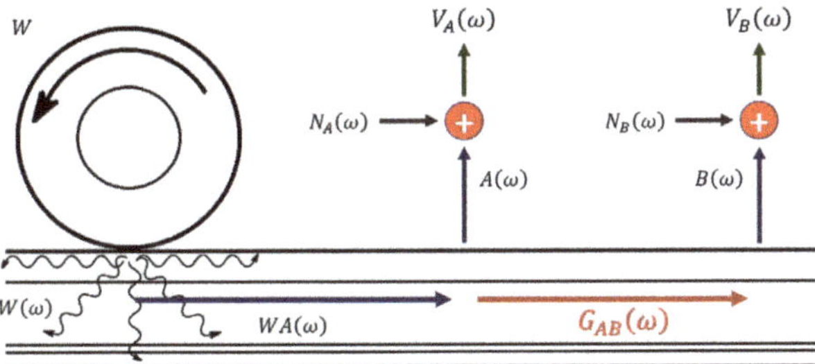

Figure 13. Representation of the linear wheel-rail interaction in the frequency domain [49].

Data are continuously recorded for the entire duration of the test [49]. A tachometer logic pulse marking the spatial position of the test car and GPS positioning are recorded, in addition to high-speed camera videos of the tested rail to verify the presence of visible discontinuities when the prototype detects an anomaly [48].

5.2.2. Defect Detection Principle

In this case, defect detection requires the estimation of the time domain *Green's function* $G_{AB}(t)$ between the two receivers A and B, without prior knowledge of the source excitation spectrum. This function represents the response of the test object measured a B from an impulse excitation at A. If the frequency domain Green's transfer function $G_{AB}(\omega)$ can be estimated, discontinuities in the rail (such as defects) can be detected by means of a procedure similar to that illustrated for the pitch–catch ultrasonic guided-wave approach. In fact, discontinuities can be perceived as a change in the structural impulse response, $G_{AB}(t)$, obtained from $G_{AB}(\omega)$ through an inverse Fourier transform. In terms of defect detection, as long as the reconstructed Green's function is stable during a test run, successful passive structural inspection becomes possible.

The estimation of the Green's function is based on the computation of the ensemble averaged *cross-power spectrum* between the output and the excitation, normalized to a modified ensemble averaged *auto-power spectrum* of the excitation. The following assumptions are made: (a) the system is linear and piecewise stationary, so that the statistics of the excitation $W(\omega)$ do not change in the observation interval; (b) the noise terms are uncorrelated, have a zero DC component, and tend to zero on average, thus enabling for separating the useful signal component from noise. The function $G_{AB}(\omega)$ is estimated by computing the deconvolution of the cross and auto power spectra. In fact, the spectra of the responses at A and B can be expressed as

$$V_A(\omega) = W(\omega) \cdot WA(\omega) \cdot A(\omega) + N_A(\omega) \tag{2}$$

and

$$V_B(\omega) = W(\omega) \cdot WA(\omega) \cdot G_{AB}(\omega) \cdot B(\omega) + N_B(\omega), \tag{3}$$

respectively; here, $V_A(\omega)$ ($V_B(\omega)$) is the response measured at receiver A (B), $W(\omega)$ is the wheel excitation spectrum, $WA(\omega)$ is the transfer function of the rail between the wheel and the transducer at A, $A(\omega)$ ($B(\omega)$) is the frequency response of the receiving sensor A (B), $N_A(\omega)$ ($N_B(\omega)$) is the uncorrelated noise originating from the environment at A (B) (see Figure 13).

Based on Equations (2) and (3), it can be shown that [49]

$$\langle \text{Cross Power} \rangle = |W(\omega)|^2 \, |WA(\omega)|^2 \, G_{AB}(\omega) \tag{4}$$

and

$$\langle \text{Auto Power} \rangle = |W(\omega)|^2 \, |WA(\omega)|^2, \tag{5}$$

in which the symbol $\langle \rangle$ represents the *ensemble average* operator.

By using Equations (4) and (5), the frequency Green's Function between the two receivers is computed as

$$\frac{\langle \text{Cross Power} \rangle}{\langle \text{Auto Power} \rangle} = \frac{|W(\omega)|^2 \, |WA(\omega)|^2 \, G_{AB}(\omega)}{|W(\omega)|^2 \, |WA(\omega)|^2} = G_{AB}(\omega). \tag{6}$$

Then, the time domain Green's function can be evaluated by taking the inverse Fourier transform from the left-hand side of the last equation, i.e., as

$$G_{AB}(t) = \frac{1}{2\pi} \int_{-\infty}^{+\infty} G_{AB}(\omega) \, e^{j\omega t} \, d\omega. \tag{7}$$

The last integral is computed in an approximate fashion through an *inverse fast Fourier transform* (IFFT).

5.2.3. Data Processing

The employed data processing is sketched in Figure 14. In each run, the recordings from receivers A and B are first amplitude-clipped to within the ± 3 standard deviations to mitigate the effects of isolated spikes in the passive reconstruction of the impulse response $G_{AB}(t)$. Then, the resulting signals undergo *fast Fourier transform* (FFT) processing, so that cross and auto power spectra and the transfer function $G_{AB}(\omega)$ can be computed. Finally, the time domain Green's function $G_{AB}(t)$ is evaluated by (a) averaging the function $G_{AB}(\omega)$ over four sensor pairs and (b) accomplishing band-pass filtering and IFFT processing. An outlier analysis is implemented in order to compute the DI related to the strength of the reconstructed transfer function; this issue has already been discussed in Section 5.1.3, when describing the active method for rail defect detection. The computation of the DI allows for normalizing the available data, thus mitigating the normal (baseline) data variability occurring in each run [48,49].

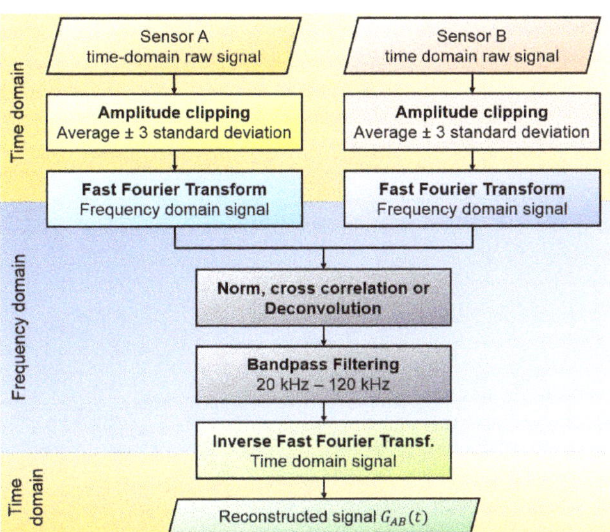

Figure 14. Signal processing steps accomplished in the passive reconstruction of the transfer function $G_{AB}(\omega)$ [48].

5.2.4. Trade-Offs

The impulse response $G_{AB}(t)$ is the result of the constructive interference of the wave modes continuously generated by the wheel excitation and propagating in the rail between the receivers. The constructive interference and, hence, the rate of convergence of the passively reconstructed transfer function (or, equivalently, its SNR) benefit from signal averaging. For this reason, time windows involving long recordings are preferred. Note, however, that, since both the transmitters and the receivers are moving along the test specimen (rail), the stationarity of the reconstructed transfer function can be guaranteed only in a fixed position. For this reason, a good trade-off needs to be achieved between the long recording time required by the averaging process and the stationarity (related to spatial localization) of the transfer function that calls for shorter observations. This explains why the test speed has to be properly selected: the higher the velocity is, the shorter the recording time is, ensuring a sufficiently accurate spatial localization [48].

Some test results are illustrated in Section 9.2.

6. Land-Based Systems Implementation-Premise: Ultrasonic Broken Rail Detector (UBRD)

In this section, we focus on the implementation of land-based systems. Since the topic is wide, the description of these systems is provided in three consecutive sections, each focusing on specific technical issues, as already mentioned at the end of Section 4.

6.1. UBRD Hardware Configuration and Generated Signals

Transmitters and receivers of the UBRD system are interleaved, as illustrated in Figure 15. Every receiver is able to acquire signals coming from both directions of the rail; to identify the exact orientation, the transmitted signals consist of different burst sequences depending on the transmitter. In fact, as it can be easily inferred from Figure 16, each transmitted sequence consists of a number of pulses spaced by a given *burst repetition interval* (BRI) and is repeated at a specific *interrogation interval* (II). This avoids the overlap at one receiver of burst trains coming from adjacent transmitters for extended periods. For such reason, receivers are also individually configured to recognize specific pulses as arriving from either the up or down direction depending on the settings of adjacent transmitters. Note also that each signal does not reach a receiver far from its transmitter because of the attenuation due to the propagation medium [50,53].

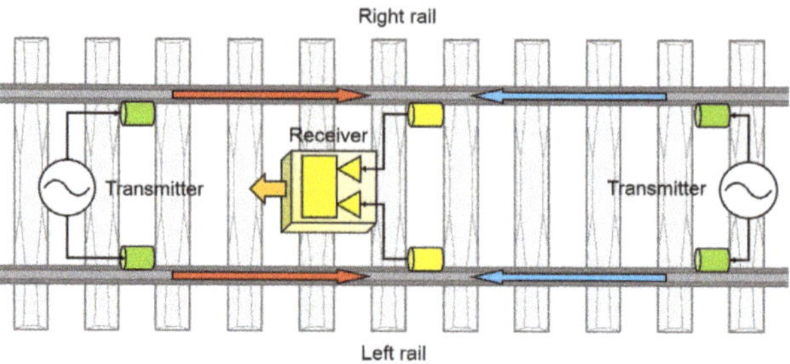

Figure 15. Block diagram of the UBRD [53].

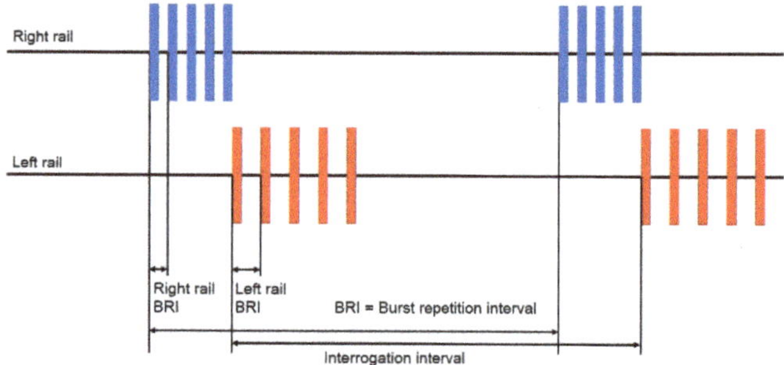

Figure 16. Burst injection scheme [50].

6.2. Defect Detection Principle and Employed Signal Processing Method

The transducers of each transmitter generate an acoustic signal that propagates through the rail; this signal travels in both directions and can reach receivers located at a significant distance from the transmitter. The received signals are filtered, amplified, and processed. Rail integrity between the transmitter and each receiver is confirmed as long as an acceptable signal is received. If a clean break occurs in the considered stretch of rail, the corresponding receiver will no longer be reached by the transmitted signal and will generate an alarm [50,53]. Receivers detect valid signals on the basis of specific criteria concerning signal frequency, burst length, burst repetition interval, and burst train continuity.

Severe continuous noise (such as that generated by an approaching train) at a receiver will affect signal detection, regardless of the efficiency of the employed detector. Under such circumstances, the receiver will generate the message "train in section", will stop processing its input signals, and will remain idle until the noise intensity decreases; then, it will return to its normal conditions.

This system is commercialized under the name "RailSonic-ultrasonic broken rail detector". The system is not classified "Fail Safe", although in Ref. [50], it is claimed that many fail safe principles have been incorporated in order to correctly detect most of equipment failures; this allows for correctly identifying these events and are, at the same time, to avoid false alarms.

6.3. UBRD Updates

To update the UBRD, a new piezoelectric transducer has been developed, thanks to numerical modeling and measurement techniques. This improvement and the use of new digital signal processing techniques have been incorporated into a new version of the UBRD system that allows for doubling the distance between transmit and receive stations [51].

Yuan et al. have proposed a method to integrate the UBRD system with a *train detection* (TD) system [54]; the resulting system is able to distinguish the following three different track conditions: *free*, *busy*, and *broken rail*. This is made possible by extracting, from the ultrasonic guided waves, the following three features: (a) *the RMS value* and (b) *the energy* in the time-domain; (c) *the frequency component with the highest amplitude* in the frequency domain. In addition, *temporal* and *spatial dependencies* of the signals are taken into consideration. High-level signal processing techniques, such as deep learning algorithms, are also employed to deal with varying environmental situations and conditions. The suitability of this approach has been assessed through experimental tests.

In Ref. [67], the same authors as Ref. [54] have proposed to exploit the *variational mode decomposition* (VMD) algorithm to de-noise and reconstruct ultrasonic guided wave signals. The VMD method is a quasi-orthogonal signal decomposition technique for non-recursively decomposing a multi-component signal into a finite number of compactly band-limited *intrinsic mode functions* (IMFs). The VMD algorithm is used to decompose ultrasonic guided wave signals into the fundamental, harmonics, interharmonics, and non-stationary disturbances, and can be used in conjunction with the method developed in Ref. [54].

7. Land-Based Systems Implementation-Evolution: Early Rail Defect Detection Capability

An experimental system able to monitor an operational rail track by means of ultrasonic guided waves has been proposed in Ref. [65]. In this section, the prototype and the methodologies adopted to process and analyze the available data are illustrated. Moreover, various technical problems and the solutions developed to solve them are briefly described.

7.1. Introduction

Numerous types of defects appear in rail tracks, and multiple parameters affect the prediction of crack growth rates and of the defect size at failure [9]. In principle, the *baseline subtraction* method could be used to detect the insurgence of a defect over time. Baseline subtraction is based on the idea of feeding a system in a given condition with a known excitation, measuring its response, and storing it as a baseline. Each and every modification in the system reflects in a change in its response to the same excitation. Therefore, in principle, the occurrence of any change potentially due to a defect can be detected by comparing each new response to the baseline. In practice, however, it is not easy to establish if such a change is caused by a defect or by a variation of other parameters affecting the system response to the considered excitation. In fact, it should be always kept in mind that, in harsh environments like rail tracks, any variation observed in the propagation conditions of ultrasonic signals can also originate from changes in *environmental and operating conditions* (EOCs); these conditions include, for instance, temperature, train passing, or maintenance operations [51].

The continuous monitoring of a rail track can provide better performance results in terms of defect detection probability than a single inspection, though it is influenced by the EOC. In this case, the most important technical challenges are: (a) the development of a permanently installed UGW-based monitoring system for rail defect detection at a reasonable cost per kilometer of track; (b) the development of a defect detection technique able to reliably operate in the presence of real EOCs. Of course, if a suitable facility is unavailable, tests can be performed on operative tracks. The problem is that damaged sections of rails are removed as soon as defects are detected; furthermore, rails must not be damaged or altered. Consequently, it is impossible to test the available prototypes on real cracks: their performance is assessed by using welds or glued masses, both acting as reflectors for guided waves [9].

Different problems must be addressed in order to achieve an effective automatic monitoring system for rail damage detection. In particular, the following specific technical issues should be investigated: (a) the behavior of a defect growing over time; (b) the influence of the selected transducers on the quality of the received signal; (c) the influence of the changes in rail properties on the propagation of guided waves and the parameterization of these changes; (d) the influence of time varying EOCs on wave propagation and the methods for compensating for these variations; and (e) the identification of the defect detection algorithm and the use of proper methods for assessing its performance.

7.2. Monitoring Set-Up

In this section, the hardware and software architecture of the system illustrated in Ref. [65] are briefly described.

7.2.1. Selected Modes for Guided Wave Propagation

Numerical simulations based on the *semi-analytical finite element algorithm* (SAFE) have allowed for identifying a guided wave mode suited for the detection of defects in rails. This mode propagates with relatively small attenuation and dispersion, and it is strongly reflected by transverse defects in the railhead. It can be excited by a transducer attached under the railhead. The identified mode, used for both transmission and reception, is a symmetric mode that has motion in the vertical axis while propagating along the longitudinal direction moreover with the energy mainly concentrated in the railhead [65]. The SAFE is a popular method developed to solve wave propagation problems in waveguides. It can be employed to compute the dispersion curves of the modes that propagate along a waveguide characterized by an arbitrary cross-section. It has been shown that, in the analysis of waveguides that are infinitely long in one direction, SAFE achieves better accuracy than *finite element* (FE) methods at a lower computational cost. A detailed description of this method can be found in Ref. [70].

7.2.2. Monitoring System Hardware and Software

Two piezoelectric transducers, forming an array and attached under the railhead, are employed to perform pulse-echo measurements [9]; the sandwich piezoelectric transducers have been designed to effectively excite the selected mode of propagation around 35 kHz [65]. The distance between the axes of the transducers has to be equal to one quarter of the wavelength of the excitation signal.

Two specific technical problems have been faced in system design [56]. The first problem is represented by the slightly different resonant frequencies of the two transducers; this is due to the tolerances characterizing the manufacturing process. Moreover, the transducer performance may change differently over time and, in the presence of temperature variations, thus causing asymmetric changes in the measured reflections and replicas. A possible solution to this problem is to independently scale the reflections from the positive direction and from the negative one, so that the difference in their amplitudes can be compensated for; this requires the estimation of the attenuation in both directions. The second problem is represented by the presence of multiple replicas originating from reflections; phased array processing can be exploited to solve this problem.

The experiments were performed on a section of tracks consisting of rails joined by aluminothermic welds. Transducers and the artificial defects mentioned in Section 7.2.3 have been placed far from the joints.

7.2.3. Behavior of a Defect during Time

The behavior of a defect over time can be analyzed by using a deteriorating defect emulator, such as a glued mass. The reflection of a glued mass deteriorates quickly over time, thanks to the stresses caused by trains and to the progressive corrosion of the rail under the glue. A monotonically increasing defect, similar to a growing crack, can be simulated if the reflected signals recorded during the time window starting at the installation of the defect emulator and ending at its detachment are time reversed [9]. The size of the mass representing the crack has to be selected to provide a realistic reflection, such as a small transverse crack in the railhead. Reflections from rail welds can provide a useful reference during testing; furthermore, if these welds can be detected, it is believed that a crack can be found well before it reaches a critical size [44].

7.3. Signal Pre-Processing

The employed excitation signal is a Hanning-windowed tone burst signal whose center frequency is equal to 35 kHz. The measurements are acquired by exciting one piezoelectric transducer, acquiring the response of both transducers of the array, and then repeating the process by exciting the other transducer. This cycle is repeated k times, and the average of the k acquired measurements is stored as a new signal in the measurement set, which consists of four time domain signals.

Noise from passing trains has to be avoided since it would overwhelm the useful signal. For this reason, the system first senses the presence of train noise and proceeds with the acquisition only in its absence; otherwise, it delays its acquisition. In the observation interval m, distinct sets of measurements are recorded.

Before defect detection or monitoring, the acquired data must be pre-processed to compensate for unwanted effects. The pre-processing steps are illustrated in Figure 17 and described in the following sub-sections. Data pre-processing involves (a) *phased array processing*; (b) *dispersion compensation*; (c) *signal stretching and scaling*; (d) *signal reordering*. These steps are required to discriminate the direction of the reflections, to reduce the influence of dispersion and to compensate for some of the changes occurring in EOC, respectively. Signal reordering accounts for the use of a mass as a defect emulator: the mass, in fact, behaves in a reversed time order with respect to a crack (see Section 7.2.3).

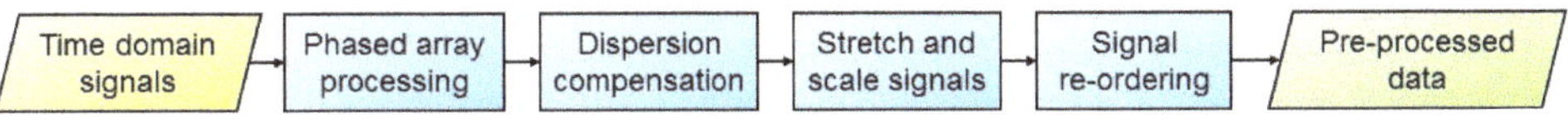

Figure 17. Pre-processing steps of the acquired signals [65].

7.3.1. Acquired Signals

The impact of noise in the received signal can be reduced by band-pass filtering, making easier the task of detecting the reflections in the steps to follow. The center frequency and bandwidth of the filter can be set to be the same of the excitation signal. During pre-processing, the signals are converted to the frequency domain.

7.3.2. Phased Array Processing

An array of transducers allows for exciting a specific mode in a desired direction of propagation, even if the array is composed of two transducers only. This result is obtained by feeding both transducers with excitation signals characterized by a small difference in delay or phase. The phase difference depends on the wavelength of the propagation mode selected at the transmission frequency and on the axial distance between the two transducers. The same idea can be exploited at the receive side; in fact, if a phase shift is applied to the signal acquired by a transducer, the energy associated with the desired mode and originating from the desired direction can be captured. It is worth noting that a perfect cancellation is impossible, since other modes characterized by different wavelengths are transmitted and received in both directions. This problem can be mitigated by increasing the overall number of transducers forming the array.

Phased array processing can be applied to the full matrix of signals acquired from an array of transducers, as explained by Wilcox in Ref. [71]. The acquired time signals are converted to the frequency domain and, after processing, are converted back to the time domain. When the first (second) transducer is excited, the responses acquired by both transducers are stored in the first (second) column of the *measured displacement matrix*

$$\mathbf{V}(\omega) = \begin{bmatrix} V_{11}(\omega) & V_{12}(\omega) \\ V_{21}(\omega) & V_{22}(\omega) \end{bmatrix};$$

(8)

Here, $V_{ij}(\omega)$ is the spectrum of the response of the i-th transducer in the array (with $i = 1$ and 2, if the array consists of two transducers only) in response to the excitation injected through the j-th transducer (with $j = 1$ and 2, if the array consists of two transducers only).

The analysis of different combinations of transmitted and received propagation modes requires the computation of *mode shape matrices*. Each column of the mode-shape matrix refers to the displacement of a mode at each of the transducer locations. The mode shape matrices for an array of transducers in a 1D waveguide (such as a rail) contains both mode shape and phase information according to the axial position of the associated transducer, for each mode in every possible direction [71]. Since it may happen that the modes selected for receive processing are different from those chosen for transmit processing, two mode shape matrices are required, one for reception and the other for transmission (denoted $\mathbf{R}(\omega)$ and $\mathbf{T}(\omega)$, respectively). If the same modes are selected for reception and transmission, then $\mathbf{R}(\omega) = \mathbf{T}(\omega)$. The coefficients of these matrices are computed through a SAFE analysis of the rail.

If the mode shape matrices $\mathbf{R}(\omega)$ and $\mathbf{T}(\omega)$, and the measured displacement matrix $\mathbf{V}(\omega)$ (8) are known, the contributions of the different modes to the received signal can be computed. The spectrum $\alpha(\omega)$ of the overall response is evaluated as

$$\alpha(\omega) = \mathbf{R}(\omega)^{-1}\,\mathbf{V}(\omega)\,\mathbf{T}(\omega)^{*^{-1}} = \left[\begin{array}{cc} \alpha_{r_1 t_1}(\omega) & \alpha_{r_1 t_2}(\omega) \\ \alpha_{r_2 t_1}(\omega) & \alpha_{r_2 t_2}(\omega) \end{array}\right]; \qquad (9)$$

Here, the spectrum $\alpha_{r_1 t_1}(\omega)$ ($\alpha_{r_2 t_2}(\omega)$) describes the transmission and reception in the forward (backward) direction, whereas $\alpha_{r_2 t_1}(\omega)$ ($\alpha_{r_1 t_2}(\omega)$) is the transmission in the forward (backward) direction and the reception in the backward (forward) direction. Note that (a) $\alpha_{r_1 t_1}(\omega)$ is ideally equal to zero; (b) the presence of reflections from features (defects, welds, and any other element reflecting back guided waves towards the array of transducers) in the forward (backward) direction are inferred from $\alpha_{r_2 t_1}(\omega)$ ($\alpha_{r_1 t_2}(\omega)$); (c) proper calibration factors can be introduced to compensate for the different sensitivities of the two transducers (see Section 7.2.2).

7.3.3. Dispersion Compensation

Even if a mode characterized by a small dispersion is selected, the effects of this phenomenon can be appreciable in the presence of a long propagation range. Dispersion compensation can be performed by exploiting the algorithm developed by Wilcox [72] and the dispersion data evaluated by means of the SAFE method in the considered scenario. The compensation process allows also for converting the received signals from the time domain to the distance domain. The correctness of this conversion can be verified by comparing the location of the welds on field and that inferred from the reflections appearing in the compensated signals.

7.3.4. Signal Stretching and Scaling

Time varying EOCs influence the acquired signals, even in a defect-free rail: for this reason, compensation techniques not affecting the influence of defects on the processed signal are required. The group velocities of guided wave modes are influenced by temperature variations; in fact, the reflections appear to be shifted in time (or distance) if measurements are acquired at a different temperature. This effect can be compensated for by stretching the dispersion-compensated signals, in time or in distance domain [65]. Since dispersion compensation converts time domain signals in the distance domain, stretching in Ref. [65] has been performed in the last domain.

The influence of temperature variations cannot be represented only through a phase shift, since the wave envelope can also be distorted. A piecewise linear stretch can be exploited to compensate for the envelope distortion; in doing so, the knowledge about the location of the welds appearing along the considered span of rails is used. If this method is adopted, the distorted envelope is stretched in a way that the reflection peaks corresponding to the welds are aligned with their known locations. In this phase, however, the phase information is discarded. The received signal might also be affected by variations in their amplitude during the monitoring period. These can originate from a change in the sensitivity of transmitters or in the attenuation of the rail with temperature, or from the rail itself gradually sinking into the ballast. To compensate for this effect, signal scaling based on the peaks originating from weld reflection can be performed. Guided waves are attenuated during propagation; for this reason, distant weld reflections appear smaller than those due to welds closer to the transducer array. To compensate for this effect, an energy-based normalization can be applied to each signal envelope by adopting the technique proposed by Moustakidis et al. in Ref. [73]. The energy attenuation employed in normalization is computed by exploiting a moving average technique.

The phase of the compensated signals is discarded, and only their envelope remains. Unluckily, the envelope is affected by a large *direct current* (DC) component, affecting the performance of the defect detection algorithms (and, in particular, of the ICA; see Section 7.4). To reintroduce phase information, the wave envelopes of scaled signals are multiplied by a single-frequency sine wave. However, the employed method for defect detection is based only on the analysis of signal envelopes [65].

7.3.5. Signal Reordering to Simulate the Monotonic Growth of a Defect

The reflection of a glued mass deteriorates quickly over time because of the stresses caused by trains and of the progressive corrosion of the rail under the glue. A monotonically increasing defect, similar to a growing crack, can be simulated if the reflected signals recorded during the time window starting at the installation of the defect emulator and ending at its detachment are time reversed [9]. A steady decrease in the amplitude of the reflection due to the artificial defect amplitude is expected, with some fluctuation due to temperature changes. Test results illustrated in Refs. [9,65] have proven that the reflection peak due to an artificial defect does not have a simple correlation with temperature; no satisfactory explanation for this behavior has been provided, although it is known that measurements may have been affected by the complicated resonant conditions producing the large reflections [9].

7.4. Defect Detection

The processing steps accomplished in defect detection are summarized in Figure 18. Because of the EOCs-induced variations in the propagation conditions of UGW signals, the simple baseline subtraction is not effective [65]. Some promising results in structural health monitoring of plates and pipes have been obtained by means of two unsupervised machine learning techniques, namely *singular value decomposition* (SVD) [74] and *independent component analysis* (ICA) [75]. These techniques outperform baseline subtraction in the detection of simulated defect signatures superimposed on measured data from a pipe subject to time varying EOCs [76]. Therefore, it makes sense to apply these techniques to rails too.

Ideally, if EOCs are perfectly compensated for, the signal obtained from pre-processing can be seen as the additive contribution of two or more distinct components: (a) one resembling the baseline ultrasonic signal containing weld reflections, with a constant weight over the duration of the experiment; (b) other components resembling the ultrasonic signature of the defect, with a weight increasing over time and corresponding to the deterioration of the defect [9].

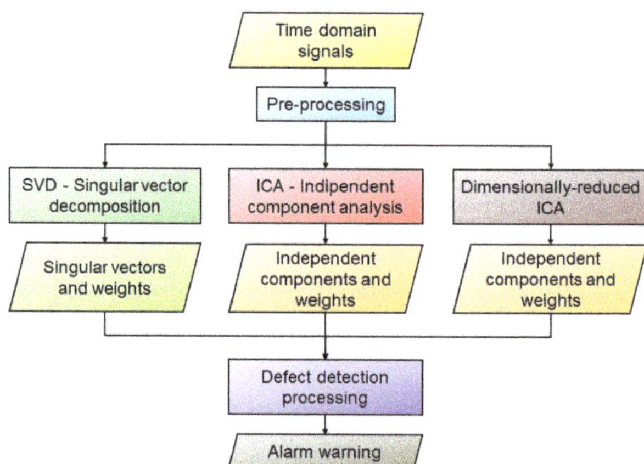

Figure 18. Processing steps in defect detection.

Data-Driven Defect Detection Techniques

The pre-processed measurements are stored in the $m \times n$ matrix $\mathbf{X}$; here, m represents the overall number of performed measurements over the monitoring time, whereas n is the number of samples acquired in each measurement. The matrix $\mathbf{X}$ can be factored as

$$\mathbf{X} = \mathbf{AC}. \tag{10}$$

The matrix $\mathbf{C}$ contains the additive components of the acquired signal (baseline and defect signatures) in the distance domain, whereas the matrix $\mathbf{A}$ is the corresponding weights in the observation interval.

The *singular value decomposition* (SVD) technique can be exploited to evaluate the factorization expressed by Equation (10) [74]. In fact, applying this technique to the real data matrix $\mathbf{X}$ produces

$$\mathbf{X} = \mathbf{USV}^T, \tag{11}$$

where $\mathbf{U}$ and $\mathbf{V}$ are the left and right *singular vector matrices*, respectively, and $\mathbf{S}$ is the diagonal *singular value matrix*. The columns of the matrix $\mathbf{U}$ are related to the slow signal behavior evolution over the monitoring period, while the columns of the matrix $\mathbf{V}$ are related to the fast signal behavior evolution over the same observation interval. $\mathbf{S}$ allows for quantifying the amount of information contained in the associated left and right singular vectors. In fact, the main diagonal of $\mathbf{S}$ collects the singular values, sorted in descending order. The size of the given dataset can be reduced by discarding small singular vectors, since these convey a negligible amount of information.

The *independent component analysis* (ICA) technique is employed to extract the additive components of a multivariable signal and analyze their relative trends. More specifically, this technique allows for decomposing the matrix $\mathbf{X}$ into a given number of independent components forming the columns of the matrix $\mathbf{C}$ in Equation (10); such components have minimal statistical correlation. In addition, the matrix $\mathbf{A}$ contains the weights of the independent components. In our case, the ICA method can be used to separate the contributions of different sources. In fact, we are interested in separating the contribution of the defect signatures from the system background, including the baseline signal and EOC-related variations. To efficiently implement ICA, the iterative FastICA algorithm can be used. This algorithm is fed by a normalized version of the data set $\mathbf{X}$, and computes the principal values and associated components of the input data by calculating the eigenvalues and eigenvectors, respectively, of the covariance matrix $\mathbf{XX}^T$. Repeatable results are generated if the initial value is selected for the vector of weights.

If the presence of data overfitting is detected while using the ICA algorithm, dimensionality reduction techniques, like the *principal component analysis* (PCA) and can be employed to mitigate this problem [77]. The combination of the ICA algorithm with principal component analysis (PCA) is called ICA *with dimension reduction* in Ref. [65]. During ICA processing, the eigenvalues (principal values) are ordered according to their magnitude, and only the largest values and their associated components are kept.

Some test results are shown in Section 9.4.

7.5. Adaptive SAFE Model for Rail Parameter Estimation

Temperature is not the only EOC variation affecting signal propagation. For instance, the propagation environment of ultrasonic guided waves in rails is modified by any change occurring in rail geometry and in the parameters of the employed materials. Compensating for these additional effects allows for achieving better performance.

7.5.1. Problem Statement

Since the performance of any monitoring system that makes use of ultrasonic guided waves is influenced by some characteristics of the propagating environment, it would be useful to estimate these characteristics from the signals acquired by the system itself. In fact, the dispersion curves of a rail depend on the properties of its material, as well as on its geometry, neither of which are known with a sufficient level of accuracy. Moreover, the rail geometry changes over time because of wear and regular maintenance operations (including rail grinding of the crown) [52].

7.5.2. Possible Solutions

Material and geometric parameters are implicit in the computation of the dispersion curves. For this reason, inferring these parameters from such curves requires solving an inverse problem through an iterative approach. Solving this inverse problem becomes easier if the corresponding forward problem, consisting of the computation of dispersion curves, can be solved efficiently and if the number of rail parameters to be estimated can be reduced.

The use of the *semi-analytical finite element* (SAFE) method for efficiently solving the forward problem is investigated in Refs. [52,57]. In these manuscripts, a set of parameters describing the geometry and material properties of a worn rail has been determined. This set is employed to develop a SAFE model for computing the propagation characteristics of worn rails. Wear and grinding of the railhead can be represented through the three geometric parameters shown in Figure 19. An adaptive mesh model has also been developed by means of parametric equations to represent material removal. This allows an efficient automatic modification of the geometry and the mesh of the considered rail.

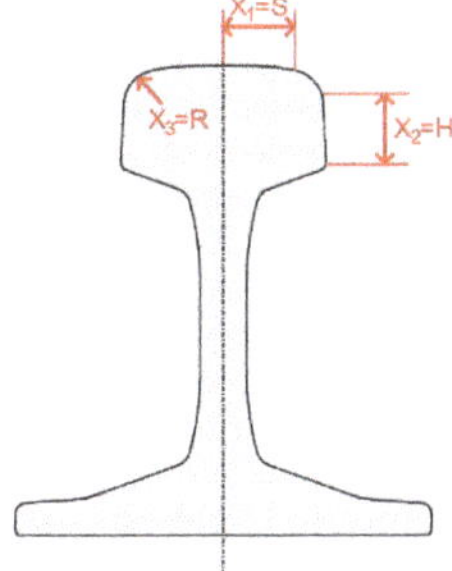

Figure 19. Representation of the parameters useful to identify the presence of a worn rail.

The use of the SAFE method requires the knowledge of different properties of the employed material, such as its *elastic modulus E* and its *density ρ*, or its *Poisson's ratio ν*. The evaluation of the dispersion curves of a waveguide through this method requires solving an eigenvalue problem. Material properties are usually unknown to an acceptable accuracy, and their value is dependent on environmental conditions. In the case of an isotopic elastic medium, it is shown that E and ρ can be combined in a single parameter, namely the *longitudinal speed of sound in the medium, $c = \sqrt{E/\rho}$*. Therefore, only two parameters, c and v, are required instead of three (i.e., E, ρ, and v): this allows for saving computation time. Computational efficiency can be further improved by solving the eigenvalue problem in terms of the parameter $\beta = \omega/c$, i.e., of the ratio between the frequency ω and the longitudinal speed of sound c. Thus, only a single set of dispersion curves needs to be computed for each value of the couple (c, v): if c changes, but v remains constant; the dispersion curves need only to be scaled according to the variation of the parameter β.

Experimental results have proven that various propagating modes could be detected on field. This has allowed for achieving a proper tracking of mode shapes computed through the SAFE method by using different material and rail geometric properties. The *modal assurance criterion* (MAC) has been adopted to accomplish the tracking task. More specifically, the mode shapes produced by the SAFE method in the presence of different input parameters have been correlated with three reference groups of the mode shapes identified manually on the basis of field measurements, as illustrated in Figure 20 [57].

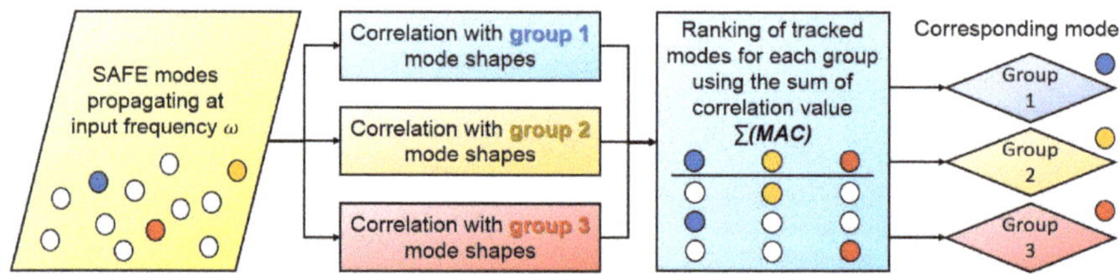

Figure 20. Representation of the mode shape tracking process that employs selected mode shape groups [57].

Finally, it is worth mentioning that the technique developed in Ref. [57] allows for identifying which set of simulated dispersion curves best fits the experimental spectrograms computed in the absence of prior knowledge about the distance of the reflectors from the transducer. Experimentation results on the field have proven that this technique performs well.

7.5.3. Conclusions

The estimation technique illustrated in Section 7.5.2 has been developed specifically for the considered application; hence, the involved inverse problem appears to be well-posed. Further research work is required to establish which conditions need to be satisfied to ensure that this technique works, or to determine when the associated inverse problem becomes ill-posed. The inverse problem has been solved by simply finding the SAFE model that provides the best fit in the sample space. The use of a response surface to interpolate the sample space should also be investigated.

8. Land-Based Systems Implementation—Other Projects

Other land-based systems and products based on similar working principles as Rail-Sonic by IMT-CSIR have been or are being developed. A brief introduction to these systems is provided in this section.

8.1. RailAcoustic by Enekom

The Broken Rail Detection System-RailAcoustic by ENEKOM is a system based on vibrations and not on ultrasonic guided waves. In Ref. [58], it is claimed that the operation of this system is exclusively based on electro-mechanical methods. A transmitter injects a vibration into the rail; this excitation is detected simultaneously by a receiver close to the transmitter and by a second receiver placed 2 km apart. The received signal is de-noised thanks to a synchronization signal, and the status of the rail is inferred from the difference between transmitted and received vibration signals. The receiver close to the transmitter allows the system to check if the excitation signal matches some specifications. Enekom claims that: (a) the modules of this system can be installed or dismantled without any damage to the involved rails; (b) this system works reliably and responsively under all weather conditions. Successful performance testing has been done on the Konya-Ankara high-speed railway.

8.2. Technical Contributions Provided by the Beijing Jiaotong University

A research group working at the Beijing Jiaotong University is investigating the propagation characteristics of rails and, in particular, the most suitable modes for detecting their defects [59–61]. They do so by analyzing the diagrams representing the strain energy distribution and the phase velocity dispersion curves of different propagation modes. Moreover, the researchers of the above-mentioned institution are trying to discover the best way to excite a specific mode or, alternatively, the best way to detect a given mode in a multi-mode scenario. As a matter of fact, their investigation has led to the development of a method for locating rail defects [62]; this method is called the single modal extraction algorithm (SMEA) and is briefly described in the next paragraph.

Single Modal Extraction Algorithm

The single modal extraction algorithm is described in Ref. [62]. It relies on the observation that: (a) the vibration displacement measured at any point on the rail is the superposition of the contributions of all the propagation modes at that point; (b) the vibration displacement due to each mode at any point can be extracted from the total displacement of the point.

A flow chart describing this method for defect location is shown in Figure 21. The principles on which this method is based can be summarized as follows. The primary task of defect location is to select the mode, the frequency, and the excitation conditions under which defect detection can be accomplished. The excitation response of the rail is analyzed through ANSYS (an engineering simulation software), that allows for simulating a three-dimensional model of the rail that also includes some defects. This allows for evaluating the vibration displacements of a series of points on the rail, thus providing a realistic representation of the signals observed in an experiment. Information about the reflected signals corresponding to the selected modes is provided by the results of the modal identification and SMEA. Defect location is based on the estimation of the group velocity and propagation time of the reflected modes.

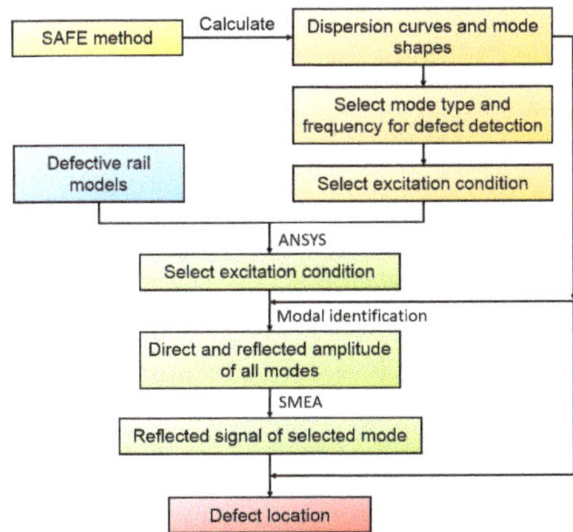

Figure 21. Flow chart describing the defect location algorithm described in Reference [62].

8.3. Technical Contributions Provided by the Xi'an University of Technology

A research group working at the Xi'an University of Technology has provided relevant contributions to the development of the electronics employed to drive the piezoelectric transducers that excite ultrasonic guided waves in rails. Their specific contributions include:

- The design of a high-voltage pulser achieving an improved transmission efficiency [63].
- The development of a tracking method able to estimate the optimal excitation frequency, i.e., the one maximizing the received signal power [64].
- The use of Baker coded UGW signals for enhancing the SNR of the received signal and the development of an adaptive peak detection algorithm [78].

As far as the last point is concerned, it is worth mentioning that Baker coding, albeit simple, is characterized by a autocorrelation with lower sidelobes than those observed in the autocorrelation of binary codes having the same length; moreover, its simplicity enables minimizing hardware complexity at the transmit side. It is also worth mentioning that the proposed peak detection algorithm offers the advantage of a low energy consumption and achieves better performance than other more sophisticated techniques, like those based on the discrete wavelet transform [79].

9. Performance Analysis

The target of a rail diagnostics system is to detect most of the defects affecting rails, while minimizing the number of false alarms [47]. Its performance can be assessed through the so-called *receiver operating characteristic curves* (ROC), which represent the trade-off between the *probability of detection* (PD) and the *probability of false alarm* (PFA) achievable by varying the threshold level adopted in defect detection. A related performance indicator is the *area under the receiver operating characteristic curve* (AUC) that provides an overall indication of the "goodness" of detection. In fact, AUC = 1 means that perfect detection is achieved for the given threshold (i.e., PD = 1 and PFA = 0); conversely, if AUC = 0, detection is jeopardized by false alarms, so that PD = 0 and PFA = 1.

In this section, various numerical results referring to the on-board and land-based systems described in the previous sections are illustrated.

9.1. On-Board Active System: Performance Analysis

Some representative DI traces acquired along a test track and at various testing speeds by means of the prototype of on-board active defect detection system described in Section 5.1 are illustrated in Figure 22; note that the SNR of the DI traces degrades as the train speed increases. The plots of Figure 22 are generated on the basis of the data available in Ref. [47].

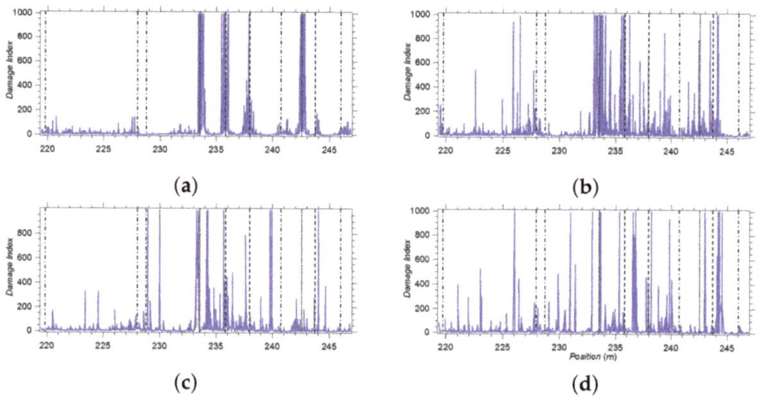

Figure 22. Example of damage index traces collected during tests (dashed lines = locations of defects; dashed-dotted lines = locations of welds; dotted lines = locations of joints). Testing speeds: (**a**) 1.6 km/h; (**b**) 8 km/h; (**c**) 16 km/h; (**d**) 24 km/h.

The performance of the above-mentioned prototype, expressed in terms of ROC, are illustrated in Figure 23 (its plots have been generated on the basis of the data available in Ref. [47]).

The probabilities of detection computed for four different PFAs on the basis of the cumulative curves shown in Figure 23 are listed in Table 2. From Figure 23 and Table 2, it is easily inferred that the performance of the considered defect detection system gets worse as the speed at which the test is accomplished increases [47]. Moreover, the experimental campaigns accomplished for this system have proven that it is sensitive to both transverse-type defects and mixed-mode cracks (vertical split heads or compound fractures). However, its performance is limited by the fact that air-coupled ultrasound transduction in steel suffers from a loss of energy due to the large impedance mismatch between air and steel, both in transmission and in reception. This explains the low SNR characterizing received signals [40] and limits the test speed [47].

Table 2. Significant PD and PFA values extracted from the cumulative ROC curves of Figure 23 [47].

Test Speed (km/h)	PD (%) Achievable for Specific PFAs				AUC (%)
	0% PFA	1% PFA	5% PFA	10% PFA	
1.6	65.38	76.92	86.54	92.31	97.58
8	7.08	47.79	78.76	85.84	95.03
16	0.00	29.23	47.69	56.92	77.90
24	1.92	9.62	28.85	32.69	69.05

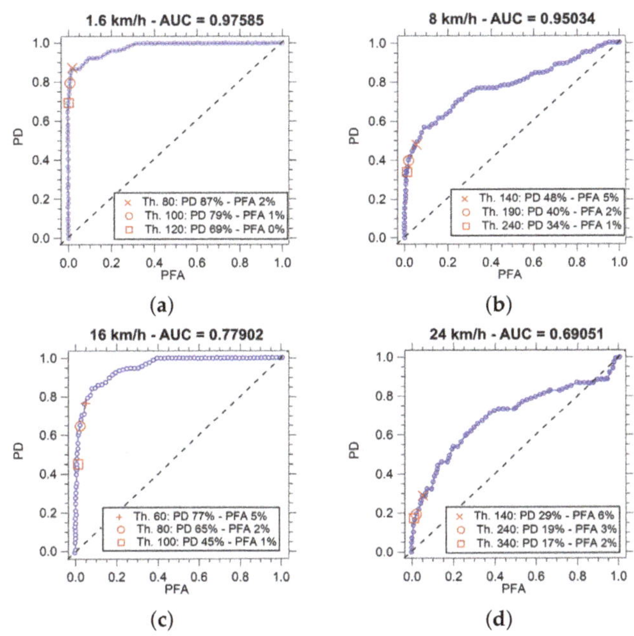

Figure 23. Cumulative ROC curves for runs executed at four different speeds: (**a**) 1.6 km/h; (**b**) 8 km/h; (**c**) 16 km/h; (**d**) 24 km/h. The results associated with three different values of the detection threshold are identified by specific labels.

9.2. On-Board Passive System: Performance Analysis

Some DI traces obtained for the on-board passive defect detection system described in Section 5.2 are illustrated in Figure 24 (its plots have been generated generated on the basis of the data available in Ref. [49]). Additional results illustrated in Ref. [66] have evidenced that:

1. The zones in which the received signal was strong enough for defect detection were primarily localized in sections of curved tracks, where wheels were flanging, so generating stronger excitation signals.
2. As speed decreases, a larger portion of the run becomes sub-optimal, since the presence of a defect cannot be easily detected.
3. At high speeds (say, higher than 96 km/h), an optimal source excitation energy is generated; this allows for achieving a stable Green's function.
4. Poor results were found in areas where the source excitation energy was too low to be detected by the receivers, e.g., in most of the tangent portion of the track. affecting the performance of the defect detection algorithms (and, in particular, of the ICA; see Section 7.4).

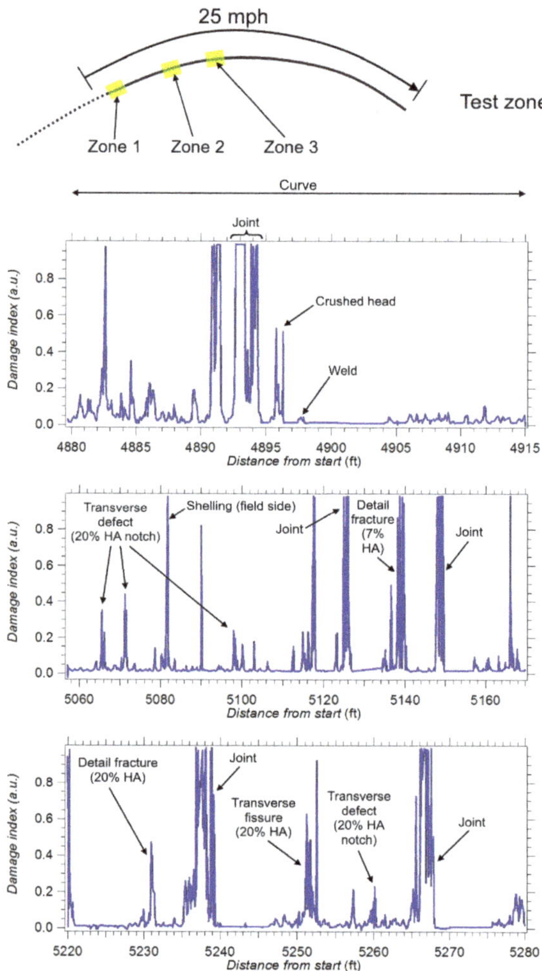

Figure 24. Damage index traces referring to the RDTF test track at 25 mph (40 km/h) in three selected test zones.

The performed tests did not allow for assessing the true repeatability of the achieved results; for this reason, further tests have been planned to investigate the reliability of the considered system [49]. Moreover, it has been found that, as long as acoustic excitation is detectable, the strength of the received signal does not play an important role in the reconstruction stability. However, instability is found when the signal intensity is comparable to the noise floor of the receiver sensors.

Based on the considerations illustrated in Section 5.2.4, concerning the conflicting needs of selecting a proper test speed and achieving an accurate localization, some experiments have been made; the obtained results are illustrated in Ref. [48]. In these experiments, various lengths for the recording time have been tested when the prototype was running at three different test speeds (30 mph, 50 mph, and 80 mph, corresponding to 48.28 km/h, 80.46 km/h, and 128.75 km/h, respectively). Such results prove that, in general, the SNR increases when the recording time increases; however, this leads to progressively losing spatial localization and to generating a non-stationary transfer function. Moreover, the SNR decreases as test speeds get larger. The duration of the time window has been selected in a way to achieve an accuracy equal to 8 in. (corresponding to 20.3 cm) in spatial localization at the various test speeds. The values evaluated for this duration are listed in Table 3.

Table 3. Compromise values between test speeds, recording time window and SNR of the passively reconstructed transfer function achieved in field tests.

Test Speed	Recording Time Window	SNR of Reconstructed Transfer Function
30 mph (48.28 km/h)	~15.25 ms	~12
50 mph (80.46 km/h)	~9.15 ms	~9
80 mph (128.75 km/h)	~5.7 ms	~4.5

Additional work is in progress to quantify and improve the reliability of the defect detection prototype by (a) determining the optimal value of the threshold for the DI; (b) minimizing the effects of poor signal reconstruction; (c) improving the strength of the source excitation signal at low speeds [66].

9.3. Land-Based Ultrasonic Broken Rail Detection System Test Results and Performance Analysis

The performance of the UBRD system version 4 is illustrated in Ref. [80]. After having solved some initial problems (equipment failures and GSM network reliability in some areas), a preliminary analysis of this system has been accomplished. In the first two months of operation, four broken rails and approximately four cracks have been detected before complete fracture occurred. It is believed that the system has prevented at least one derailment, the cost of which is similar to the cost of installing the system on the entire 840 km line. Detailed performance results have not been published yet.

9.4. Land-Based UBRD System with Early Rail Defect Detection Capability: Performance Analysis

Various experimental results referring to the land-based UBRD system with early defect detection capability are analyzed in Ref. [65]. As already mentioned in Section 7.4, the reordered set of pre-processed data has been processed by three algorithms (namely, SVD, ICA, and dimensionally-reduced ICA), whose performance in defect detection has been evaluated. Since the outcomes generated by these algorithms are highly dependent on the input dataset, we do not illustrate here an accurate analysis of the achieved results. We comment briefly on the performance achieved by such algorithms.

It is expected that, when a monotonically growing defect is found, its weight in ICA or SVD also grows monotonically. To establish if a weight exhibits a monotonic increase, the *Mann–Kendall* (M–K) test can be exploited, as suggested in Ref. [76]. This test is based on computing the difference between the number of increments and that of decrements between each pair of data points forming the input sequence; normalizing this difference with respect to the overall number of data points produces the normalized test statistic, denoted Z_{mk}. If the value of Z_{mk} is greater than 1.96, the probability of a monotonic trend is equal to 0.95; larger values of Z_{mk} indicate that this probability is even higher. The M–K test has been performed on the weights computed by all the considered algorithms.

9.4.1. Defect Detection in the Presence of Large Defects

From test results, it can be inferred that:

- The SVD, ICA, and dimensionally-reduced ICA techniques are able to separate the signature of artificial defects from the baseline.
- Their performance is appreciably affected by the EOCs and the multi-mode nature of ultrasonic guided waves propagation in rails. As far as the last issue is concerned, it is important to stress that mode conversion is one of the possible causes of phantom reflections appearing when ultrasonic guided waves impinge on a defect; the multi-mode nature of propagation may cause the defect signature to spread over different components. Moreover, the EOCs can have a different influence on each mode.
- In general, the independent components produced by the ICA algorithm appear to be less noisy than the singular vectors generated by the SVD technique.
- The presence of relatively large monotonic trends is confirmed by the analysis of the M–K test scores over the component weights.
- A good match between the components achieving a large M–K test score and the components related to the defect signature is found in the SVD, ICA, and dimensionally-reduced ICA techniques.

9.4.2. Defect Detection in the Presence of Small Defects

The performance of the SVD, ICA, and dimensionally-reduced ICA algorithm was tested also by considering a dataset composed only of the measurements corresponding to the initial evolution of the defect. When the defect appears, it is supposed to reflect only a small amount of the impinging UGW. Therefore, the corresponding dataset is composed of the measurements in which the peak of reflection caused by the defect is small in amplitude. In this case:

- From the outcomes of SVD analysis, it has been inferred that the presence of an artificial defect cannot be easily detected. Moreover, selecting the singular values characterized by the highest M–K score does not guarantee defect detection; furthermore, in these conditions, the defect cannot be located by analyzing the components of the singular vectors.
- The analysis of the results generated by the ICA have proven that the defect signature is spread over multiple independent components, whose associated weights exhibit a monotonic increase, proportional to defect growth.

All of these results have led to the conclusion that ICA with dimension reduction performs better than the SVD algorithm in the presence of EOCs when attempting to detect the growth of small defects [65].

9.4.3. Concluding Analysis on the Evolved UBRD System

The cancellation of unwanted modes through phased array processing can be improved by using a larger number of transducers in the array employed for transmission and reception. The resulting system is able to accurately estimate the distance and direction of reflectors along the considered span of rails. However, since EOCs are not exclusively due to temperature variations, an important role in the achievable performance is played by techniques for compensating EOCs other than temperature. Defect detection has been performed through SVD and ICA; test results have led to the conclusion that ICA with dimension reduction performs better than SVD. Nevertheless, it has been found that the signature of a given defect spreads over different independent components; this might be due to the multi-modality of defect reflection signature. The last issue deserves further investigation.

Based on the results obtained in Ref. [65], it can be stated that the detection of relatively small defects in the railhead is possible even if an array consisting of two transducers only is used and that, despite this positive result, the development of an automatic monitoring system, capable of autonomous operation with a low rate of false alarms, requires substantial effort. For this reason, the authors of that manuscript suggest that, during the design of an autonomous machine learning algorithm, the decision to stop trains or to ask for an on field inspection should be made by an expert able to interpret its output. A possible workflow for the development of a defect detection algorithm is also suggested: this could start from the analysis of the size of M–K scores, looking for values characterized by a significantly larger trend than the others. Then, it can be verified if the associated independent component resembles a single reflection from a defect. An appropriate level of alarm could be assigned on the basis of magnitude of the considered weight and its rate of increase. In any case, decision thresholds for these steps should be selected only after analyzing a large dataset.

10. Discussion and Conclusions

In this manuscript, different types of rail defects have been described and the main techniques for their detection have been outlined; the main features of these techniques are summarized in Table 4. Then, our attention has focused on the diagnostic technique based on ultrasonic guided waves. This technique is expected to solve the problem of the high-speed (or quasi-real-time) inspection of both the surface and the inner part of rails. In our analysis, two possible approaches, namely an on-board approach and a land-based approach, have been taken into consideration. In the on-board approach, the ultrasonic instrumentation is installed on an inspection car for high-speed scanning. In the land-based approach, instead, the ultrasonic equipment is attached to the rail in a fixed position for quasi-real-time monitoring. Moreover, in our analysis of the on-board approach, two different technical methods have been described: an active method and a passive one. In the on-board active method, ultrasonic guided waves are generated and sensed by the instrumentation installed on the inspection vehicle. On the other hand, the on-board passive method exploits the ultrasonic waves generated by the rail-wheel interaction of the moving train; consequently, the instrumentation placed in the lead car of the train only measures the response of the rail. A land-based system for ultrasonic broken rail detection is already active in South Africa; the system is being evolved to detect early-stage cracks too.

Table 4. NDT techniques for rail defect detection.

Technique	Approach	Detected Defects	Scanning Speed	Performance	Open Problems
Eddy current (EC)	Manual inspection, on-board, non-contact	Surface defects and near-surface internal railhead defects.	$\leq$70 km/h	Reliable detection of surface defects; adversely affected by grinding marks. It is sensitive to probe lift-off.	Mature technology
Magnetic flux leakage (MFL)	On-board, non-contact	Surface defects and near-surface internal railhead defects. Unsuitable for longitudinal fissures.	$\leq$35 km/h	Reliable detection of surface and near-surface defects in the railhead. Cracks smaller than 4 mm are not detected. Performance degradation observed at high speeds.	Mature technology, research aims at increasing its scanning speed
Visual inspection	Manual inspection, on-board, non-contact	Surface breaking defects, rail head profile, corrugation, missing parts, defective ballast. No internal defects.	From 4 to 320 km/h	Reliable detection on the surface of the rail and track. Internal cracks are not detected. Sensitive to lighting conditions (these require proper compensation).	Mature technology
Radiography	Manual system	Welds and all types of known internal defects.	Static	Reliable detection of internal defects in rails and welds that are difficult to inspect by means of other techniques. Some transverse defects can be missed. Safety hazard, costly and time-consuming.	Mature technology
Conventional ultrasound	Manual inspection, on-board, contact	Surface defects, railhead, web and foot internal defects.	<70 km/h in tests; in practice around 15 km/h	Reliable manual inspection, but rail foot defects can be missed. In high-speed inspection, surface defects smaller than 4 mm, as well as internal defects (especially in the rail foot) can be missed. Limit on the maximum speed achievable. Surface shallow RCF defects can mask severe internal cracks.	Mature technology

Technique	Approach	Detected Defects	Scanning Speed	Performance	Open Problems
Active on-board guided waves ultrasound	Active on-board, non-contact	Surface defects and rail head web and foot internal defects. Best for transverse and mixed mode defects.	$\leq$16 km/h laser; $\leq$24 km/h air-coupled piezoelectric	Reliable on-board system able to detect surface and internal defects. Affected by sensors lift-off variations. Difficult to be deployed at high speeds. The employed laser is costly and hard to maintain.	In case of piezoelectric excitation, the airborne signal reverberation limits the achievable test speed because of acoustic mismatch between air and steel.
Passive on-board guided waves ultrasound	Passive, on-board	Surface defects and rail head web and foot internal defects. Best for transverse and mixed mode defects.	>96 km/h	The system reliably detects defects when the signal strength is high, i.e., where the rail-wheel interaction is stronger (e.g., in curves). As long as the quality of the received signal is good, its strength is not so important. System under development.	Quantification and improvement of defect detection reliability by selecting appropriate thresholds. Improvement of source excitation signals at low speeds or in tangent tracks.

Technique	Approach	Detected Defects	Scanning Speed	Performance	Open Problems
Ultrasonic broken rail detector (UBRD)	Active, land-based	Broken rails	Fixed system (1 km sections), quasi real-time operation	The system is designed to reliably detect clean breakages in rails. It is already working in operating conditions.	Mature technology. Evolution to reliably detect defects in rails and to increase section length.
Land-based guided waves ultrasound rail defect detection	Active, land-based	Transverse and mixed mode internal defects	Fixed system (1 km sections), quasi real-time operation	The technology is under development. The aim is to detect and monitor a growing defect, and raise an alarm when it reaches a critical size.	Study of the behavior of a defect during time, compensation of resonant transducers variations, compensation of varying environmental and operating conditions which modify the propagation of guided waves.

10.1. Advantages and Disadvantages of the Considered Systems

The systems reviewed in this manuscript present both advantages and disadvantages. Due to the approaches followed being so different, a direct comparison is impossible. However, an attempt can be made to assess their pros and cons by considering their main characteristics, range of application, and ease of development.

The main disadvantage of the active on-board inspection system is its limited inspection speed; in these terms, the system does not achieve a significant improvement over the conventional ultrasonic inspection systems. On the contrary, if the passive on-board system is considered, the opposite problem is found: in fact, the system is able to provide good results only when the diagnostic vehicle is moving at high speed or along curved tracks. This limitation can be circumvented by exploiting the ability of this system to reconstruct the waveguide transfer function (which allows defect detection) without the perfect knowledge of the excitation signal. An alternative to on-board approaches is constituted by land-based systems, such as the UBRD and the UBRD evolved to recognize small-scale defects. The main feature of the simple UBRD system, the detection of complete rail breaks, is already implemented in the track circuits, which are well known to railway engineers. The advantage of the UBRD system is that it does not perform critical operations in terms of railway traffic safety (such as train detection); therefore, it might not be subject to stringent safety constraints, unlike track circuits. The main advantages of the evolved UBRD system are represented by its ability to identify defects before a complete rail breakage and the possibility of accomplishing a continuous monitoring of the health status of rails. The last feature is not provided by on-board systems that, because of their nature, call for a periodic inspection of the infrastructure. It should be also kept in mind that a continuous monitoring system allows for acquiring a large number of measurements having limited accuracy. In fact, the process of searching for a defect, identifying it, and monitoring its evolution is repeated continuously over a long period. Basically, it does not matter that a defect is identified and reported at its onset, when it still has a small size. What really matters is that the presence of a defect is signaled and its evolution kept under control over time, so that maintenance operations for its removal can be planned before they become dangerous for the safety of railway traffic. A periodic inspection system, depending on the frequency of its tests, needs to acquire accurate measurements. A defect must be detected in time, when it is still small in size, so that its evolution can be monitored, inspection after inspection, before it reaches a critical state. This is needed since any defect can exhibit a sudden evolution.

Other advantages and disadvantages to be taken into consideration are those related to the practical usage of the considered systems. If the health of the rails of an entire railway network has to be monitored, a land-based system must be installed on all its tracks; this could be time-consuming, laborious, and expensive. In addition, system maintenance might require a relevant effort, especially when some of its devices have been installed in remote areas, which are hard to reach. Another relevant problem is represented by the fact that its equipment, being physically attached to the rails, could represent an obstacle during any operation of infrastructure renovation. On the contrary, if an on-board system is employed, it is only needed to equip an adequate number of diagnostic vehicles, and make them travel frequently and regularly throughout the network. This means that fewer devices, albeit more complicated, need to be built, installed, and maintained. Moreover, their regular maintenance may take place in workshops and not on the field, perhaps in remote areas. On the other hand, the costs associated with carrying out special diagnostic runs have to be paid; moreover, these runs could prevent the circulation of commercial service trains in certain time intervals, reducing the capacity of the network. If a system achieving a good balance between performance, installation, and maintenance costs is developed, these problems can be overcome. In fact, a part of the fleet of commercial service trains could be equipped with such a system, allowing such trains to perform diagnostic inspections while performing their normal duties.

Another relevant issue to be considered is the complexity of system testing during its development. An on-board system requires the execution of high-speed tests on field using railway vehicles in a relatively early stage of its development period. For this reason, it is required to have (a) a test vehicle capable of sustaining the envisioned inspection speeds, (b) the possibility of accessing an infrastructure suitable for tests, and finally (c) qualified personnel to drive and route the diagnostic vehicle on the selected infrastructure. All this might be quite costly. From this perspective, a land-based system has very different requirements. In fact, its initial development does not require access to particular infrastructures other than a sufficiently long stretch of rail. This is true at least until it is needed to test the system in its effective operating conditions. Even in this case, however, it is sufficient to have access to an adequate stretch of track where the system can be installed. Once the system has been set up, the execution of tests does not require to run dedicated trains or to employ specialized personnel.

10.2. Future Developments

Various possible developments of the systems described in this manuscript have been already mentioned in the previous sections. In summary, as far as the passive on-board system is concerned, research efforts should aim at improving its performance in defect detection at low speeds. Moreover, it is necessary to quantify and improve the reliability of the developed prototype in defect identification. On the other hand, to achieve a reliable operation of land-based systems, a fundamental research problem needs to be solved. In fact, it is needed to acquire a deeper understanding on the evolution over time of the defect response to UGW solicitations. In addition, the influence of variable EOCs on such a response must be studied in detail. Some improvement in defect identification and localization based on unsupervised machine learning techniques is also foreseeable.

10.3. Conclusions

All of the ultrasonic systems described in the technical literature still suffer from various technical problems, related to ultrasonic propagation and sensing in rail, to the influence of environmental and operating factors, etc. Therefore, even if various technically relevant results have already been obtained in this field, substantial research efforts are still needed. The strong interest in the above-mentioned problems is motivated by the increasing attention paid to railway transport safety in recent years.

Author Contributions: D.B. has written the first draft of the whole manuscript, whereas G.M.V. has carefully rewritten it. G.F. has revised the final manuscript. All authors have read and agreed to the published version of the manuscript.

Funding: This research received no external funding.

Institutional Review Board Statement: Not applicable.

Informed Consent Statement: Not applicable.

Data Availability Statement: Data sharing not applicable.

Acknowledgments: We would like to thank Alstom Ferroviaria S.p.A. (Bologna, Italy) for funding a PhD scholarship on the research topic analyzed in this manuscript.

Conflicts of Interest: The authors declare no conflict of interest.

Appendix A. Guided Waves

Any disturbance affecting a portion of an elastic medium propagates through the medium itself in a finite time as a mechanical sound wave (*elastic wave*). An ultrasound or ultrasonic wave is characterized by a frequency greater than 20 kHz. There are two types of ultrasonic waves [81]: *bulk* (fundamental) waves and *guided* waves. Bulk propagation refers to waves propagating without any boundary, like in an infinite medium (or in media whose boundaries does not influence wave propagation). Guided waves propagate in bounded media, like plates, rods, or tubes.

In general, an elastic wave consists of two components, which propagate independently from each other; these components are known as *longitudinal* and *transverse (shear)* waves. In longitudinal waves, the variation of the propagating quantity is observed parallel to the propagation direction. In transverse waves, instead, the variation of propagating quantity is orthogonal to the propagation direction. The existence of such waves depend on the elastic properties of the medium in which they propagate [82].

In a waveguide, interference phenomena arise from the waves bouncing back and forth inside the waveguide itself when impinging on its boundaries. Bouncing produces mode conversion because of reflection and refraction of longitudinal and shear waves. Depending on the angle and frequency, constructive, destructive, or intermediate interference takes place: this leads to hundreds of solutions of constructive interference points and, consequently, to guided wave packets traveling in the waveguide and named *modes*. The interference points can be represented through a *wave velocity dispersion curve*, which relates phase velocity to frequency. Each waveguide has its own set of dispersion curves, and every curve in the set is related to a specific mode of propagation. In the technical literature, dispersion curves are commonly employed to display all the types of waves and modes that can propagate inside a given waveguide. The ability to select, for a given dispersion curve, specific phase velocity and frequency has a significant impact on the penetration power and sensitivity of an ultrasound inspection system based on guided waves [83].

Another important feature to be taken into consideration is the wave structure of the selected mode, i.e., the *mode shape*. This shows the mode in-plane displacement, out-of-plane displacement, or actual stress distribution that varies across the thickness of the considered waveguide [70]. This knowledge is useful to establish the maximum penetration power of waves in a given structure, or to establish the maximum sensitivity to a defect located in a specific area of a test specimen [83].

When an ultrasonic wave propagates through a medium, it undergoes attenuation. The waveform and the amplitude of an ultrasonic wave are influenced by a number of factors, including ultrasonic beam spreading, energy absorption, dispersion, nonlinearity, transmission at interfaces, scattering by inclusions and defects, Doppler effect, etc. [81]. Attenuation is a fundamental factor to be always kept in mind when designing ultrasonic guided waves systems, especially if they are required to operate on a long range.

A finite body can support an infinite number of guided wave modes; these represent the solutions of a differential guided wave problem, defined by the boundary conditions. Specific solutions to guided wave problems are those related to *Rayleigh*, *Lamb*, and *Stonely* waves. Rayleigh waves are waves on the surface of a semi-infinite solid, Lamb waves are waves of plain strain occurring in a free plate, and Stonely waves are waves that occur at the interface between two media [70].

Appendix B. Characteristics of Guided Waves in Rails

Modeling the rail as a waveguide is not easy because of its complicated profile. For this reason, the accurate computation of the dispersion characteristics of guided waves traveling through rails represents a difficult problem, whose solutions are based on finite elements methods [7]. A frequently used method is the *semi-analytical finite element* (SAFE) method.

The dispersion curves of rails exhibit *mode repulsion* and *mode crossing*: these phenomena cannot be easily distinguished. The meaning of these two terms can be understood by analyzing the dispersion curves of a waveguide. When two modes (curves) approach each other, they can cross (mode crossing) or suddenly diverge without crossing (mode repulsion). The mode shapes of rails, having a symmetric profile (e.g., see Figure 1), are either symmetric or anti-symmetric. A mode is symmetric when its energy is distributed symmetrically over the waveguide cross-section, i.e., it is equally distributed between field side and gauge side in the case of rails. A mode is antisymmetric when its energy is concentrated non-symmetrically in the waveguide, i.e., on one side only if rails are considered. It has been shown that: (a) symmetric and antisymmetric modes can cross each other; (b) the modes within symmetric and antisymmetric families do not cross each other [84]. Furthermore, the introduction of even a small asymmetry in the waveguide shape produces repulsion forces that prevent mode crossings. This information is useful in the selection of a propagation mode to be employed for rail integrity inspection. Examples of wavenumber vs. frequency curves and of the concentration of energy for different modes are shown in Figure 10.

The energy of different (symmetric or antisymmetric) modes is concentrated in the head, web, or foot of the rail. This explains the possibility of locating defects in a specific portion of the rail cross-section by selecting the most appropriate mode [7].

Some concerns can originate from the influence of ties, fasteners, and absorbing pads. It has been found that, at high frequencies (such as those of ultrasonic guided waves), their presence does not influence dispersion curves, but only decay rates [85]. Moreover, experimental studies have led to the conclusion that the most effective frequency interval for long range wave propagation along railway tracks is between 20 and 40 kHz [86].

The three main technical problems observed in the use of guided ultrasonic waves are [7]:

1. *Dispersion*—If different modes are excited, they travel at different (frequency-dependent) velocities in both directions. Consequently, each mode takes a different time to travel along the employed waveguide, so compromising spatial resolution (for instance, a given reflector can generate multiple echoes). This problem can be mitigated by resorting to dispersion compensation.
2. *Coherent noise*—This noise is observed in the same frequency band of the signal of interest. It is due to: (a) the excitation and reception of unwanted modes; (b) the transmission of waves in the wrong direction along the waveguide and the reception of echoes from that direction. Therefore, mitigating coherent noise requires exciting and sensing only the selected modes. This can be accomplished by using a proper transducer or excitation signal, and by suppressing unwanted modes.
3. *Changes in temperature or in material properties with age*—These changes, even if small, affect the above-mentioned dispersion phenomenon. This problem has to be taken into account when comparing signals received at different instants.

Appendix B.1. Excitation of Guided Modes

A mode is most efficiently excited when the harmonic force applied by the employed transducer to the considered waveguide is well coupled to the displacement associated with the mode itself. Therefore, a transducer should be possibly placed on the rail surface where the displacement of the selected mode shape is large; the polarization of mode shape and that of the transducer must be equal. These considerations also apply to efficient reception [42]. Mode control is achieved by choosing an appropriate transducer and a suitable excitation signal [7].

Important alternatives for the transducers are represented by piezoelectric transducers and *electro-magnetic acoustic transducers* (EMATs). An EMAT generates a wave in a given structure via the Lorentz force and/or magnetostriction (see Figure A1a). The phase velocity c_p of such a wave is given by $c_p = f \cdot \lambda$, where f is the frequency of the driving signal and λ is the meander coil spacing (i.e., the wavelength imposed by the EMAT). Direction control can be achieved by employing a second coil overlapping the first one, but displaced from it along the structure by a quarter wavelength. A piezoelectric transducer generates compression waves towards the employed structure via a coupling medium, as shown in Figure A1a (where θ_t denotes the orientation angle of the transducer with respect to the structure). The wavelength λ_p of the waves generated in the structure is related to the wavelength λ_c of the compression waves traveling through the coupling medium by $\lambda_p = \lambda_c / \sin \theta_i$, where θ_i is the angle of incidence. The phase velocity in the structure is given by $c_p = v / \sin \theta_i$, where v is the velocity of the compression waves in the coupling medium. Therefore, a mode at a given frequency can be excited by properly orienting the employed transducer.

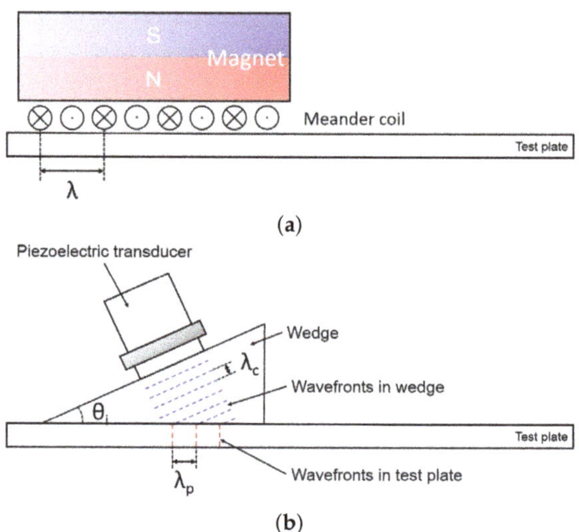

Figure A1. Representation of the excitation of guided waves through (**a**) an EMAT or (**b**) a piezoelectric transducer [7].

The size of the transducer and the generated excitation signal influence the achieved degree of modal selectivity. On the one hand, the effective wavelength bandwidth (the effective phase velocity bandwidth) depends on the transducer size if an EMAT (a piezoelectric transducer) is employed; on the other hand, the frequency bandwidth depends on the excitation signal. If the transducer has a diameter of 3–5 times the selected wavelength, satisfactory mode control is achieved: moreover, an array of point sources has to be preferred in long-range testing with respect to a monolithic transducer. A fundamental contribution to the improvement of the signal to coherent noise ratio is provided by the signal processing techniques employed at the receive side; in fact, these should allow for extracting the desired input mode-received mode combination only, while rejecting the others [7].

Appendix B.2. Selection of Guided Modes

As it can be easily inferred from the dispersion curves shown in Figure 10a, various propagating modes can be excited in rails in the frequency range of interest (25 kHz–45 kHz). However, not all of these modes are suitable for long-range propagation due to the attenuation they experience (e.g., in the foot of the rail) or their high dispersivity [44].

As already mentioned in Appendix A, a specific mode is highly sensitive to defects at positions of the rail cross section where the energy in the mode shape is most concentrated. The following considerations have to always be kept in mind when selecting the mode best suited for detecting a defect in pulse-echo operation. When a uni-modal wave impinges on a defect (or on another feature), a portion of its energy is reflected, but the reflected wave is no longer uni-modal because of modal conversion. Therefore, it is important to acquire data about reflection coefficients for each combination of incident and reflected modes; such data can be organized in a matrix of reflection coefficients for each defect or feature. A useful way of representing the information contained in such a matrix is through a color or grayscale map of the amplitudes of each element [42]; an example of such a map is shown in Figure A2 [44]. Based on this map, it can be shown that, on the one hand, a symmetric mode with energy concentrated in the railhead (e.g., mode 1 of Figure A2) is well suited to distinguish between cracks in the railhead and welds. On the other hand, a mode with energy concentrated in the web (e.g., mode 4 of Figure A2) can be exploited to detect welds and damages in the rail web, but not to differentiate them. Finally, it is worth noting that detecting cracks at thermite welds (through a land-based approach; see Section 6) or in the foot of the rail can be very difficult.

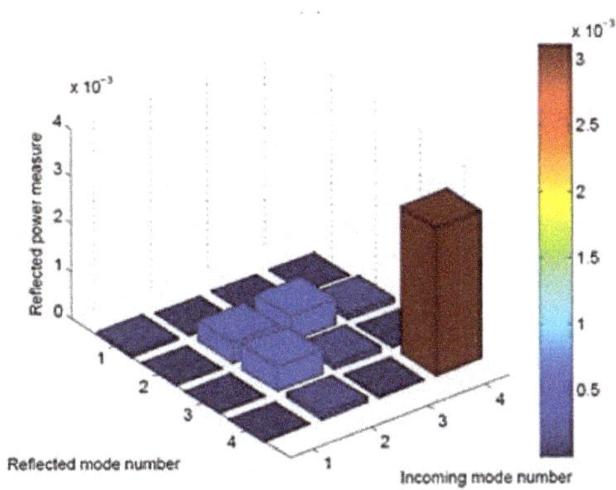

Figure A2. Reflection map for a thermite weld with a 6 mm thick weld cap (picture taken from Ref. [44]).

References

1. Zumpano, G.; Meo, M. A new damage detection technique based on wave propagation for rails. *Int. J. Solids Struct.* **2006**, *43*, 1023–1046. [CrossRef]
2. Ferreira, L.; Murray, M. Modelling rail track deterioration and maintenance: Current practices and future needs. *Transp. Rev.* **1997**, *17*, 207–221. [CrossRef]
3. European Union Agency of Railways. *Railway Safety in the European Union-Safety Overview 2017*; Publications Office of the European Union: Luxembourg, 2017.
4. Papaelias, P.; Roberts, M.C.; Davis, C.L. A Review on Non-Destructive Evaluation of Rails: State-of-the-Art and Future Development. *Proc. Inst. Mech. Eng. Part F J. Rail Rapid Transit* **2008**, *222*, 367–384. [CrossRef]
5. Rizzo, P. Sensing solutions for assessing and monitoring railroad tracks. In *Sensor Technologies for Civil Infrastructures*; Wang, M.L., Lynch, J.P., Sohn, H., Eds.; Woodhead Publishing: Cambridge, UK, 2014; Volume 56, pp. 497–524.
6. Fadaeifard, F.; Toozandehjani, M.; Mustapha, F.; Matori, K.; Mohd Ariffin, M.K.A.; Zahari, N.; Nourbakhsh, A. Rail inspection technique employing advanced nondestructive testing and Structural Health Monitoring (SHM) approaches—A review. In Proceedings of the Malaysian International NDT conference and Exhibition (MINDTCE 13), Kualalampur, Malaysia, 16–18 June 2013.
7. Cawley, P.; Lowe, M.J.S.; Alleyne, D.N.; Pavlakovic, B.; Wilcox, P. Practical long range guided wave testing: Applications to pipes and rail. *Mater. Eval.* **2003**, *61*, 66–74.
8. Ghofrani, F.; Pathak, A.; Mohammadi, R.; Aref, A.; He, Q. Predicting rail defect frequency: An integrated approach using fatigue modeling and data analytics. *Comput. Aided Civ. Infrastruct. Eng.* **2019**, *35*, 1–15. [CrossRef]
9. Loveday, P.; Taylor, R.M.C.; Long, C.; Ramatlo, D. Monitoring the reflection from an artificial defect in rail track using guided wave ultrasound. *AIP Conf. Proc.* **2018**, *1949*, 090003.
10. Sadeghi, J.M.; Askarinejad, H. Development of track condition assessment model based on visual inspection. *Struct. Infrastruct. Eng.* **2011**, *7*, 895–905. [CrossRef]
11. Muravev, V.V.; Boyarkin, E.V. Nondestructive testing of the structural-mechanical state of currently produced rails on the basis of the ultrasonic wave velocity. *Russ. J. Nondestruct. Test.* **2003**, *39*, 24–33. [CrossRef]
12. Abbaszadeh, K.; Rahimian, M.; Toliyat, H.A.; Olson, L.E. Rails defect diagnosis using wavelet packet decomposition. *IEEE Trans. Ind. Appl.* **2003**, *39*, 1454–1461.
13. Jeffrey, B.D.; Peterson, M.L.; Gutkowski, R.M. Assessment of rail flaw inspection data. *AIP Conf. Proc.* **2000**, *509*, 789–796.
14. Vadillo, E.G.; Tarrago, J.A.; Zubiaurre, G.G.; Duque, C.A. Effect of sleeper distance on rail corrugation. *Wear* **1998**, *217*, 140–146. [CrossRef]
15. Bohmer, A.; Klimpel, T. Plastic deformation of corrugated rails—A numerical approach using material data of rail steel. *Wear* **2002**, *253*, 150–161. [CrossRef]
16. Cannon, D.F.; Edel, K.O.; Grassie, S.L.; Sawley, K. Rail defects: An overview. *Fatigue Fract. Eng. Mater. Struct.* **2003**, *26*, 865–886. [CrossRef]
17. Cannon, D.F.; Pradier, H. Rail rolling contact fatigue research by the European Rail Research Institute. *Wear* **1996**, *191*, 1–13. [CrossRef]

18. Grassie, S.; Nilsson, P.; Bjurstrom, K.; Frick, A.; Hansson, L.G. Alleviation of rolling contact fatigue on Swedens heavy haul railway. *Wear* **2002**, *253*, 42–53. [CrossRef]
19. Australian Rail Track Corporation. *Rail Defects Handbook-Some Rail Defects, Their Characteristics, Causes and Control. Document RC2400*; Australian Rail Track Corporation: Mile End, Australia, 2006.
20. Shull, P.J. *Nondestructive Evaluation: Theory, Techniques, and Applications*, 1st ed.; CRC Press: New York, NY, USA, 2002.
21. Bray, D.E.; Stanley, R.K. *Nondestructive Evaluation. A Tool in Design, Manufacturing, and Service*, 2nd ed.; CRC Press: New York, NY, USA, 1997.
22. Mix, P. *Introduction to Nondestructive Testing, A Training Guide*, 2nd ed.; John Wiley and Sons: Hoboken, NJ, USA, 2005.
23. Clark, R. Rail flaw detection: Overview and need for future developments. *NDT&E Int.* **2004**, *37*, 111–118.
24. Rajamäki, J.; Vippola, M.; Nurmikolu, A.; Viitala, T. Limitations of eddy current inspection in railway rail evaluation. *Proc. Inst. Mech. Eng. Part F J. Rail Rapid Transit* **2018**, *232*, 121–129. [CrossRef]
25. Xu, P.; Zhu, C.; Zeng, H.; Wang, P. Rail crack detection and evaluation at high speed based on differential ECT system. *Measurement* **2020**, *166*, 108152. [CrossRef]
26. Wang, J.; Dai, Q.; Lautala, P. *Real-Time Rail Defect Detection with Eddy Current (EC) Technique: Signal Processing and Case Studies of Rail Samples*; Final Rep. of the Project NURail2020-MTU-R18; National University Railway Center: Urbana, IL, USA, 2020.
27. Nichoha, V.; Storozh, V.; Matiieshyn, Y. Results of the development and research of information-diagnostic system for the magnetic flux leakage defectoscopy of rails. In Proceedings of the 2020 IEEE 15th International Conference on Advanced Trends in Radioelectronics, Telecommunications and Computer Engineering (TCSET), Lviv-Slavske, Ukraine, 25–29 February 2020; pp. 852–857.
28. Deutschl, E.; Gasser, C.; Niel, A.; Werschonig, J. Defect detection on rail surfaces by a vision based system. In Proceedings of the IEEE Intelligent Vehicles Symposium, Parma, Italy, 14–17 June 2004; pp. 507–511.
29. Singh, M.; Singh, S.; Jaiswal, J.; Hempshall, J. Autonomous rail track inspection using vision based system. In Proceedings of the IEEE International Conference on Computational Intelligence for Homeland Security and Personal Safety, Alexandria, VA, USA, 16–17 October 2006; pp. 56–59.
30. Molina, L.F.; Resendiz, E.; Edwards, J.R.; Hart, J.M.; Barkan, C.P.L.; Ahuja, N. Condition monitoring of railway turnouts and other track components using machine vision. In Proceedings of the Transportation Research Board 90th Annual Meeting, Washington, DC, USA, 23–27 January 2011; pp. 11–1442.
31. Molina, L.F.; Edwards, J.R.; Barkan, C.P.L. Emerging condition monitoring technologies for railway track components and special trackwork. In Proceedings of the 2011 Joint Rail Conference, Pueblo, CO, USA, 16–18 March 2011; pp. 151–158.
32. Oukhellou, L.; Côme, E.; Bouillaut, L.; Aknin, P. Combined use of sensor data and structural knowledge processed by Bayesian network: Application to a railway diagnosis aid scheme. *Transp. Res. Part C* **2008**, *16*, 755–767. [CrossRef]
33. Rete Ferroviaria Italiana. *Diagnostics Services Catalogue 2018*; Rete Ferroviaria Italiana: Rome, Italy, 2018.
34. Vandone, A.; Rizzo, P.; Vanali, M. Two-stage algorithm for the analysis of infrared images. *Res. Nondestruct. Eval.* **2012**, *23*, 69–88. [CrossRef]
35. Peng, D.; Jones, R. Lock-in thermographic inspection of squats on rail steel head. *Infrared Phys. Technol.* **2013**, *57*, 89–95. [CrossRef]
36. Usamentiaga Fernández, R.; Sfarra, S.; Fleuret, J.; Yousefi, B.; García Martínez, D.F. Rail inspection using active thermography to detect rolled-in material. In Proceedings of the 14th Quantitative InfraRed Thermography (QIRT) Conference, Berlin, Germany, 25–29 June 2018.
37. Bayissa, W.L.; Dhanasekar, M. High speed detection of broken rails, rail cracks and surface faults. In *Advanced Project Management Training Needs*; CRC for Rail Innovation: Brisbane, Australia, 2011.
38. Marais, J.J.; Mistry, K.C. Rail integrity management by means of ultrasonic testing. *Fatigue Fract. Eng. Mater. Struct.* **2003**, *26*, 931–938. [CrossRef]
39. Olympus. *Advances in Phased Array Ultrasonic Technology Applications*; Olympus: Waltham, MA, USA, 2011; p. 280.
40. Mariani, S.; Nguyen, T.; Phillips, R.R.; Kijanka, P.; Lanza di Scalea, F.; Staszewski, W.J.; Fateh, M.; Carr, G. Noncontact ultrasonic guided wave inspection of rails. *Struct. Health Monit.* **2013**, *12*, 539–548. [CrossRef]
41. Coccia, S.; Bartoli, I.; Salamone, S.; Phillips, R.; Lanza di Scalea, F.; Fateh, M.; Carr, G. Noncontact Ultrasonic Guided Wave Detection of Rail Defects. *Transp. Res. Rec.* **2009**, *2117*, 77–84. [CrossRef]
42. Wilcox, P.; Pavlakovic, B.; Evans, M.; Vine, K.; Cawley, P.; Lowe, M.; Alleyne, D. Long Range Inspection of Rail Using Guided Waves. *AIP Conf. Proc.* **2003**, *657*, 236–243.
43. Ryue, J.; Thompson, D.J.; White, P.R.; Thompson, D.R. Wave Propagation in Railway Tracks at High Frequencies. In *Noise and Vibration Mitigation for Rail Transportation Systems. Notes on Numerical Fluid Mechanics and Multidisciplinary Design*; Schulte-Werning, B., Thompson, D., Gautier, P.-E., Hanson, C., Hemsworth, B., Nelson, J., Maeda, T., de Vos, P., Eds.; Springer: Berlin/Heidelberg, Germany, 2008; Volume 99, pp. 440–446.
44. Long, C.; Loveday, P. Prediction of Guided Wave Scattering by Defects in Rails Using Numerical Modelling. *AIP Conf. Proc.* **2014**, *1581*, 240–247.
45. Rizzo, P.; Cammarata, M.; Bartoli, I.; Lanza di Scalea, F.; Salamone, S.; Coccia, S.; Phillips, R. Ultrasonic Guided Waves-Based Monitoring of Rail Head: Laboratory and Field Tests. *Adv. Civ. Eng.* **2010**, *2010*, 1–13. [CrossRef]
46. Mariani, S.; di Scalea, F.L. Predictions of defect detection performance of air-coupled ultrasonic rail inspection system. *Struct. Health Monit.* **2018**, *17*, 684–705. [CrossRef]

47. Mariani, S.; Nguyen, T.; Zhu, X.; Lanza di Scalea, F. Field Test Performance of Noncontact Ultrasonic Rail Inspection System. *J. Transp. Eng. Part A Syst.* **2017**, *143*, 04017007. [CrossRef]
48. Lanza Di Scalea, F.; Zhu, X.; Capriotti, M.; Liang, A.; Mariani, S.; Sternini, S. Passive Extraction of Dynamic Transfer Function From Arbitrary Ambient Excitations: Application to High-Speed Rail Inspection From Wheel-Generated Waves. *ASME J. Nondestruct. Eval. Diagn. Progn. Eng. Syst.* **2018**, *1*, 011005. [CrossRef]
49. Liang, A.; Sternini, S.; Capriotti, M.; Lanza di Scalea, F. High-Speed Ultrasonic Rail Inspection by Passive Noncontact Technique. *Mater. Eval.* **2019**, *77*, 941–950.
50. Ultrasonic Broken Rail Detector Overview. Available online: http://www.railsonic.co.za/pdf_files/UBRD_Overview_V1.pdf (accessed on 25 December 2020).
51. Loveday, P.W.; Burger, F.A.; Long, C.S. Rail Track Monitoring in SA using Guided Wave Ultrasound. In Proceedings of the Conference & Exhibition of the South African Institute for NDT (SAINT-2018), Johannesbourg, South Africa, 17–18 February 2018.
52. Setshedi, I.; Long, C.; Loveday, P.; Wilke, D. Adaptive SAFE model of a rail for parameter estimation. In Proceedings of the South African Conference on Computational and Applied Mechanics (SACAM), Vaal University of Technology, Vanderbijlpark, South Africa, 17–19 September 2018.
53. Steyn, B.M.; Pretorius, J.F.W.; Burger, F. Development, testing and operation of an Ultrasonic Broken Rail Detector System. In *Computers in Railways IX*; Allan, J., Brebbia, C.A., Hill, R.J., Sciutto, G., Sone, S., Eds.; WIT Press: Ashurst Lodge, UK; Southampton, UK, 2004; Volume 74, pp. 351–358.
54. Yuan, L.; Yang, Y.; Hernández, Á.; Shi, L. Feature Extraction for Track Section Status Classification Based on UGW Signals. *Sensors* **2018**, *18*, 1225. [CrossRef] [PubMed]
55. Loveday, P.W.; Dhuness, K.; Long, C.S. Experimental development of electromagnetic acoustic transducers for measuring ultraguided waves. In Proceedings of the Eleventh South African Conference on Computational and Applied Mechanics (SACAM 2018), Vanderbijlpark, South Africa, 17–19 September 2018; pp. 693–702.
56. Loveday, P.; Long, C. Influence of resonant transducer variations on long range guided wave monitoring of rail track. *AIP Conf. Proc.* **2016**, *1706*, 150004-1–150004-6.
57. Setshedi, I.; Loveday, P.; Long, C.; Wilke, D. Estimation of rail properties using semi-analytical finite element models and guided wave ultrasound measurements. *Ultrasonics* **2019**, *96*, 240–252. [CrossRef] [PubMed]
58. Enekom. *Enekom RCFSS UsersManual V4*; Enekom: Ankara, Turkey, 2019.
59. Xining, X.; Lu, Z.; Bo, X.; Zujun, Y.; Liqiang, Z. An Ultrasonic Guided Wave Mode Excitation Method in Rails. *IEEE Access* **2018**, *6*, 60414–60428. [CrossRef]
60. Shi, H.; Zhuang, L.; Xu, X.; Yu, Z.; Zhu, L. An Ultrasonic Guided Wave Mode Selection and Excitation Method in Rail Defect Detection. *Appl. Sci.* **2019**, *9*, 1170. [CrossRef]
61. Xu, X.; Xing, B.; Zhuang, L.; Shi, H.; Zhu, L. A Graphical Analysis Method of Guided Wave Modes in Rails. *Appl. Sci.* **2019**, *9*, 1529. [CrossRef]
62. Xing, B.; Yu, Z.; Xu, X.; Zhu, L.; Shi, H. Research on a Rail Defect Location Method Based on a Single Mode Extraction Algorithm. *Appl. Sci.* **2019**, *9*, 1107. [CrossRef]
63. Wei, X.; Yang, Y.; Yao, W.; Zhang, L. Design of full bridge high voltage pulser for sandwiched piezoelectric ultrasonic transducers used in long rail detection. *Appl. Acoust.* **2019**, *149*, 15–24. [CrossRef]
64. Wei, X.; Yang, Y.; Yao, W.; Zhang, L. An automatic optimal excitation frequency tracking method based on digital tracking filters for sandwiched piezoelectric transducers used in broken rail detection. *Measurement* **2019**, *135*, 294–305. [CrossRef]
65. Loveday, P.W.; Long, C.S.; Ramatlo, D.A. Ultrasonic guided wave monitoring of an operational rail track. *Struct. Health Monit.* **2019**, *19*, 1666–1684. [CrossRef]
66. Liang, A.; Sternini, S.; Capriotti, M.; Zhu, P.X.; Lanza di Scalea, F.; Wilson, R. Passive extraction of Green's function of solids and application to high-speed rail inspection. In *Proceedings of the SPIE 10970, Sensors and Smart Structures Technologies for Civil, Mechanical, and Aerospace Systems, SPIE Smart Structures + Nondestructive Evaluation, Denver, CO, USA, 27 March 2019*; Lynch, J.P., Huang, H., Sohn, H., Wang, K.-W., Eds.; SPIE: Bellingham, WA, USA, 2019; Volume 10970, pp. 109700R-1–109700R-10.
67. Yuan, L.; Yang, Y.; Alonso, I.H.; Li, S. Application of VMD Algorithm in UGW-based Rail Breakage Detection System. In Proceedings of the IEEE International Conference on Vehicular Electronics and Safety (ICVES), Madrid, Spain, 12–14 September 2018; pp. 1–6.
68. Benzeroual, H.; Khamlichi, A.; Zakriti, A. Detection of Transverse Defects in Rails Using Noncontact Laser Ultrasound. *Proceedings* **2020**, *42*, 43. [CrossRef]
69. Teidj, S. Defect Indicators in a Rail Based on Ultrasound Generated by Laser Radiation. *Procedia Manuf.* **2020**, *46*, 863–870. [CrossRef]
70. Rose, J.L. *Ultrasonic Guided Waves in Solid Media*; Cambridge University Press: Cambridge, UK, 2014.
71. Wilcox, P.D. Guided-Wave Array Methods. In *Encyclopedia of Structural Health Monitoring*; Boller, C., Chang, F.-K., Fujino, Y., Eds.; John Wiley & Sons: Hoboken, NJ, USA, 2009.
72. Wilcox, P.D. A rapid signal processing technique to remove the effect of dispersion from guided wave signals. *IEEE Trans. Ultrason. Ferroelectr. Freq. Control* **2003**, *50*, 419–427. [CrossRef]

73. Moustakidis, S.; Kappatos, V.; Karlsson, P.; Selcuk, C.; Gan, T.H.; Hrissagis, K. An intelligent methodology for railways monitoring using ultrasonic guided waves. *J. Nondestruct. Eval.* **2014**, *33*, 694–710. [CrossRef]
74. Liu, C.; Harley, J.B.; Bergés, M.; Greve, D.W.; Oppenheim, I.J. Robust ultrasonic damage detection under complex environmental conditions using singular value decomposition. *Ultrasonics* **2015**, *58*, 75–86. [CrossRef]
75. Dobson, J.; Cawley, P. Independent component analysis for improved defect detection in guided wave monitoring. *Proc IEEE* **2016**, *104*, 1620–1631. [CrossRef]
76. Liu, C.; Dobson, J.; Cawley, P. Efficient generation of receiver operating characteristics for the evaluation of damage detection in practical structural health monitoring applications. *Proc. R. Soc.* **2017**, *437*, 1–26. [CrossRef]
77. Hyvarinen, A.; Oja, E. Independent component analysis: Algorithms and applications. *Neural Netw.* **2000**, *13*, 411–430. [CrossRef]
78. Wei, X.; Yang, Y.; Ureña, J.; Yan, J.; Wang, H. An Adaptive Peak Detection Method for Inspection of Breakages in Long Rails by Using Barker Coded UGW. *IEEE Access* **2020**, *8*, 48529–48542. [CrossRef]
79. Yuan, L.; Yang, Y.; Hernández, Á.; Shi, L. Novel adaptive peak detection method for track circuits based on encoded transmissions. *IEEE Sens. J.* **2018**, *18*, 6224–6234. [CrossRef]
80. Burger, F.A.; Loveday, P.; Long, C. Large scale implementation of guided wave based broken rail monitoring. *AIP Conf. Proc.* **2015**, *1650*, 771–776.
81. Ihara, I. Ultrasonic Sensing: Fundamentals and its Applications to Nondestructive Evaluation. In *Sensors. Lecture Notes Electrical Engineering*; Mukhopadhyay, S., Huang, R., Eds.; Springer: Berlin/Heidelberg, Germany, 2008; Volume 21, pp. 287–305.
82. Mazzoldi, P.; Nigro, M.; Voci, C. *Fisica Vol. 1-Meccanica-Termodinamica*, 1st ed.; EdiSES: Napoli, Italy, 1991; Chapter 8, pp. 215–232.
83. Rose, J.L. A Baseline and Vision of Ultrasonic Guided Wave Inspection Potential. *ASME. J. Press. Vessel Technol.* **2002**, *124*, 273–282. [CrossRef]
84. Loveday, P.; Long, C.; Ramatlo, D. Mode Repulsion of Ultrasonic Guided Waves in Rails. *Ultrasonics* **2018**, *84*, 341–349. [CrossRef]
85. Ryue, J.; Thompson, D.; White, P.; Thompson, D.R. Investigations of propagating wave types in railway tracks at high frequencies. *J. Sound Vib.* **2008**, *315*, 157–175. [CrossRef]
86. Ryue, J.; Thompson, D.; White, P.; Thompson, D.R. Decay rates of propagating waves in railway tracks at high frequencies. *J. Sound Vib.* **2009**, *320*, 955–976. [CrossRef]

Article

Mooring-Failure Monitoring of Submerged Floating Tunnel Using Deep Neural Network

Do-Soo Kwon [1,2], Chungkuk Jin [1,*], MooHyun Kim [1] and Weoncheol Koo [2,*]

[1] Department of Ocean Engineering, Texas A&M University, Haynes Engineering Building, 727 Ross Street, College Station, TX 77843, USA; dskwon7752@tamu.edu (D.-S.K.); m-kim3@tamu.edu (M.K.)

[2] Department of Naval Architecture and Ocean Engineering, Inha University, Incheon 22212, Korea

* Correspondence: jinch999@tamu.edu (C.J.); wckoo@inha.ac.kr (W.K.); Tel.: +1-979-204-3454 (C.J.); +82-032-860-7348 (W.K.)

Received: 21 August 2020; Accepted: 18 September 2020; Published: 21 September 2020

Abstract: This paper presents a machine learning method for detecting the mooring failures of SFT (submerged floating tunnel) based on DNN (deep neural network). The floater-mooring-coupled hydro-elastic time-domain numerical simulations are conducted under various random wave excitations and failure/intact scenarios. Then, the big-data is collected at various locations of numerical motion sensors along the SFT to be used for the present DNN algorithm. In the input layer, tunnel motion-sensor signals and wave conditions are inputted while the output layer provides the probabilities of 21 failure scenarios. In the optimization stage, the numbers of hidden layers, neurons of each layer, and epochs for reliable performance are selected. Several activation functions and optimizers are also tested for the present DNN model, and Sigmoid function and Adamax are respectively adopted to enhance the classification accuracy. Moreover, a systematic sensitivity test with respect to the numbers and arrangements of sensors is performed to find the appropriate sensor combination to achieve target prediction accuracy. The technique of confusion matrix is used to represent the accuracy of the DNN algorithms for various cases, and the classification accuracy as high as 98.1% is obtained with seven sensors. The results of this study demonstrate that the DNN model can effectively monitor the mooring failures of SFTs utilizing real-time sensor signals.

Keywords: submerged floating tunnel; deep neural network; machine learning; sensor optimization; failure monitoring accuracy; mooring line; sigmoid function; Adamax; categorical cross-entropy

---

1. Introduction

SFT (Submerged floating tunnel) is an alternative infrastructure to conventional/floating bridges and immersed tunnels for deep-sea crossing. It is balanced underwater by its buoyancy, weight, and constraint forces by means of mooring systems [1]. Feasibility studies of SFT have been performed by many researchers worldwide based on its potential advantages such that SFT can be safe in both waves and earthquakes since wave loads exponentially decay with submergence depth and seismic excitations are indirectly transmitted through flexible moorings [2]. However, the real SFT construction is not realized since its safety and feasibility are not fully guaranteed yet. Since the structure is deeply submerged, its structural health monitoring is another big challenging area. In this regard, the present paper focuses on smart structural health monitoring to design safer and more reliable SFTs in the future.

Two major investigations are essential in the design and operation of SFT for its safety and reliability. First, in the design stage, dynamic and structural analyses are needed under various environmental conditions and scenarios. Environmental loads include waves, earthquakes, currents, and tsunamis. Among them, waves tend to be the most important environmental loading if the submergence depth is not sufficiently large [3–7]. For SFT with a large diameter, it was found through hydro-elastic analysis that large static and dynamic mooring tensions were critical issues [7].

To solve this, various design concepts were suggested, e.g., variable-span mooring design [7] and suspension-cable type [8]. However, despite the effort to reduce mooring tension, there still exist several uncertainties, such as nonlinear SFT behaviors and related snap loading that can lead to unexpectedly large tension. In such a case, unexpected mooring failure can potentially happen. Similarly, collisions and underwater explosions can break mooring lines. Second, when any mooring failure happens, it has to be rapidly detected to avoid further problems. Since SFTs are not visible, the detection is not that simple. Diverse monitoring systems have been suggested for underwater flexible systems. The most direct method is to use ROVs (remotely operated vehicles); however, they are expensive and continuous real-time monitoring is not possible. The beam-theory-and-sensor-based monitoring systems with accelerometers/strain gauges [9] and inclinometers/GPS were developed [10]. These methods need mode shapes or structural parameters in advance. Other researchers also used analytical, transfer-function, and mode-matching methods for riser monitoring [11]. However, the above methods still require many sensors to predict global behaviors and local failures.

Recently, ML (machine learning) has newly been employed for the smart monitoring of marine structures. While the above conventional methods require a lot of sensors to monitor potential failures, machine-learning-based algorithms can accurately detect problems with the fewer number of sensors. For example, Chung et al. [12] performed analyses to detect the damage of TLP (tension leg platform) mooring by using DNN (deep neural networks). Jaiswal and Ruskin [13] presented the development of a machine learning algorithm for mooring-failure detection using measured vessel positions and 6-DOF acceleration data. Sidarta et al. [14] developed an ANN (artificial neural network) algorithm for detecting broken mooring lines for FPSO (floating production storage and offloading). Sidarta et al. [15] also developed a machine-learning-based algorithm for detecting mooring-line failures of a semi-submersible. The above examples utilized floaters' motions to detect mooring-line failures so that much fewer sensors can be used compared to the conventional deterministic methods. They [12–15] also assumed that floating structures are rigid to simplify the analyses of motion-sensor signals.

In this study, we developed a ML-based mooring-failure-monitoring system for a long and highly flexible SFT with DNN algorithms to detect mooring-line failures in real time without employing human effort or any devices. The effects of variable number of sensors (accelerometers) and their locations were analyzed. The training data for the developed ML algorithm was produced by running the author-developed time-domain SFT simulation program [7] using the commercial software, OrcaFlex [16]. The simulations of SFT's wave-induced motions and mooring tensions were partly validated through comparisons with an independent SFT hydro-elastic simulation program [7,17] and a series of experimental results with small SFT sections [18]. To validate the developed smart monitoring system, various intact/failure scenarios and wave conditions were considered for training and testing of the algorithm. In the algorithm optimization stage, not only the numbers of the hidden layers, neurons, and epochs but also the activation function and optimizer were properly utilized to enhance the detection accuracy. Contrary to previous researches [12–15], in the case of highly elastic SFT, a single sensor cannot cover the entire motion, so several sensors are required, as presented in [19]. In this regard, in the testing stage, we checked the detection accuracy of the developed algorithm with different numbers and arrangements of sensors. Machine learning algorithms for the failure detection of submerged deformable structures like SFT have rarely been investigated in the open literature. In this regard, this study using DNN will help other researchers to investigate similar problems in the future.

2. Numerical Model

2.1. Submerged Floating Tunnel

We considered SFT with 28 mooring lines as an example, as shown in Figure 1. Material properties and design parameters are presented in Table 1. The tunnel, 20 m in diameter and 800 m in length, is made of high-density concrete. BWR (buoyancy weight ratio) is set as 1.3. We assume that the tunnel has fixed stations at both ends and water depth is constant at 100 m. The submergence depth, which is a vertical distance between the MWL (mean water level) and the tunnel centerline, is set at 61.5 m. As can be seen in Figure 1, four 60-degree inclined mooring lines made of studless chains are installed with 100-m interval along the longitudinal length. The lengths of the mooring lines are 50.2 m for lines #1 and #2 and 38.7 m for lines #3 and #4. In addition, accelerometers are located along the tunnel's centerline, as shown in Figure 1c. We conduct a systematic sensitivity test with respect to the numbers and arrangements of accelerometers, as discussed in Section 4.3.

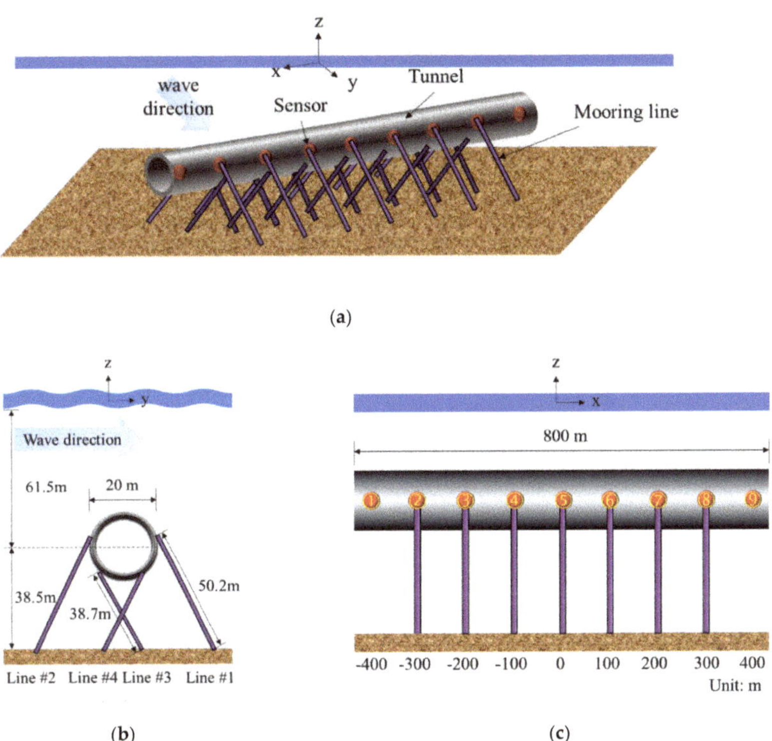

Figure 1. 2D and 3D schematic drawings (accelerometers are marked with red dots in (**a**); mooring line numbers are given in (**b**); sensor numbers (1–9) are given in (**c**)).

Table 1. Design parameter of the SFT (E is Young's modulus, I is area moment of inertia, A is cross-sectional area).

Parameter	Value	Unit
Tunnel length	800	m
Tunnel diameter	20	m
Interval of mooring lines	100	m
End boundary condition	Fixed-fixed	-
Material of tunnel	High-density concrete	-
Length of mooring lines	50.2 (line #1,2), 38.7 (line #3,4)	m
Added mass coefficient	1.0	-
Nominal diameter of mooring lines	0.18	m
Drag coefficient	0.55 (tunnel), 2.4 (mooring lines)	-
Bending stiffness (EI)	1.34×10^{11} (tunnel), 0 (mooring lines)	kN·m^2
Axial stiffness (EA)	3.23×10^{9} (tunnel), 2.77×10^{6} (mooring lines)	kN

2.2. Time-Domain Numerical Simulation

Authors developed a numerical model [7] that can perform tunnel-mooring-coupled time-domain simulations using OrcaFlex [16] to produce data-sets for machine learning. The SFT was modeled by the long-beam model, which consists of nodes (lumped masses) and segments (massless linear and rotational springs). The mass, drag, and other important properties are all lumped at the nodes, and stiffness properties are modeled in the segments. The time-domain equation of motion for SFT can be expressed as:

$$\mathbf{M}\ddot{\mathbf{x}} + \mathbf{K}\mathbf{x} = \mathbf{F_M} + \mathbf{w} + \mathbf{F_C}\delta(\mathbf{x}_i - \xi) \tag{1}$$

where $\mathbf{M}$ and $\mathbf{K}$ are system's mass and structural stiffness matrices, respectively, $\mathbf{x}$ is the displacement vector, $\mathbf{x}_i$ is the longitudinal position vector of tunnel nodes, ξ is the vector representing longitudinal locations of mooring lines, $\mathbf{F_M}$ is the hydrodynamic force vector, $\mathbf{w}$ is the wet-weight vector (i.e., net sum of weight and buoyancy), $\mathbf{F_C}$ is the constraint force vector at the tunnel-mooring connection locations (i.e., coupling force induced by mooring lines on the tunnel and vice versa), and δ is the Dirac delta function. Upper dot in the equations means time derivative of a variable.

The line's elastic behaviors are considered by the $\mathbf{K}\mathbf{x}$ term with axial, bending, and torsional springs. The axial and torsional springs located at the center of two neighboring nodes evaluate the tension force and torsional moment while the rotational springs located at either side of the node estimate the shear force and bending moment.

The hydrodynamic force at nodes' instantaneous locations was evaluated by the Morison equation for a moving body, which can be written for a cylindrical object as:

$$\mathbf{F_M} = -C_A\rho V\ddot{\mathbf{x}}^n + C_M\rho V\dot{\boldsymbol{\eta}}^n + \frac{1}{2}C_D\rho A\left|\boldsymbol{\eta}^n - \dot{\mathbf{x}}^n\right|\left(\boldsymbol{\eta}^n - \dot{\mathbf{x}}^n\right) \tag{2}$$

where C_A, C_M, and C_D are the added mass, inertia, and drag coefficients, respectively, V and A stand for the displaced volume and projected area, ρ is density of water, η is velocity of a fluid particle, and superscript n means the normal direction. More details of the numerical model can be found in [7]. The developed SFT dynamics simulation program was partly validated through comparisons with a series of experimental results for a small SFT segment with similar mooring set-up [18].

2.3. Big-Data Generation under Various Environmental Conditions and Failure Scenarios

Big data were generated and collected by using the developed time-domain simulation program under various wave conditions and intact/failure scenarios, as summarized in Tables 2 and 3. As shown in Figure 1c, we considered accelerometers as sensors. However, we collected displacement signals directly assuming that real-time double integration is feasible with an appropriate bandwidth filter. Then, we conducted a systematic sensitivity test with respect to the arrangements of accelerometers to discover effective sensor combinations. Note that sensors #1 and #9 were not used for the analysis due to the fixed-fixed boundary conditions. As presented in Table 2, 21 intact/failure cases were simulated for 1800 sec with a time interval of 0.2 sec at each environmental condition. The failure cases were simulated by disconnecting the target mooring lines. The wave heading is assumed to be normal to the longitudinal direction of SFT. Based on symmetry, failure scenarios on the half domain were considered. For the random wave-elevation generation, a JONSWAP wave spectrum was utilized, and 100 regular-wave components were superposed to generate the random wave signals. Signal repetition in time histories was avoided through the equal energy method in which each regular wave component has equal spectral energy. Ten wave conditions were considered, as shown in Table 3. The enhancement parameter (γ) of the wave spectrum was fixed at 2.14. The total data points and time were, therefore, 210 (21 scenarios times 10 environmental conditions) and 378,000 sec (210 simulations times 1800 s). As given in Table 3, we employed 80% of data collected from eight environmental conditions for training, and thus the total data points for training were 134 while the number of data points for testing was 76. Data from the first eight environmental conditions were utilized for studies of optimization and failure-detection performance with different arrangements of the sensors. Later, we further tested the feasibility of the algorithm by employing data from the last two environmental conditions not used for training.

Table 2. Mooring-failure scenarios (mooring line number is based on the line number in Figure 1b, In = intact case, All = failure of all mooring lines (# 1, 2, 3, and 4) at their tunnel location).

Item	Failure Case																				
	1	2	3	4	5	6	7	8	9	10	11	12	13	14	15	16	17	18	19	20	In
Tunnel Location (m)	−300				−200				−100				0				−300	−200	−100	0	-
Mooring Line Number	1	2	3	4	1	2	3	4	1	2	3	4	1	2	3	4	All	All	All	All	-

Table 3. Wave conditions (Tr is training, Te is testing, percentage in bracket denotes the percentage of data used for training or testing).

Significant Wave Height (Hs) (m)	Peak Period (Tp) (s)	Data Usage
1	6	Tr (80%), Te (20%)
1	10.5	Tr (80%), Te (20%)
1	13	Tr (80%), Te (20%)
5	6	Tr (80%), Te (20%)
5	10.5	Tr (80%), Te (20%)
5	13	Tr (80%), Te (20%)
11.7	10.5	Tr (80%), Te (20%)
11.7	13	Tr (80%), Te (20%)
3	8	Te (100%)
7	12	Te (100%)

3. Deep Neural Network

3.1. Artificial Neural Network

ANN is a parallel computational model consisting of adaptive processing units that are densely connected to each other [20]. ANN is a biologically inspired computing method based on the neural structures of the brain. Neurons and layers are basic elements to design a neural network structure. To be specific, a layer consists of a certain number of neurons while neurons in a layer are connected with neurons in neighboring layers linearly. There are three types of layers, i.e., input, hidden, and output layers [21]. When there exist multiple hidden layers, it is called DNN.

Learning in DNN is a process of determining the weights and biases in each layer so that the error between the final output value and the actual value can be minimized. Figure 2 shows the mechanism of the learning process. This learning procedure is called feedforward because it flows from the input nodes a to the last output $\hat{y}$ through the intermediate layer where the activation function $f(x)$ exists. The weighted sum of the inputs x is firstly estimated by weights ω, a, and bias b, and activation function decides whether outside connections consider this neuron as activated or not. Finally, the network produces the predicted value $\hat{y}$ [22]. In this sense, the error of the network is defined as the correct answer y minus the predicted value $\hat{y}$.

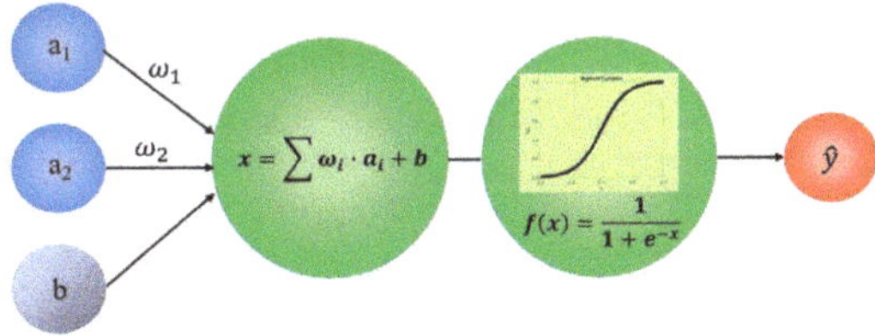

Figure 2. Layout of feedforward algorithm.

Classification learning needs to update the weights based on the error values to enhance accuracy. In this regard, the backpropagation algorithm was adapted. The backpropagation algorithm trains the neural network by backpropagating the error of the output layer to the hidden layers while updating weights and bias. In this study, a delta rule was utilized, which can be expressed as [23,24]:

$$\omega_{i+1} = \omega_i + \eta \cdot \delta \cdot a \tag{3}$$

where η and δ are the learning rate and delta. The learning rate is a parameter of the optimization algorithm that determines the step size at each epoch, moving towards the minimum loss function [24]. Delta is first calculated by multiplying the error value by the derivative of the activation function, and error is then calculated for the next layer as:

$$\delta = f'(x) \cdot e \tag{4}$$

$$e = W^T \delta \tag{5}$$

where W^T is the transpose of weight matrix $\{\omega_1, \omega_2, \dots, \omega_n\}$. Delta of the output node is backpropagated to calculate the delta of the hidden nodes while repeating this process of backpropagation to the leftmost hidden layer. The neural network learning of feedforward and backpropagation is repeated until the error converges.

3.2. Neural Network Model Architecture

Figure 3 shows the present DNN model for mooring-failure monitoring. In this study, mooring-failure locations are to be predicted by the change of tunnel motion-sensor signals under given wave conditions. In other words, the input layer consists of the tunnel motions from sensors and wave conditions, and the output layer provides the probabilities of 21 failure scenarios. We optimized the number of hidden layers and neurons, and the selected numbers of hidden layers and neurons per each layer were three and 200, respectively, which is discussed in Section 4.2 in detail. Therefore, the proposed architecture consists of an input layer, three hidden layers, and an output layer.

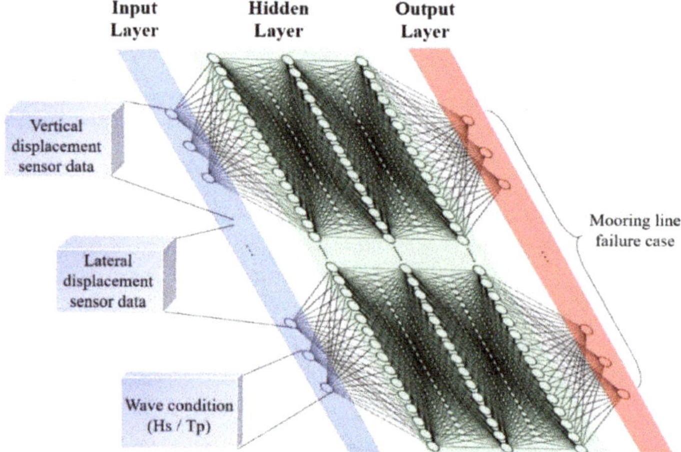

Figure 3. Deep neural network architecture with three hidden layers (although 3 hidden layers are not that deep, it is called DNN in this paper for convenience).

3.3. Building Neural Network for Classification Tasks

To build the neural networks, the first step is importing the data and defining the input and target variables as illustrated in Section 3.2. To the next, we need to create the structure of ANN that is composed of the input, hidden, and output layers. To be specific, the number of hidden layers and the size of input and output are defined. Hidden and output layer neurons have activation functions that can learn and perform more complex tasks by performing a nonlinear transformation on the input. In this study, while four types of activation functions were investigated, Sigmoid function was selected as the best activation function for the hidden layers, and Softmax was used as the activation function of the output layer. Model compiling is the next stage to build the neural network. In the compile process, optimizer and loss function need to be specified. Optimizer is the part of the machine learning process that actually updates parameters such as weights to decrease losses. On the other hand, the loss is a prediction error in the neural network, and the method of calculating the loss is called the loss function, also referred to as the cost function. In this study, Adamax was used as optimizer after comparing four different optimizers, and Cross-entropy was adopted to define a loss function in ANN [23]. Table 4 summarizes the characteristics of those activation functions and optimizers, and Figure 4 represents the layout of DNN model applied in the present research. The optimization process to select the activation function and optimizer is discussed in Section 4.2.2.

Table 4. Characteristics of activation functions and optimizers.

Parameter		Characteristic	Advantage	Disadvantage
Activation function	Sigmoid	• Sigmoid takes a real value as input and outputs another value between 0 and 1. It's easy to work with and has all the nice properties of activation functions [25]	• Smooth gradient • Good for a classifier • Have activations bound in a range	• Vanishing gradient problem • Not zero centered • Sigmoid saturates and kills gradients
	ReLU	• ReLU stands for rectified linear unit. Mathematic form: $y = max(0, x)$ [26,27]	• Biological plausibility • Sparse activation • Better gradient propagation • Efficient computation	• Non-differentiable at zero • Not zero-centered • Unbounded • Dying ReLU problem
	Tanh	• The range of the Tanh function is from −1 to 1. Tanh is also Sigmoidal (s-shaped) [25]	• Derivatives are steeper than Sigmoid	• Vanishing gradient problem
	SELU	• SELU stands for Scaled Exponential Linear Unit. It is self-normalizing the neural network [27]	• Network converges faster. • No problem of Vanishing and exploding gradient	• Relatively new activation function – needs more papers on architectures
Optimizer	Adam	• Using moving average of the gradient instead of gradient itself [28]	• Only requires first-order gradients with little memory requirement	• Generalization issue [29,30]
	Adamax	• It is a variant of Adam based on the infinity norm [28]	• Infinite-order norm makes the algorithm surprisingly stable	
	RMSProp	• A gradient-based optimization technique used in training neural networks [31,32]	• Differentiation between parameters is maintained and prevent convergence to zero [31]	• Zero initialization bias problem
	AdaGrad	• An optimizer with parameter-specific learning rates, which are adapted relative to how frequently a parameter gets updated during training [32]	• Well-suited for dealing with sparse data • Lesser need to manually tune learning rate	• Accumulates squared gradients in denominator • Causes the learning rate to shrink

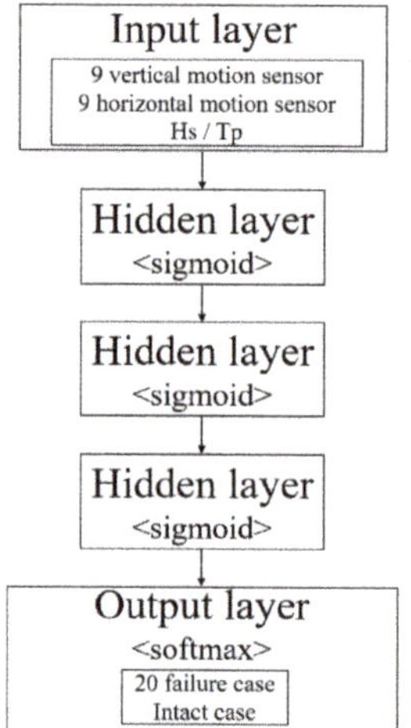

Figure 4. Layout of the present neural network model.

Finally, model fitting is required to execute the model. The training process will run for a fixed number of iterations over the dataset called epochs. The number of dataset rows is also arranged before the model weights are updated within each epoch, called the batch size. The number of epochs means the number of times that the algorithm repeatedly learns the entire training dataset. One epoch means that each sample of the training dataset can be used to update internal model parameters. The size of epochs is usually large up to hundreds to thousands to minimize errors sufficiently. The batch size is the number of samples processed before the model is updated. The numbers of epochs and batch sizes can be determined experimentally through trial and error. The model must be sufficiently trained to well map the rows of input data to the output classification. There are always some errors in the model, but after a certain point in time for a given model configuration, the model converges, and the amount of error is reduced. In this study, 2000 epochs and 200 batch sizes were used. Figure 5 shows the learning process of the neural network. Loss is a sum of errors that occurred for each example in the training set. Loss values indicate how well a particular model works after each optimization iteration. Ideally, the loss is expected to decrease after each or multiple iterations. The accuracy of the model is usually determined after training and correction of model parameters. The test sample is then provided to the model and compared to the actual target, and the number of times that the model makes a mistake is memorized.

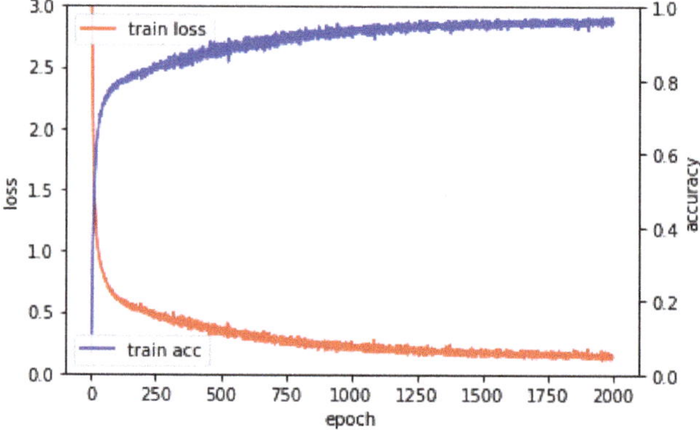

Figure 5. Learning process of the network model.

4. Results and Discussions

4.1. Failure Identification Examples

While the mooring-failure-detection algorithm is designed, finding appropriate measurement parameters is the first important task to provide high-accuracy detection. Using the tension sensor is the most direct method to detect the failure. However, it is unrealistic to install the tension sensor to all mooring lines because tension sensors need regular calibrations and are hard to be repaired and replaced especially for deeply submerged structures such as SFT. In this regard, accelerometers, which can easily be installed inside the tunnel along its longitudinal direction, are chosen. Furthermore, it can be directly connected to the electric wire to continuously function in real time. Being inside in dry space, it is easy to be calibrated and checked up. For the present monitoring algorithm, we directly use the lateral and vertical displacements of SFT, which can be obtained by integrating the acceleration signals twice. We confirmed that appropriate bandwidth filter and baseline correction can accurately recover the displacements from accelerations during time integrations even if there is noise. Zheng et al. [33] showed the feasibility of real-time displacement monitoring using double integration of acceleration with measurement noise based on a recursive baseline correction and recursive high-pass filter. However, when mean displacements cannot be captured and/or noise cannot be handled during time integrations, real-time displacements can alternatively be obtained from inclinometers instead of accelerometers, as developed by 3rd author's research group [34].

Figures 6 and 7 show the time histories of lateral and vertical displacements of the tunnel at its mid-length (sensor No. 5, x = 0 m) in intact and failure scenarios. While simulation time is 1800 sec (9000 steps), the results for the first 400 sec are presented there. The corresponding significant wave height and peak period are 11.7 m and 13 sec. Cases 1 and 13 in Figure 6 are the results when one of the mooring lines is broken at x = −300 m and x = 0 m, respectively (see Table 2). There are noticeable differences between the intact case and Case 13. For the lateral motion, Case 13 has lager motions in the negative direction since one-mooring failure leads to less stiffness of the system there. The larger fluctuations and static shift of the vertical displacement are observed for Case 13. These results demonstrate that failure can more easily be detected by motion changes when sensors are close to the failure locations. On the other hand, for Case 1, since the failure location (x = −300 m) is far away from the sensor location (x = 0 m), there is no visible difference in displacements between intact and failure cases. In addition, as shown in Figure 7, depending on the left- and right-side mooring failure, the corresponding asymmetrical bias can be observed in the lateral motion trends. However, the corresponding difference in the vertical response is small, as can be seen from Figure 7b. By observing the pattern of those sensor signals, the monitoring algorithm can figure out the best guess for the failure incidence. If we place several accelerometers along the longitudinal direction of the tunnel, the prediction accuracy can significantly be improved while much more detailed comparisons of multiple sensor signals have to be made. As was seen in Figure 7, two-directional (horizontal and vertical) signals will be beneficial to better classify the failure location.

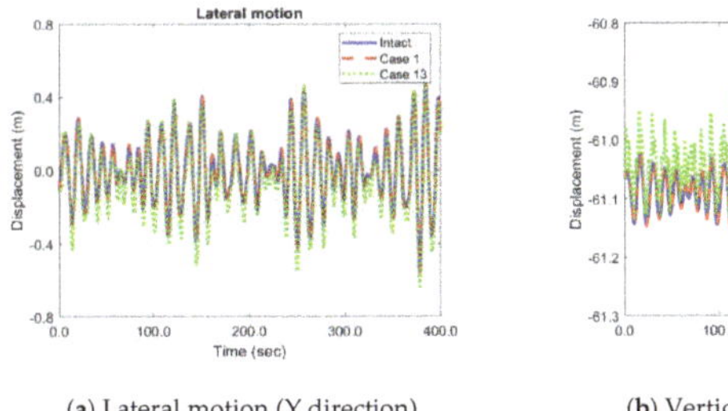

(a) Lateral motion (Y direction)   (b) Vertical motion (Z direction)

Figure 6. Time histories of lateral and vertical displacements of the tunnel at its mid-length (sensor No. 5) in intact and failure conditions.

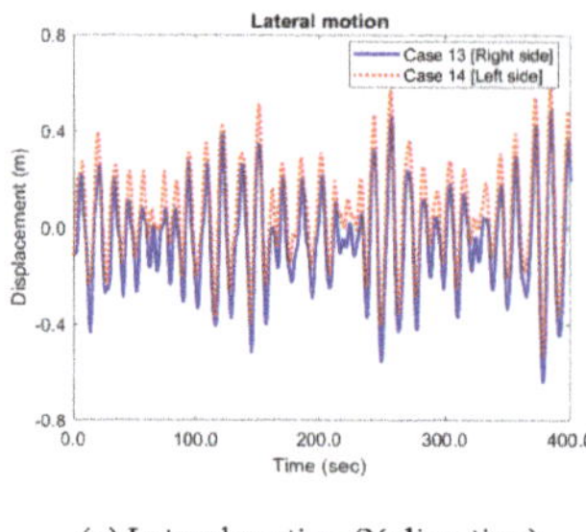 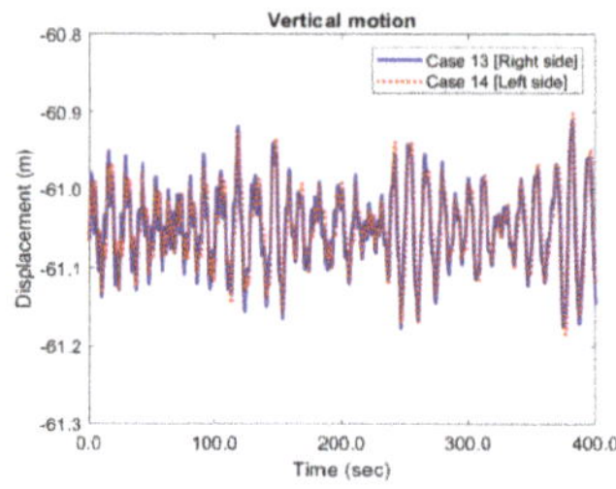

(a) Lateral motion (Y direction)   (b) Vertical motion (Z direction)

Figure 7. Time histories of lateral and vertical displacements of the tunnel at its mid-length (numerical sensor No. 5) in two different mooring-failures in the opposite sides.

4.2. Optimization

4.2.1. Hidden Layer and Neuron

Several optimizations are processed before evaluating the performance of the machine-learning algorithm. The optimized parameters are determined based on the accuracy and loss function defined in Section 3. Accuracy is a way to measure the performance of a classification model. The higher the accuracy, the better the algorithm. The accuracy of the perfect failure-detection algorithm is 1. On the other hand, the loss function considers the probability or uncertainty of the prediction depending on how different the prediction is from the actual value. The loss is the sum of the errors that occurred for each sample in the training. The perfect failure-detection algorithm has zero loss.

The performance tests are firstly conducted with respect to the numbers of the hidden layers and neurons in each layer. Generally, ANN cannot analytically determine the best numbers of layers and neurons of each layer. In addition, using a single hidden layer makes it difficult to express the relationship between inputs and outputs since they are linearly linked. Therefore, the number of hidden layers and neurons should be optimized through the performance tests, as presented in Figure 8. In the optimization process, we utilized the signals from seven sensors (#2–8) that are equally spaced. The algorithm randomly selects 80% of the entire dataset for training from the first eight environmental conditions given in Table 3 and uses the remaining 20% of the dataset to check the accuracy and loss. While the epoch increases from 1 to 2000, accuracy and loss are evaluated. These procedures are also adopted in the next optimization process. As shown in Figure 8, the highest accuracy and lowest loss can be acquired with three hidden layers and 200 neurons, and these values are used in the ensuing study. In general, the higher the epoch, the higher the accuracy. Interestingly, increasing the hidden layers and neurons more than three and 200 does not provide higher accuracy.

4.2.2. Activation Function and Optimizer

Next, the activation function and optimizer are to be selected through this optimization process. Sigmoid, Tanh, ReLU, and SELU are compared for the activation-function optimization while RMSProp, AdaGrad, Adam, and Adamax are compared for the optimizer performance. Their characteristics and advantages/disadvantages are summarized in Table 4 [25–32].

As shown in Figure 9, the Sigmoid function, one of the most classic activation function, provides the best results than any other activation functions. This indicates that the Sigmoid function operates nicely as a classifier for the present goal. As for optimizer, Adamax outperforms other optimizers. In this regard, Sigmoid and Adamax are selected as the activation function and optimizer.

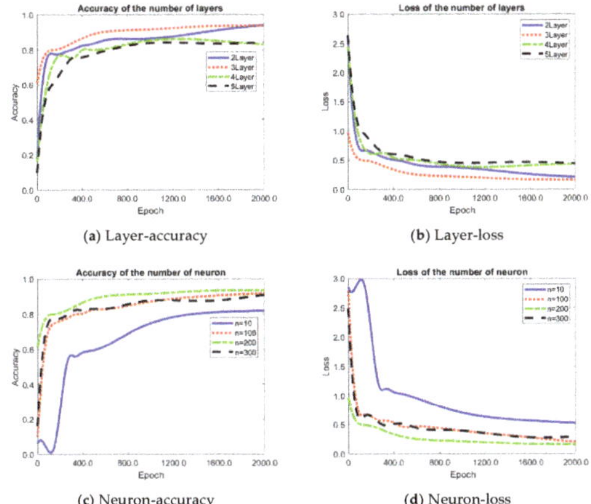

(a) Layer-accuracy

(b) Layer-loss

(c) Neuron-accuracy

(d) Neuron-loss

Figure 8. Performance test for the number of hyperparameters.

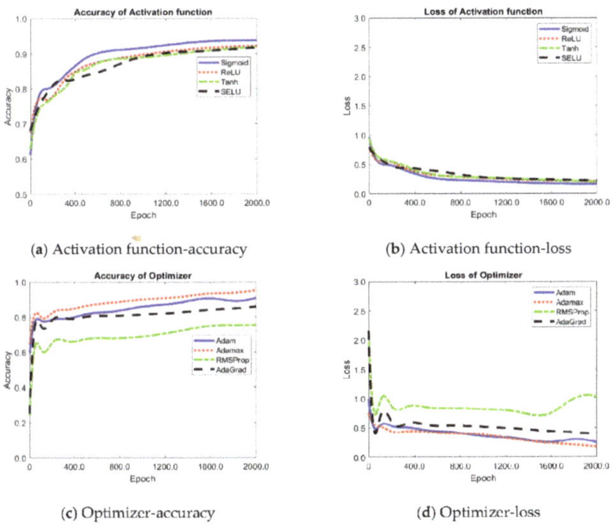

(a) Activation function-accuracy

(b) Activation function-loss

(c) Optimizer-accuracy

(d) Optimizer-loss

Figure 9. Performance test for the selection of the activation function and optimizer.

4.3. Failure-Detection Performance

Based on the previous optimization results, the numbers of hidden layers and neurons are three and 200, and the selected activation function and optimizer are Sigmoid and Adamax, respectively. The epoch is set as 2000. These parameters and functions are further utilized to check the failure-detection accuracy of the present algorithm. Again, the algorithm randomly selects 80% of the entire dataset for training from the first eight environmental conditions given in Table 3 and uses the remaining 20% of the dataset for testing by checking the corresponding accuracy and loss.

Figure 10 shows the correlation between numerical sensors for lateral and vertical motions. Correlation analysis is a technique that analyzes the linear relationship between two variables measured continuously. For example, in Figure 10a, the figure in the first row and the second column shows the relationship between the sensor No. 2 and sensor No. 3 (see Figure 1 for sensor numbers). We can see their strong relationship since the figure looks like a very slender ellipse along the diagonal.

For the combination of first row and last column representing sensors 2 and 8, a less strong relationship can be observed with a wider ellipse. In other words, the longer the distance between the sensors, the less the correlation. The relationship is less strong in vertical displacements under different failure scenarios since vertical motions are smaller having 60-degree mooring toe angle and relatively large BWR (=1.3), and thus less sensitive to certain mooring-line failure. If the BWR is decreased, the vertical displacements are increased [7,17], then they are to play a more important role. As a result, the distance between sensors should be short enough to acquire high detection accuracy but large enough to minimize cost and complexity. Therefore, the optimization of the sensor interval and number in real monitoring applications is important.

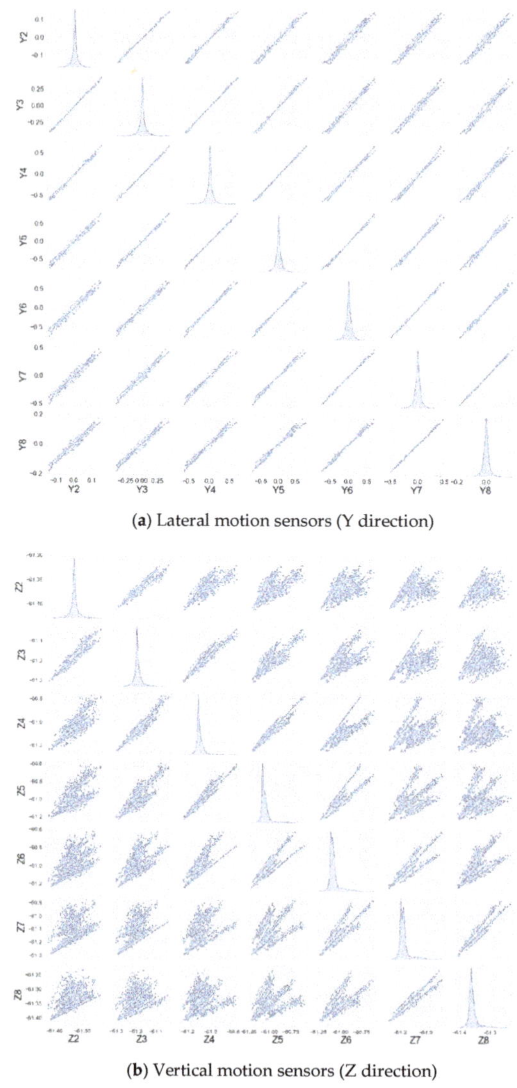

(**a**) Lateral motion sensors (Y direction)

(**b**) Vertical motion sensors (Z direction)

Figure 10. Correlation of numerical sensor data for lateral (**a**) and vertical (**b**) motions.

Figures 11–16 show the confusion matrices at different sensor combinations. The confusion matrix represents the performance of the classifier in a series of test data of which actual values are known [35]. The confusion matrices are a good option for reporting the performance results in multi-class classification problems because it is possible to observe the relations between the classifier outputs and the true ones. Table 5 summarizes the accuracy of classification with different numbers of sensors. The elements n_{ij} in the confusion matrix, where i and j are row and column identifiers, indicate the cases belonging to i that is classified as j. Hence, the elements in the diagonal (n_{ii}) are the elements correctly classified, while the off-diagonal elements in the matrix are misclassified. The total number of cases is expressed as:

$$N = \sum_{i=1}^{M} \sum_{j=1}^{M} n_{ij} \tag{6}$$

where M is the number of failure/intact scenarios. The confusion matrices show the ratio of detected cases to total cases in a range [0,1]. An asymmetric confusion matrix can reveal a biased classifier. The accuracy is the probability of performing the correct classification. The accuracy can be calculated from the confusion matrix as:

$$\text{Accuracy} = \frac{\sum_{i=1}^{M} n_{ii}}{N} \tag{7}$$

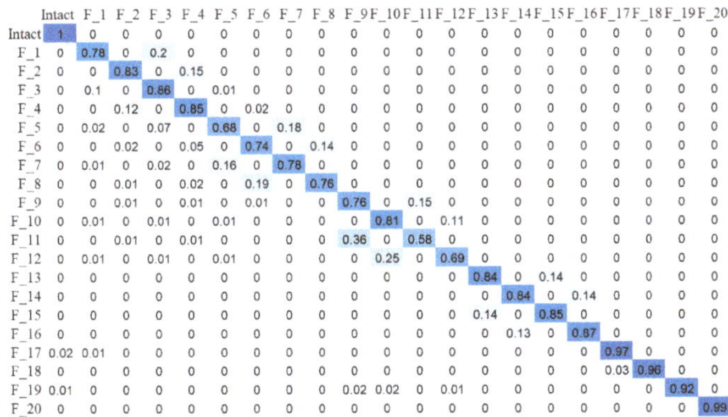

	Intact	F_1	F_2	F_3	F_4	F_5	F_6	F_7	F_8	F_9	F_10	F_11	F_12	F_13	F_14	F_15	F_16	F_17	F_18	F_19	F_20
Intact	1	0	0	0	0	0	0	0	0	0	0	0	0	0	0	0	0	0	0	0	0
F_1	0	0.78	0	0.2	0	0	0	0	0	0	0	0	0	0	0	0	0	0	0	0	0
F_2	0	0	0.83	0	0.15	0	0	0	0	0	0	0	0	0	0	0	0	0	0	0	0
F_3	0	0.1	0	0.86	0	0.01	0	0	0	0	0	0	0	0	0	0	0	0	0	0	0
F_4	0	0	0.12	0	0.85	0	0.02	0	0	0	0	0	0	0	0	0	0	0	0	0	0
F_5	0	0.02	0	0.07	0	0.68	0	0.18	0	0	0	0	0	0	0	0	0	0	0	0	0
F_6	0	0	0.02	0	0.05	0	0.74	0	0.14	0	0	0	0	0	0	0	0	0	0	0	0
F_7	0	0.01	0	0.02	0	0.16	0	0.78	0	0	0	0	0	0	0	0	0	0	0	0	0
F_8	0	0	0.01	0	0.02	0	0.19	0	0.76	0	0	0	0	0	0	0	0	0	0	0	0
F_9	0	0	0.01	0	0.01	0	0.01	0	0	0.76	0	0.15	0	0	0	0	0	0	0	0	0
F_10	0	0.01	0	0.01	0	0.01	0	0	0	0	0.81	0	0.11	0	0	0	0	0	0	0	0
F_11	0	0	0.01	0	0.01	0	0	0	0	0.36	0	0.58	0	0	0	0	0	0	0	0	0
F_12	0	0.01	0	0.01	0	0.01	0	0	0	0	0.25	0	0.69	0	0	0	0	0	0	0	0
F_13	0	0	0	0	0	0	0	0	0	0	0	0	0	0.84	0	0.14	0	0	0	0	0
F_14	0	0	0	0	0	0	0	0	0	0	0	0	0	0	0.84	0	0.14	0	0	0	0
F_15	0	0	0	0	0	0	0	0	0	0	0	0	0	0.14	0	0.85	0	0	0	0	0
F_16	0	0	0	0	0	0	0	0	0	0	0	0	0	0	0.13	0	0.87	0	0	0	0
F_17	0.02	0.01	0	0	0	0	0	0	0	0	0	0	0	0	0	0	0	0.97	0	0	0
F_18	0	0	0	0	0	0	0	0	0	0	0	0	0	0	0	0	0	0.03	0.96	0	0
F_19	0.01	0	0	0	0	0	0	0	0	0.02	0.02	0	0.01	0	0	0	0	0	0	0.92	0
F_20	0	0	0	0	0	0	0	0	0	0	0	0	0	0	0	0	0	0	0	0	0.99

Figure 11. Failure detected location when there are 3 sensors at points 2, 5, and 8 (Accuracy: 84.1%).

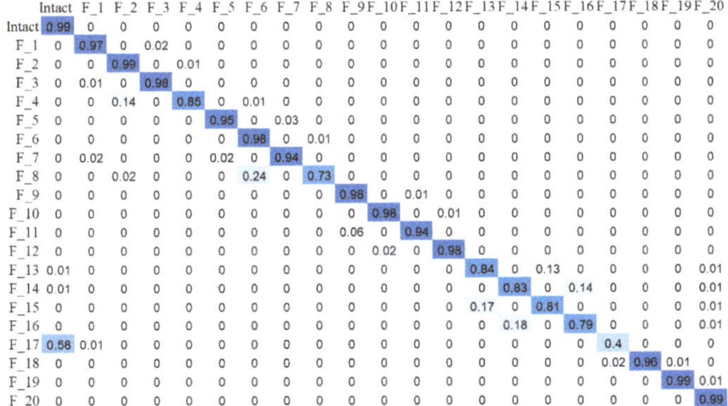

	Intact	F_1	F_2	F_3	F_4	F_5	F_6	F_7	F_8	F_9	F_10	F_11	F_12	F_13	F_14	F_15	F_16	F_17	F_18	F_19	F_20
Intact	0.99	0	0	0	0	0	0	0	0	0	0	0	0	0	0	0	0	0	0	0	0
F_1	0	0.97	0	0.02	0	0	0	0	0	0	0	0	0	0	0	0	0	0	0	0	0
F_2	0	0	0.99	0	0.01	0	0	0	0	0	0	0	0	0	0	0	0	0	0	0	0
F_3	0	0.01	0	0.98	0	0	0	0	0	0	0	0	0	0	0	0	0	0	0	0	0
F_4	0	0	0.14	0	0.85	0	0.01	0	0	0	0	0	0	0	0	0	0	0	0	0	0
F_5	0	0	0	0	0	0.95	0	0.03	0	0	0	0	0	0	0	0	0	0	0	0	0
F_6	0	0	0	0	0	0	0.98	0	0.01	0	0	0	0	0	0	0	0	0	0	0	0
F_7	0	0.02	0	0	0	0.02	0	0.94	0	0	0	0	0	0	0	0	0	0	0	0	0
F_8	0	0	0.02	0	0	0	0.24	0	0.73	0	0	0	0	0	0	0	0	0	0	0	0
F_9	0	0	0	0	0	0	0	0	0	0.98	0	0.01	0	0	0	0	0	0	0	0	0
F_10	0	0	0	0	0	0	0	0	0	0	0.98	0	0.01	0	0	0	0	0	0	0	0
F_11	0	0	0	0	0	0	0	0	0	0.06	0	0.94	0	0	0	0	0	0	0	0	0
F_12	0	0	0	0	0	0	0	0	0	0	0.02	0	0.98	0	0	0	0	0	0	0	0
F_13	0.01	0	0	0	0	0	0	0	0	0	0	0	0	0.84	0	0.13	0	0	0	0	0.01
F_14	0.01	0	0	0	0	0	0	0	0	0	0	0	0	0	0.83	0	0.14	0	0	0	0.01
F_15	0	0	0	0	0	0	0	0	0	0	0	0	0	0.17	0	0.81	0	0	0	0	0.01
F_16	0	0	0	0	0	0	0	0	0	0	0	0	0	0	0.18	0	0.79	0	0	0	0.01
F_17	0.58	0.01	0	0	0	0	0	0	0	0	0	0	0	0	0	0	0	0.4	0	0	0
F_18	0	0	0	0	0	0	0	0	0	0	0	0	0	0	0	0	0	0.02	0.96	0.01	0
F_19	0	0	0	0	0	0	0	0	0	0	0	0	0	0	0	0	0	0	0	0.99	0.01
F_20	0	0	0	0	0	0	0	0	0	0	0	0	0	0	0	0	0	0	0	0	0.99

Figure 12. Failure detected location when there are 4 sensors at points 2, 4, 6, and 8 (Accuracy: 90.4%).

	Intact	F_1	F_2	F_3	F_4	F_5	F_6	F_7	F_8	F_9	F_10	F_11	F_12	F_13	F_14	F_15	F_16	F_17	F_18	F_19	F_20
Intact	1	0	0	0	0	0	0	0	0	0	0	0	0	0	0	0	0	0	0	0	0
F_1	0	0.74	0	0.25	0	0	0	0	0	0	0	0	0	0	0	0	0	0	0	0	0
F_2	0	0	0.94	0	0.05	0	0	0	0	0	0	0	0	0	0	0	0	0.01	0	0	0
F_3	0	0.04	0	0.95	0	0	0	0	0	0	0	0	0	0	0	0	0	0	0	0	0
F_4	0	0	0.25	0	0.74	0	0	0	0	0	0	0	0	0	0	0	0	0	0	0	0
F_5	0	0	0	0.01	0	0.92	0	0.07	0	0	0	0	0	0	0	0	0	0	0	0	0
F_6	0	0	0	0	0.01	0	0.98	0	0.01	0	0	0	0	0	0	0	0	0	0	0	0
F_7	0	0	0	0	0	0.01	0	0.98	0	0	0	0	0	0	0	0	0	0	0	0	0
F_8	0	0	0	0	0	0	0.08	0	0.92	0	0	0	0	0	0	0	0	0	0	0	0
F_9	0	0	0	0	0	0	0	0	0	0.98	0	0.01	0	0	0	0	0	0	0	0	0
F_10	0	0	0	0	0	0	0	0	0	0	0.97	0	0.02	0	0	0	0	0	0	0	0
F_11	0	0	0	0	0	0	0	0	0	0.04	0	0.96	0	0	0	0	0	0	0	0	0
F_12	0	0	0	0	0	0	0	0	0	0	0.01	0	0.98	0	0	0	0	0	0	0	0
F_13	0	0	0	0	0	0	0	0	0	0	0	0	0	0.87	0	0.13	0	0	0	0	0
F_14	0	0	0	0	0	0	0	0	0	0	0	0	0	0	0.93	0	0.06	0	0	0	0
F_15	0	0	0	0	0	0	0	0	0	0	0	0	0	0.08	0	0.91	0	0	0	0	0
F_16	0	0	0	0	0	0	0	0	0	0	0	0	0	0	0.15	0	0.85	0	0	0	0
F_17	0.09	0.01	0	0	0	0	0	0	0	0	0	0	0	0	0	0	0	0.89	0	0	0
F_18	0	0	0	0	0	0	0	0	0	0	0	0	0	0	0	0	0	0	1	0	0
F_19	0	0	0	0	0	0	0	0	0	0	0	0	0	0	0	0	0	0	0	0.99	0
F_20	0	0	0	0	0	0	0	0	0	0	0	0	0	0	0	0	0	0	0	0	0.99

Figure 13. Failure detected location when there are 5 sensors at points 3, 4, 5, 6, and 7 (Accuracy: 92.9%).

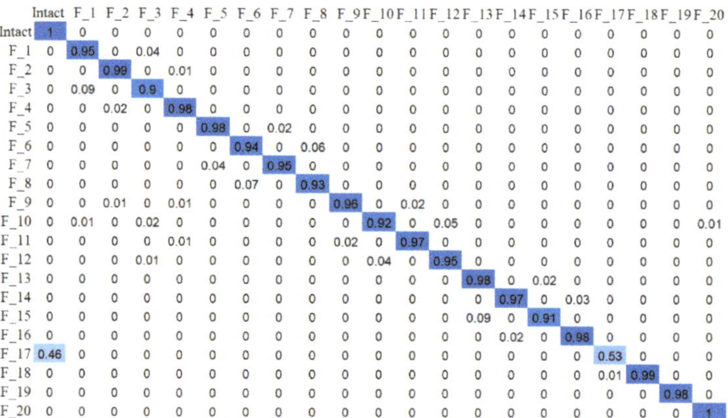

	Intact	F_1	F_2	F_3	F_4	F_5	F_6	F_7	F_8	F_9	F_10	F_11	F_12	F_13	F_14	F_15	F_16	F_17	F_18	F_19	F_20
Intact	1	0	0	0	0	0	0	0	0	0	0	0	0	0	0	0	0	0	0	0	0
F_1	0	0.95	0	0.04	0	0	0	0	0	0	0	0	0	0	0	0	0	0	0	0	0
F_2	0	0	0.99	0	0.01	0	0	0	0	0	0	0	0	0	0	0	0	0	0	0	0
F_3	0	0.09	0	0.9	0	0	0	0	0	0	0	0	0	0	0	0	0	0	0	0	0
F_4	0	0	0.02	0	0.98	0	0	0	0	0	0	0	0	0	0	0	0	0	0	0	0
F_5	0	0	0	0	0	0.98	0	0.02	0	0	0	0	0	0	0	0	0	0	0	0	0
F_6	0	0	0	0	0	0	0.94	0	0.06	0	0	0	0	0	0	0	0	0	0	0	0
F_7	0	0	0	0	0	0.04	0	0.95	0	0	0	0	0	0	0	0	0	0	0	0	0
F_8	0	0	0	0	0	0	0.07	0	0.93	0	0	0	0	0	0	0	0	0	0	0	0
F_9	0	0	0.01	0	0.01	0	0	0	0	0.96	0	0.02	0	0	0	0	0	0	0	0	0
F_10	0	0.01	0	0.02	0	0	0	0	0	0	0.92	0	0.05	0	0	0	0	0	0	0	0.01
F_11	0	0	0	0	0.01	0	0	0	0	0.02	0	0.97	0	0	0	0	0	0	0	0	0
F_12	0	0	0	0.01	0	0	0	0	0	0	0.04	0	0.95	0	0	0	0	0	0	0	0
F_13	0	0	0	0	0	0	0	0	0	0	0	0	0	0.98	0	0.02	0	0	0	0	0
F_14	0	0	0	0	0	0	0	0	0	0	0	0	0	0	0.97	0	0.03	0	0	0	0
F_15	0	0	0	0	0	0	0	0	0	0	0	0	0	0.09	0	0.91	0	0	0	0	0
F_16	0	0	0	0	0	0	0	0	0	0	0	0	0	0	0.02	0	0.98	0	0	0	0
F_17	0.46	0	0	0	0	0	0	0	0	0	0	0	0	0	0	0	0	0.53	0	0	0
F_18	0	0	0	0	0	0	0	0	0	0	0	0	0	0	0	0	0	0.01	0.99	0	0
F_19	0	0	0	0	0	0	0	0	0	0	0	0	0	0	0	0	0	0	0	0.98	0
F_20	0	0	0	0	0	0	0	0	0	0	0	0	0	0	0	0	0	0	0	0	1

Figure 14. Failure detected location when there are 5 sensors at points 2, 3, 5, 7, and 8 (Accuracy: 94.2%).

	Intact	F_1	F_2	F_3	F_4	F_5	F_6	F_7	F_8	F_9	F_10	F_11	F_12	F_13	F_14	F_15	F_16	F_17	F_18	F_19	F_20
Intact	1	0	0	0	0	0	0	0	0	0	0	0	0	0	0	0	0	0	0	0	0
F_1	0	0.98	0	0.02	0	0	0	0	0	0	0	0	0	0	0	0	0	0	0	0	0
F_2	0	0	0.97	0	0.03	0	0	0	0	0	0	0	0	0	0	0	0	0	0	0	0
F_3	0	0.04	0	0.96	0	0	0	0	0	0	0	0	0	0	0	0	0	0	0	0	0
F_4	0	0	0.02	0	0.98	0	0	0	0	0	0	0	0	0	0	0	0	0	0	0	0
F_5	0	0	0	0	0	0.98	0	0.02	0	0	0	0	0	0	0	0	0	0	0	0	0
F_6	0	0	0	0	0	0	0.9	0	0.1	0	0	0	0	0	0	0	0	0	0	0	0
F_7	0	0	0	0	0	0.12	0	0.87	0	0	0	0	0	0	0	0	0	0	0	0	0
F_8	0	0	0	0	0	0	0.01	0	0.98	0	0	0	0	0	0	0	0	0	0	0	0
F_9	0	0	0	0	0	0	0	0	0	0.98	0	0.02	0	0	0	0	0	0	0	0	0
F_10	0	0	0	0.01	0	0	0	0	0	0	0.98	0	0.02	0	0	0	0	0	0	0	0
F_11	0	0	0	0	0	0	0	0	0	0.01	0	0.99	0	0	0	0	0	0	0	0	0
F_12	0	0	0	0	0	0	0	0	0	0	0.01	0	0.99	0	0	0	0	0	0	0	0
F_13	0	0	0	0	0	0	0	0	0	0	0	0	0	0.96	0	0.04	0	0	0	0	0
F_14	0	0	0	0	0	0	0	0	0	0	0	0	0	0	1	0	0	0	0	0	0
F_15	0	0	0	0	0	0	0	0	0	0	0	0	0	0.01	0	0.99	0	0	0	0	0
F_16	0	0	0	0	0	0	0	0	0	0	0	0	0	0	0.07	0	0.93	0	0	0	0
F_17	0.08	0	0	0	0	0	0	0	0	0	0	0	0	0	0	0	0	0.91	0	0	0
F_18	0	0	0	0	0	0	0	0	0	0	0	0	0	0	0	0	0	0.01	0.98	0.01	0
F_19	0	0	0	0	0	0	0	0	0	0	0	0	0	0	0	0	0	0	0	0.99	0
F_20	0	0	0	0	0	0	0	0	0	0	0	0	0	0	0	0	0	0	0	0	1

Figure 15. Failure detected location when there are 5 sensors at points 2, 4, 5, 6, and 8 (Accuracy: 96.7%).

	Intact	F_1	F_2	F_3	F_4	F_5	F_6	F_7	F_8	F_9	F_10	F_11	F_12	F_13	F_14	F_15	F_16	F_17	F_18	F_19	F_20
Intact	1	0	0	0	0	0	0	0	0	0	0	0	0	0	0	0	0	0	0	0	0
F_1	0	0.98	0	0.02	0	0	0	0	0	0	0	0	0	0	0	0	0	0	0	0	0
F_2	0	0	0.96	0	0.04	0	0	0	0	0	0	0	0	0	0	0	0	0	0	0	0
F_3	0	0.02	0	0.98	0	0	0	0	0	0	0	0	0	0	0	0	0	0	0	0	0
F_4	0	0	0.02	0	0.97	0	0	0	0	0	0	0	0	0	0	0	0	0	0	0	0
F_5	0	0	0	0	0	1	0	0	0	0	0	0	0	0	0	0	0	0	0	0	0
F_6	0	0	0	0	0	0	0.96	0	0.03	0	0	0	0	0	0	0	0	0	0	0	0
F_7	0	0	0	0	0	0.07	0	0.93	0	0	0	0	0	0	0	0	0	0	0	0	0
F_8	0	0	0	0	0	0	0	0	1	0	0	0	0	0	0	0	0	0	0	0	0
F_9	0	0	0	0	0	0	0	0	0	0.98	0	0.02	0	0	0	0	0	0	0	0	0
F_10	0	0	0	0	0	0	0	0	0	0	0.97	0	0.02	0	0	0	0	0	0	0	0
F_11	0	0	0	0	0	0	0	0	0	0.01	0	0.99	0	0	0	0	0	0	0	0	0
F_12	0	0	0	0	0	0	0	0	0	0	0.01	0	0.99	0	0	0	0	0	0	0	0
F_13	0	0	0	0	0	0	0	0	0	0	0	0	0	0.97	0	0.03	0	0	0	0	0
F_14	0	0	0	0	0	0	0	0	0	0	0	0	0	0	0.99	0	0.01	0	0	0	0
F_15	0	0	0	0	0	0	0	0	0	0	0	0	0	0.01	0	0.99	0	0	0	0	0
F_16	0	0	0	0	0	0	0	0	0	0	0	0	0	0	0.05	0	0.95	0	0	0	0
F_17	0.01	0	0	0	0	0	0	0	0	0	0	0	0	0	0	0	0	0.99	0	0	0
F_18	0	0	0	0	0	0	0	0	0	0	0	0	0	0	0	0	0	0.01	0.99	0	0
F_19	0	0	0	0	0	0	0	0	0	0	0	0	0	0	0	0	0	0	0	0.99	0
F_20	0	0	0	0	0	0	0	0	0	0	0	0	0	0	0	0	0	0	0	0	1

Figure 16. Failure detected location when there are 7 sensors at points 2, 3, 4, 5, 6, 7, and 8 (Accuracy: 98.1%).

Table 5. Failure-detection accuracy at different sensor combinations.

	Sensor Location					
	2, 5, 8	2, 4, 6, 8	3, 4, 5, 6, 7	2, 3, 5, 7, 8	2, 4, 5, 6, 8	2, 3, 4, 5, 6, 7, 8
Accuracy (%)	84.1	90.4	92.9	94.2	96.7	98.1

Here, we set a rule; if the classification probability of a certain failure case is less than 50%, there is no classification at each time step. For instance, at nth time step, the probabilities of Case 3 and Case 5 are 42.6% and 45.0%, respectively, then, they are regarded as no classification. Base on this rule, the summation of values is less than 1 in several rows in Figures 11–16.

As shown in Figure 11, with three sensors (No. 2, 5, and 8), the overall classification accuracy is low (84.1% accurate), except for the intact case and cases when all mooring lines are broken i.e., Cases 17–20. Mostly, misclassification occurs when the adjacent mooring is broken, but, the probability of correct classification is always higher than that of misclassification, even if the classification was not good enough. Figure 12 shows the results with four sensors (No. 2, 4, 6, and 8). After adding one more sensor, the overall classification accuracy is increased to 90.4% i.e., failure detection performance is significantly enhanced. However, there are some scenarios where accuracy is lower than the three-sensor case, which supports the necessity of sensor-location optimization. We can further check the importance of sensor arrangement in the longitudinal direction, as shown in Figures 13–15. With the same sensor number of five, quite different detection accuracies can be obtained. Widely spread sensors (Figures 14 and 15) provide higher accuracy than centrally packed sensors (Figure 13). When comparing Figure 14 with Figure 15, a sufficient number of sensors should be located near central locations where the largest motion variations occur as a result of different mooring failures. Placing three sensors in the central areas (Figure 15) results in higher detection accuracy than the case with one sensor in the center (Figure 14). Finally, as shown in Figure 16, with seven sensors, the highest detection accuracy of 98.1% is obtained. Small misclassifications recognizing the mooring-failure case of F7 as F5 occurred, but their failure locations were the same i.e., x = −200 m. Judging from the trend, with more sensors, the mooring failure prediction can be even higher than 98.1%. The improvement from Figure 15 with five sensors (96.7%) to Figure 16 with 7 sensors (98.1%) is only marginal. Therefore, when 97% prediction accuracy is good enough, we can install five sensors instead of seven sensors to lower the relevant cost and complexity. Even lower failure-prediction accuracy than 97% may still be acceptable since with the warning sign, we can always double-check the actual failure by sending divers or underwater drones.

In order to further verify model accuracy in various marine environments, two more wave conditions (Hs = 3 m, Tp = 8 sec; Hs = 7 m, Tp = 12 sec, i.e., the last two environmental conditions given in Table 3) were considered for testing to evaluate the failure detection performance, which was not considered in training. Same as the previous cases, 80% of the dataset from the first eight environmental conditions were used as a training set, and the data from seven sensors (#2–8) that showed the best performance were used. As shown in Figures 17 and 18, the results show that even if different environmental conditions were tested, sufficient accuracy of over 97% was obtained. Therefore, the reliability of the model is secured by showing very high accuracy even for the environmental conditions not used as training data.

	Intact	F_1	F_2	F_3	F_4	F_5	F_6	F_7	F_8	F_9	F_10	F_11	F_12	F_13	F_14	F_15	F_16	F_17	F_18	F_19	F_20
Intact	1	0	0	0	0	0	0	0	0	0	0	0	0	0	0	0	0	0	0	0	0
F_1	0	0.97	0	0.03	0	0	0	0	0	0	0	0	0	0	0	0	0	0	0	0	0
F_2	0	0	0.99	0	0.01	0	0	0	0	0	0	0	0	0	0	0	0	0	0	0	0
F_3	0	0.01	0	0.99	0	0	0	0	0	0	0	0	0	0	0	0	0	0	0	0	0
F_4	0	0	0.02	0	0.97	0	0	0	0	0	0	0	0	0	0	0	0	0	0	0	0
F_5	0	0	0	0	0	0.99	0	0.01	0	0	0	0	0	0	0	0	0	0	0	0	0
F_6	0	0	0	0	0	0	0.97	0	0.02	0	0	0	0	0	0	0	0	0	0	0	0
F_7	0	0	0	0	0	0.02	0	0.98	0	0	0	0	0	0	0	0	0	0	0	0	0
F_8	0	0	0	0	0	0	0.01	0	0.99	0	0	0	0	0	0	0	0	0	0	0	0
F_9	0	0	0	0	0	0	0	0	0	0.97	0	0.03	0	0	0	0	0	0	0	0	0
F_10	0	0	0	0	0	0	0	0	0	0	0.97	0	0.02	0	0	0	0	0	0	0	0
F_11	0	0	0	0	0	0	0	0	0	0.01	0	0.99	0	0	0	0	0	0	0	0	0
F_12	0	0	0	0	0	0	0	0	0	0	0.01	0	0.99	0	0	0	0	0	0	0	0
F_13	0	0	0	0	0	0	0	0	0	0	0	0	0	0.98	0	0.02	0	0	0	0	0
F_14	0	0	0	0	0	0	0	0	0	0	0	0	0	0	1	0	0	0	0	0	0
F_15	0	0	0	0	0	0	0	0	0	0	0	0	0	0.02	0	0.98	0	0	0	0	0
F_16	0	0	0	0	0	0	0	0	0	0	0	0	0	0	0.3	0	0.7	0	0	0	0
F_17	0.04	0	0	0	0	0	0	0	0	0	0	0	0	0	0	0	0	0.96	0	0	0
F_18	0	0	0	0	0	0	0	0	0	0	0	0	0	0	0	0	0	0	1	0	0
F_19	0	0	0	0	0	0	0	0	0	0	0	0.01	0	0	0	0	0	0	0.01	0.98	0
F_20	0	0	0	0	0	0	0	0	0	0	0	0	0	0	0	0	0	0	0	0	1

Figure 17. Confusion matrix of the different environmental condition (Hs = 3 m, Tp = 8 sec) (Accuracy: 97.0%).

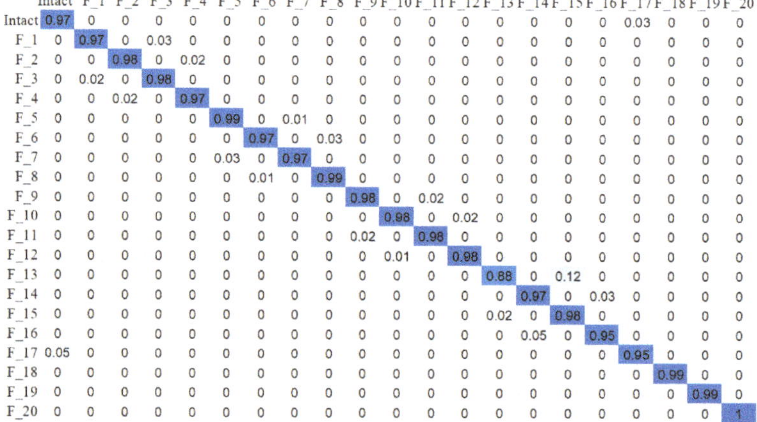

	Intact	F_1	F_2	F_3	F_4	F_5	F_6	F_7	F_8	F_9	F_10	F_11	F_12	F_13	F_14	F_15	F_16	F_17	F_18	F_19	F_20
Intact	0.97	0	0	0	0	0	0	0	0	0	0	0	0	0	0	0	0	0.03	0	0	0
F_1	0	0.97	0	0.03	0	0	0	0	0	0	0	0	0	0	0	0	0	0	0	0	0
F_2	0	0	0.98	0	0.02	0	0	0	0	0	0	0	0	0	0	0	0	0	0	0	0
F_3	0	0.02	0	0.98	0	0	0	0	0	0	0	0	0	0	0	0	0	0	0	0	0
F_4	0	0	0.02	0	0.97	0	0	0	0	0	0	0	0	0	0	0	0	0	0	0	0
F_5	0	0	0	0	0	0.99	0	0.01	0	0	0	0	0	0	0	0	0	0	0	0	0
F_6	0	0	0	0	0	0	0.97	0	0.03	0	0	0	0	0	0	0	0	0	0	0	0
F_7	0	0	0	0	0	0.03	0	0.97	0	0	0	0	0	0	0	0	0	0	0	0	0
F_8	0	0	0	0	0	0	0.01	0	0.99	0	0	0	0	0	0	0	0	0	0	0	0
F_9	0	0	0	0	0	0	0	0	0	0.98	0	0.02	0	0	0	0	0	0	0	0	0
F_10	0	0	0	0	0	0	0	0	0	0	0.98	0	0.02	0	0	0	0	0	0	0	0
F_11	0	0	0	0	0	0	0	0	0	0.02	0	0.98	0	0	0	0	0	0	0	0	0
F_12	0	0	0	0	0	0	0	0	0	0	0.01	0	0.98	0	0	0	0	0	0	0	0
F_13	0	0	0	0	0	0	0	0	0	0	0	0	0	0.88	0	0.12	0	0	0	0	0
F_14	0	0	0	0	0	0	0	0	0	0	0	0	0	0	0.97	0	0.03	0	0	0	0
F_15	0	0	0	0	0	0	0	0	0	0	0	0	0	0.02	0	0.98	0	0	0	0	0
F_16	0	0	0	0	0	0	0	0	0	0	0	0	0	0	0.05	0	0.95	0	0	0	0
F_17	0.05	0	0	0	0	0	0	0	0	0	0	0	0	0	0	0	0	0.95	0	0	0
F_18	0	0	0	0	0	0	0	0	0	0	0	0	0	0	0	0	0	0	0.99	0	0
F_19	0	0	0	0	0	0	0	0	0	0	0	0	0	0	0	0	0	0	0	0.99	0
F_20	0	0	0	0	0	0	0	0	0	0	0	0	0	0	0	0	0	0	0	0	1

Figure 18. Confusion matrix of the different environmental condition (Hs = 7 m, Tp = 12 sec) (Accuracy: 97.2%).

5. Conclusions

This paper presents the development of a DNN (deep neural network) model for detecting and classifying mooring-failure locations for SFT using lateral and vertical motion-sensor signals. This task can be considered as pattern-recognition and classification problems, and DNN is appropriate for the task. The input variables of the algorithm are sea state (Hs and Tp) and lateral and vertical displacement data from three to seven sensors. The hydro-elastic tunnel-mooring coupled time-domain simulations were performed under different wave conditions and failure/intact scenarios, and the big dataset of numerical-sensor signals was acquired. Through simulations, we confirmed that trends and magnitudes of displacements were changed under different failure (or intact) scenarios. The entire data set basically consists of 80% training data and 20% test data not used for training from eight wave conditions. Two wave conditions not used for training were additionally employed for further testing. The optimization process was progressively conducted with respect to the numbers of hidden layers and neurons of each layer, activation function, and optimizer. The selected optimal numbers of hidden layers and neurons were three and 200, respectively. The Sigmoid activation function and Adamax optimizer were selected for high classification accuracy. We also tested the failure-detection performance with different numbers and combinations of motion sensors. Misclassification largely decreases as the number of sensors is increased with proper intervals. With seven motion sensors, the classification accuracy of 98.1% was achieved. The prediction accuracy higher than 90% can also be achieved even with four sensors. Besides, the model displays the accuracy of over 97% with the different wave conditions that were not used in the training data set. The results show good feasibility of the developed machine-learning-based monitoring system for the mooring-failure detection of future SFTs or similar deformable submerged structures.

Author Contributions: All authors have equally contributed to the publication of this article with respect to the design of the target model, validation of the numerical model, analysis, and writing. All authors have read and agreed to the published version of the manuscript.

Funding: This research was supported by the MOTIE (Ministry of Trade, Industry, and Energy) in Korea, under the Fostering Global Talents for Innovative Growth Program (P0008750) supervised by the Korea Institute for Advancement of Technology (KIAT). This work was supported by the National Research Foundation of Korea (NRF) grant funded by the Korea government (MSIT) (No. 2017R1A5A1014883).

Conflicts of Interest: The authors declare no conflict of interest.

References

1. Di Pilato, M.; Perotti, F.; Fogazzi, P. 3D dynamic response of submerged floating tunnels under seismic and hydrodynamic excitation. *Eng. Struct.* **2008**, *30*, 268–281. [CrossRef]
2. Faggiano, B.; Landolfo, R.; Mazzolani, F. The SFT: An innovative solution for waterway strait crossings. In *IABSE Symposium Report*; International Association for Bridge and Structural Engineering: Lisbon, Portugal, 2005; Volume 90, pp. 36–42.
3. Seo, S.-I.; Mun, H.-S.; Lee, J.-H.; Kim, J.-H. Simplified analysis for estimation of the behavior of a submerged floating tunnel in waves and experimental verification. *Marine Struct.* **2015**, *44*, 142–158. [CrossRef]
4. Chen, Z.; Xiang, Y.; Lin, H.; Yang, Y. Coupled vibration analysis of submerged floating tunnel system in wave and current. *Appl. Sci.* **2018**, *8*, 1311. [CrossRef]
5. Long, X.; Ge, F.; Hong, Y. Feasibility study on buoyancy–weight ratios of a submerged floating tunnel prototype subjected to hydrodynamic loads. *Acta Mech. Sin.* **2015**, *31*, 750–761. [CrossRef]
6. Lu, W.; Ge, F.; Wang, L.; Wu, X.; Hong, Y. On the slack phenomena and snap force in tethers of submerged floating tunnels under wave conditions. *Marine Struct.* **2011**, *24*, 358–376. [CrossRef]
7. Jin, C.; Kim, M.-H. Time-domain hydro-elastic analysis of a SFT (submerged floating tunnel) with mooring lines under extreme wave and seismic excitations. *Appl. Sci.* **2018**, *8*, 2386. [CrossRef]
8. Won, D.; Seo, J.; Kim, S.; Park, W.-S. Hydrodynamic Behavior of Submerged Floating Tunnels with Suspension Cables and Towers under Irregular Waves. *Appl. Sci.* **2019**, *9*, 5494. [CrossRef]
9. Ma, Z.; Sohn, H. Structural Displacement Estimation by FIR Filter Based Fusion of Strain and Acceleration Measurements. In Proceedings of the 29th International Ocean. and Polar Engineering Conference, Honolulu, HI, USA, 16–21 June 2019; International Society of Offshore and Polar Engineers: Cupertino, CA, USA, 2019.

10. Choi, J.; Kim, J.M.-H. Development of a New Methodology for Riser Deformed Shape Prediction/Monitoring. In Proceedings of the ASME 2018 37th International Conference on Ocean., Offshore and Arctic Engineering, Madrid, Spain, 17–22 June 2018; American Society of Mechanical Engineers Digital Collection: New York, NY, USA, 2018.

11. Mercan, B.; Chandra, Y.; Maheshwari, H.; Campbell, M. Comparison of Riser Fatigue Methodologies Based on Measured Motion Data. In Proceedings of the Offshore Technology Conference, Houston, TX, USA, 2–5 May 2016; Offshore Technology Conference, Inc.: Houston, TX, USA, 2016.

12. Chung, M.; Kim, S.; Lee, K.; Shin, D.H. Detection of damaged mooring line based on deep neural networks. *Ocean Eng.* **2020**, *209*, 107522. [CrossRef]

13. Jaiswal, V.; Ruskin, A. Mooring Line Failure Detection Using Machine Learning. In Proceedings of the Offshore Technology Conference, Houston, TX, USA, 6–9 May 2019; Offshore Technology Conference, Inc.: Houston, TX, USA, 2019.

14. Sidarta, D.E.; Lim, H.-J.; Kyoung, J.; Tcherniguin, N.; Lefebvre, T.; O'Sullivan, J. Detection of Mooring Line Failure of a Spread-Moored FPSO: Part 1—Development of an Artificial Neural Network Based Model. In Proceedings of the International Conference on Offshore Mechanics and Arctic Engineering, Glasgow, Scotland, UK, 9–14 June 2019; American Society of Mechanical Engineers: New York, NY, USA; Volume 58769, p. V001T01A042.

15. Sidarta, D.E.; O'Sullivan, J.; Lim, H.-J. Damage detection of offshore platform mooring line using artificial neural network. In Proceedings of the International Conference on Offshore Mechanics and Arctic Engineering, Madrid, Spain, 17–22 June 2018; American Society of Mechanical Engineers: New York, NY, USA; Volume 51203, p. V001T01A058.

16. Orcina. OrcaFlex Manual. Available online: https://www.orcina.com/SoftwareProducts/OrcaFlex/index.php (accessed on 17 October 2018).

17. Jin, C.; Kim, M.-H. Tunnel-mooring-train coupled dynamic analysis for submerged floating tunnel under wave excitations. *Appl. Ocean. Res.* **2020**, *94*, 102008. [CrossRef]

18. Cifuentes, C.; Kim, S.; Kim, M.; Park, W. Numerical simulation of the coupled dynamic response of a submerged floating tunnel with mooring lines in regular waves. *Ocean Syst. Eng.* **2015**, *5*, 109–123. [CrossRef]

19. Kim, H.; Jin, C.; Kim, M.; Kim, K. Damage detection of bottom-set gillnet using Artificial Neural Network. *Ocean Eng.* **2020**, *208*, 107423. [CrossRef]

20. Hassoun, M.H. *Fundamentals of Artificial Neural Networks*; MIT Press: Cambridge, MA, USA, 1995.

21. Anderson, D.; McNeill, G. Artificial neural networks technology. *Kaman Sci. Corp.* **1992**, *258*, 1–83.

22. Rumelhart, D.; McCelland, G.; Williams, T. Training Hidden Units: Generalized Delta Rule. *Explor. Parallel Distrib. Process.* **1986**, *2*, 121–160.

23. Goodfellow, I.; Bengio, Y.; Courville, A. *Deep Learning*; MIT Press: Cambridge, MA, USA, 2016.

24. Murphy, K.P. *Machine Learning: A Probabilistic Perspective*; MIT Press: Cambridge, MA, USA, 2012.

25. Karlik, B.; Olgac, A.V. Performance analysis of various activation functions in generalized MLP architectures of neural networks. *Int. J. Artif. Intell. Expert Syst.* **2011**, *1*, 111–122.

26. Agarap, A.F. Deep Learning Using Rectified Linear Units (relu). *arXiv* **2018**, arXiv:1803.08375.

27. Ramachandran, P.; Zoph, B.; Le, Q.V. Searching for Activation Functions. *arXiv* **2017**, arXiv:1710.05941.

28. Kingma, D.P.; Ba, J. Adam: A Method for Stochastic Optimization. *arXiv* **2014**, arXiv:1412.6980.

29. Wilson, A.C.; Roelofs, R.; Stern, M.; Srebro, N.; Recht, B. The marginal value of adaptive gradient methods in machine learning. In Proceedings of the 31st Conference on Neural Information Processing Systems, Long Beach, CA, USA, 4–9, December 2017; Neural Information Processing Systems: San Diego, CA, USA; pp. 4148–4158.

30. Keskar, N.S.; Socher, R. Improving Generalization Performance by Switching from Adam to Sgd. *arXiv* **2017**, arXiv:1712.07628.

31. Tieleman, T.; Hinton, G. Lecture 6.5-rmsprop: Divide the gradient by a running average of its recent magnitude. *COURSERA Neural Netw. Mach. Learn.* **2012**, *4*, 26–31.

32. Mukkamala, M.C.; Hein, M. Variants of Rmsprop and Adagrad with Logarithmic Regret Bounds. *arXiv* **2017**, arXiv:1706.05507.

33. Zheng, W.; Dan, D.; Cheng, W.; Xia, Y. Real-time dynamic displacement monitoring with double integration of acceleration based on recursive least squares method. *Measurement* **2019**, *141*, 460–471. [CrossRef]

34. Choi, J.; Kim, J.M.-H. Development of a New Methodology for Riser Deformed Shape Prediction/Monitoring. In Proceedings of the International Conference on Offshore Mechanics and Arctic Engineering, Madrid, Spain, 17–22 June 2018; American Society of Mechanical Engineers: New York, NY, USA, 2018; Volume 51241, p. V005T04A041.
35. Kohavi, R.; Provost, F. Confusion matrix. *Mach. Learn.* **1998**, *30*, 271–274.

 *applied sciences*

Article

Damping of Beam Vibrations Using Tuned Particles Impact Damper

Mateusz Żurawski * and Robert Zalewski

Institute of Machine Design Fundamentals, Warsaw University of Technology, 02-524 Warsaw, Poland; Robert.Zalewski@pw.edu.pl
* Correspondence: Mateusz.Zurawski@pw.edu.pl; Tel.: +48-535-278-333

Received: 22 July 2020; Accepted: 9 September 2020; Published: 11 September 2020

Abstract: The presented paper reveals an innovative device which is the Tuned Particle Impact Damper (TPID). The damper enables the user change the dynamical features of the vibrating system thanks to rapidly tuning the volume of the container where the grains are locked. The effectiveness of proposed semi-active damping methodology was confirmed in experiments on vibrations of a cantilever beam excited by kinematic rule. Various damping characteristics captured for different volumes of the grains container and mass of granular material are presented. It is confirmed that the proposed TPID device allowed for efficient attenuation of the beam's vibration amplitude in the range of its resonant frequency vibrations.

Keywords: Particle Impact Damper; adaptive-passive damping; damping of vibrations; experiments

---

1. Introduction

Devices that are able to attenuate unwanted mechanical vibrations have interested researches and engineers for several decades. In the everyday engineering applications we can distinguish various types of vibration damper designs. The most popular passive shock absorbers constructions consist of viscous properties of elastomers or more or less sophisticated fluids or gels. Nowadays, it seems that so called Particles Impact Dampers (PID) start to play an increasingly important role in the field of vibrations damping. The classical PID principle of operation lies in inelastic collisions of loose particles enclosed in the specially designed container or cavities. The Energy dissipation mechanism is additionally related to the frictional contact of grains and the container casing. The PID device connected to the vibrating element or structure absorbs a part of global kinetic energy of the investigated system and reduces its vibrations amplitudes. The effectiveness of such a damping strategy mainly results from coupling of friction and impact phenomenon. The energy dissipation process in the most commonly used damping devices is generally related either to the friction phenomenon or previously mentioned visco-elastic material deformations. Such mechanisms increase the operating temperature and also the wear of material resulting in decreasing of damping properties. Considering the aforementioned limitations, the use of granular dampers in which the influence of temperature and material degradation is less important seems to be particularly interesting. The additional advantages of PIDs are: they can perform in a very wide range of frequencies, they can be made of recycled materials, economical aspect, the global mass of the system with grains moving in the internal cavity is generally lower than the initial. All the previously mentioned features of PIDs cause that such devices are very often applied in dynamic systems working in harsh environments. The adjustment of damping properties in a considered group of shock absorbers is limited and consists of choosing the right kind, consistency, shape and size of grains for the given application.

In world literature a significant number of studies have been carried out in the field of granular dampers analysis. They are mainly focused on efficiency ([1–4]), optimization ([5–7]), applications ([8–11]) and modeling ([12–14]). An interesting problem is discussed in Reference [15] where Particles Impact Damper is treated as a double pendulum striking element. It is confirmed that such an object is especially effective to attenuate vibrations of various slender elements. Various excitation rules and PID parameters are taken into account in experimental studies. Other papers suggest, that the PID can be a part of alternative energy harvesting device [16]. In this case the classical grains are replaced by a moving magnet forming the previously mentioned original energy harvester. Taking into account damping of vibrations of elements operating in an extreme environment, the reliability and durability of devices are extremely important. The typical example of such an environment is aerospace engineering, where PIDs are especially popular. It is presented in Reference [17] that attenuation of vibrations of highly sensitive electric circuits can be achieved by PIDs. Application of granular damping seems to also be optimal solution in the case of taking-off or landing of spacecrafts. Various modifications of classical PID designs are also investigated in recently published papers. In Reference [18], the authors propose an innovative solution for the semi-active Tuned Liquid Column Damper (sTLCD). Such a device is connected to the base structure using the spring with controlled stiffness. The efficiency of the considered system is discussed and different control algorithms are taken into consideration in the paper. Applied control strategies enabled the authors to effectively tune the frequency of a structure. The proposed solution is especially interesting in civil engineering where the increased danger of earthquakes exists. A lot of papers dealing with granular damping discuss various modeling approaches and problems encountered in the laboratory tests. Also experimentally determined damping characteristics are presented and discussed in detail ([19–21]).

Despite the apparent simplicity of the PIDs fundamental working rule it can be concluded, that capturing the real behavior of the discussed group of devices is rather complex. Previously mentioned papers generally simplify the PIDs modelling process to the 1 or 2 Degrees of Freedom (DOF) problems incorporating additionally external forces or equivalent damping parameters. In Reference [22] or Reference [23] the complex dynamics of vibrating sphere particles is compared to the dense gas flow. Such a solution allows to represent the damping characteristics by formulas dependent on: a mass ratio, volume ratio, restitution coefficient and excitation frequency. The different approach to the PIDs modelling problems can be observed in Reference [24] or Reference [25], where the sophisticated numerical software is used to capture the real laboratory test results. In Reference [24] an interesting original technique to predict the harmonic response of particle damping structure based on ANSYS software is discussed. Other approach basing on Discreet Element Method (DEM) is investigated in Reference [25] where the energy dissipation problems of harmonically excited elements are solved numerically. To solve the classical inverse problem consisting of calibrating the selected model parameters basing on the real response of the PID device various numerical optimization techniques are taken into consideration. In Reference [26] the damper parameters are captured using Artificial Neural Networks. The efficiency of proposed methodology was compared with two other popular parameters identification methods - Feed Forward Back Propagation Network and Radial Basis Function. An interesting practical engineering application of PIDs in the automotive industry is presented in Reference [27] where a granular damper is used as a sound absorber. It is confirmed that such a device is also suitable for increasing the driving comfort of the passengers. Granular dampers are also used in suppression of vibrations in centrifugal field of gear transmission ([28]). In such a case the tightly packed loose material is placed in specially designed holes in the gears. The reliability of PIDs is validated by the comparison of real response of the system with the numerical DEM simulations. The other grain based damping of vibration techniques being the alternative to the discussed PIDs are so called Chain Dampers ([29,30]) or Vibro-Impact Nonlinear Energy Sinks ([31–33]). Application of various PIDs seems to be especially interesting in the controlling nonlinear dynamics of slender and light objects resulted from the unknown disturbances. Typical examples of such structures are wind turbines ([34,35]).

Reference [36] describes the effective damping of rail vibrations by adding a TPID consisting of Tuned Absorber Mass and Impact Ball. The mathematical model, experiments and simulations are introduced to establish the optimal damping parameters for the system. In the mentioned paper, the authors analyze the effectiveness of the structure by separating the components of the damper. Reference [37] provides the literature review over PID dampers. It presents various technical solutions and modelling methods. They describe typical applications of the TPIDs, which however allow to change the parameters of the attenuator, but require a significant human involvement aspect. In Reference [38] a simple mathematical model of the vibrating system including TPID damper described as a 2-DOF system is presented. The shock absorber model is mainly described by two coupled parameters: equivalent damping and equivalent mass. Reference [39] presents the construction of a damper named TPID. The gap (free volume of the system) can be manually changed by the threaded lid. The research takes into account experiments, simulations and focuses on comparing three types of dampers: Particle Dampers, Tuned Particles Dampers and Tuned Mass Dampers. Determining the effectiveness of attenuation ability depending on the various excitation frequency and grains mass is also discussed.

In the previous paper of the authors [40], the initial TPID concept was discussed. Preliminary study is presented and discussed. In contrast to Reference [40], in this paper, the authors present a significantly greater number of experimental results conducted for free and kinematically excited vibrations. Based on the recent results it possible to determine the effectiveness of the TPID damper to attenuate vibrations depending on the application of various grains material, variable mass of particles and container volume and the excitation frequency. The main novelty of this paper is the presentation of several different TPID intermediate states during a dynamic strategy of changing the container volume. A mathematical model based on a nonlinear–exponential control function has been proposed. A verification of the real system response and numerical simulations is additionally provided.

The authors of this paper state, that Tuned Particle Impact Dampers can be competitive to other more sophisticated and expensive control schemes for industrial wind turbines such as proportional integral derivative regulators. The overview of PID literature presented in the Introduction section includes mainly passive damping strategies. The dynamics of attenuated by loose materials structures can be basically controlled by grains volume ratio, mass of the granular material or the coefficient of restitution. Although it was shown that PIDs are competitive to other damping techniques, up to this point granular dampers could not be used as a part of semi-active damping strategies. In this paper the authors propose an innovative Tuned Particles Impact Damping device. Its novelty consists of the possibility to rapidly change the volume of the compartment where the grains are encapsulated. This solution enables the implementation of a completely new adaptive-passive strategy of semi-active vibration damping. The paper is divided into several sections. In the next one the idea of the Tuned Particles Impact Damper is presented. In the third and fourth Sections the original test stand and experimental results are discussed respectively. The paper is completed with a Conclusion Section where the summary and the perspective research are revealed.

2. Idea of the Tuned Particles Impact Damper

Popular granular vibrations attenuators are devices introducing additional damping to the system. Dynamic properties of such structures depend on many parameters, where the most important are: size, shape and volume of the container, filling ratio of granular media, grains features like density, shape and roughness. All previous studies related to the application of PIDs in dynamic systems assumed that changing of damping properties can be achieved by passive adjusting of previously mentioned parameters. Taking into consideration the advantages of semi-active methods of damping of vibrations over classical passive strategies the authors decided to improve the classical PIDs and introduce the Tuned Particles Impact Damper (TPID) enabling the user to adjust in the damping properties.

The present paper focuses on the revealing the extraordinary features of the TPID device prototype, where rapid changes of the volume container allow for adjusting the damping characteristics of the system aimed in reducing the effects of temporary external excitations. The simplified idea of investigated TPID prototype and scheme of experimental setup is presented in Figure 1.

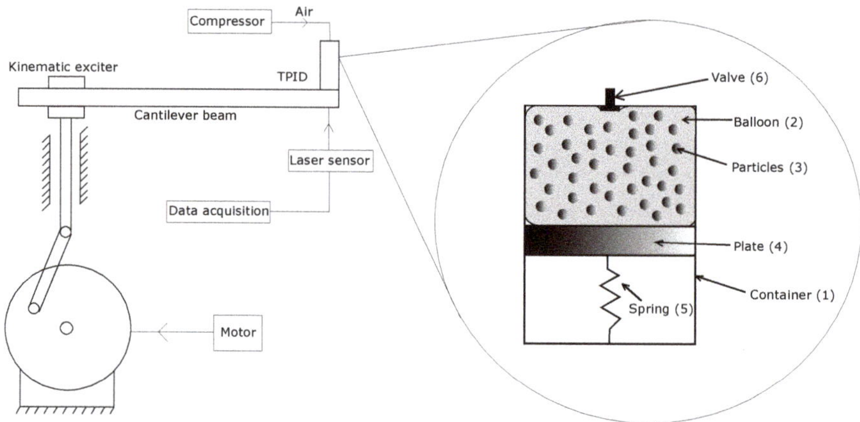

Figure 1. Scheme of the tests stand and the Tuned Particles Impact Damper (TPID) prototype.

The TPID can be treated as an additional device that can be mounted onto (eventually into) various mechanical systems to control their dynamical features, mainly to reduce their mechanical vibrations. The TPID prototype considered in this paper consists of the container (1) having a varying volume of the range (V_{min} = 0.2 [dm^3]; V_{max} = 0.79 [dm^3]). This volume can be controlled by the balloon (2) located inside the container. The balloon in its initial state is deflated and closely encapsulates the loose granular material (3) placed inside of it. In such a case the grains can be treated as a concentrated mass located in a selected place of the system. At the same time discs (4) connected with springs (5) limit the container volume to V_{min} value. The balloon is additionally equipped with a valve (6), the end of which protrudes from the container, enabling connection of the external pump. Changing the pressure inside the balloon expands it causing the movement of the disc and increasing the operating volume of the container. At this stage the granular material starts to vibrate which results in two main phenomenon: inelastic collisions between single grains and interaction with the casing. A further inflating of the balloon causes reaching the V_{max} volume of the device. The rapid changes of the TPID operating volume of the TPID device is a novel mechanism enabling adaptive-passive or in future semi-active damping of vibrations strategies. The additional advantages of the proposed TPID prototype are its uncomplicated design, high strength of its components and finally environmental friendliness.

3. Test Stand

To reveal the extraordinary features of the TPID prototype, the special laboratory test stand was designed (Figure 2). It was aimed to conduct the research on the kinematically excited cantilever beam equipped with the TPID device. Various TPID damper parameters were taken into consideration during the experiments. In the first stage of empirical investigations, to fully understand the features of the investigated system the values of TPID such as the volume of container, mass of the grains or type of granular material were systematically changed. Subsequently, the tip of the vibrated beam amplitudes for the resonant range were measured for the various parameters configuration.

The features of the investigated system, including the geometry of the grains and the dimensions of the test stand, are presented in Table 1. The range of laboratory tests included three different granular media (Figure 3). The spherical shape of particles having diameter $D = 6$ [mm] were tested in laboratory tests. Additionally, five different filling ratios (grains volume to total volume) were taken into account. At the current stage of investigations the authors did not concern the wear of grains material process and its influence on the damping properties of the device. The whole research plan is presented in Figure 4.

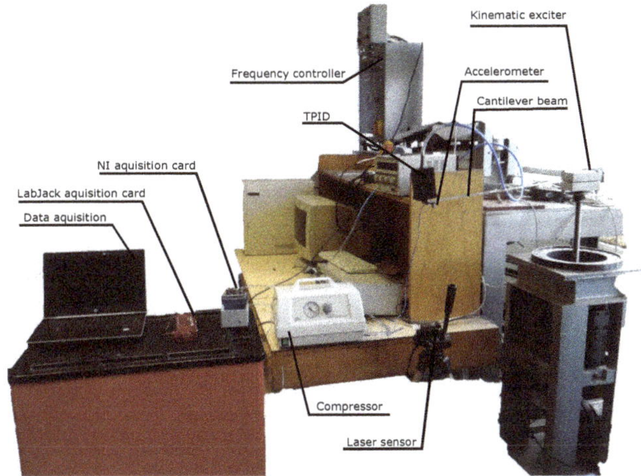

Figure 2. Test stand.

Table 1. Parameters used in experiments.

Component	Parameter	Value
Cantilever beam	Size	Length: 600 [mm], Width: 35 [mm], High: 2 [mm]
	Material	Steel
PID	Size	Length: 63 [mm], Width: 63 [mm], High: 200 [mm]
	Material	Plastic and steel
Grains-case 1	Size	$D = 6$ [mm]
	Material	Plastic
	Density	$\rho_p = 1000 \left[\frac{\text{kg}}{\text{m}^3}\right]$
Grains-case 2	Size	$D = 6$ [mm]
	Material	Steel
	Density	$\rho_s = 7860 \left[\frac{\text{kg}}{\text{m}^3}\right]$
Grains-case 3	Size	$D = 6$ [mm]
	Material	Mixed (plastic and steel)
Balloon	Size	$Volume_1 = 0.20$ [dm^3]
		$Volume_2 = 0.35$ [dm^3]
	Material: rubber	$Volume_3 = 0.50$ [dm^3]
		$Volume_4 = 0.65$ [dm^3]
		$Volume_5 = 0.79$ [dm^3]

The measuring system consisted of LabJack and National Instruments data acquisition cards. The beam tip displacements were measured by Emron laser sensor and additional accelerometer located at the end of the beam. The signals were recorded on a personal computer with dedicated recording and analyzing software. The previously mentioned beam was clamped in a special handle and mounted in the individually designed exciter (Figure 2). A special arrangement of gears, belt transmissions and eccentric disks were transforming the rotary movement of the electric rotor into a reciprocating movement of the exciter. The kinematic excitation rule including variable frequencies (up to 60 Hz) and fixed amplitude level (±5 mm) were considered. The data was sampled at 0.5 kHz.

(a)   (b)   (c)

Figure 3. The investigated grains. (**a**) Plastic particles; (**b**) Steel particles; (**c**) Mixed particles.

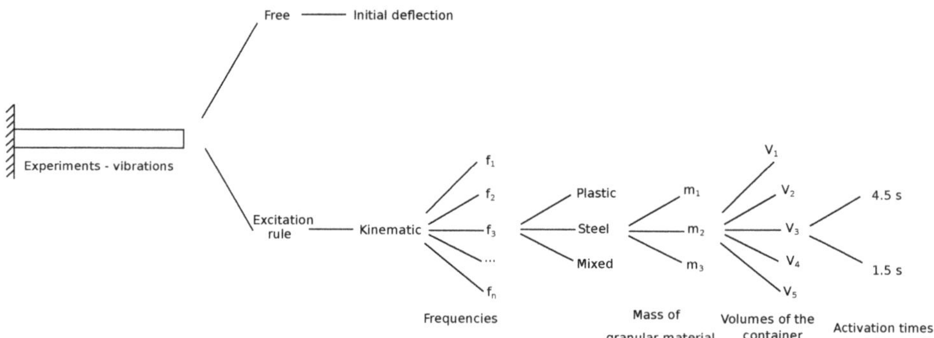

Figure 4. Research plan.

For rapid changes of the rubber balloon volume the compressor equipped with pressure sensors was used. The volume changing of the TPID compartment consisted of inflating and deflating the balloon that tightly filled the working chamber of the granular shock absorber. Recorded data were analyzed in the SageMath environment based on Phyton programing language.

4. Experiments

In the laboratory tests the authors focused on determining original damping characteristics for various operating parameters of the TPID device presented in Table 2. The impact of changeable volume of the container is specially underlined. The main objectives of laboratory tests were to confirm the usefulness of introduced granular damping strategy and to reveal the innovative adaptive-passive method for attenuation of the investigated beam vibrations. Several excitation frequencies were taken into account in experimental research. The whole variety of applied frequencies covered the range before and after the resonance. Various vibration acceleration spectra acquired for different kinds of granular materials were determined. The effectiveness of the proposed solution is especially visible for the resonant frequency where the activation and deactivation of the TPID device caused reasonable changes in the recorded amplitudes of the vibrated beam.

Table 2. Root Mean Square (RMS) of the displacement for various masses of the grains.

Volume [dm^3]	RMS [mm]		
	$M_p = 0.1M_s$	$M_p = 0.2M_s$	$M_p = 0.05M_s$
0.20	42.4	46.3	43.9
0.35	37.4	43.5	43.8
0.50	36.8	36.5	43.1
0.65	29.7	32.2	42.1
0.79	22.5	31.5	38.8

4.1. The Influence of Container Volume and Mass of Particles

Five various volumes of the container were investigated in the laboratory tests. The lowest value corresponds to the fully compacted grains. Such a case results in loading of the system by the concentrated mass of the grains. The dissipative properties of TPID element is then reduced to the minimum. For the maximum value of volume the vibrating grains can freely collide between each other. Also the impacts of particles and walls of the container is observable, which results in increased dissipative properties of the granular damper. Three different compositions of the particles were tested. Consequently, three various masses were included in the described experiments. Typical laboratory tests results are depicted in Figures 5 and 6. Root Mean Square (RMS) values of the beam tip displacements (1) for various operating parameters were determined to reveal the recorded response of the considered system. Additionally the RMS data are presented in Table 2.

$$D_{RMS} = \sqrt{\frac{\sum_{i=1}^{n} a_i(f_j)^2}{n}},\tag{1}$$

where a_i—maximum displacement amplitude in "i" cycle of vibration, f_j—frequency of excitation, j—case number, n—number of amplitude values in each case.

Diagrams depicted in Figure 5a,c,e illustrate the effect of variable container volume for three types (masses) of granular materials. The selection of appropriate mass of particles is of first importance. This parameter can not be easily changed in the presented TPID prototype. At this stage of the research it was assumed, that the total mass of the particles (M_p) was 5%, 10% and 20% of the whole system mass (M_s). The second and the most important aspect is the volume of the container, controlled by the volume of the inflatable balloon placed inside the casing.

Data depicted in Figure 5 confirm that changing the volume of the container may lead to reduction of the resonance amplitudes. Lower number of particles (lower mass) is related to limited number of "intergranular" collisions what obviously decreases the dissipative properties of the TPIDs (Figure 5c,e). In Figure 5a,b it can be observed that the RMS amplitude for the case where the particles mass is 10% of the total mass of the system and for the lowest container volume (V_{min}) was 42.5 mm. The value of the considered parameter for the maximum volume of the container (V_{max}) was 22.5 mm, which can be described as a reduction of vibrations by 53%. Lower damping properties were observed for the remaining two cases of the particles mass. In experimental studies, in which the granular mass was 20% of the mass of the entire system, the recorded RMS amplitudes for minimal and maximal volumes were 46.3 mm and 31.5 mm respectively. Analyzing the presented data, it turns out that the system vibration amplitude decreased by 32%. In a situation where the TPID damper worked with the use of granules with a mass equal to only 5%, the reduction of system vibrations was much lower than in the first two cases and amounted to 12%. Analyzing the RMS amplitude for all cases also including the intermediate values of the damper volume, it can be observed that the amplitudes of the system response change nonlinearly, and thus the vibration damping effectiveness is also nonlinear as a function of the TPID damper working space variation. Switching "on" and "off" processes, which involves increasing the initial value of the operating volume slightly changes the natural frequencies of the considered system. This phenomenon is presented in Figure 5b,d,f. Nonlinear contact forces occur during collisions between the granules and the damper container. In the considered mechanical systems they are coupled with changes in the dynamics of the structure. The authors of Reference [7] show that changing (increasing) the gap volume in the PID damper results in two main phenomena: initially the reduction of the system's frequency is observable, but after exceeding some specific volume threshold value the increase in the natural frequency of the system is noticed. The reason is the response of the grains movement, the occurrance contact forces and their duration in relation to the moving damped system. In the previously mentioned paper such an effect of changing the natural frequency is explained by a change of the effective mass. In the first range, the effective mass of the system increases, and after exceeding the specified value of the gap, the effective mass decreases. The structure which we proposed is evidently operating in the first described range. Such an observation shows some similarities between TPID and Tuned Mass Dampers. It can be stated that the proposed design of novel TPID device ensures all the advantages of classical PIDs [12], [41] or [3].

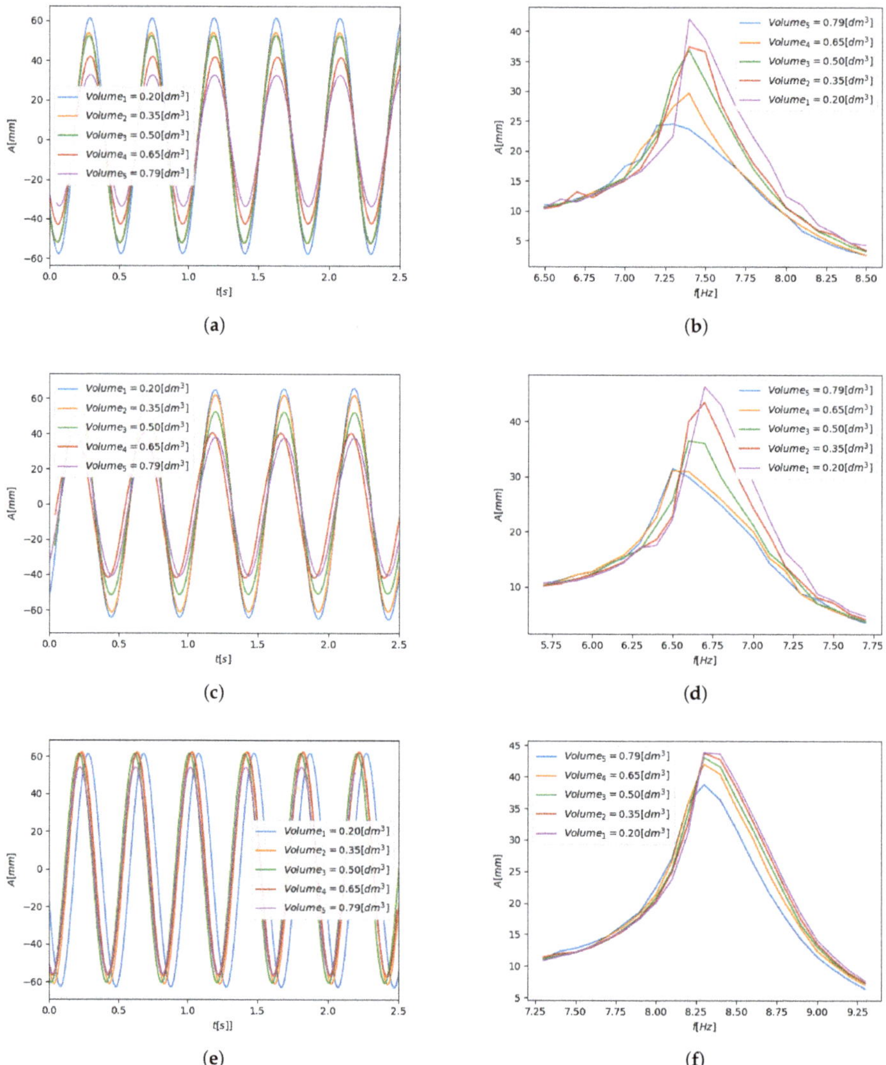

Figure 5. Displacement waveforms (**a,c,e**) and RMS values (**b,d,f**) for various granular grains compositions. (**a**) Displacements of the beam—$M_p = 0.1M_s$; (**b**) RMS of beam displacements—$M_p = 0.1M_s$; (**c**) Displacements of the beam—$M_p = 0.2M_s$; (**d**) RMS of beam displacements—$M_p = 0.2M_s$; (**e**) Displacements of the beam—$M_p = 0.05M_s$; (**f**) RMS of beam displacements—$M_p = 0.05M_s$.

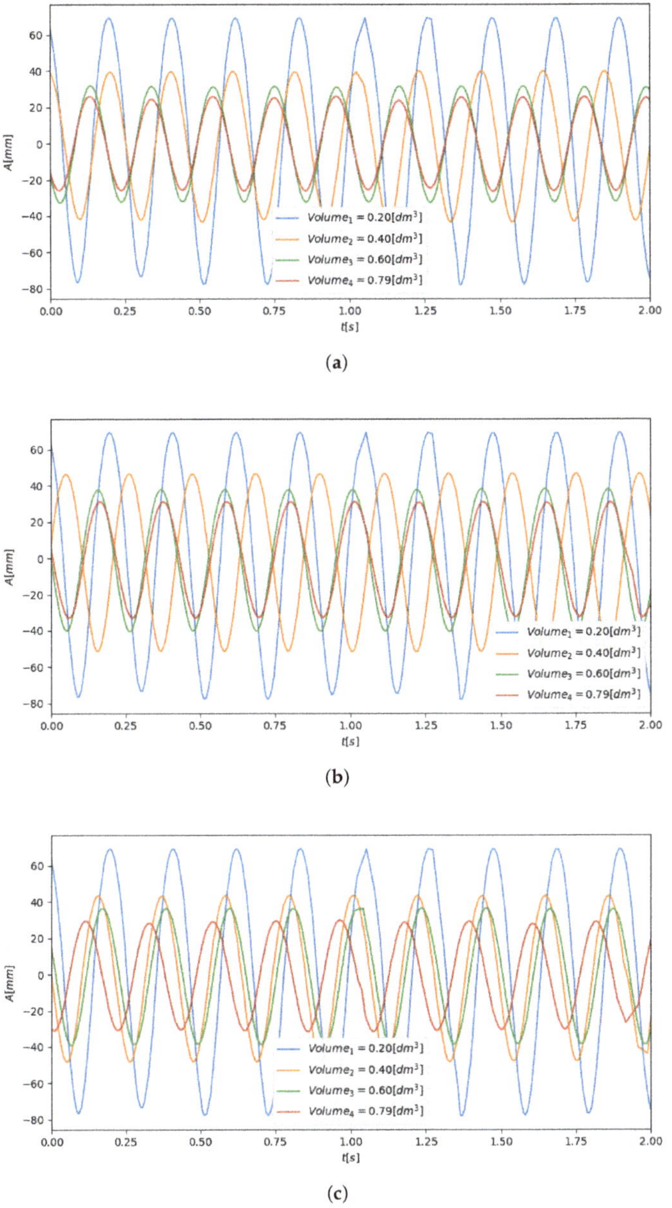

Figure 6. Displacement waveforms for various composition of the grains. (**a**) Beam displacements—steel grains; (**b**) Beam displacements—plastic grains; (**c**) Beam displacements—mixed grains.

4.2. The Influence of the Grains Material

In this section, the authors focus on the influence of the grains material on the response of investigated system. In Figure 6a–c various displacement waveforms in the resonant range of the beam vibrations are depicted. At this stage four different volumes of the TPID compartment and three compositions of grains (steel, plastic and particles mixture) were taken into account. In the experiments the constant mass of the grains equal 10% of the dampened system was assumed. Such an approach resulted in slightly different volume of the loose material applied in TPID and in consequences the changeable number of particles involved in the collisions.

Analyzing the data presented in Figure 6a it can be stated, that for the steel particles the energy dissipation process is effective (more than 50% efficiency) even for small volumes of the damper compartments. The main disadvantage of this type of material is significantly reduced controlling process of the system dynamics. Changing the volume from the V_{min} to V_{max} causes a reduction of the displacement amplitudes by approximately 20%. A different phenomenon was observed for plastic grains (Figure 6b). In this case even the slightly changed volume of the damper compartment resulted in changing the damping properties of the system. Increasing the volume value from $V_{min} = 0.2$ [dm^3] to $V = 0.35$ [dm^3] causes decreasing of the displacement amplitude by app. 30%. Additionally, inflating the balloon to $V = 0.50$ [dm^3] is connected to further decreasing of the amplitudes up to 50%. Applying the V_{max} volume of the compartment reduces the value of recorded amplitudes by up to 70%. Plastic grains seem to be a much more interesting material for the purpose of controlling the dynamics of the investigated beam. Pumping the balloon, resulting in a gradual change in the volume of the operating space of the damper allows for a smooth change of its damping properties and allows for the application of this type of material in an effective strategy of semi-active vibration attenuation. For the steel-plastic composition of the grains (Figure 6c) experimental results are quite similar to the previously mentioned case illustrated in Figure 6b. Recorded amplitudes for the $V = 0.35$ [dm^3] volume are slightly lower for grains mixture than for the pure plastic particles. In the contrary the amplitudes observed for $V = 0.50$ [dm^3] are to some extent higher for the granular material combination. Although the dissipative properties of the TPID device in the case of fully inflated balloon (V_{max} of the compartment) seem to be independent on the type of used granular material (Figure 7), the plastic particles appear to be the optimal solution from the controlling point of view.

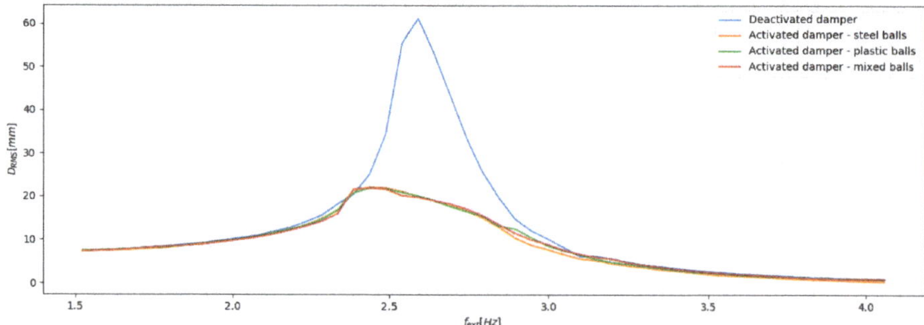

Figure 7. RMS of displacements for various grains materials.

4.3. Spectrum Analysis

In the literature, it is hard to find the spectral analysis for the vibrated objects dampened by the PIDs. Such a situation results from the fact, that generally PIDs are passively calibrated to attenuate vibrations in the resonant range. In this paper the authors present an innovative Tuned Particles Impact Damping strategy, therefore the spectrum analysis seems to provide important information about the dynamics of the considered system. In Figure 8 the typical spectral analysis obtained for the constant mass of particles (10% of the whole system) and various granular materials is depicted. The maximal volume of the TPID compartment corresponding to the best damping properties of the device was taken into consideration. As the reference result the spectrum for the switched "off" state (minimal volume of the device) is presented.

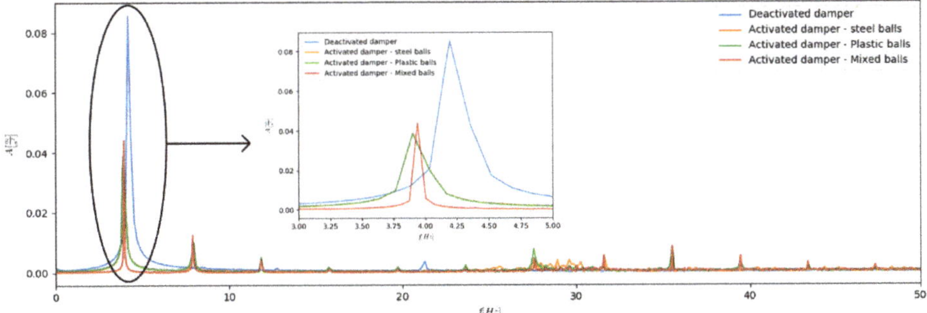

Figure 8. Exemplary spectrum analysis.

Detailed analysis of the data depicted in Figure 8 revealed the significant decreasing of the acceleration amplitudes of the first component. Additionally in the zoomed region the shifting phenomenon of natural frequencies is observable. Similar phenomenon is observed for Tuned Mass Damper class of devices. Moreover, the complex character of intergranular collisions and multiple contacts between the grains and walls of the TPID casing results in insignificant increases of subsequent component amplitudes. This observation seems to be the only disadvantage of the proposed damper prototype. Nevertheless, it is worth emphasizing that the proposed solution is mainly intended for systems operating close to the first resonance frequency range. In Figure 8 the authors presented the whole spectrum including the higher frequencies. The basic intention of the authors was to focus on the first harmonic as the TPID devices are developed to work effectively in devices that will be subjected to such frequencies. The presented spectrum shows that the change of the material and the filling ratio also affect the higher harmonics but it is less important from the engineering point of view.

5. Modelling

At the initial stage of TPID investigation the 2-DOF linear model (Figure 9), described by equations of motion (2) has been selected to capture the real behavior of the considered system. The authors want to emphasize the dynamics of the tested object in the resonant range for the TPID being in switched "on" and "off" states.

$$M\ddot{X} + C_1\dot{X} + K_1 X + C_2\dot{u} + K_2 u = 0. \tag{2}$$

$$M = \begin{pmatrix} M & 0 \\ 0 & M_p \end{pmatrix} \qquad K_1 = \begin{pmatrix} K + K_p & -K_p \\ -K_p & K_p \end{pmatrix} \qquad C_1 = \begin{pmatrix} C + C_p & -C_p \\ -C_p & C_p \end{pmatrix}$$

$$C_2 = \begin{pmatrix} -C & 0 \\ 0 & 0 \end{pmatrix} \qquad K_2 = \begin{pmatrix} -K & 0 \\ 0 & 0 \end{pmatrix},$$

where M—reduced mass of the system (mass of the container included), M_p—mass of particles, K—beam stiffness, C—beam damping, K_p—TPID stiffness, C_p—controllable TPID damping.

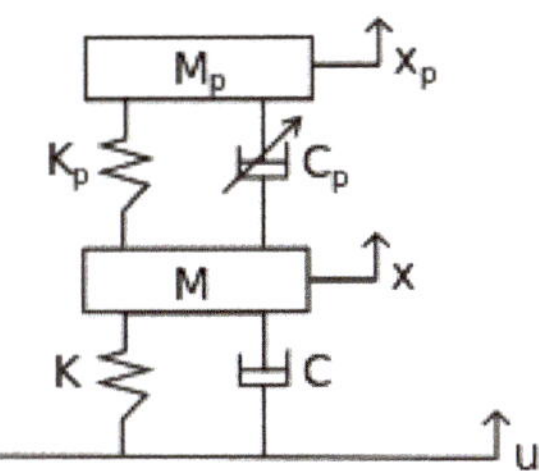

Figure 9. Mathematical model.

K, C and M parameters were identified for free vibrations of the investigated cantilever beam in separate experimental tests. It was assumed that the value of K_p parameter is constant during laboratory tests and results from the attachment of the container to the beam. It was calibrated basing on the experiments conducted for the fully compacted particles that is, for the lowest volume value (V_{min}) of the TPID container. The C_p parameter values depend on the type of inelastic collisions between single particles and the character of their mutual contact with the container. In the paper it is assumed that variations of this parameter are the function of the TPID operation volume and can be described as:

$$C_p = C_{TPID} \cdot A(t) + C_{p_0}, \tag{3}$$

where C_{TPID}-damping parameter for the fully opened TPID device (volume of the container is V_{max}), $A(t)$ activation function (4), C_{p_0} -initial damping in the system (considered as the "intergranular" friction forces acting in the switched "off" state).

The C_{TPID} parameter was determined for the test conducted in a fully turned "on" state. It involves the maximal number of collisions between loose material and container. On the contrary, the C_{p_0} parameter value was captured for the minimal value of the container volume. Although the mathematical model (Figure 2) seems to be simple, its complexity lies in the previously mentioned complex phenomenon related to not fully understood interactions between internal particles. The confirmation of this fact is the data depicted in Figure 10 where the logarithmic damping decrement for the investigated beam free vibrations is illustrated. In Figure 10a,b the damping characteristics for various particles masses and different container volumes are presented respectively. The characteristics have been obtained on the basis of supplementary experiments. They consisted of measurements of free vibrations of the considered beam initially deflected at the free right end by 30 mm. The release of the deflected beam was achieved by burning out a thin thread attached to the tip of the element.

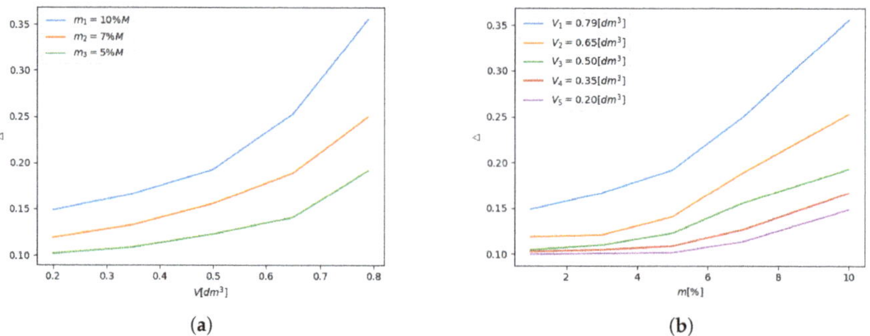

Figure 10. Logarithmic damping decrement. (**a**) Variable particles mass; (**b**) Variable container volume.

Basing on data presented in Figure 10 it can be stated that the higher mass of grains results in a more efficient energy dissipation process. Similarly, the higher volume of the TPID compartment is related to the more effective vibrations attenuation. The next problem is encountered during the volume changing of the compartment, that is, inflating and deflating of the internal balloon. These transition processes are not fully recognized yet, therefore the authors proposed exponential functions to reproduce them (Figure 11). The function (4) takes 0 value for the deflated balloon (the grains are compacted and can be treated as a concentrated mass). In the contrary, for the V_{max} volume of the container the activation function takes a value of 1.

$$A(t) = \begin{cases} 0 & \Leftrightarrow 0 < t < t_1 \\ (e^{a \cdot \frac{(t-t_1)^\alpha}{\Delta_t^\alpha}} - 1) \cdot (H(t-t_1) - H(t-t_2)) & \Leftrightarrow t_1 < t < t_2 \\ 1 & \Leftrightarrow t_2 < t < t_3 \\ (e^{b \cdot \frac{(t_3-t)^\alpha}{\Delta_t^\alpha}} - 1) \cdot (H(t-t_3) - H(t-t_4)) & \Leftrightarrow t_3 < t < t_4 \\ 0 & \Leftrightarrow t_4 < t < \infty \\ , \end{cases}$$

$$(4)$$

where α—arbitrary coefficient ($\alpha = 3$ was assumed during numerical simulations), $\Delta_t = t_2 - t_1 = t_4 - t_3$—the activation and deactivation time (at this stage of investigations it was assumed that inflating and deflating times of the balloon were the same) and coefficients $a = b = 0.69$ because simplifying the formula (4) for cases $t_1 < t < t_2$ or $t_3 < t < t_4$, $e^a - 1 = 1$.

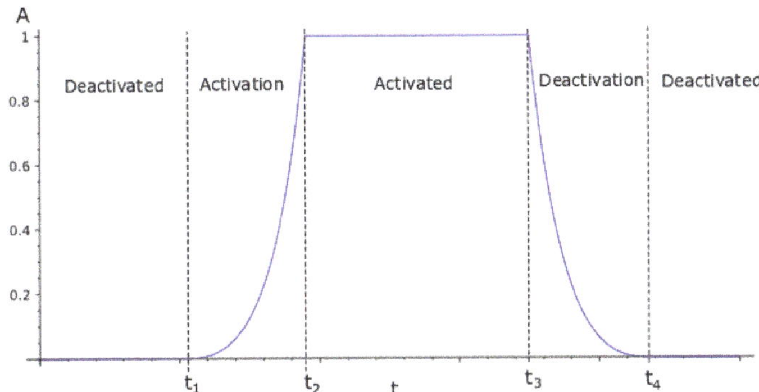

Figure 11. Activation function $A(t)$.

In the experiments, two various forms of activation functions were taken into consideration (Figure 12a,b). They corresponded to changes in the compartment volumes from V_{min} to V_{max} in $\Delta_{t_1} = 4.5$ s and $\Delta_{t_2} = 1.5$ s respectively. The investigated beam tip waveforms for the resonant range of excitations are depicted in Figure 13. For all described cases the plastic particles were involved. Two various numbers of grains corresponding to 10% and 5% of the total mass of the beam were taken into account. Additionally, different activation times, according to the data depicted in Figure 12, were applied in the experiments. Moreover, in Figure 13 the numerical simulation results are verified with the real response of the system.

The data depicted in Figure 13 reveal interesting information. It is evident that the activation of the TPID device significantly reduced the resonant vibration amplitudes. The efficiency of granular damper is strongly dependent on the number (mass) of particles being involved in the damping process. On the basis of Figure 12 it can be observed that the larger amount of granular media increased the dissipative properties of the system. Changing the mass of the grains from 5% of the total mass into 10% increased the damping properties of considered device by 35%. This phenomenon is apparent for both damper activation times. The detailed analysis of data depicted in Figure 13 also revealed some damping inertia of the system. For the case of larger amount of particles and applied activation time $\Delta_{t_1} = 4.5$ s (resulted from inflating the balloon to the maximal volume) the $\Delta_{t_{11}}$ time necessary for the recorded waveforms stabilization was 5.4 s (Figure 13d). Similarly to stabilize the vibrations of the beam dampened by lower number of particles (5% of the total mass) the same activation function $\Delta_{t_{12}} = 6.4$ s was required. Such observations were also valid for the shorter activation time $\Delta_{t_2} = 1.5$ s. The damping inertia for this case was $\Delta_{t_{21}} = 1.9$ s and $\Delta_{t_{22}} = 3.5$ s for larger and smaller amount of particles respectively. It is worth emphasizing that the transition from the TPID deactivated into fully activated state was relatively smooth. There were no uncontrolled increases in vibration amplitudes during the balloon pumping process. Figure 14, which depicts the envelopes of registered waveforms for various parameters of the investigated TPID prototype, can be a transparent summary of the experimental work carried out in the paper.

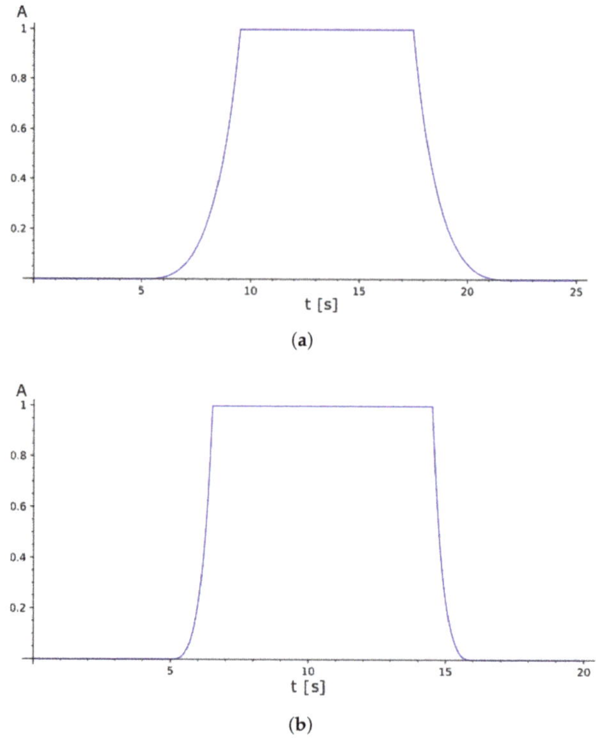

Figure 12. A proposition of activation functions. (a) $\Delta_{t_1} = 4.5$ s; (b) $\Delta_{t_2} = 1.5$ s.

It is worth emphasizing that the proposed mathematical model of the system is quite reliable. The verification between real experimental responses of the vibrated beam and numerical simulations being carried out on the basis of Equations (2)–(4) revealed a satisfactory convergence (Figure 13). The differences between real and theoretical waveforms for the "turned on" and "turned off" states are negligible.

In the next stage of a research the authors examined responses of the kinematically excited cantilever beam for various volumes of the TPID container (excluding turned "on" and "off" states). Three various volumes of the device corresponding to volumes equal to 0.35, 0.50 and 0.65 [dm^3] were taken into account. Plastic grains having the mass equal 10% of the total mass of the system were involved in the tests. The registered waveforms are presented in Figure 15. In Figure 15 also the numerical simulation results, derived from Equations (2)–(4) are presented. Knowing the temporary volume value of the TPID device the corresponding damping coefficient value C_{theo} was calculated Equation (3). Basing on data depicted in Figure 15 it can be stated that the model is capable to capture the real behavior of the dampened system. To calculate the quantitative difference between the theoretical and experimental responses the E_{error} parameter has been introduced (5)

$$E_{error_i} = (1 - \frac{X_{theo_i}}{X_{exp_i}}) \cdot 100\%. \tag{5}$$

The E_{error} parameter value decreases for higher values of the container volume (Figure 15). The list of values of the considered parameter is presented in Table 3.

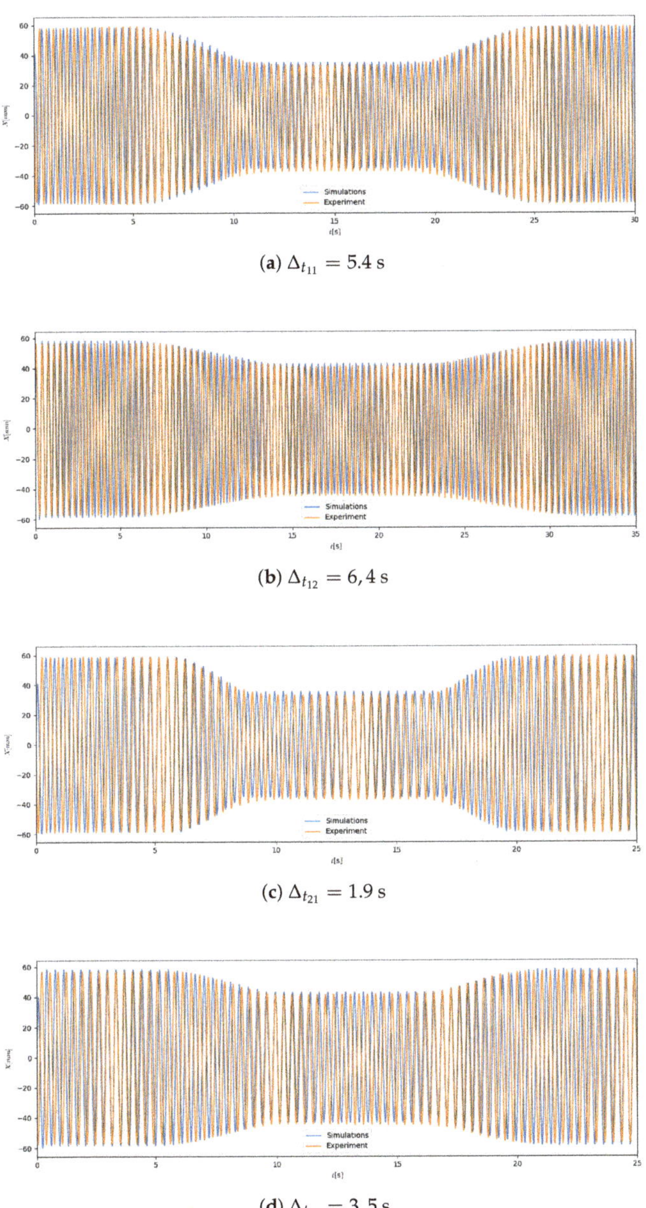

(**a**) $\Delta_{t_{11}} = 5.4$ s

(**b**) $\Delta_{t_{12}} = 6,4$ s

(**c**) $\Delta_{t_{21}} = 1.9$ s

(**d**) $\Delta_{t_{22}} = 3,5$ s

Figure 13. Verification of experimental and numerical data for various operational parameters: (**a**) 10% mass of grains, 4.5 s activation time; (**b**) 5% mass of grains, 4.5 s activation time; (**c**) 10% of grains, 1.5 s activation time; (**d**) 5% mass of grains, 1.5 s activation time.

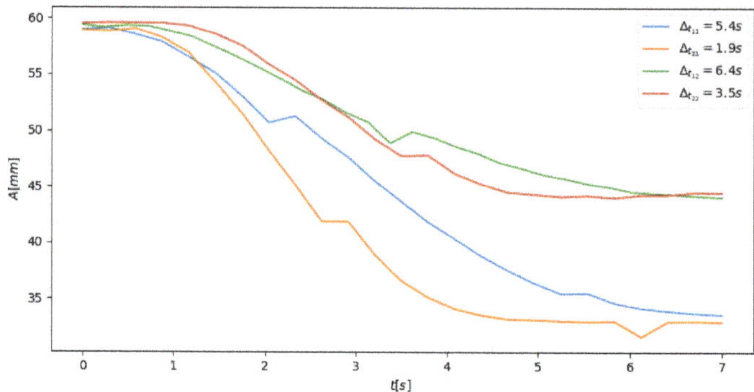

Figure 14. Envelopes of recorded waveforms for various operating parameters.

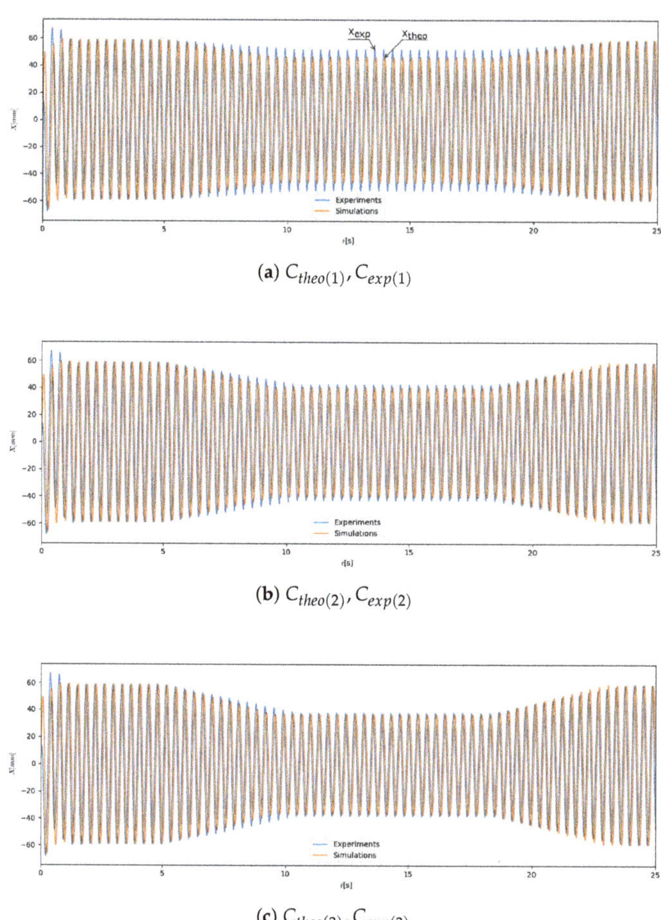

(**a**) $C_{theo(1)}, C_{exp(1)}$

(**b**) $C_{theo(2)}, C_{exp(2)}$

(**c**) $C_{theo(3)}, C_{exp(3)}$

Figure 15. Waveforms for various values of the container volume: (**a**) 0.35 [dm^3]; (**b**) 0.50 [dm^3]; (**c**) 0.65 [dm^3].

Table 3. Error parameter values for various volumes of the TPID compartment.

V	0.35 [dm^3]	0.50 [dm^3]	0.65 [dm^3]
Error	6.5%	2.5%	0.5%

6. Conclusions

The novelty of the proposed TPID device is the application of a special balloon, that has variable volume and is filled with granules. Through inflating and deflating the previously mentioned balloon, the TPID device volume is shifted in a controlled manner, enabling the rapid change of its damping parameters—the volume ratio of particles and container. Damping properties of granules made of plastic, steel and their combinations have been experimentally tested. In addition, tests were carried out for various cases of the container filling ratio and grains mass for the full range of excitation frequencies (before, in resonance and after resonance ranges). Experiments confirmed that the proposed TPID damper structure retains the advantages of standard PID devices ([3,5,7,22]). It was also observed that the type of grains material strongly affects the damping properties of the granular damper. The particles are characterized by various physical parameters, for example, Young's modulus. Differences in values of the particles restitution coefficient cause changes in the grains movement and collisions character. They also result in different values of contact forces. Basing on detailed analyses of experimental data the authors concluded that the plastic particles allow for a wide range of controllability and tuning of the system response.

Another parameter that significantly influences the dissipative properties of investigated TPID prototype is the mass of the particles. Laboratory tests revealed that increasing the mass of granular material being involved in inelastic collisions inside the PID container improved the damping features of the prototype. Obviously some specific threshold value for the number of particles being involved in the damping process exist. Exceeding this value will cause the limitations of free space for intergranular impacts and consequently will lead to decreasing the dissipative properties of the system. This complex phenomenon, dependent not only on the filling ratio of the TPID, but also on the shape of grains and container and materials involved in contacts, was not discussed in the paper. Assuming too low a mass of granular materials (insufficient number of particles) will not provide the desired vibrations attenuation effect.

The authors assumed that the Young's modulus is the main parameter of granular material that affects the damping properties. Defining in an empirical way the energy dissipation components is a challenging task. At the current stage of investigations the authors did not conducted such tests. The shape and changeable in time volume of the container make the described problem more complicated. The TPID container was developed by the 3D printing technique. The applied filament material is much softer than the grains material. The problem of investigating the container's material influence on the global damping properties of the system (TPID effectiveness) is important and requires carrying out additional and time-consuming studies. Also crucial from the operational point of view effects of grains degradation or temperature changes were not discussed in the paper.

The biggest advantage of the proposed TPID prototype is the possibility of adjusting its damping characteristics according to various types of temporary excitations. Discussed proposition by simply adding the expandable balloon to the classical PID device enabled the user to fully control the volume of the grains casing (by the external pump) and in fact transformed the passive PID device into the innovative semi-active TPID damper.

In the modeling section the simple 2DOF linear mathematical model was adopted. To capture nonlinearities of the real system resulted from the inelastic character of collisions inside the granular media the exponential activation function was proposed. The function confirmed its reliability not only for the "on" and "off" states of the TPID but also allowed us to achieve satisfactory results for various transient states of the device being inflated. The biggest differences between the model and the real object did not exceed 6.5%.

In the authors' opinion, the proposed TPID construction opens new perspectives for the innovative semi-active, inexpensive and environmental friendly damping of vibrations strategies.

Author Contributions: Conceptualization, M.Ż. and R.Z.; methodology, M.Ż. and R.Z.; software, M.Ż.; validation, M.Ż. and R.Z.; formal analysis, M.Ż.; investigation, M.Ż.; resources, R.Z.; data curation, M.Ż.; writing–original draft preparation, M.Ż.; writing–review and editing, R.Z.; visualization, M.Ż.; supervision, R.Z.; project administration, M.Ż. and R.Z.; funding acquisition, M.Ż. and R.Z. All authors have read and agreed to the published version of the manuscript.

Funding: This research received no external funding. The APC was funded by GRANT WEWNĘTRZNY DLA PRACOWNIKÓW POLITECHNIKI WARSZAWSKIEJ WSPIERAJĄCY PROWADZENIE DZIAŁALNOŚCI NAUKOWEJ W DYSCYPLINIE INŻYNIERIA MECHANICZNA, grant number 504/04555/1150/43.090048.

Conflicts of Interest: The authors declare no conflict of interest.

References

1. Fu, B.; Jiang, H.; Wu, T. Comparative studies of vibration control effects between structures with particle dampers and tuned liquid dampers using substructure shake table testing methods. *Soil Dyn. Earthq. Eng.* **2019**, *121*, 421–435, doi:10.1016/j.soildyn.2019.03.031. [CrossRef]

2. Gnanasambandham, C.; Stender, M.; Hoffmann, N.; Eberhard, P. Multi-scale dynamics of particle dampers using wavelets: Extracting particle activity metrics from ring down experiments. *J. Sound Vib.* **2019**, *454*, 1–13, doi:10.1016/j.jsv.2019.04.009. [CrossRef]

3. Marhadi, K.S. Particle impact damping: Effect of mass ratio, material, and shape. *J. Sound Vib.* **2005**, *283*, 433–448, doi:10.1016/j.jsv.2004.04.013. [CrossRef]

4. Wang, Y.; Liu, B.; Tian, A.; Wei, D.; Jiang, X. Prediction methods for the damping effect of multi-unit particle dampers based on the cyclic iterations of a single-unit particle damper. *J. Sound Vib.* **2019**, *443*, 341–361, doi:10.1016/j.jsv.2018.10.035. [CrossRef]

5. Snoun, C.; Trigui, M. Design parameters optimization of a particles impact damper. *Int. J. Interact. Des. Manuf.* **2018**, *12*, 1283–1297, doi:10.1007/s12008-018-0463-y. [CrossRef]

6. Zhang, R.; Zhang, Y.; Zheng, Z.; Mo, L.; Wu, C. Parametrical optimization of particle dampers based on particle swarm algorithm. *Appl. Acoust.* **2020**, *160*, 107083, doi:10.1016/j.apacoust.2019.107083. [CrossRef]

7. Zhang, K.; Chen, T.; Wang, X.; Fang, J. Rheology behavior and optimal damping effect of granular particles in a non-obstructive particle damper. *J. Sound Vib.* **2016**, *364*, 30–43, doi:10.1016/j.jsv.2015.11.006. [CrossRef]

8. Zhang, C.; Zhao, Z.; Chen, T.; Liu, H.; Zhang, K. Discrete element method model of electromagnetic particle damper with a ferromagnetic end cover. *J. Sound Vib.* **2019**, *446*, 211–224, doi:10.1016/j.jsv.2019.01.034. [CrossRef]

9. Junling, C.; Christos, G. Tuned rolling-ball dampers for vibration control in wind turbines. *J. Sound Vib.* **2013**, *332*, 5271–5282, doi:10.1016/j.jsv.2013.05.019. [CrossRef]

10. Xiao, W.; Lu, D.; Song, L.; Guo, H.; Yang, Z. Influence of particle damping on ride comfort of mining dump truck. *Mech. Syst. Signal Process.* **2020**, *136*, 106509, doi:10.1016/j.ymssp.2019.106509. [CrossRef]

11. Zhang, Z.; Staino, A.; Basu, B.; Nielsen, S.R. Performance evaluation of full-scale tuned liquid dampers (TLDs) for vibration control of large wind turbines using real-time hybrid testing. *Eng. Struct.* **2016**, *126*, 417–431, doi:10.1016/j.engstruct.2016.07.008. [CrossRef]

12. Du, Y.; Wang, S. Modeling the fine particle impact damper. *Int. J. Mech. Sci.* **2010**, *52*, 1015–1022, doi:10.1016/j.ijmecsci.2010.04.004. [CrossRef]

13. Yüzbasioglu, T.; Aramendiz, J.; Fidlin, A. On the numerical simulations of amplitude-adaptive impact dampers. *J. Sound Vib.* **2020**, *468*, 115023, doi:10.1016/j.jsv.2019.115023. [CrossRef]

14. Vinayaravi, R.; Kumaresan, D.; Jayaraj, K.; Asraff, A.; Muthukumar, R. Experimental investigation and theoretical modelling of an impact damper. *J. Sound Vib.* **2013**, *332*, 1324–1334, doi:10.1016/j.jsv.2012.10.032. [CrossRef]

15. Egger, P.; Caracoglia, L. Analytical and experimental investigation on a multiple-mass-element pendulum impact damper for vibration mitigation. *J. Sound Vib.* **2015**, *353*, 38–57, doi:10.1016/j.jsv.2015.05.003. [CrossRef]

16. Afsharfard, A. Application of nonlinear magnetic vibro-impact vibration suppressor and energy harvester. *Mech. Syst. Signal Process.* **2018**, *98*, 371–381, doi:10.1016/j.ymssp.2017.05.010. [CrossRef]

17. Veeramuthuvel, P.; Shankar, K.; Sairajan, K. Application of particle damper on electronic packages for spacecraft. *Acta Astronaut.* **2016**, *127*, 260–270, doi:10.1016/j.actaastro.2016.06.003. [CrossRef]

18. Sonmez, E.; Nagarajaiah, S.; Sun, C.; Basu, B. A study on semi-active Tuned Liquid Column Dampers (sTLCDs) for structural response reduction under random excitations. *J. Sound Vib.* **2016**, *362*, 1–15, doi:10.1016/j.jsv.2015.09.020. [CrossRef]

19. Sánchez, M.; Manuel Carlevaro, C. Nonlinear dynamic analysis of an optimal particle damper. *J. Sound Vib.* **2013**, *332*, 2070–2080, doi:10.1016/j.jsv.2012.09.042. [CrossRef]

20. Sánchez, M.; Pugnaloni, L.A. Effective mass overshoot in single degree of freedom mechanical systems with a particle damper. *J. Sound Vib.* **2011**, *330*, 5812–5819, doi:10.1016/j.jsv.2011.07.016. [CrossRef]

21. Lei, X.; Wu, C. Optimizing parameter of particle damping based on (Leidenfrost) effect of particle flows. *Mech. Syst. Signal Process.* **2018**, *104*, 60–71, doi:10.1016/j.ymssp.2017.10.037. [CrossRef]

22. Djemal, F.; Chaari, R.; Gafsi, W.; Chaari, F.; Haddar, M. Passive vibration suppression using ball impact damper absorber. *Appl. Acoust.* **2019**, *147*, 72–76, doi:10.1016/j.apacoust.2017.09.011. [CrossRef]

23. Yao, B.; Chen, Q.; Xiang, H.; Gao, X. Experimental and theoretical investigation on dynamic properties of tuned particle damper. *Int. J. Mech. Sci.* **2014**, *80*, 122–130, doi:10.1016/j.ijmecsci.2014.01.009. [CrossRef]

24. Zhang, R.; Wu, C.; Zhang, Y. A novel technique to predict harmonic response of Particle-damping structure based on ANSYS secondary development technology. *Int. J. Mech. Sci.* **2018**, *144*, 877–886, doi:10.1016/j.ijmecsci.2017.10.035. [CrossRef]

25. Yin, Z.; Su, F.; Zhang, H. Investigation of the energy dissipation of different rheology behaviors in a non-obstructive particle damper. *Powder Technol.* **2017**, *321*, 270–275, doi:10.1016/j.powtec.2017.07.090. [CrossRef]

26. Veeramuthuvel, P.; Shankar, K.; Sairajan, K.; Machavaram, R. Prediction of Particle Damping Parameters Using RBF Neural Network. *Procedia Mater. Sci.* **2014**, *5*, 335–344, doi:10.1016/j.mspro.2014.07.275. [CrossRef]

27. Hu, L.; Shi, Y.; Yang, Q.; Song, G. Sound reduction at a target point inside an enclosed cavity using particle dampers. *J. Sound Vib.* **2016**, *384*, 45–55, doi:10.1016/j.jsv.2016.08.016. [CrossRef]

28. Xiao, W.; Li, J.; Wang, S.; Fang, X. Study on vibration suppression based on particle damping in centrifugal field of gear transmission. *J. Sound Vib.* **2016**, *366*, 62–80, doi:10.1016/j.jsv.2015.12.014. [CrossRef]

29. Gharib, M.; Ghani, S. Free vibration analysis of linear particle chain impact damper. *J. Sound Vib.* **2013**, *332*, 6254–6264, doi:10.1016/j.jsv.2013.07.013. [CrossRef]

30. Gharib, M.; Karkoub, M.; Ghamary, M. Numerical investigation of Linear Particle Chain impact dampers with friction. *Case Stud. Mech. Syst. Signal Process.* **2016**, *3*, 34–40, doi:10.1016/j.csmssp.2016.03.001. [CrossRef]

31. Farid, M.; Gendelman, O. Response regimes in equivalent mechanical model of strongly nonlinear liquid sloshing. *Int. J.-Non Mech.* **2017**, *94*, 146–159, doi:10.1016/j.ijnonlinmec.2017.04.006. [CrossRef]

32. Gendelman, O. Analytic treatment of a system with a vibro-impact nonlinear energy sink. *J. Sound Vib.* **2012**, *331*, 4599–4608, doi:10.1016/j.jsv.2012.05.021. [CrossRef]

33. Gendelman, O.; Alloni, A. Forced System with Vibro-impact Energy Sink: Chaotic Strongly Modulated Responses. *Procedia IUTAM* **2016**, *19*, 53–64, doi:10.1016/j.piutam.2016.03.009. [CrossRef]

34. Fitzgerald, B.; Sarkar, S.; Staino, A. Improved reliability of wind turbine towers with active tuned mass dampers (ATMDs). *J. Sound Vib.* **2018**, *419*, 103–122, doi:10.1016/j.jsv.2017.12.026. [CrossRef]

35. Hussan, M.; Rahman, M.S. Multiple tuned mass damper for multi-mode vibration reduction of offshore wind turbine under seismic excitation. *Ocean. Eng.* **2018**, *160*, 449–460, doi:10.1016/j.oceaneng.2018.04.041. [CrossRef]

36. Jin, J.; Yang, W.; Koh, H.I.; Park, J. Development of tuned particle impact damper for reduction of transient railway vibrations. *Appl. Acoust.* **2020**, *169*, 107487, doi:10.1016/j.apacoust.2020.107487. [CrossRef]

37. Lu, Z.; Wang, Z.; Masri, S.F.; Lu, X. Particle impact dampers: Past, present, and future. *Struct. Control. Health Monit.* **2018**, *25*, e2058, doi:10.1002/stc.2058. [CrossRef]

38. Zhang, K.; Chen, T.; Wang, X.; Fang, J. A model of Tuned Particle Damper. In Proceedings of the 22nd International Congress on Sound and Vibration, Florence, Italy, 12–16 July 2015.

39. Zhang, K.; Xi, Y.; Chen, T.; Ma, Z. Experimental studies of tuned particle damper: Design and characterization. *Mech. Syst. Signal Process.* **2018**, *99*, 219–228, doi:10.1016/j.ymssp.2017.06.007. [CrossRef]

40. Zurawski, M.; Zalewski, R.; Chilinski, B. Concept of an Adaptive-Tuned Particles Impact Damper. *J. Theor. Appl. Mech.* **2020**, *58*, 811–816, doi:10.15632/jtam-pl/122431. [CrossRef]
41. Dehghan-Niri, E.; Zahrai, S.M.; Rod, A.F. Numerical studies of the conventional impact damper with discrete frequency optimization and uncertainty considerations. *Sci. Iran.* **2012**, *19*, 166–178, doi:10.1016/j.scient.2012.01.001. [CrossRef]

 *applied sciences*

Article

Health and Structural Integrity of Monitoring Systems: The Case Study of Pressurized Pipelines

Vladimír Chmelko *, Martin Garan, Miroslav Šulko and Marek Gašparík

Faculty of Mechanical Engineering, Institute of Applied Mechanics and Mechatronics, Slovak University of Technology, Námestie Slobody 17, 81231 Bratislava, Slovakia; martin.garan@stuba.sk (M.G.); miroslav.sulko@stuba.sk (M.Š.); marek.gasparik@stuba.sk (M.G.)
* Correspondence: vladimir.chmelko@stuba.sk; Tel.: +421-2-57296225

Received: 28 July 2020; Accepted: 28 August 2020; Published: 31 August 2020

Abstract: In the operation of some structures, particularly in energy or chemical industry where pressurized pipeline systems are employed, certain unexpected critical situations may occur, which must be definitely avoided. Otherwise, such situations would result in undesirable damage to the environment or even the endangerment of human life. For example, the occurrence of such nonstandard states can significantly affect the safety of high-pressure pipeline systems. The following paper discusses basic physical prerequisites for assembling the systems that can sense loading states and monitor the operational safety conditions of pressure piping systems in the long-run. The appropriate monitoring system hardware with cost-effective data management was designed in order to enable the real-time monitoring of operational safety parameters. Furthermore, the paper presents the results obtained from the measurements of existing real-time safety monitoring systems for selected pipeline systems.

Keywords: monitoring system; pipeline; health and structural integrity

---

1. Introduction

Obviously, the operational loading for which a real structure is designed can differ from the real loading in many cases. Based on statistical records, the majority of damages and collapses of structures in operation are caused merely by a change in operating conditions, except for those caused by a human factor [1,2]. In the case of pressurized pipelines (e.g., gas transit systems, oil pipelines, or other liquid pipeline systems under pressure), the piping systems may be subjected to a quasi-static loading or of a predominantly variable loading due to a varying pressure of the operating medium. An expected design loading may change only in some locations, which are really difficult to predict in the design stage. In the case of compressor stations, for instance, or the geometrical-fluid ratios in the closed side branches, the setting of the compressor's operating point can cause dynamic phenomena, i.e., pipeline vibrations, as pointed out in [3–9].

Other changes occurring in the operation of piping systems may occur in the subsoil of the pipeline. For example, the subsoil slide on a slope or subsoil slip after a prolonged rainfall can change the design of the pipeline stress. Most accidents of the pipeline systems may arise as a result of the following simultaneous interactions ("system synergy") of unexpected changes during the operations, e.g.,

- internal pressure with induced vibrations,
- additional bending stress from the subsoil drop loaded with an internal pressure, and
- the pipeline wall thickness decreases caused by corrosion at the point of additional bending stress with an internal pressure, etc.

The occurrence of such nonstandard and unplanned situations in a real operation is generally difficult to predict. The only possible solution for how to record these states is to develop and implement monitoring systems that can monitor not only the operating parameters, such as the pressure and temperature of transported media, but, also, the construction parameters, enabling to determine the reliability and operational safety [10]. Based on ten years of our experience with the development of several safety monitoring systems of pipeline operations in the field of gas transportation, two systems have turned out to be beneficial: the monitoring of dangerous vibrations and the monitoring of pipes with corrosion defects and additional bending.

Hence, the objective of this paper is to further analyze three important problems that arise out of the previously stated; that is, to determine:

- which quantities should be measured continuously by sensors,
- what are their permissible limit values for a safe operation of the pipeline systems, and
- how to predict the corrosion process and its synergy with additional bending loading.

2. Methodology and Results

2.1. Vibrations Induced in the Side Branches

The dangerous self-excited gas pressure oscillations may occur at the T-branch of the gas pipe system due to the following cases:

- the gas flow rate is sufficiently high,
- there are necessary fluid and geometric conditions for the generation of excited pressure oscillations, and
- there was a resonant match between the excitation frequency and the natural frequency of the closed tap volume (existence of Helmholtz standing waves in the branch—the so-called "whistle effect").

Pressure oscillations induced by the standing wave can vibrate the piping system around the T-branch. This effect is even more pronounced with several closed side branches in a row [3]. Oscillation of this type also occurs in bypasses, where the sharp radius of the arc creates conditions close to the blind side branches (see Figure 1). These operating states can be partially predicted in the design phase of the piping system.

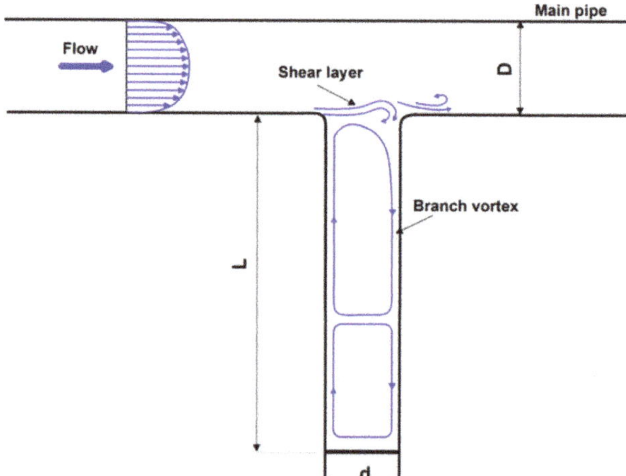

Figure 1. Principle of self-excited oscillations in closed side branches. D is the diameter of the main pipe, and L (d) is a length (diameter) of a side branch.

The critical gas velocity v_n for self-excitation can be calculated according to the relationship

$$v_n = \frac{f_{ne}d}{S_t} \tag{1}$$

where S_t is the Strouhal number for the geometric and fluid ratios in the T-branch, and f_{ne} is the oscillation frequency given as

$$f_{ne} = \frac{(0.88 \div 1)nc}{4L} \tag{2}$$

where $n = 1, 3, 5,..., c$ is the sound velocity in an environment [7].

Due to the unclear geometry of the rounding edge of the T-branch or the change in fluid conditions compared to the design state, it is not always possible to eliminate such oscillations.

A suitable device for ensuring operational safety should be installed in a monitoring system that can be based on sensing vibration parameters (accelerometers) or strain sensors. Accelerometers enable detecting quantities such as the velocity of oscillation or acceleration and frequency values at which the pipe oscillates. Allowable values for the velocity of the oscillation are stated by guidelines and industry regulations. As far as the effects of oscillation on the safe operations are concerned, it is essential to know the amplitude of the strain (distortions), which could cause the fatigue of a material. However, this type of information cannot be provided by accelerometers. The need to know the maximum displacement of the oscillating part relative to the base (branch oscillation displacement relative to the main pipe or fixed ground) would require the measuring of accelerations and calculating of displacements in multiple directions and in multiple parts of the pipeline node, as documented in Figure 2.

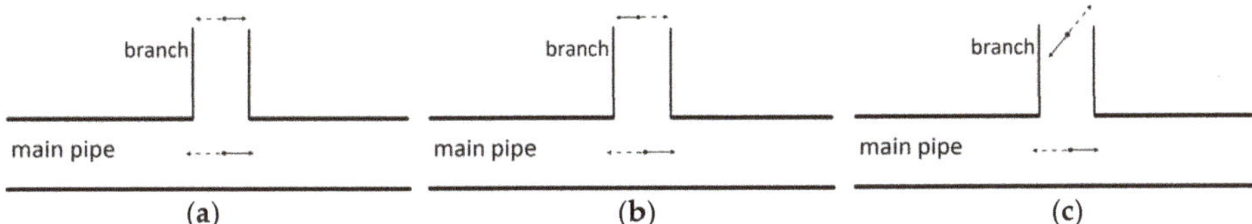

Figure 2. Vibration of the pipe and branch: (**a**) in-phase in the same direction (without the relative deflection) (**b**) out p-hase in the same direction (inducing deflection of branch), and (**c**) in a different direction (inducing deflection of the branch).

In general, the pipelines oscillate in a wide spectrum of frequencies. The integrity of pipelines is put at risk only at frequencies that cause significant magnitude of the strain amplitudes. Having compared the frequency composition of the oscillation shown in Figure 3, measured by accelerometers and bending stress detected by strain gauges on the identical section of the pipeline, it is obvious that significant deformation amplitudes occur only at frequencies up to 3 Hz. The frequency spectrum measured by accelerometer sensors (Figure 3 on the left side) does not reveal the significant magnitudes of strain that can decrease the operational lifetime of a pipe.

More reliable results can be obtained by the direct measuring of strains at the critical cross-section. The suitable arrangement of the sensors allows to calculate directly the maximum value of the stress in the most loaded cross-section of the vibrating pipeline part.

Figure 4 shows the deployment of a minimum number of strain sensors along the perimeter of the pipeline's cross-section, allowing the calculations of bending, torsion, and normal forces. The methodology of measurements and calculations is elaborated in more detail in the papers [11,12].

Due to the high mean stresses due to internal pressure, it is not possible to easily compare the measured amplitudes of the strain from vibrations with their safe limit against fatigue crack. It is necessary to separate the individual oscillation cycles and take into account the mean value of the cycle [13,14]. The real-time monitoring of dangerous vibrations can be accomplished by processing the measured time process of strains in time batches, e.g., in 30 or 60 s.

For reliable information about the operational capability of the oscillating pipeline, it is, however, necessary to solve two more problems:

- to determine cyclic properties of the critical volume of the material (often the weld joint) and
- the condition for assessing the criticality of the vibrations.

The critical volume of the weld joint material in terms of oscillation is the heat affected area or the area with defects in the weld metal. In view of the validity of the cyclic properties of the material, it is appropriate to perform cyclic tests of the welded samples, as shown in Figure 5. The cyclic tests of specimens taken along the pipeline circumference in a place of the weld joint include all the effects of the microstructure of all individual weld joint zones and its geometry, as well as possible defects.

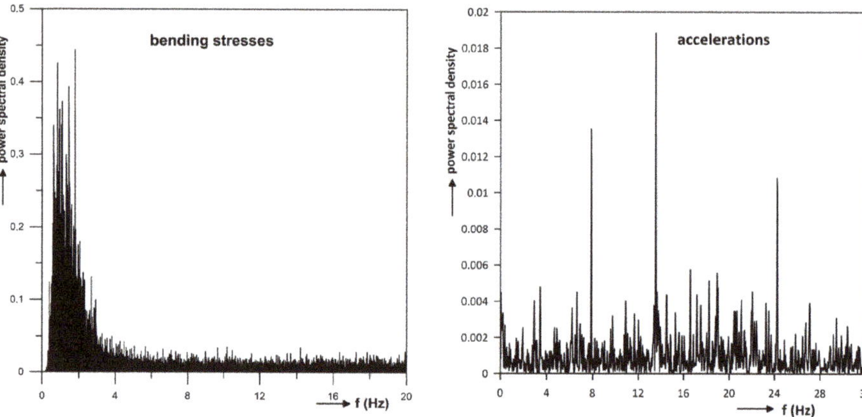

Figure 3. Frequency composition of signals measured by accelerometers and strain gauges on the vibrating pipe branch.

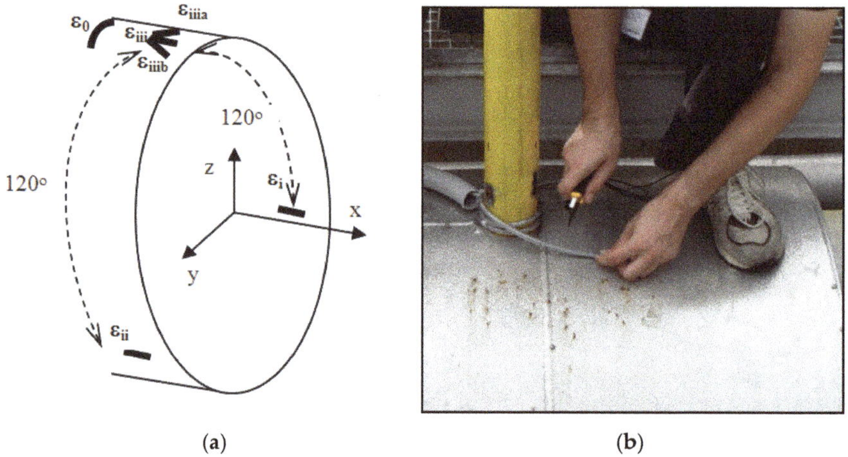

(a)   (b)

Figure 4. (**a**) The arrangement of strain sensors along the perimeter of the pipeline, and (**b**) the application of the strain sensors (gauges) on the real pipe.

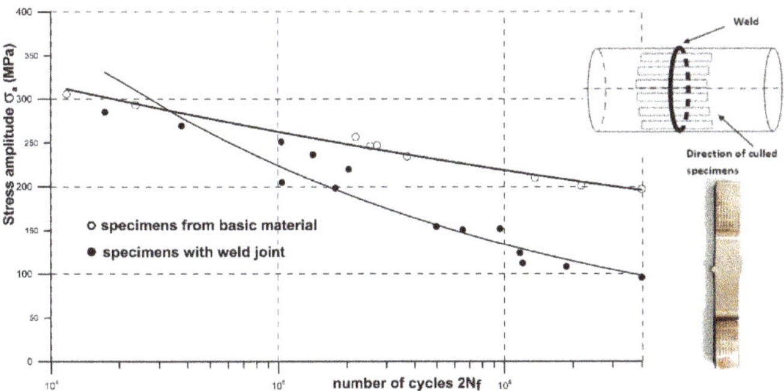

Figure 5. Dependency σ_a-$2N_f$ [15] for the base material and pipeline specimens made from the real pipeline with the weld joint.

The condition of the vibration permissibility of the T-branch of the welded pipeline is then given by the relation

$$\varepsilon_{aeq} \leq \frac{0.8\sigma_c^{5.10^6}}{E} \, , \tag{3}$$

where $\sigma_c^{5.10^6}$ is a fatigue limit for 5 million cycles to failure (for the selected probability of survival, i.e., 95%), and ε_{aeq} is an amplitude of equivalent strain in the critical cross-section, taking into account the mean value of the cycle (Goodman [16] or Morrow's [17] method). Such a cycle is mainly induced due to stresses from the internal pressure and possible combination of normal (axial plus bending loading) and shear stresses (torsional loading). E is the Young modulus of the material.

The amplitudes of strain (stress) lower than value of the cyclic anelasticity limit induced by oscillation may be considered as safe, as was presented in [18]. These amplitudes of strains below this cyclic anelasticity limit are thus neglected. Thus, the monitoring system takes into account only the amplitudes of strains above this limit. The example of putting such a pipeline monitoring system in practice for the gas pipeline courtyard compressor station is presented by its main online information screen in Figure 6. This screen shows the information about critical cross-section points of the pipeline system at the compressor courtyard. KM1-KM9 are those critical monitoring places where there are installed 5 strain gauges, according to the methodology displayed in Figure 4. These strain sensors continuously sense the deformation state of each monitoring cross-section KM1-KM9. The evaluated and visualized magnitudes are:

- safety in terms of the pressure integrity of the pipe,
- fatigue damage accumulation state [19], and
- vibration permissibility.

The long-term operation of the monitoring system places high demands on the stability of the sensors and the associated electronics, which can lead to the undesired valuating offset effect of the measured signals. Although it is possible to employ FBG (fiberglass) sensors [20], VSG (vibrating string gauges) [21], or resistive strain gauges to continuously sense strain, they are not convenient for measuring pipeline deformations caused by vibrations due to high frequencies. Resistive strain gauges frequently have an undesirable floating offset effect, which is formed by the sum of the heat drift of strain gauges, wires, and electronic circuits. Thus, it is not satisfactory to rely on the conventional thermal compensation of strain gauges as recommended by their manufacturers [22].

For long-time monitoring of strains, the extended thermal compensation concept for strain gauges also includes electronic offset compensation, as it is explained in more detail in [12,23]. The example of the long-term overall stress record in a pipeline surface based on strain sensors measured without any influence of the unwanted offset is shown in Figure 7.

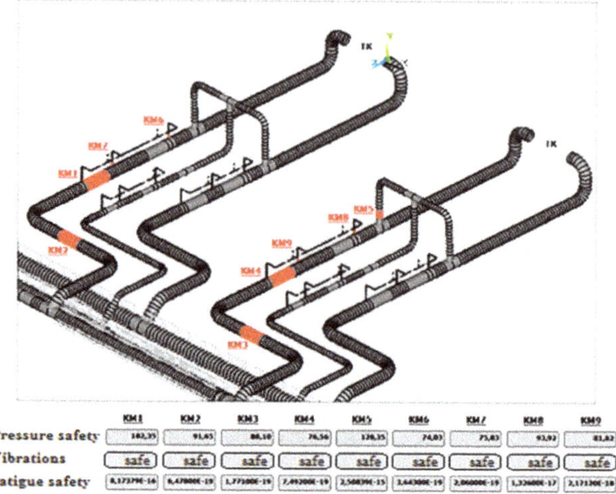

Figure 6. Main screen of the monitoring system for the super-structure of the pipeline yard above the ground installed at the compressor station.

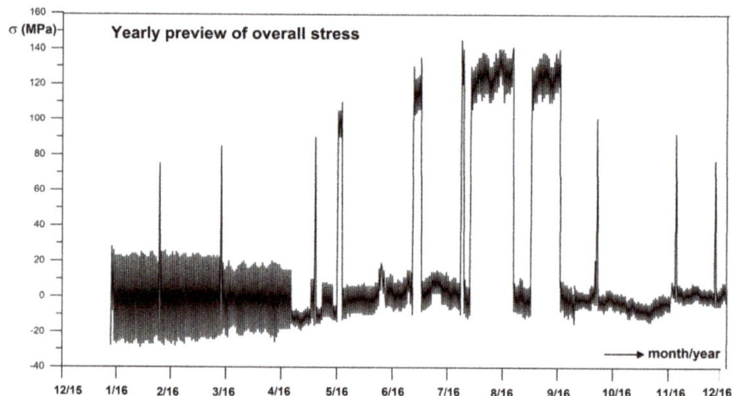

Figure 7. The progress of daily stress peaks in the cross-section of the pipe near the compressor obtained by a monitoring system with an extended thermal and device compensation concept.

2.2. The Additional Bending in Operations of Pressurized Pipelines

In the operation of line piping systems and their parts, the changing of the soil support may cause additional bending stresses in the pipeline casing loaded with an internal pressure. In practice, such a situation may occur on pipes laid on slopes. In pipelines placed perpendicular to the slope, additional downward bending stresses may occur as a result of the downward slope densification, resulting in less torsion. Similarly, a pipeline placed downstream of a downward slope can be subjected to the additional bending stress at the point of the transition curve to the horizontal section.

When placing the pipeline, it is necessary to compact the subsoil, so that the pipe is supported over its entire length. However, it is sometimes difficult to achieve the ground compaction while excavating the pipeline because of the maintenance during its operation. After heavy rains, the subsoil may drop in such sections, or the subsoil may be undermined, and the pipeline will become a beam due to its own gravity. The additional bending stress, together with the internal pressure, can endanger its integrity (Figure 8).

Figure 8. Gas pipeline accident due to the additional bending of a pipeline system laid on a slope with insufficient amounts of compacted subsoil.

The critical pipeline sections where the problems with the slope stability or subsoil stability may be expected can also be equipped with monitoring systems. The arrangement of the sensors along the circumference (see Figure 3) allows to monitor the value of the additional bending stress and possible torsional stress. Those stresses can be evaluated by the following relationships:

$$\sigma_{M_B max} = \frac{2}{3} E \sqrt{\varepsilon_i^2 + \varepsilon_{ii}^2 + \varepsilon_{iii}^2 - \varepsilon_i \varepsilon_{ii} - \varepsilon_i \varepsilon_{iii} - \varepsilon_{ii} \varepsilon_{iii}} \tag{4}$$

$$\tau_{M_t} = G(\varepsilon_{iiib} - \varepsilon_{iiia}), \tag{5}$$

where E is the Young modulus of the material in tensile/compression, and G is Young´s modulus of material in torsion.

An example of the occurrence of the additional bending stress of the pipeline section after a prolonged rainfall recorded by the monitoring system is shown in Figure 9.

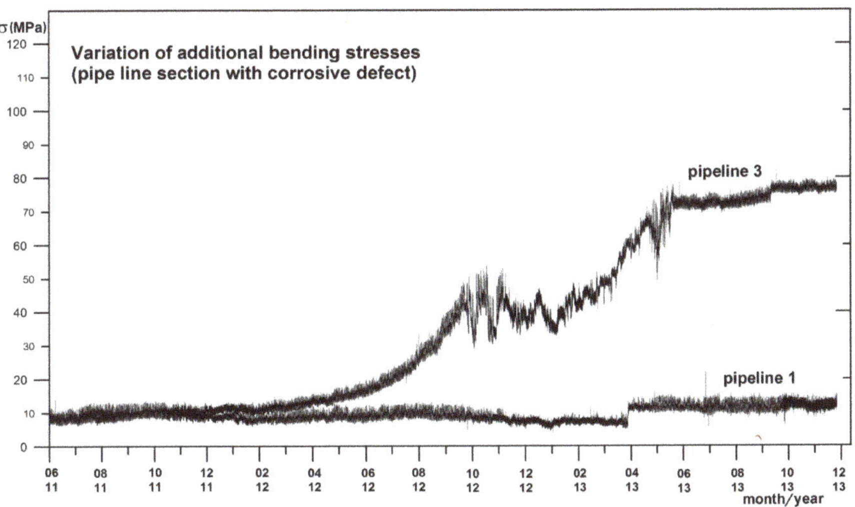

Figure 9. Gradual rise of stresses from additional bending stresses on the pipeline.

2.3. Weakening of Pipe Walls Due to Corrosion

Insufficient corrosion protection, the breakage of such protection, or its absence subsequently lead to the initiation of a corrosion process under appropriate environmental conditions. By gradually weakening the wall thickness, the safety of the operation of the pressured pipeline is reduced. Wall weakening is particularly dangerous when it is combined with additional bending stresses.

The concept of the online monitoring of the pipeline conditions with corrosion loss of the wall thickness placed in the ground at a location endangered by landslides (risk of additional bending stress on pipeline loading) loaded with an internal gas pressure is shown in Figure 10.

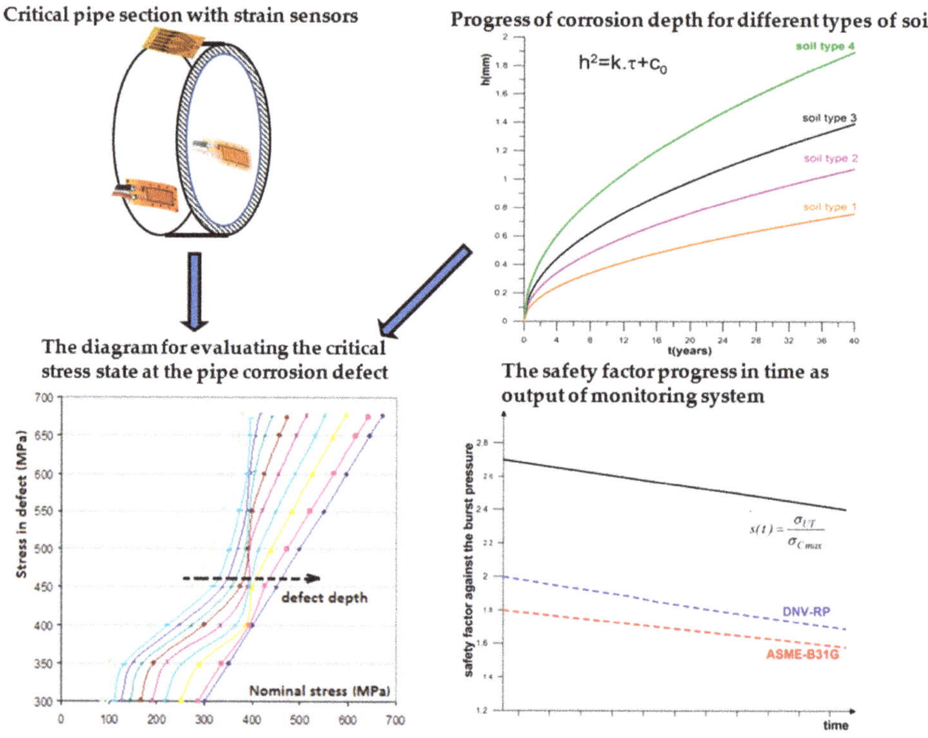

Figure 10. The schematic concept of the online monitoring of a pressurized pipeline system with corrosion defects.

The input data for such a system are the following:

- depth of existing corrosion defect found by pipeline inspection,
- kinetics of corrosion defect development,
- stress-strain state of the pipeline, and
- material properties of the pipeline.

The kinetics of corrosion layer growth is most often described by parabolic law in the form [24,25]

$$\Delta h^2 = k.\tau + c_0 \qquad \tau_0 = 0;\ c_0 \neq 0, \tag{6}$$

where Δh is the corrosion length change (mm of corrosion depth increase per year), k is the constant depending on temperature, τ is a time, and c_0 is an integration constant calculated from at least one corrosion depth.

The actual depth of the corrosion defect is therefore given by the type of corrosive environment and the time of its growth. Using the computer simulations, it is made possible to create diagrams of the stress-corrosion defect dependencies for the nominal stresses in the pipeline and the corrosion defect depth as a parameter [26]. The monitoring system thus calculates the depth of corrosion loss at the current time, according to Equation (6), calculates the current stress value in the pipeline cross-section from the strain sensors placed on the pipeline at the corrosion defect location, and, also, calculates the value of the maximum stress in the corrosion defect, as is shown in Figure 11.

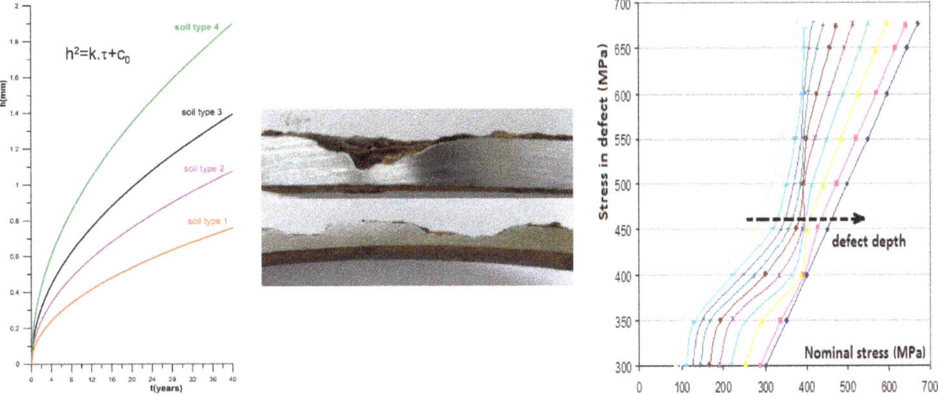

Figure 11. The progress of the corrosion depth over time for different soil types and stress diagram for different depths of corrosion defects for nominal stress values obtained by direct measurements.

As Equation (4) expresses the maximum value of tension caused by the bending stress in the cross-sectional fibers, which is subjected to the highest loading, and not the value at the point of a corrosion defect, it is necessary to optimize the location of strain sensors with regards to the location of the corrosion defect (Figure 12).

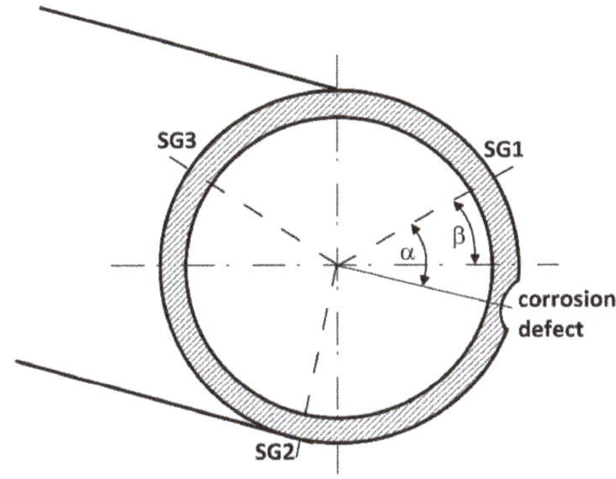

Figure 12. The location of the corrosion defect in relation to a strain gauge 1 (SG1).

The bending stress in the corrosion defect σ_{MBcor} is given by the equation:

$$\sigma_{MBcor} = \sigma_{MBmax} \sin(\alpha - \beta) \tag{7}$$

where the angle β defines the vector location of the additional bending moment with respect to strain sensor SG1 and may be expressed by the following equation:

$$tan\beta = \frac{\sqrt{3}(\varepsilon_2 - \varepsilon_3)}{2\varepsilon_1 - \varepsilon_2 - \varepsilon_3} \tag{8}$$

where the angle α defines the location of a corrosion defect in relation to the same strain sensor.

Thus the relationships derived in [12] may be elaborated to include the following form:

$$\sigma_x = E\left(\frac{\varepsilon_1 + \varepsilon_2 + \varepsilon_3}{3} + \frac{2}{3}sin(\alpha - \beta)\sqrt{\varepsilon_1^2 + \varepsilon_2^2 + \varepsilon_3^2 - \varepsilon_1\varepsilon_2 - \varepsilon_1\varepsilon_3 - \varepsilon_2\varepsilon_3}\right) \tag{9}$$

$$\sigma_y = E.\varepsilon_y = E(\varepsilon_a + \varepsilon_b - \varepsilon_x) \tag{10}$$

$$\frac{\gamma}{2} = \frac{\varepsilon_a - \varepsilon_b}{2} \tag{11}$$

The principal stresses may be calculated by employing the methodology in [12] by the following equations:

$$\sigma_1 = E.\varepsilon_1 = E\left[\frac{\varepsilon_x + \varepsilon_y}{2} + \sqrt{\left(\frac{\varepsilon_x - \varepsilon_y}{2}\right)^2 + \left(\frac{\gamma}{2}\right)^2}\right] \tag{12}$$

$$\sigma_2 = E.\varepsilon_2 = E\left[\frac{\varepsilon_x + \varepsilon_y}{2} - \sqrt{\left(\frac{\varepsilon_x - \varepsilon_y}{2}\right)^2 + \left(\frac{\gamma}{2}\right)^2}\right] \tag{13}$$

$$\sigma_3 = E.\varepsilon_3 = -p, \tag{14}$$

where p is an internal pressure.

The current safety level of a pipeline with corrosive wall loss can be calculated in-time according to the following relationship:

$$s(t) = \frac{\sigma_{UT}}{\sigma_{Cmax(t)}}, \tag{15}$$

where σ_{UT} is a true strength limit for given material of pipeline, and $\sigma_{Cmax(t)}$ is the current value of the maximum stress calculated in the place of the corrosion defect using the diagram in Figure 11 for a measured value of the nominal stress in the pipe and an actual depth of the corrosion defect.

The safety level of a pipeline with corrosive wall loss can also be calculated using ASME-B31G [26] or DNV-101RP [27] normative relationships, which are significantly more conservative than the reality is. The resulting time progress of the safety level for each of the approaches shown in Figure 9 (normative relationships or diagrams from computer simulations) provides the operator of the pipeline system with the basic information on operational safety.

3. Conclusions

The highest number of pipeline system accidents in hydrocarbon transportation occurs due to the concurrence of several adverse effects [28,29], such as:

- induced vibrations due to fluid-geometry or the work of compressors,
- additional bending loading, and
- corrosion losses of the wall thickness.

These unplanned operational situations can be evaluated using appropriately designed monitoring systems. The physical principles of the creation and functioning of such systems were presented in the article. It should be noted that an important part of such systems is their appropriate hardware assembly, which must be implemented in industrial conditions and as simple and robust as possible. The monitoring system itself is controlled by a software application that provides the collection and evaluation of data from sensors, calculation of the resulting quantities, and their visualization in-time. All the monitoring systems described in this article have been developed and employed in a real operation of pipeline systems throughout the period of several years. They helped to detect the processes of self-excited vibrations using strain sensors and additional bending stresses in the operation of pipeline gas transport systems. The credibility of the values provided by the tested systems thus results from the direct measurements of strain as a key quantitie for assessing the health and operational safety of such structures. The diagrams in Figures 3, 7 and 9 shown in the article were diagrams recorded by real, presented monitoring systems.

Author Contributions: Conceptualization, methodology (V.C.); measurement (M.Š.); software M.G. (Martin Garan); electronic M.G. (Marek Gašparík). All authors have read and agreed to the published version of the manuscript.

Funding: This work was supported by the Slovak Research and Development Agency under the contract No. APVV-17-0666 and by the Research & Development Operational Program funded by the ERDF ITMS: 26240220084 Science, Bratislava.

Conflicts of Interest: The authors declare no conflict of interest.

References

1. Ricker, E.R. Analysis of Pipeline Steel Corrosion Data from NBS (NIST) Studies. *J. Res. Natl. Inst. Stand. Technol.* **2010**, *115*, 373–392. [CrossRef] [PubMed]
2. Restrepo, C.E.; Simonoff, J.S.; Zimmerman, R. Causes, cost consequences, and risk implications of accidents in US hazardous liquid pipeline infrastructure. *Int. J. Crit. Infrastruct. Prot.* **2009**, *2*, 38–50. [CrossRef]
3. Wachal, J.C.; Smith, D.R. *Vibration Troubleshooting of Existing Piping Systems*; EDI Paper No.59; Engineering Dynamics Inc.: San Antonio, TX, USA, 1991.
4. Radavich, P.M.; Selamet, A.; Novak, J.M. A Computational Approach for Flow-acoustic Coupling in Closed Side Branches. *J. Acoust. Soc. Am.* **2001**, *109*, 1343–1353. [CrossRef] [PubMed]
5. Jungowski, W.M.; Botros, K.K.; Studzinski, W. Cylindrical Side-branch as Tone Generator. *J. Sound Vib.* **1989**, *131*, 265–285. [CrossRef]
6. Rogers, L.E. Design Stage Acoustic Analysis of Natural Gas Piping Systems in Centrifugal Compressor Stations. *J. Eng. Gas Turbines Power* **1992**, *114*, 727–736. [CrossRef]
7. Arnulfi, G.L.; Giannattasio, P.; Giusto, C.; Massardo, A.F.; Micheli, D.; Pinamonti, P. Multistage Centrifugal Compressor Surge Analysis: Part I—Experimental Investigation. *J. Turbomach.* **1999**, *121*, 305–311. [CrossRef]
8. Arnulfi, G.L.; Giannattasio, P.; Giusto, C.; Massardo, A.F.; Micheli, D.; Pinamonti, P. Multistage Centrifugal Compressor Surge Analysis: Part II—Numerical Simulation and Dynamic Control Parameters Evaluation. *J. Turbomach.* **1999**, *121*, 312–320. [CrossRef]
9. Gravdahl, J.T.; Willems, F.; De Jager, B.; Egeland, O. Modeling for Surge Control of Centrifugal Compressors: Comparison with experiment. *IEEE* **2000**, *2*, 1341–1346.
10. Farrar, R.C.; Worden, K. *Structural Health Monitoring: A Machine Learning Perspective*; John Wiley & Sons: Hoboken, NJ, USA, 2012.
11. Chmelko, V.; Garan, M. Long-term monitoring of strains in a real operation of structures. In Proceedings of the 14th IMEKO TC10 Workshop on Technical Diagnostics, Milano, Italy, 27–28 June 2016; pp. 333–336.
12. Chmelko, V.; Garan, M.; Šulko, M. Strain measurement on pipelines for long-term monitoring of structural integrity. *Measurement* **2020**, *163*, 107863. [CrossRef]
13. Amzallag, C.; Gerey, P.; Robert, J.L.; Bahuaudt, J. Standardization of the rainflow counting method for fatigue analysis. *Int. J. Fatigue* **1994**, *16*, 287. [CrossRef]
14. Papuga, J.; Vízková, I.; Lutovinov, M.; Nesládek, M. Mean stress effect in stress-life fatigue prediction re-evaluated. *MATEC Web Conf.* **2018**, *165*, 10018. [CrossRef]
15. Basquin, O.H. The exponential law of endurance tests. *Am. Soc. Test. Mater. Proc.* **1910**, *10*, 625–630.
16. Morrow, J.D. Fatigue properties of metals, section 3.2. In *Fatigue Design Handbook*; Pub. No. AE-4; SAE: Warrendale, PA, USA, 1968; p. 21.
17. Goodman, J. *Mechanics Applied to Engineering*; Longmans, Green and Co.: London, UK, 1919; pp. 631–636.
18. Chmelko, V. Cyclic anelasticity of metals. *Met. Mater.* **2014**, *52*, 353–359. [CrossRef]
19. Chmelko, V.; Kliman, V.; Garan, M. In-time monitoring of fatigue damage. *Procedia Eng.* **2015**, *101*, 93–100. [CrossRef]
20. Tennyson, R.C.; Don, W.; Cherpillod, T. Monitoring Pipeline Integrity Using Fiber Optic Sensors. In Proceedings of the CORROSION 2005, NACE International, Houston, TX, USA, 3–7 April 2005.

21. Glisic, B. Comparative study of distributed sensors for strain monitoring of pipelines. *Geotech. Eng.* **2019**, *50*, 28–35.

22. Hoffmann, K. *Eine Einfuh-rung in die Technik des Messens mit Dehnungsmessstreifen*; HBM: Darmstadt, Germany, 2012.

23. Šulko, M.; Garan, M.; Chmelko, V. *PUV 5017-2015, G01L 1/22*; Industrial Property Office: Banská Bystrica, Slovakia, 2015.

24. National Research Council Canada. Metallic corrosion. In Proceedings of the 9th International Congress on Metallic Corrosion, Toronto, ON, Canada, 3–7 June 1984.

25. Chmelko, V.; Berta, I. Analytical Solution of the Pipe Burst Pressure Using Bilinear Stress-Strain Model and Influence of Corrosion Defects on it. *Procedia Struct. Integr.* **2019**, *18*, 600–607. [CrossRef]

26. ASME B31.G. *Manual for Determining the Remaining Strength of Corroded Pipelines*; ASME: New York, NY, USA, 2009.

27. *DNV RP F101-Parts A & B: Corroded Pipelines*; Det Norske Veritas AS: Oslo, Norway, 2015.

28. Cosham, A.; Hopkins, P. *The Pipeline Defect Assessment Manual (PDAM)*; Penspen Ltd.: London, UK, 2003.

29. Nanney, S. *Pipeline Safety Update*; NAPCA Workshop: Houston, TX, USA, 2012.

Article

Strain Response Characteristics of RC Beams Strengthened with CFRP Sheet Using BOTDR

Ki-Nam Hong [1], Won-Bo Shim [1,*], Yeong-Mo Yeon [1] and Kyu-San Jeong [2]

[1] Department of Civil Engineering, Chungbuk National University, 1 Chungdae-ro, Seowon-Gu, Cheongju, Chungbuk 28644, Korea; hong@chungbuk.ac.kr (K.-N.H.); yym235@chungbuk.ac.kr (Y.-M.Y.)
[2] Sustainable Infrastructure Research Center, Korea Institute of Civil Engineering and Building Technology (KICT), 283, Goyang-daero, Ilsanseo-gu, Goyang-si, Gyeonggi-do 10223, Korea; jungkyusan@kict.re.kr
*   Correspondence: firstice@chungbuk.ac.kr; Tel.: +82-10-9950-2363

Received: 30 July 2020; Accepted: 27 August 2020; Published: 29 August 2020

Abstract: This paper presents the structural behaviors of reinforced concrete (RC) beams that have been strengthened with carbon-fiber-reinforced polymer (CFRP) sheets experimentally and numerically. Test specimens were subjected to four-point bending, and structural behavior was observed using a strain gauge and a Brillouin optical time domain reflectometer (BOTDR) sensor. Non-linear finite element analysis was conducted to examine the applicability and reliability of numerical models using the commercial finite element code, LS-DYNA. In the results, the de-bonded section between the beam substrate and CFRP sheet affected the initial crack in the structure, while the ultimate load, which is related to structural failure, was unaffected. The predicted results correlated well with the experimental observations in terms of the trend of the load-displacement curve, initial crack load, ultimate load and failure mode. Additionally, it is shown that the de-bonding behaviors in the interface were examined using the strain distributions for the CFRP sheets through the experiment and numerical simulations.

Keywords: BOTDR; CFRP sheet; un-bonded position; cover delamination; interfacial de-bonding

---

1. Introduction

Reinforced concrete (RC) has recently become one of the most common building materials for social infrastructures. However, RC structures have drawbacks that include a significant reduction to their structural performance over time [1,2]. In the civil engineering community, repairing or strengthening deteriorated RC structures using carbon-fiber-reinforced polymer (CFRP) composites is often done because it is economical and can be constructed in a short amount of time. One popular method for strengthening deteriorated RC structures is attaching CFRP sheets on the external surface of a structure. This method is widely used to improve the loading carrying capacity of the structure with its straight-forward constructability and applicability; it also only causes slight changes in geometry. Additionally, it is applicable to bulk bridge structures, which need continuous strengthening, or a damaged RC structure, since it constrains its damaged parts.

During the last few years, many experimental studies have been performed on the structural behavior of RC beam structures that have been strengthened with external CFRP sheets [3–5]. The structural behavior of the beam structures was affected by the bonded CFRP sheets. The flexural strength and stiffness of the beam structures that were well bonded with CFRP sheets were enhanced [3,4]. However, the RC beam structures that were strengthened with CFRP sheets contained a separation between the concrete surface and the sheet during use. This separation is dangerous in a practical sense, because it can degrade the performance of the composite sheet on the structure and affect the load carrying capacity of the beam structure all while being difficult to sight during inspection [2]. Therefore, it is imperative to set up an accurate instrument system to monitor the structural behavior in order to prepare countermeasures.

To measure the deformation of the RC structure and reinforcement member, a conventional electrical resistance strain sensor is widely used. When the conventional strain sensor is used to observe structural deformations in beam structures, multiple sensors or instrumentations are needed because of their short range. The physical quantities, which are measured with the use of electrical devices, are dependent on the surface condition and are unreliable for long term monitoring. Fiber optic sensors (FOS) have been introduced to solve the limitations of the conventional electrical devices. FOS are able to measure physical quantities such as strain and temperature with a high degree of sensitivity, resolution, accuracy and signal-to-noise ratio in a wide range of temperatures [6–9]. These sensors are not only capable of being used in service for long term monitoring [10,11] but can also be applied to the smart instrument technique due to their light weight and small size [12–15]. Woods et al. [16] measured the dynamic behavior of fiber-reinforced polymer strengthened RC shear walls using a distributed fiber optic sensor (DFOS) in the form of the fiber bragg grating (FBG) technique under a cyclic loading condition. It is shown that the DFOS can observe the structural behavior of the concrete member and fiber reinforced polymer. Wu et al. [17] studied the flexural strengthening of a full-scale pre-cast girder that was reinforced with pre-stressed PBO (poly-phenylene benzobisoxazole) sheets using a Brillouin optical time domain reflectometer (BOTDR) under various loading conditions. The BOTDR is one of the fiber optical instruments used, and has been shown to effectively observe the structural behavior of RC beams [18]. While the BOTDR has relatively lower accuracy than the FBG technique, it is effective when applied to the practical bulk structures because of its convenient installation and choices of monitoring point [18].

The objective of this study is to assess the applicability of BOTDR fiber sensors to measure the de-bonding behavior of CFRP sheets on externally strengthened RC structures for their structural integrity. Four-point bending tests were conducted to examine the structural behavior of RC beams that were strengthened with CFRP sheets. The test results were compared with the predicted results, which were provided from a finite element analysis using LS-DYNA. Additionally, the applicability and reliability of the numerical model was examined.

2. Experiment

2.1. Experimental Variable and Manufacturing

The specimens were manufactured in the form of a wide slab, which provided a large selection of variables and an ease for reinforcement. Figure 1 shows the configuration and dimension of a specimen. With a 30 mm concrete cover, the specimen was reinforced with 5 SD400-D16 steel bars in a longitudinal direction and SD400-D10 was used for the erection bars and shear stubs with a 200 mm spacing. Ready mix concrete was used to cast the beam specimens. After the cast was removed, the beam specimens were kept moist for 28 days as a curing period.

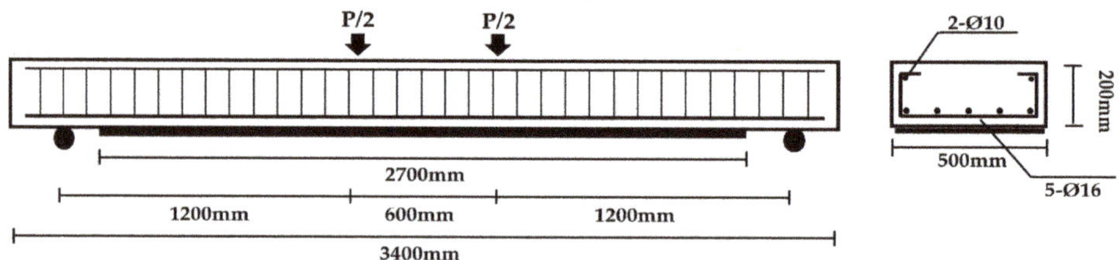

Figure 1. Configuration and dimension of a specimen.

Primer was spread on the concrete surface to improve the adhesion of the epoxy resin. Plastic films were used to simulate the separation of a CFRP sheet on the specimen. CFRP sheets were attached in two layers using epoxy resin to the tension region of the specimen, which included plastic films. A BOTDR sensor was set up to monitor the behavior of the CFRP sheets on the places where reinforcement was included. Figure 2 shows the geometry of the BOTDR fiber sensor placement. Since signal distortion and breaking can arise when placing a fiber optic sensor on a CFRP sheet, the sensor was attached using a nylon net. Nylon net is lightweight and has low strength, and it can be expected that it will not affect the structural behavior of RC structures. The BOTDR fiber sensor was placed in the arrangement seen in Figure 2, and the total length of the fiber was 27,000 mm. The fiber sensor was fixed on the nylon net using an industrial adhesive tape, and then the nylon net was attached on the concrete substrate using epoxy resin. Finally, the specimen was stored where there was an adequate moisture level and temperature for a curing period of more than 7 days. The experimental variables in this study are presented in Table 1.

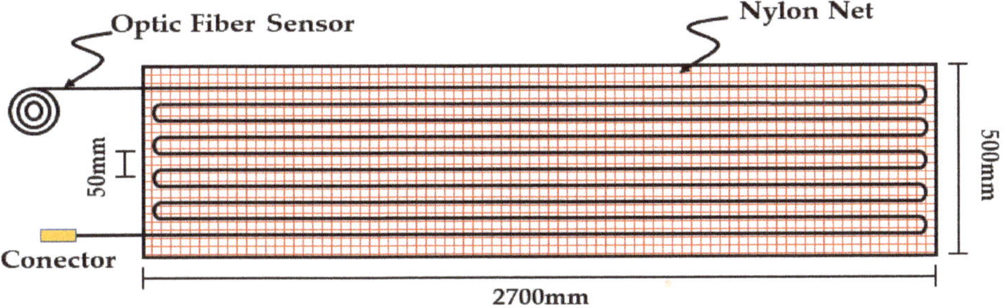

Figure 2. Geometry of the Brillouin optical time domain reflectometer (BOTDR) sensor placement.

2.2. Loading and Data Acquisition

The test specimens were subjected to four-point bending using a 2000 kN actuator with displacement control at 1.5 mm/min. The test specimens were subjected to four-point bending using a 2000 kN actuator with displacement control at 1.5 mm/min. Since BOTDR fiber sensors can measure the strain in CFRP sheets at a static loading point, the loading was progressively applied and paused at 20 kN, 60 kN, 80 kN, 95 kN, 110 kN, 125 kN and 140 kN. The static load points were chosen based on the crack load of the concrete substrate and elastic/plastic behavior of the steel bar. After the load reached the ultimate load, it was gradually decreased due to compressive failure in the concrete member and de-bonding/ripping off in the interface. When the load decreased to 80% of the ultimate load, the tests were terminated. As seen in Figure 3, 2 displacement meters were placed in the middle of the lower surface in order to measure the deflection. In addition, conventional electrical strain gauges were placed in the middle of the specimen. The applied load, deflection and strain values were acquired using a data logger system.

Table 1. Experimental variable details.

Specimen	CFRP Bonded Level (%)	Unbonded Position
C-0	0	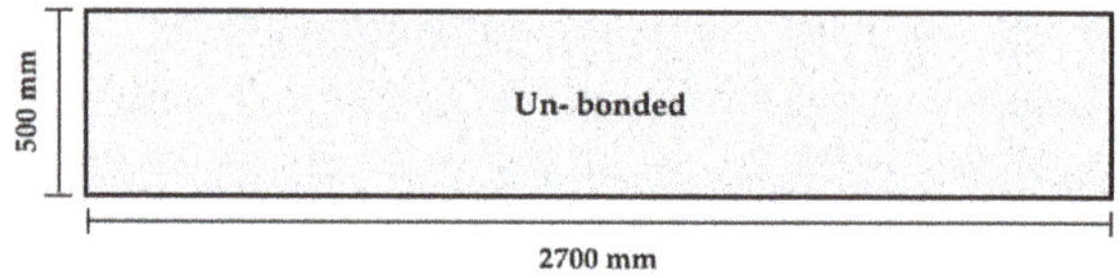 500 mm (height) × 2700 mm — Un-bonded
A-100	100	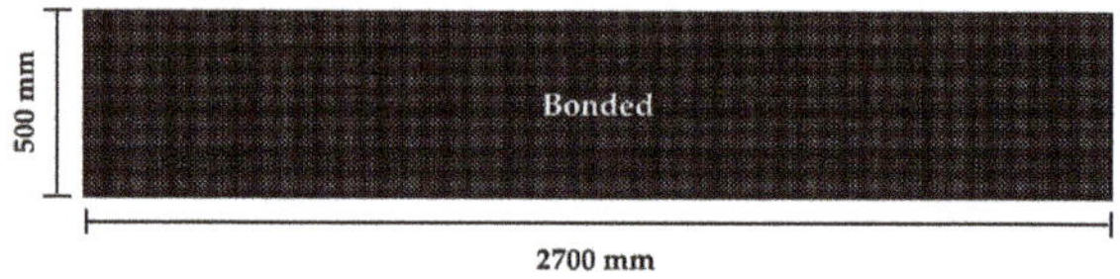 500 mm (height) × 2700 mm — Bonded
M-50	50	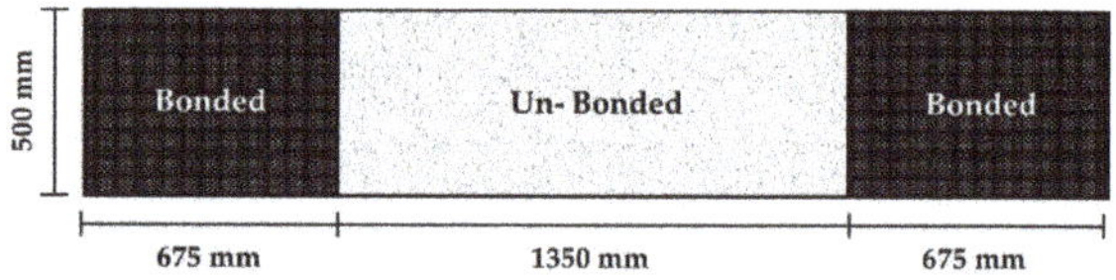 500 mm (height); Bonded 675 mm — Un-Bonded 1350 mm — Bonded 675 mm
M-30	30	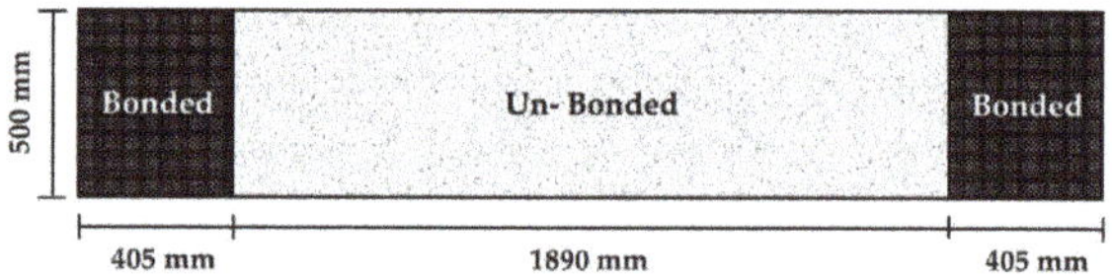 500 mm (height); Bonded 405 mm — Un-Bonded 1890 mm — Bonded 405 mm
E-50	50	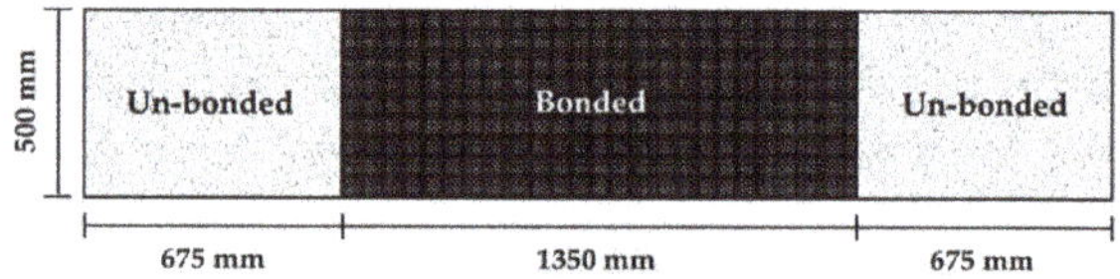 500 mm (height); Un-bonded 675 mm — Bonded 1350 mm — Un-bonded 675 mm
E-30	30	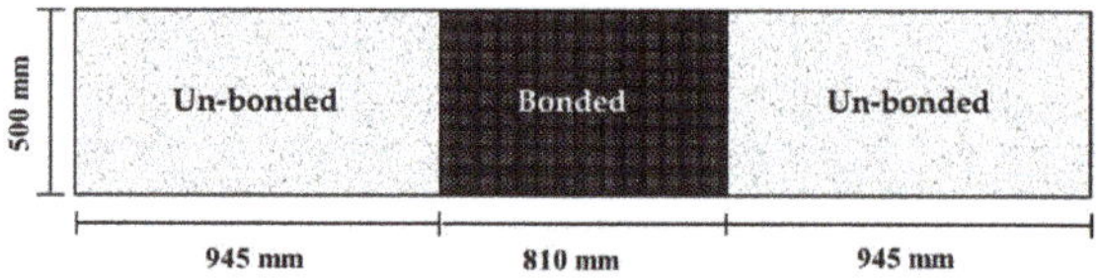 500 mm (height); Un-bonded 945 mm — Bonded 810 mm — Un-bonded 945 mm
LE-50	50	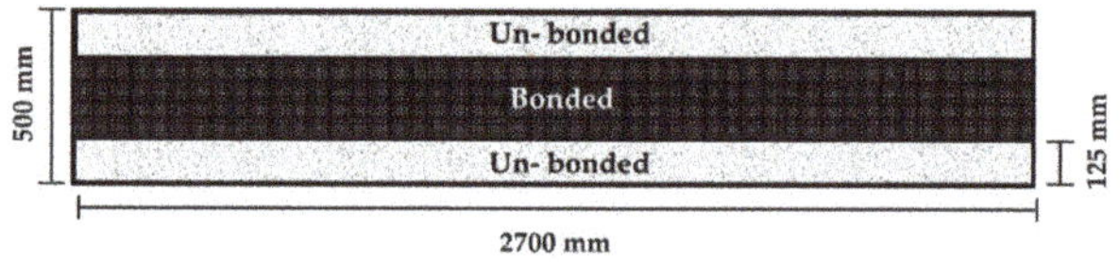 500 mm (height), 125 mm bands; Un-bonded / Bonded / Un-bonded — 2700 mm
LH-50	50	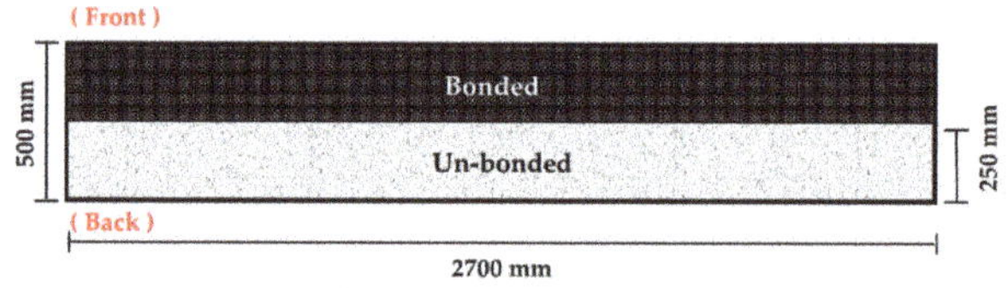 500 mm (height), 250 mm bands; (Front) Bonded / Un-bonded (Back) — 2700 mm

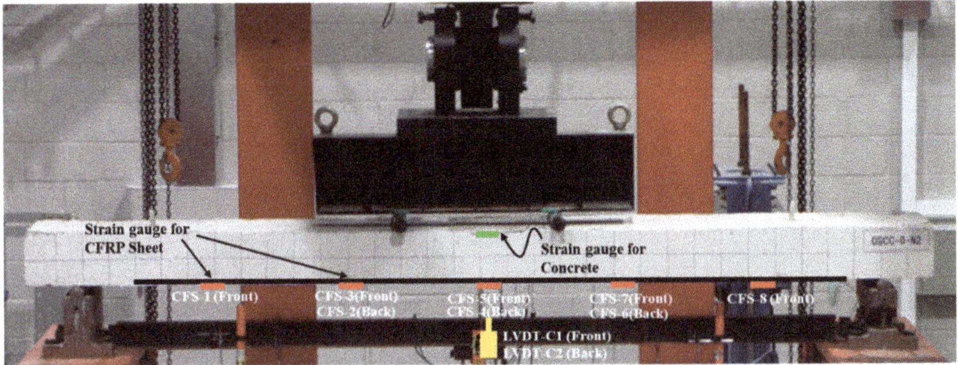

Figure 3. Test setup.

3. Finite Element Modeling

3.1. Analysis

LS-DYNA, a commercial finite element code, was employed to numerically analyze the behavior of the test specimens. In LS-DYNA, a large selection of material models for concrete, steel and unidirectional composites are equipped and the de-bonding failure along the interface can be simulated using material models and contact modeling [19–21]. In particular, the concrete model reasonably predicts the compressive-stress-adopting element erosion option at the ultimate strength and prevents any stress locking in the post peak region [20,21].

Figure 4 displays the finite element model used in this study. The concrete part was modeled using eight-node volumetric hexagonal solid elements. Two-node Hughes-Liu beam elements were utilized to model the main and transverse steel bars. The CFRP sheet, which was attached to the concrete for strengthening, was modeled using four-node Belytschko-Tsay shell elements. The solid beam and shell elements were approximately 20 mm in size. The concrete part and steel bars were considered as being perfectly bonded to each other using the LAGRANGIAN_IN_SOLID option.

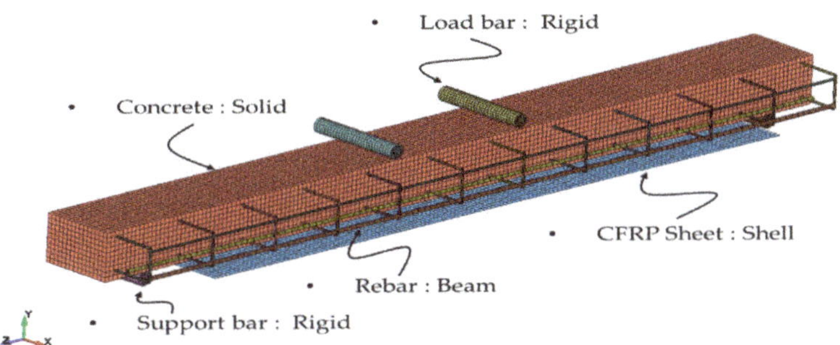

Figure 4. Finite element modeling.

The loading was applied in displacement control and finished when the displacement became 50 mm. Support and loading rods were modeled using rigid elements while avoiding the load concentration around the loading point when the load was directly applied to the element of the concrete material. The contact behavior between the concrete part and rods was considered using the AUTOMATIC_SURFACE_TO_SURFACE contact option.

3.2. Material Modeling

The material properties considered in this study are summarized in Table 2. In general, the concrete material, which is a non-homogenous anisotropic material, represents the nonlinear behavior at a low load level and exhibits different responses in the tension and compression fields. The material model 145, MAT_SCHWER_MURRAY_CAP, was adopted to model the anisotropic nonlinear behavior of the concrete material. Additional strength tests were conducted to determine the compressive strength of the concrete considered in this study using cylindrical nominal specimens. The material parameters for the concrete material were found in Jiang et al. [22].

The material properties for the CFRP sheet and steel bars were taken from the standard material test results. The material model 24, MAT_PIECEWISE_LINEAR_PLASTICITY, was used to model the steel bar. In this material model, the steel exhibits elasto-plastic behavior depending on the arbitrary stress and strain relation. Since the failures in the specimens occurred in the concrete material and the interface, the unidirectional CFRP composite material was considered as an orthotropic elastic material using the material model known as MAT_ORTHOTROPIC_ELASTIC.

Table 2. Material properties.

Material	Parameter	Magnitude
Concrete	Material model	MAT_145_SCHWER_MURRAY_CAP_MODEL
	Density (kg/m^3)	2320
	Compressive strength (MPa)	26.0
	Max aggregate size (mm)	25
Steel bar	Material model	MAT_024_PIECEWISE_LINEAR_PLASTICITY
	Density (kg/m^3)	7850
	Young's modulus (MPa)	200,000
	Poisson's ratio	0.3
	Yield stress (MPa)	460
	Tangent modulus (MPa)	2094
CFRP Sheet	Material model	MAT_002_ORTHOTROPIC_ELASTIC
	Thickness per layer (mm)	0.1
	Density (MPa)	2400
	Longitudinal modulus (MPa)	138,000
	Transverse modulus (MPa)	9650
	In-plane shear modulus (MPa)	5240
	Out of plane shear modulus (MPa)	2240
	Poisson's ratio	0.4
	Minor Poisson's ratio	0.021

3.3. Interface Modeling

When a bending load is applied to an RC beam strengthening with CFRP, de-bonding failure is one major failure that affects the structural strength. In order to simulate the failure along the interface between the concrete part and CFRP sheet, a large amount of interface modeling was presented [19–21]. In this study, the de-bonding failure between the concrete part and CFRP sheets was modeled using the tiebreak contact option. The tiebreak contact option, TIEBREAK_SURFACE_TO_SURFACE, is one tie contact option in LS-DYNA and works the same as general tie contact options before satisfying the failure criterion. The failure criterion used in this study is presented in Equation (1).

$$\left(\frac{|\sigma_n|}{NFLS}\right)^2 + \left(\frac{|\sigma_s|}{SFLS}\right)^2 \geq 1 \tag{1}$$

where σ_n is the normal stress, σ_s is the shear stress, NFLS is the normal failure stress and SFLS is the shear failure stress. Under tensile and shear loads, the normal and shear stresses in the interface are calculated. When the failure criterion is satisfied, the tie contact is released along the interface.

After failure arises, the interface behaves as a surface to surface contact. In order to simulate the interfacial separation accurately, it is necessary to determine the normal failure stress and shear failure stress through empirical models or proper tensile and shear tests. The failure stresses used in this study were determined through Equations (2)–(4) [23,24]:

$$NFLS = 0.447(f')^{0.55} \quad (\text{MPa}) \tag{2}$$

$$SFLS = 1.5\beta_w NFLS \quad (\text{MPa}) \tag{3}$$

$$\beta_w = \sqrt{\frac{2.25 - b_f/b_c}{1.25 + b_f/b_c}} \tag{4}$$

where f' is compressive strength of the concrete material, b_c is center to center spacing of CFRP strips and b_f is width of CFRP strip.

4. Experimental and Numerical Results

4.1. Failure Behavior

The failure behavior of the test specimens under a four-point bending condition was examined using experiments and numerical simulations. Figure 5 presents the failure behavior of the A-100 specimen from the experiment and analysis. The failure occurred in the form of de-bonding failure along the interface between the concrete substrate and the CFRP sheets. This type of failure behavior appeared in the test specimen for M-30, M-50, LE-50 and LH-50. Figure 6 shows the failure behavior of the E-30 specimen. De-bonding failure initiated from the edge of the CFRP sheet along the interface and developed to cover the delamination in the concrete substrate. The failure propagated to the vicinity of where the tensile steel bars were placed. In Figures 5b and 6b, the predicted failure behaviors of the test specimens for A-100 and E-30 correlated well with the test results. Additionally, the strain energy distributions in the concrete substrates progressed in accordance with the failure behaviors.

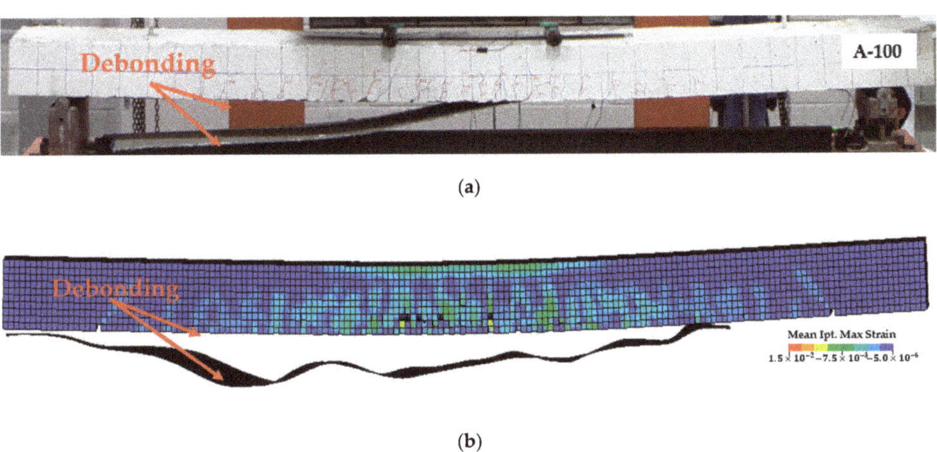

Figure 5. De-bonding failure mode (A-100). (**a**) Experiment. (**b**) Analysis.

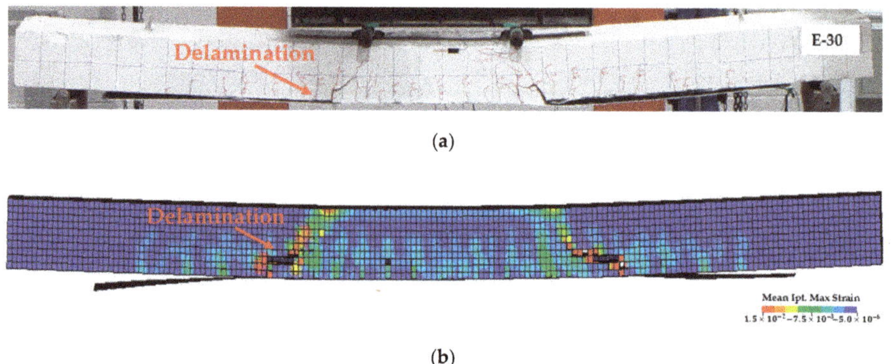

(a)

(b)

Figure 6. Delamination failure mode (E-30). (**a**) Experiment, (**b**) Analysis.

4.2. Failure Load and Load-Deflection Response

The test results are summarized in Table 3. The initial crack loads for the A-100, M-50, E-50, E-30 and LE-50 specimens were higher than the initial crack load for the C-0 specimen. The M-30 and LH-30 specimens were shown to be vulnerable to cracking, because the initial crack loads for the M-30 and LH-50 specimens decreased by 2.01% and 15.35%, respectively. Additionally, while the initial crack load for the LH-50 specimen, which was strengthened in the front, increased 120% when compared to the initial crack load for C-0, the initial crack load for the LH-50 specimen, which was strengthened in the back, decreased 15%. This is because the strengthening done using CFRP sheet asymmetrically caused a reduction of load capacity with load eccentricity. However, the ultimate load for all of the specimens varied over 10% in comparison to the one for the C-0 specimen. It is thought that the un-bond area did not affect the ultimate load of the structure, while the initial crack dominated the un-bonded area of the structure. Nevertheless, it is considered in order to carefully monitor the failures in the RC beam structures because the failure resulted in a reduction of structural performance over a long-term period.

Table 3. Comparison of loads and failure mode *.

Specimen	$P_{cr.EX}$ (kN)	R_{cr} (%)	$P_{u.EX}$ (kN)	$P_{u.FE}$ (kN)	$P_{u.FE}$ /$P_{u.EX}$	R_u (%)	$\Delta_{u.EX}$ (mm)	$\Delta_{u.FE}$ (mm)	$\Delta_{u.FE}$ /$\Delta_{u.EX}$	Failure Mode
C-0	14.46	0.0	101.62	100.52	1.01	0.0	50.02	43.51	0.87	Flexural
A-100	35.43	145.02	145.74	150.62	0.97	43.42	45.78	35.75	0.78	Interface De-bonding
M-50	19.74	36.51	128.98	134.42	0.96	26.92	44.43	44.95	1.01	Interface De-bonding
M-30	14.17	−2.01	129.9	133.35	0.97	27.83	41.87	32.99	0.79	Interface De-bonding
E-50	27.66	91.29	122.54	125.41	0.98	20.59	50.70	51.87	1.02	Cover Delamination
E-30	20.00	38.31	102.92	106.67	0.96	1.28	50.06	47.72	0.95	Cover Delamination
LE-50	16.69	15.42	123.54	127.82	0.97	21.57	43.19	33.35	0.77	Interface Debonding
LH-50	Front: 32.12 Back: 12.24	Front: 122.13 Back: −15.35	114.08	128.40	0.89	12.26	34.21	36.21	1.06	Interface Debonding

* $P_{cr.EX}$: initial crack load; $P_{u.EX}$: ultimate load of experiment; $P_{u.FE}$: ultimate load of FEM; R_{cr}: ration of crack load increase compared to the control specimen (C-0); R_u: ration of ultimate load increase compared to the control specimen (C-0); $\Delta_{u.EX}$: midspan deflection of experiment at ultimate load, $\Delta_{u.EX}$: midspan deflection of FEM at ultimate load.

Figure 7 shows the comparison of the load-deflection curves between the experiments and numerical simulations. While the behavior of the beam structure during the stage where the load was distributed to the steel bars and the CFRP sheet slightly differed, the load-deflection behaviors between the experiment and simulation in terms of the initial slope of the curve, global trend and steel bar yielding point matched well. In the simulation, the ultimate loads were over-predicted by 2% to 4%. This is reasonably acceptable when it is considered that the beam structure exhibits the anisotropic non-linear behavior. In the LH-50 specimen, the predicted ultimate load was approximately 10% higher than the one from the experiment and the failure behaviors from the experiment and the simulation differed. While the failure occurred along the interface before the steel bars yielded in the experiment, the interface failure appeared after the steel bars yielded. However, it is an insignificant discrepancy since the ultimate loads when the interface failure initiated are comparable in the experiment and simulation. In the case of the E-50 and E-30 specimens, which had concrete ripping off, the maximum deflections in the experiment and the simulation varied within 5% and the failure behaviors were similar.

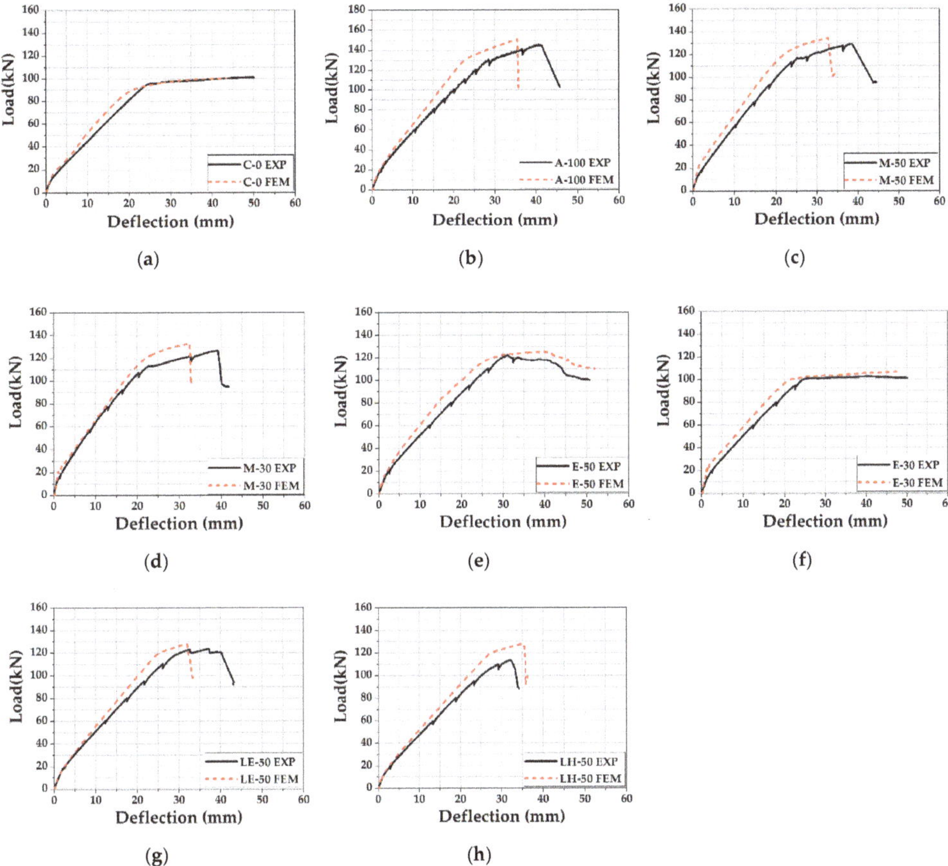

Figure 7. Comparison of load-deflection curves between experiment and analysis. (**a**) C-0, (**b**) A-100, (**c**) M-50, (**d**) M-30, (**e**) E-50, (**f**) E-30, (**g**) LE-50, and (**h**) LH-50.

4.3. Strain Behavior of CFRP Sheet

The ultimate strain results in the middle of the CFRP sheets from the experiments and numerical simulations are summarized in Table 4. In the experiment, the strain was monitored using two types of instruments: a conventional electrical strain gauge and a BOTDR sensor. When using the strain gauge, the ultimate strain for the M-30 specimen, which had an unbound area in the middle of the beam structure, was 30% lower than the one for the A-100 specimen, which was strengthened perfectly. It is shown that placing the strengthening CFRP sheets in the middle of the beam structure results in a more advantageous performance of the CFRP sheets on the structural behavior. In the simulation, the ultimate strains were slightly over-predicted by 15% in contrast to those from the strain gauges. The ultimate strains that were monitored using the BOTDR sensors were 17% lower than those from the strain gauges.

Figure 8 presents the comparison of the load-strain curves in the experiment and analysis in the CFRP sheet. In most cases, the predicted load-strain behaviors corresponded well with the observed loads and strains when using the strain gauges and BOTDR sensors. The slopes of the load-strain curves in the numerical simulation were relatively lower than those in the experiments. It is thought that the difference of the slopes in the numerical simulation and experiments was a result of the measuring conditions in the experiments. The strain from the strain gauge affected the surface condition where the strain gauge was placed and could be roughly measured due to the adhesive epoxy. In the case of the BOTDR sensor, the strain was considered to be more stable than the one that was from the strain gauge because the sensor was perfectly bonded with the CFRP sheet and the strain result's average value was in the range of 500 mm. Therefore, it was found that the numerical results were relatively close to the experimental results using the BOTDR sensor. The load-strain response seen in Figure 8e from the BOTDR sensor shows the difference with other curves. Considering the fact that it happened only in the E-30 specimen and the load-deflection response for the E-30 specimen were reasonable, it is thought to be caused by the slip of the BOTDR fiber sensor which needed to be fully constrained in the epoxy during the experiment.

Table 4. Comparison of the strains in the carbon-fiber-reinforced polymer (CFRP) sheet.

Specimen	CFRP Strain (με)			Gauge/FEM	Gauge/BOTDR
	Gauge	FEM	BOTDR		
A-100	6035.70	6114.95	4261.06	0.99	1.42
M-50	2867.61	4113.02	3190.17	0.70	0.90
M-30	1896.72	2810.13	459.20	0.67	4.13
E-50	2407.52	2952.68	2106.83	0.82	1.14
E-30	3405.64	2998.07	2482.14	1.14	1.37
LE-50	4594.38	5347.05	3056.74	0.86	1.50
LH-50	2877.01	3908.65	3967.69	0.74	0.73

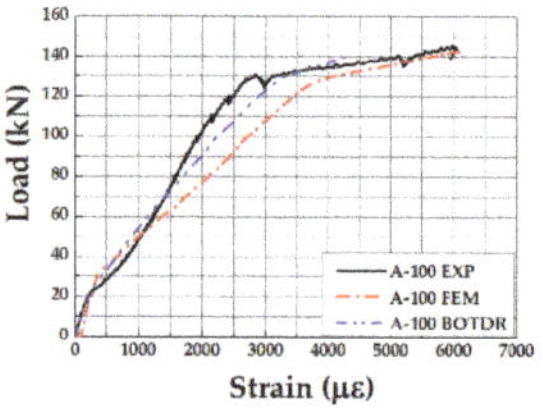

(a) 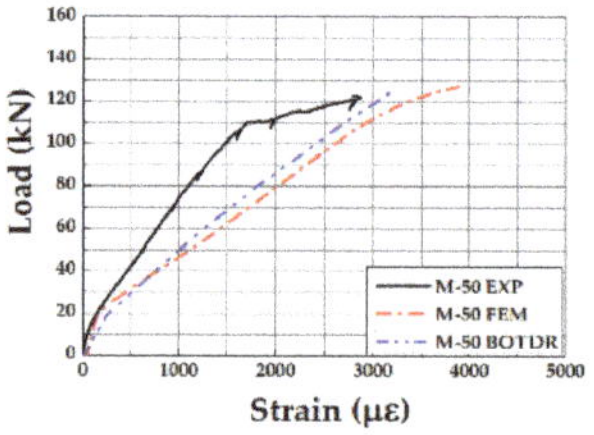 (b) 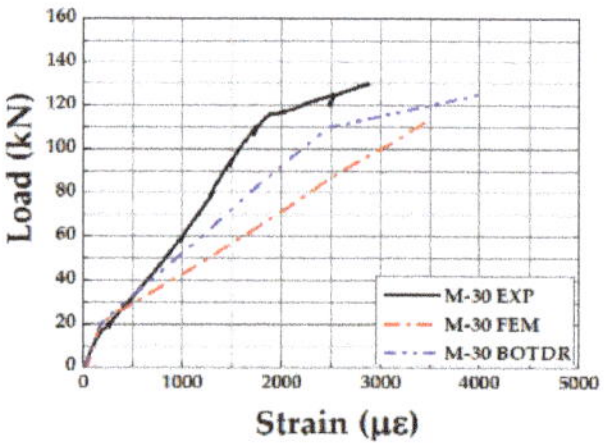 (c)

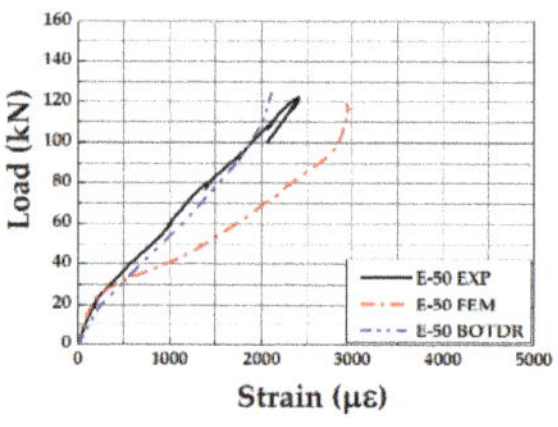

(d) 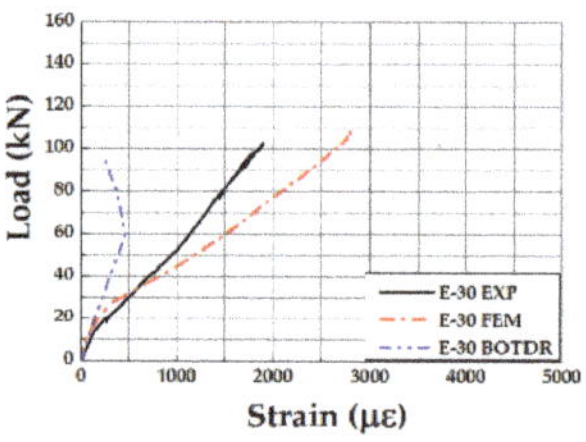 (e) 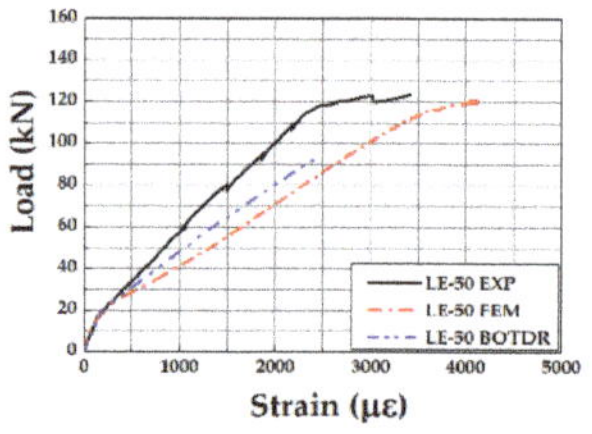 (f)

Figure 8. *Cont.*

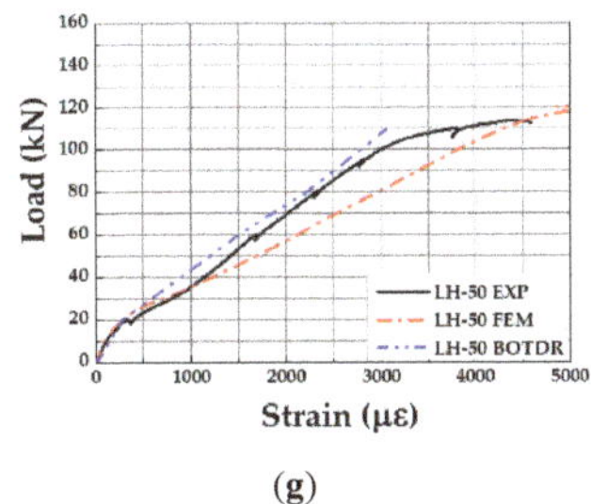

(**g**)

Figure 8. Comparison of load-strain curves between experiment and analysis in the CFRP sheet.
(**a**) A-100, (**b**) M-50, (**c**) M-30, (**d**) E-50, (**e**) E-30, (**f**) LE-50, and (**g**) LH-50.

In addition, the strain distribution for the CFRP sheets was investigated using the strain results through the BOTDR sensors and numerical simulations. The strain distributions that used experimental strain results were visualized using Origin, a numerical analysis program. In the visualization, the strain distributions were classified in 8 levels following the order of the magnitude and a thin plate spline algorithm was utilized to improve the visibility. Figure 9 shows the strain distribution of the CFRP sheet in the A-100 specimen from the experiment and simulation. As seen in Figure 9, the predicted strain distribution matched well with the experimental result in terms of the magnitude of the strain and the shape of the strain distribution. In the experimental result, the high strain region was relatively narrower than the one from the simulation. It is thought that the resolution of the experimental results is insufficient to present the accurate strain distribution. Figure 10 presents the strain distribution of the CFRP sheet in the M-50 specimen. The numerical result correlated well with the experimental result. While the shape of the developed strain in the CFRP is similar to the one in the A-100 specimen, the magnitude of the strain in the middle of the CFRP sheet was relatively lower. This is because the M-50 specimen included the un-bonded region between the CFRP and concrete substrate in the middle of the beam structure.

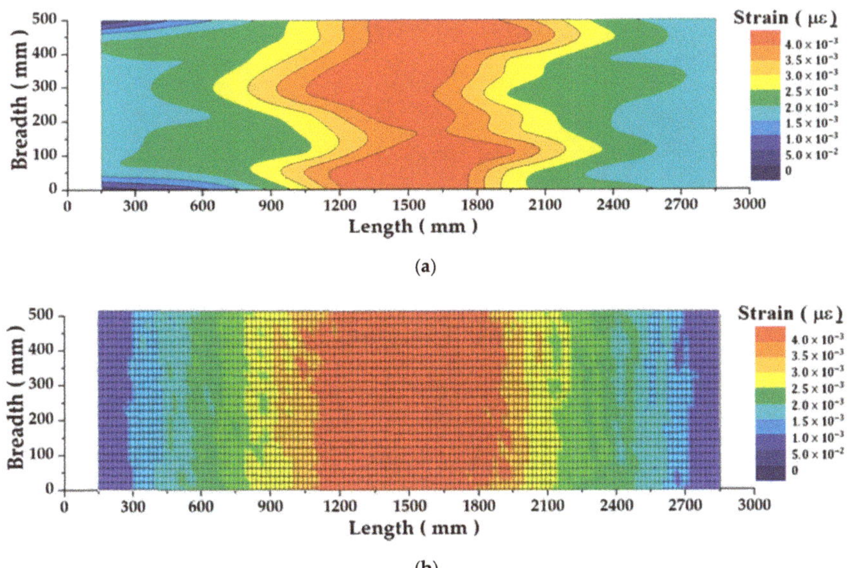

(**a**)

(**b**)

Figure 9. Comparison of CFRP strain distributions using BOTDR sensor and analysis (A-100).
(**a**) BOTDR, (**b**) FEM.

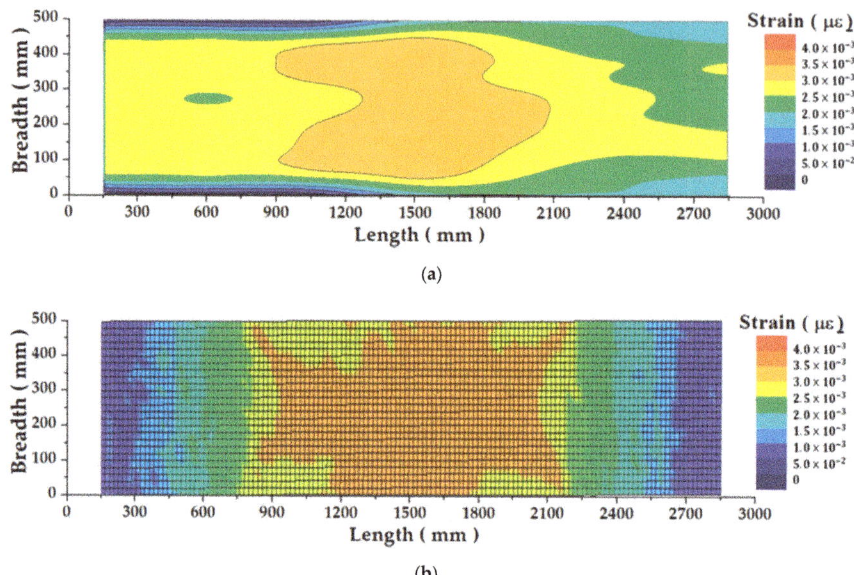

Figure 10. Comparison of CFRP strain distributions using BOTDR sensor and analysis (M-50). (**a**) BOTDR, (**b**) FEM.

The strain distribution of the CFRP sheet in the E-30 specimen is displayed in Figure 11. The strain did not develop at the end of the CFRP sheet and grew in the middle where the CFRP sheet and the concrete substrate were bonded. As with the two cases described above, the numerical simulation predicted the strain distribution of the CFRP sheet reasonably.

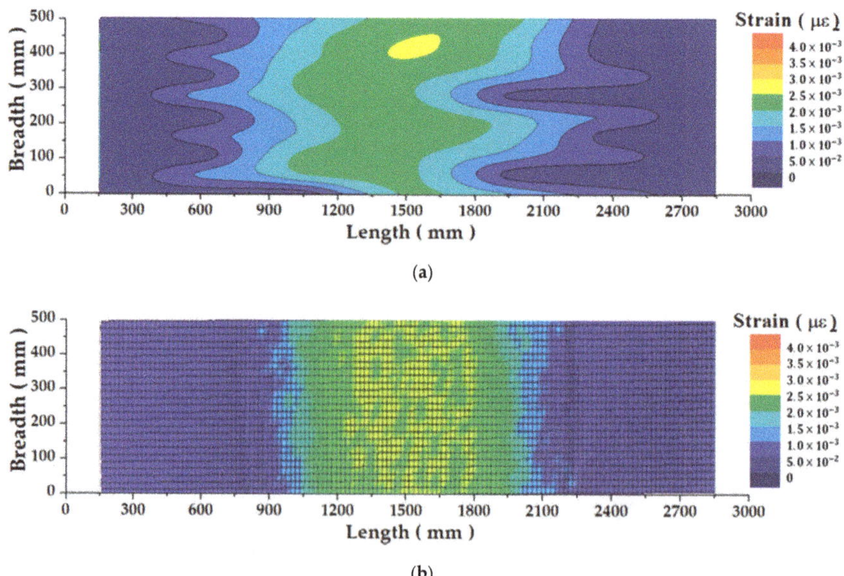

Figure 11. Comparison of CFRP strain distributions using BOTDR sensor and analysis (E-50). (**a**) BOTDR, (**b**) FEM.

5. Conclusions

In this study, the structural behavior of RC beams that had been strengthened with CFRP sheets was monitored using a BOTDR fiber sensor experimentally. Additionally, a numerical simulation was performed using LS-DYNA, a commercial non-linear finite element analysis program, and compared with the simulation from the experiment. The conclusions are described below:

(1) The de-bonding behavior between the concrete substrate and CFRP sheets affected the initial crack in the structure, while the ultimate load, which was related to the structural failure, was unaffected. Nonetheless, it is necessary to cautiously monitor the failures in RC beam structures because any failure could lead to a reduction in structural performance over a long period of time.

(2) The strain distribution for the CFRP sheets can be visualized using the strain results through the BOTDR sensor, and it allows for a decent investigation of the de-bonding behavior of the CFRP sheets on the concrete substrate.

(3) The numerical simulation correlated well with the experimental results in terms of the failure mode, failure pattern and the position where the de-bonding occurred. Therefore, it is possible to utilize the numerical simulation to study the de-bonding behavior of RC structures that are strengthened with CFRP sheets.

(4) The strain results from the BOTDR fiber sensor and numerical simulation are considered to be stable and well matched, while the strain result from the strain gauge is relatively unstable because the strain gauge affected the surface condition where the gauge was placed. Therefore, it is thought that the BOTDR fiber sensor is more reliable when monitoring the RC structure than the conventional strain gauge.

Author Contributions: K.-N.H. and W.-B.S. conceptualized the study; W.-B.S. obtained field data; W.-B.S. and Y.-M.Y. implemented data processing under the supervision of K.-N.H.; the original draft of the manuscript was written by K.-N.H. and W.-B.S. with editorial contributions from Y.-M.Y. and K.-S.J. All authors have read and agreed to the published version of the manuscript.

Funding: This paper work was financially supported by Ministry of the Interior and Safety as Human Resource Development Project in Disaster Management.

Conflicts of Interest: The authors declare no conflict of interest.

References

1. Lim, D.H. An Experimental Study of Flexural Strengthening Method of Reinforced Concrete Beams with Near Surface Mounted CFRP Strips. *J. Korean Soc. Civ. Eng.* **2013**, *33*, 131–136. (In Korean) [CrossRef]

2. Park, J.-S.; Kim, B.-C.; Park, K.-T.; Jung, K.-S. Analytical Study on the Bond Performance of Fiber Optic Sensor Embedded Carbon Fiber Sheets for Strengthening Concrete Structures. *J. Korean Soc. Adv. Compos. Struct.* **2018**, *9*, 31–36. (In Korean) [CrossRef]

3. Toutanji, H.; Zhao, L.; Zhang, Y. Flexural behavior of reinforced concrete beams externally strengthened with CFRP sheets bonded with an inorganic matrix. *Eng. Struct.* **2006**, *28*, 557–566. [CrossRef]

4. Li, Y.-F.; Xie, Y.-M.; Tsai, M.-J. Enhancement of the flexural performance of retrofitted wood beams using CFRP composite sheets. *Constr. Build. Mater.* **2009**, *23*, 411–422. [CrossRef]

5. Khan, A.U.R.; Fareed, S. Behaviour of Reinforced Concrete Beams Strengthened by CFRP Wraps with and without End Anchorages. *Procedia Eng.* **2014**, *77*, 123–130. [CrossRef]

6. Lee, D.G.; Mitrovic, M.; Friedman, A.; Carman, G.P.; Richards, L. Characterization of Fiber Optic Sensors for Structural Health Monitoring. *J. Compos. Mater.* **2002**, *36*, 1349–1366. [CrossRef]

7. Todd, M.D.; Nichols, J.M.; Trickey, S.T.; Seaver, M.; Nichols, C.J.; Virgin, L.N. Bragg grating-based fiber optic sensors in structural health monitoring. *Philos. Trans. R. Soc. A* **2007**, *365*, 317–343. [CrossRef]

8. Ansari, F. Practical Implementation of Optical Fiber Sensors in Civil Structural Health Monitoring. *J. Intell. Mater. Syst. Struct.* **2007**, *18*, 879–889. [CrossRef]

9. Deng, L.; Cai, C.S. Applications of fiber optic sensors in civil engineering. *Struct. Eng. Mech.* **2007**, *25*, 577–596. [CrossRef]

10. Casas, J.; Cruz, P.J.S. Fiber Optic Sensors for Bridge Monitoring. *J. Bridg. Eng.* **2003**, *8*, 362–373. [CrossRef]
11. Betz, D.C.; Staudigel, L.; Trutzel, M.N.; Kehlenbach, M. Structural Monitoring Using Fiber-Optic Bragg Grating Sensors. *Struct. Health Monit.* **2003**, *2*, 145–152. [CrossRef]
12. Zhan-Feng, G.; Yan-Liang, D.; Bao-Chen, S.; Xiu-Mei, J. Strain monitoring of railway bridges using optic fiber sensors. *J. Qual. Maint. Eng.* **2007**, *13*, 186–197. [CrossRef]
13. Doyle, C.; Staveley, C.; Henderson, P. Structural health monitoring using optical fiber strain sensing systems. In Proceedings of the 4th International Workshop on Structural Health Monitoring, Stanford, CA, USA, 1 July 2003; Volume 1, pp. 1–8.
14. Maalej, M.; Rizkalla, S.H. Fiber optic sensing and intelligent processing of sensor data for structural health monitoring. In Proceedings of the 14th International Conference on Optical Fiber Sensors, Island of San Giorgio Maggiore, Venice, Italy, 11–13 October 2000; Volume 4185, pp. 238–241.
15. Li, S.; Wu, Z. Development of Distributed Long-gage Fiber Optic Sensing System for Structural Health Monitoring. *Struct. Health Monit.* **2007**, *6*, 133–143. [CrossRef]
16. Woods, J.; Lau, D.; Bao, X.; Li, W. Measuring strain fields in FRP strengthened RC shear walls using a distributed fiber optic sensor. *Eng. Struct.* **2017**, *152*, 359–369. [CrossRef]
17. Wu, Z.; Xu, B.; Hayashi, K.; Machida, A. Distributed optic fiber sensing for a full-scale PC girder strengthened with prestressed PBO sheets. *Eng. Struct.* **2006**, *28*, 1049–1059. [CrossRef]
18. Kim, B.-C.; Jung, K.-S.; Park, J.-S.; Park, K.-T. Brillouin-OTDR Strain Response Analysis of Optical Fiber-embedded Carbon Fiber Sheet. *J. Korean Soc. Adv. Compos. Struct.* **2018**, *9*, 1–8. [CrossRef]
19. Hussein, M.E.; Tarek, H.A.; Saleh, H.A.; Yousef, A.A. Experimental and FE study on RC one-way slabs upgraded with FRP composites. *J. Korean Soc. Civil. Eng.* **2011**, *31*, 399–408.
20. Yoo, D.-Y.; Min, K.-H.; Lee, J.-Y.; Yoon, Y.-S. Evaluating Impact Resistance of Externally Strengthened Steel Fiber Reinforced Concrete Slab with Fiber Reinforced Polymers. *J. Korea Concr. Inst.* **2012**, *24*, 293–303. [CrossRef]
21. Li, L.; Lee, J.Y.; Min, K.H.; Yoon, Y.-S. Evaluating Local Damages and Blast Resistance of RC Slabs Subjected to Contact Detonation. *J. Korea Inst. Struct. Maint. Insp.* **2013**, *17*, 37–45. [CrossRef]
22. Jiang, H.; Zhao, J. Calibration of the continuous surface cap model for concrete. *Finite Elem. Anal. Des.* **2015**, *97*, 1–19. [CrossRef]
23. Lu, X.; Teng, J.; Ye, L.; Jiang, J. Bond-slip models for FRP sheets/plates bonded to concrete. *Eng. Struct.* **2005**, *27*, 920–937. [CrossRef]
24. Lu, X.; Teng, J.; Ye, L.P.; Jiang, J.J. Intermediate Crack Debonding in FRP-Strengthened RC Beams: FE Analysis and Strength Model. *J. Compos. Constr.* **2007**, *11*, 161–174. [CrossRef]

Article

Bayesian Calibration of Hysteretic Parameters with Consideration of the Model Discrepancy for Use in Seismic Structural Health Monitoring

Rosario Ceravolo, Alessio Faraci and Gaetano Miraglia *

Politecnico di Torino, Department of Structural, Geotechnical and Building Engineering & Responsible Risk Resilience interdepartmental Centre, Corso Duca degli Abruzzi, 24-10129 Turin, Italy; rosario.ceravolo@polito.it (R.C.); alessio.faraci@polito.it (A.F.)
* Correspondence: gaetano.miraglia@polito.it

Received: 7 July 2020; Accepted: 15 August 2020; Published: 22 August 2020

Abstract: Bayesian model calibration techniques are commonly employed in the characterization of nonlinear dynamic systems, as they provide a conceptual and effective framework to deal with model uncertainties, experimental errors and procedure assumptions. This understanding has resulted in the need to introduce a model discrepancy term to account for the differences between model-based predictions and real observations. Indeed, the goal of this work is to investigate model-driven seismic structural health monitoring procedures based on a Bayesian uncertainty quantification framework, and thus make relevant considerations for its use in the seismic structural health monitoring, focusing on masonry structures. Specifically, the Bayesian inference has been applied to the calibration of nonlinear hysteretic systems to both provide: (i) most probable values (MPV) of the parameters following the calibration; and (ii) estimates of the model discrepancy posterior distribution. The effect of the model discrepancy in the calibration is first illustrated recurring to a single degree of freedom using a Bouc–Wen type oscillator as a numerical benchmark. The model discrepancy is then introduced for calibrating a reference nonlinear Bouc–Wen model derived from real data acquired on a monitored masonry building. The main novelty of this study is the application of the framework of uncertainty quantification on models representing data measured directly on masonry structures during seismic events.

Keywords: Bayesian inference; uncertainty quantification; masonry structures; seismic structural health monitoring; Bouc–Wen model; model calibration; hysteretic system identification

---

1. Introduction

Models play a key role in simulating the behavior of engineering structures, though even very detailed models may fail to represent critical mechanisms. The variety of schemes and uncertainties that are typical of civil structures makes the prediction of the actual mechanical behavior and structural performance a difficult task. All this being said, computer simulations are useful engineering tools to design complex systems and to assess their performance [1–3]. These simulations aim at reproducing the underlying physical phenomena in question providing a solution for the governing equations. However, accurate modelling of the structural systems requires them to be calibrated and validated with direct observations and measured experimental data [4].

Most calibration methods are essentially regression techniques that estimate the model parameters based on the outputs, and eventually on the input, by means of various optimization algorithms. Some of the classical approaches of parameter estimation include weighted least-squares estimation, best linear unbiased estimation, etc. [4]. Basically, they are optimization problems of minimizing the difference between computed model output and experimental measured data. However, such deterministic approaches bump into some common problems: they are prone to be ill-conditioned, extremely susceptible to errors, and affected by stability issues. As a consequence, a single optimal parameter vector is not sufficient to specify the structural model, but rather a family of all plausible values of the model parameters needs to be identified that are consistent with the observations [5]. Moreover, a common situation in practice is to have limited acquired data and hence sparse datasets. This introduces epistemic uncertainty (lack of knowledge) alongside aleatory uncertainty (natural variability) [6]. All these factors result in uncertainty on model prediction. Therefore, addressing the uncertainty in the model parameters and studying how it influences the uncertainty in the response of the system is crucial and necessary to get accurate predictions with realistic levels of confidence. This evidence naturally leads to the consideration of the problem from a probabilistic perspective. The existing literature on probabilistic calibration is extensive [7–9] and focuses particularly on Bayesian approaches [10–16], Bayesian approaches with model class selection [17] and Bayesian filtering methods [18,19].

A consequence of using imperfect models in the calibration is that the identified parameters may not correspond to their physical namesakes [1]. The resulting posteriors do not necessarily give estimates of the true value of physical parameters, but rather give values which lead the model to best explain the data. In order to estimate the true physical value of parameters, careful thought needs to be taken when considering which parameters to include in the calibration. In addition, the model discrepancy term must be precisely specified to connect model prediction to the observations [6].

In the civil engineering research field, the Bayesian methodology has been used especially for the structural model updating based on experimental dynamic data [5,20] to update the uncertainty of the mechanical parameters of reinforced concrete (RC) structures [21,22] and masonry structures [23–25]. The novelty of this study is the application of this framework on models representing data measured directly on masonry structures during seismic events. The goal is to investigate model-driven seismic structural health monitoring (SHM) procedures by incorporating the posterior uncertainty linked to updated model discrepancy in order to propose its use for the seismic SHM of buildings. This points to the importance of correlating the choice of the discrepancy model function to the degradation amount and to the characteristics of the external seismic input. In this context, the use of simplified physics-based models is made herein, to capture the general nonlinear dynamical behavior of the structure at an acceptable time-computational cost (e.g., to support real-time applications), without resorting to complex FE models that would have required the introduction of surrogate modelling with a consequent increase in uncertainty. Specifically, the well-known Bayesian inversion framework (Section 2) has been applied to the calibration of nonlinear hysteretic systems. This approach clearly defines the errors and considers the uncertainties present in the model explicitly, providing the full multivariate distribution of the calibrated parameters and insights into a variety of quantities of interest (e.g., expected values, maximum a posteriori estimates). The uncertainty quantification (UQ) framework developed in [6] is first applied to a numerical benchmark and then on a case study, whose data came from a building monitored within the Italian Seismic Observatory of Structures (OSS) [26] (Section 3). The calibration of the latter is conducted exploring different parametric and non-parametric degrading models, quantifying the uncertainties that arise from each of them using the same discrepancy function. Finally, the effect of different levels of degradation on the inference of model parameters is also investigated in Section 3 and discussed in Section 4. Conclusions are then drawn in Section 5.

2. Materials and Methods

In this section some theoretical concepts are recalled about uncertainty quantification that are deemed necessary in order to understand the main concepts and terminology in view of the presented applications to SHM.

In the context of Bayesian statistics, it is assumed that all model parameters are random variables. This randomness describes the degree of uncertainty related to their realizations. The inference goal is to draw conclusions making probability statements about the hypermeters x given the evidence of actual observations y, through Bayes' theorem:

$$\pi(x|y) = \frac{\pi(x)\,\mathcal{L}(x;y)}{Z},$$ (1)

where $\pi(x|y)$ is the posterior distribution of the hyperparameters, $\pi(x)$ is their prior distribution and $\mathcal{L}(x;y)$ is the likelihood function. Z stands for normalizing factor, named evidence or marginal likelihood, that ensures that the distribution $\pi(x)$ integrates to 1. Z is defined by the integral $\int_{D_x} \pi(x)\,\mathcal{L}(x;y)\,dx$. In this way, the solution of the inverse problem coincides with the posterior probability distribution.

2.1. Discrepancy Model

In engineering problems, the analysis of physical systems is often performed by means of computational models which are commonly based on analytical equations governing the system or numerical methods. A computational forward model $\mathcal{M}$ is usually defined as a function that maps a set of model input parameters, x, governing the system to predict certain output quantities of interest (QoI) y. However, forward models are only a mathematical representation of the real system. As a consequence, a discrepancy term shall be introduced to connect model predictions $\mathcal{M}(x)$ to the experimental observations y [6]:

$$y = \mathcal{M}(x) + \varepsilon.$$ (2)

In this discrepancy term, the effects of measurement error on $y \in Y$ and model inaccuracy are gathered. The assumption of additive Gaussian discrepancy with zero mean and residual covariance matrix Σ has been made in this work:

$$\varepsilon \sim N(\varepsilon|0, \Sigma).$$ (3)

$\Sigma = \Sigma(x_\varepsilon)$, where its parameters x_ε are additional unknowns to infer jointly with the input parameters. $x_\mathcal{M}$ of $\mathcal{M}(x)$ [6] is a quantity not known a priori, this issue being overcome by parametrizing the residual covariance matrix. A diagonal covariance matrix with unknown residual variances σ^2 is assumed, specifically $\Sigma = \sigma^2 I$. This results in reducing the discrepancy parameter vector in a single scalar ($x_\varepsilon \equiv \sigma^2$) and in setting the parameter vector as $x = \left(x_\mathcal{M}, \sigma^2\right)$. Assuming then a prior distribution $\pi\left(\sigma^2\right)$ for the unknown variance of ε, as well as independency on the priors of the uncertain model and the discrepancy, the joint prior distribution can be drawn out:

$$\pi(x) = \pi(x_\mathcal{M})\,\pi\left(\sigma^2\right).$$ (4)

With these assumptions, a particular measurement point $y_i \in Y$ is a realization of a Gaussian distribution with mean value $\mathcal{M}(x)$ and variance σ^2. In this way the likelihood function reads:

$$\mathcal{L}\left(x_\mathcal{M}, \sigma^2; Y\right) = N\left(y|\mathcal{M}(x), \sigma^2\right).$$ (5)

Specifically:

$$\mathcal{L}\left(x_\mathcal{M}, \sigma^2; Y\right) = \prod_{i=1}^{N} \frac{1}{\sqrt{(2\pi\sigma^2)^N}} e^{-\frac{1}{2\sigma^2}(y_i - \mathcal{M}(x_\mathcal{M}))^T (y_i - \mathcal{M}(x_\mathcal{M}))},$$ (6)

where N is the number of measurement points. Hence, the posterior distribution can be computed as:

$$\pi\left(x_\mathcal{M}, \sigma^2|Y\right) = \frac{1}{Z}\pi(x_\mathcal{M})\pi\left(\sigma^2\right)L\left(x_\mathcal{M}, \sigma^2; Y\right),$$ (7)

which summarizes the updated information about the unknowns as $x = \left(x_{\mathcal{M}}, \sigma^2\right)$ after conditioning on the observation y.

2.2. Bayesian Calibration Procedure

In this section, we introduce the key steps of the entire procedure used for the calibration of the stochastic systems examined below (for the numerical benchmark and for the case study). Specifically, a three-phase approach was used as follows:

- Phase 1: definition of the seismic input excitation and of the computational model. First, a ground earthquake acceleration record must be selected as seismic input excitation of the system. Then, one specifies general options for the physical model and its governing laws. In this work the system of Ordinary Differential Equations (ODEs) governing the Bouc–Wen hysteretic oscillator (Section 3) are implemented and solved numerically with the explicit Runge–Kutta method.
- Phase 2: definition of the probabilistic prior information on the model parameters. Once the computational model $\mathcal{M}(x)$ is defined, one has to select carefully which model parameters $x_{\mathcal{M}}$ to include in the calibration in order to get a reliable set of physical values from the resulting posteriors estimates after the Bayesian updating. Prior information on possible values of the hyperparameters are set by setting their prior probability distributions $\pi(x_{\mathcal{M}})$, defining for each hyperparameter the type of univariate distribution (i.e., uniform, Gaussian, lognormal distributions, etc.) and its statistical moments. The prior information is obtained making some considerations about the amount of the dissipated energy during the hysteresis.
- Phase 3: Bayesian model updating. At this stage, the Bayesian model updating can be carried out using the experimental data Y inferring the posterior distributions of the hyperparameters. However, in many practical applications, a closed form of Equation (7) does not exist. For this reason, Markov Chain Monte Carlo (MCMC) simulations have been conducted herein, allowing for an approximate expectation in Equation (8).

Specifically, the posterior distribution can also be seen as an intermediate quantity used to compute the conditional expectation of a certain quantity of interest (QoI) $h(x) : D_x \mapsto \mathbb{R}$ [27], so that:

$$\mathbb{E}\left[h(X)\big|Y\right] = \int_{D_x} h(x)\, \pi(x|Y)\, dx \approx \frac{1}{T} \sum_{t=1}^{T} h\left(x^{(t)}\right), \tag{8}$$

where $x^{(t)}$ is the step of the chain at iteration t, whereas T is the total number of the generated MCMC sample point. In this way, the posterior is explored by realizing appropriate Markov chains over the prior support. Specifically, the affine invariant ensemble sampler (AIES) solver algorithm [28] has been used in this work, and an additive Gaussian discrepancy with zero mean and residual diagonal covariance matrix $\Sigma = \sigma^2 I$ with unknown residual variances σ^2 has been introduced to connect the model response to the experimental observations. To infer σ^2, a uniform prior distribution was assumed:

$$\pi\left(\sigma^2\right) \sim \mathcal{U}\left(0, max|Y|^2\right), \tag{9}$$

whose standard deviation was set equal to the maximum of the absolute value over the time of the experimental observation Y at the beginning of the calibration. The Bayesian updating is then conducted by minimizing this value (i.e., maximizing the likelihood function of Equation (6)) until the convergence of the Markov chains is reached. The obtained sample can be used to estimate output statistics by drawing samples from the posterior. Finally, having obtained model parameters posterior distributions, a deterministic model response prediction can be estimated choosing the maximum a posteriori (MAP) or the mean of the distributions as a point estimate. The aforementioned approach is summarized by the flowchart in Figure 1.

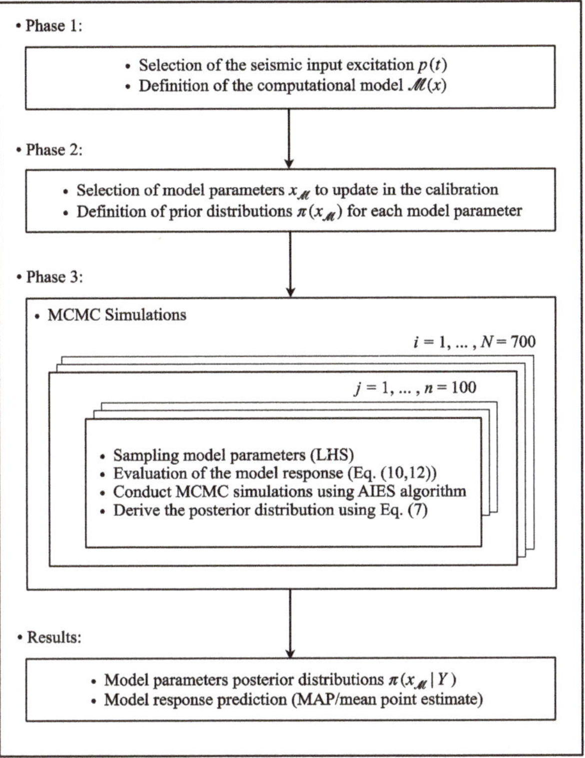

Figure 1. Flowchart of the Bayesian model calibration procedure.

3. Results

3.1. Numerical Benchmark: Calibration of a SDoF Bouc–Wen Type Hysteretic System

The Bouc–Wen model of hysteresis is widely used in structural engineering [29–34]. The model was proposed by Bouc [35], and thereafter modified by Wen [36]. The differential equation governing the motion of a single degree of freedom (SDoF) is written in the form:

$$m\ddot{u} + c\dot{u} + f_r(u(t), z(t)) = p(t), \tag{10}$$

where m is the mass of the system, c is the viscous linear damping coefficient, $u(t)$ is the displacement, $f_r(u,z)$ is the restoring force, and $p(t)$ is the external excitation. $\dot{u}(t)$ and $\ddot{u}(t)$ denote derivations in time. According to the Bouc–Wen model, the restoring force is given by the following expression:

$$f_r(u,z) = f_r^{el}(u,z) + f_r^{h}(u,z) = \alpha k_i u(t) + (1-\alpha)k_i z(t), \tag{11}$$

where $f_r^{el}(u,z)$ represents the elastic component whereas $f_r^{h}(u,z)$ is the hysteretic component which depends on the past history of the system response, k_i is the initial stiffness of the system, α is the ratio between the final stiffness and initial one and $z(t)$ is the hysteretic displacement as defined by Baber and Noori [37] to enhance the capacity of the original model to represent hysteretic cycles:

$$\dot{z}(t) = \frac{A(\varepsilon)\dot{u}(t) - \nu(\varepsilon)\left[\beta|\dot{u}(t)|\,|z(t)|_{N-1}z(t) + \gamma\dot{u}(t)|z(t)|^{N}\right]}{\eta(\varepsilon)}. \tag{12}$$

In Equation (12) the parameters β, γ and N control the shape of the cycles, while the additional parameters $A(\varepsilon)$, $\eta(\varepsilon)$ and $v(\varepsilon)$ are degradation functions in terms of the dissipated hysteretic energy $\varepsilon^h(t)$ and take into account the stiffness and strength degradation:

$$A(\varepsilon) = A_0(\varepsilon) - \delta_A \varepsilon^h(t), v(\varepsilon) = v_0(\varepsilon) - \delta_v \varepsilon^h(t), \eta(\varepsilon) = \eta_0(\varepsilon) - \delta_\eta \varepsilon^h(t), \tag{13}$$

where the constant values of A_0, v_0, η_0 are usually set to unity [38]. Whereas the values δ_A, δ_v, δ_η are constant terms which specify the amount of stiffness and strength degradation [39].

Finally, the dissipated hysteretic energy $\varepsilon^h(t)$ is given by:

$$\varepsilon^h(t) = \int_{u(0)}^{u(t)} f_r^h(u,z)du = (1-\alpha)k_i \int_0^t z(\tau)\dot{u}(\tau)d\tau. \tag{14}$$

A record of the Montenegro earthquake (1979) was selected from the PEER Ground Motion Database (PEER, 2019) as seismic excitation (Figure 2) for the benchmark.

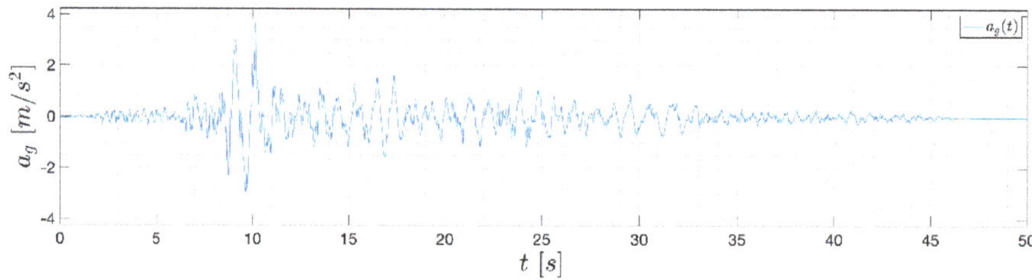

Figure 2. Montenegro (1979) ground motion record with peak ground acceleration (PGA) = 3.59 m/s^2.

This input excitation was used to simulate the response of a SDoF Bouc–Wen–Baber–Noori (BWBN) nonlinear system with mass of $m = 1200$ kg and monitored by a dynamic monitoring system composed of one accelerometer installed on the system lumped mass. The initial stiffness k_i of the system was calibrated with the objective of obtaining a plausible system frequency for a masonry structure of approximately 4 Hz. A damping ratio of $\xi = 3\%$ was adopted according to relevant literature on masonry structures. Finally, the remaining BWBN parameters used to generate the simulated record are presented in Table 1, while Figure 3 depicts the simulation record data. The Bayesian calibration was then carried out on the simulated velocity response. The latter was chosen because it is simpler to describe such a system by a single ODE with respect to velocity. Deterministic values were set for the following parameters of the BWBN model: $m = 1200$ kg, $A = 1$, $N = 1$, and also for the damping ratio, $\xi = 3\%$. The latter assumption was made to release uncertainties related to the identification of ξ from the calibration process. Instead, the remaining ones were considered independent random variables with associated distributions given in Table 2 and constitute the vector of uncertain model parameters x_M. It is worth specifying that the degradation effect of the BWBN model has been neglected on purpose, in order to introduce a discrepancy in the model used for emulating the reference response of the system. Besides, no uncertainties in input excitation have been considered herein, in order to address them only on the model side. The first moments of the prior probability density functions (PDFs) of model parameters k_i and α, namely $\mu_{k_i} = 7.42$ and $\mu_\alpha = 0.063$, were selected by two linear regressions on the reference restoring force curve for small and high displacements. The standard deviations of these parameters were chosen in order to take into account an amount of variation from these values of the 10%. It is worth noticing that in Table 2 a new random variable $\gamma*$ was introduced in order to guarantee the bounded-input, bounded-output (BIBO) stability [40]. Indeed, assuming as prior knowledge on the parameter γ the required BIBO stability condition (i.e., γ being uniformly distributed between the bounds $-\beta \leq \gamma \leq \beta$), the conditional distribution $\pi(\gamma|\beta)$ must be uniform.

Table 1. Parameters of the BWBN model adopted for simulating the record data.

k_i (N/m)	β (m^{-1})	γ	N	δ_A	δ_v	δ_η
7.6	63	63	1	0	2.43	6.5

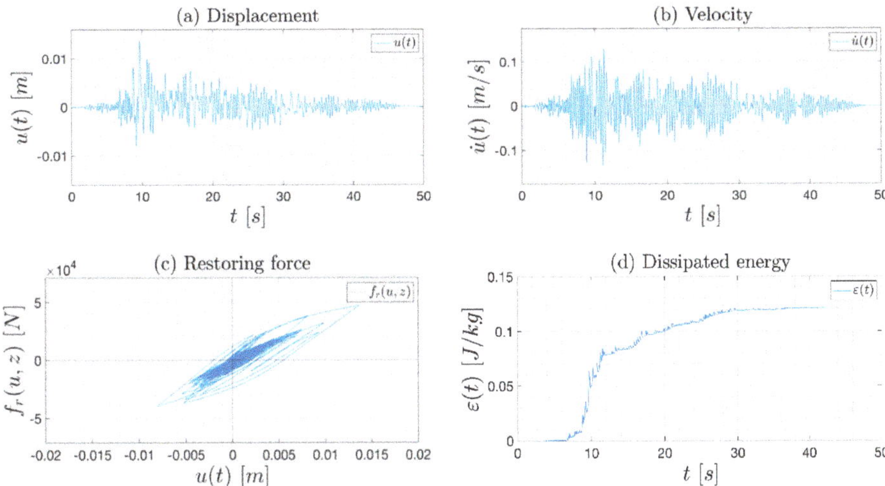

Figure 3. Simulated reference data: (**a**) response displacement of the Bouc–Wen–Baber–Noori (BWBN) oscillator; (**b**) response velocity of the BWBN oscillator; (**c**) hysteresis loop of the BWBN system; (**d**) total dissipated energy of the BWBN system (both elastic and hysteretic components of the restoring force are gathered), normalized with respect to the mass of the oscillator.

Table 2. Uncertain parameters of the BWBN benchmark model.

Parameter	Distribution	Support	Mean	Std. Dev.
k_i(N/m)	Gaussian	[4.27, 11.85]	7.42	0.1
$\beta\left(\text{m}^{-1}\right)$	Uniform	[55, 65]	60	2.8
$\gamma*$	Uniform	[−1, 1]	0	0.57
α	Lognormal	[0, 1]	0.063	0.01

This can be achieved by transforming the input variables, namely removing the parameter γ and introducing an auxiliary variable $\gamma* \sim U(-1,1)$. Hence, for a joint realization of the parameters β and $\gamma*$, the actual parameter reads: $\gamma = \gamma * \beta$ for $\beta \neq 0$. The standard deviations of the latter parameters were computed trivially once the supports of the uniform distributions were defined. To infer the unknown residual variances σ^2 (Equation (9)), a uniform prior distribution was assumed:

$$\pi\left(\sigma^2\right) \sim U\left(0, max|\dot{u}|^2\right), \tag{15}$$

whose standard deviation was set equal to the maximum of the absolute value over the time of the experimental observation $\dot{u}$. Markov Chain Monte Carlo (MCMC) simulations have been conducted using MATLAB-based Uncertainty Quantification framework UQLab-V1.3-113® [6] with 100 chains, 700 steps and AIES solver algorithm [28]. The number of the BWBN forward model $\mathcal{M}^{BW}(x)$ calls in MCMC was 70′000. The system of ordinary differential equation (ODEs; Equations (10) and (12)) was solved with the MATLAB solver ode45 (explicit Runge–Kutta method with relative error tolerance 1×10^{-3}). The Latin hypercube sampling (LHS) method was used to get the parameters prior distributions (Table 2) of the BWBN model, which are shown in Figure 4.

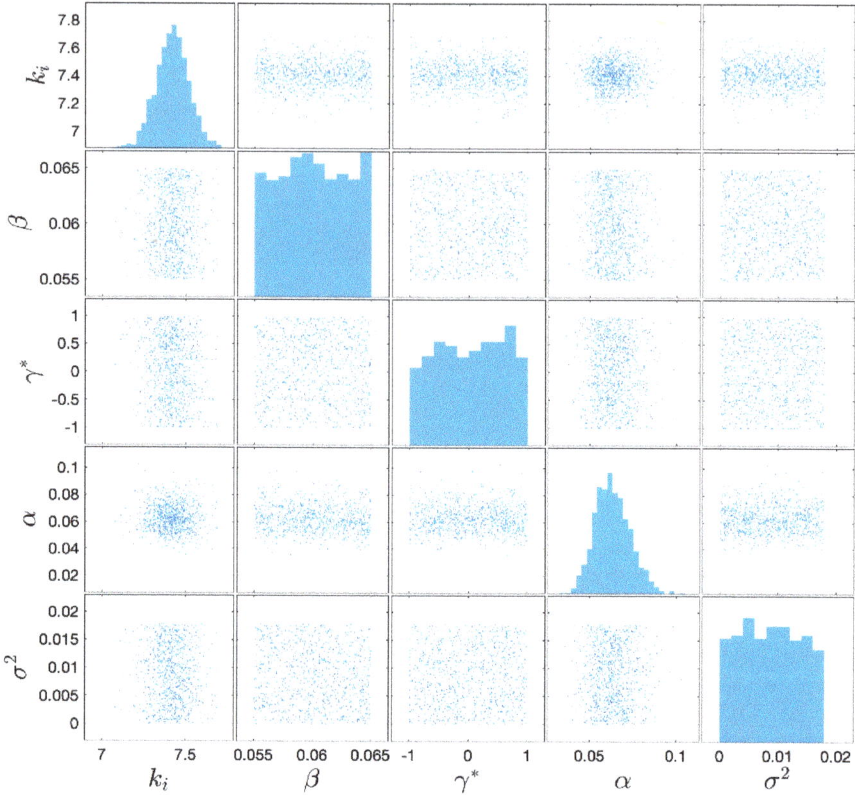

Figure 4. Prior samples of the BWBN benchmark model parameters.

The evolutions of the Markov chains are plotted instead in Figure 5 for each model parameter $x = \{k_i, \beta, \gamma*, \alpha, \sigma^2\}$. These give valuable insights about convergence of the chains. Indeed, it can be clearly seen that 700 steps are sufficient to reach the steady state. Consequently, samples generated by the chains follow the posterior distributions. However, the sample points generated by the AIES MCMC algorithm have been post-processed carrying out the burn-in of the first half of sample points before convergence to avoid the pollution of the estimate of posterior properties. In our case, the quality of the generated MCMC chains can be judged satisfactory. Table 3 reports the result of the Bayesian inversion analysis: mean and standard deviation of the posterior 5%–95% quantiles of the distribution and Maximum A Posteriori (MAP) point estimate. The MAP can be assumed as the most probable parameter value following calibration. This value has been used as a best-fit parameter.

Table 3. Marginal posterior distribution of the BWBN benchmark model parameters.

Parameter	Mean	Std. dev.	(0.05–0.95) Quant.	MAP
k_i (N/m)	4.5	0.18×10^{-1}	(4.4–4.5)	4.4
$\beta \left(\text{m}^{-1} \right)$	65	0.36	(64–65)	64
$\gamma*$	-0.38	0.25	$(-0.4\text{--}-0.34)$	-0.40
α	0.035	0.53×10^{-2}	(0.027–0.044)	0.031
σ^2	2×10^{-4}	4.7×10^{-6}	$(1.9\text{--}2.1) \times 10^{-4}$	60×10^{-3}

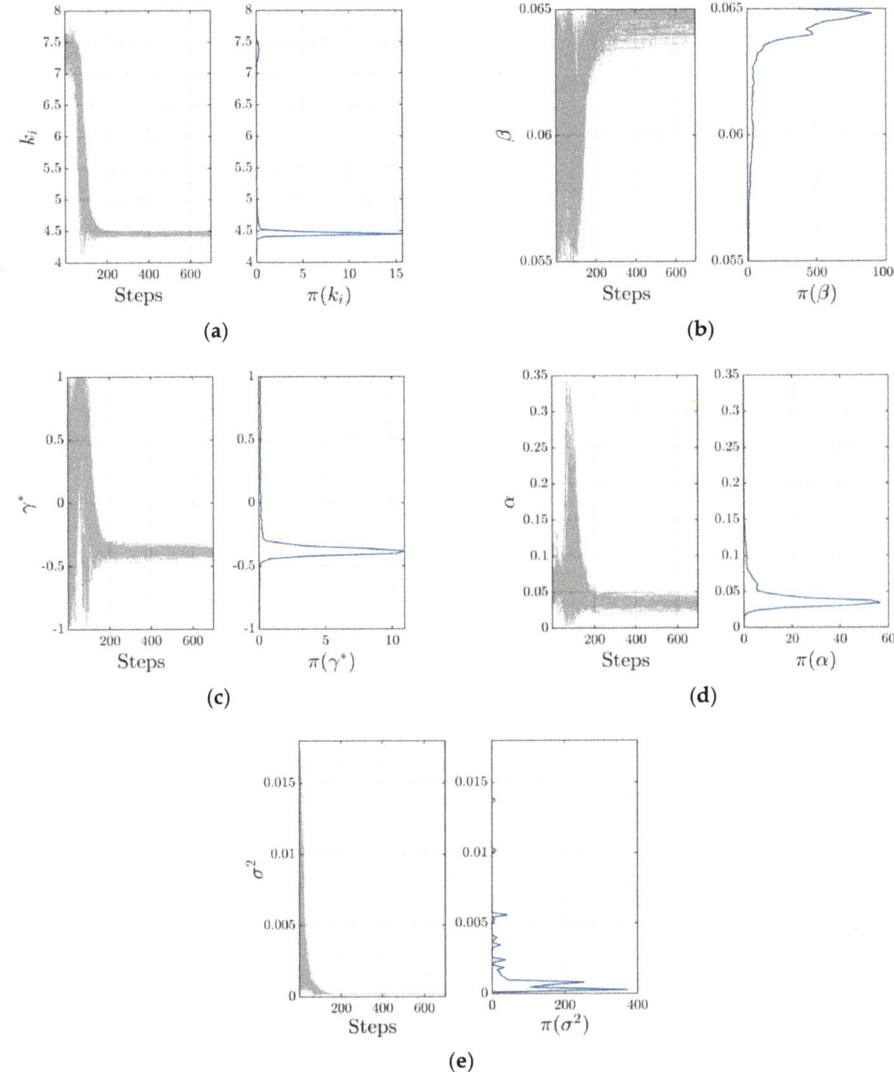

Figure 5. Trace plots of the Markov Chains and corresponding kernel density estimation (KDE) for each BWBN model parameter [6]: (**a**) k_i; (**b**) β; (**c**) $\gamma*$; (**d**) α; (**e**) σ^2.

The model response prediction using MAP as a point estimate of the posterior distributions is plotted in Figure 6. It can be clearly seen how the inferred response reproduces the experimental record quite accurately, with a mean square error (MSE) over the time of 2.007×10^{-4} (i.e., half of the average of the MSE computed on all the post-predictive model runs, MSE $= 4.0415 \times 10^{-4}$). In fact, we should not dwell on this specific prediction; rather, we should consider its confidence intervals. The latter tells us how uncertainties on model input propagate through the model.

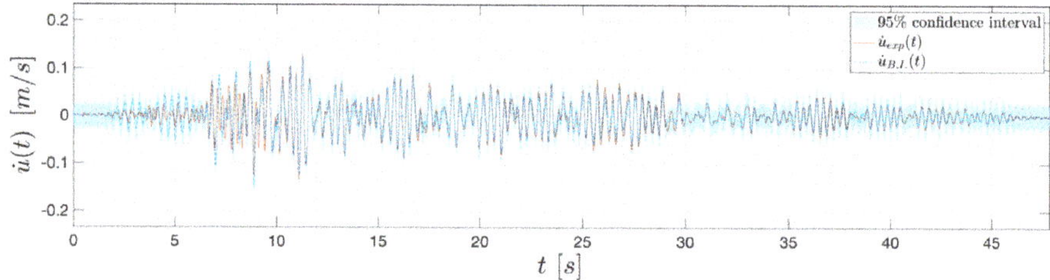

Figure 6. Velocity response prediction of the BWBN benchmark model.

From inspection of Figure 6 it can be concluded that uncertainties on the input parameters produce higher uncertainty of the response prediction in the lower amplitude regions rather than in the peak values. However, this is a satisfying result in earthquake engineering, where one is more interested in predicting maximum response quantities. The posterior distributions of the calibrated parameter are plotted in Figure 7.

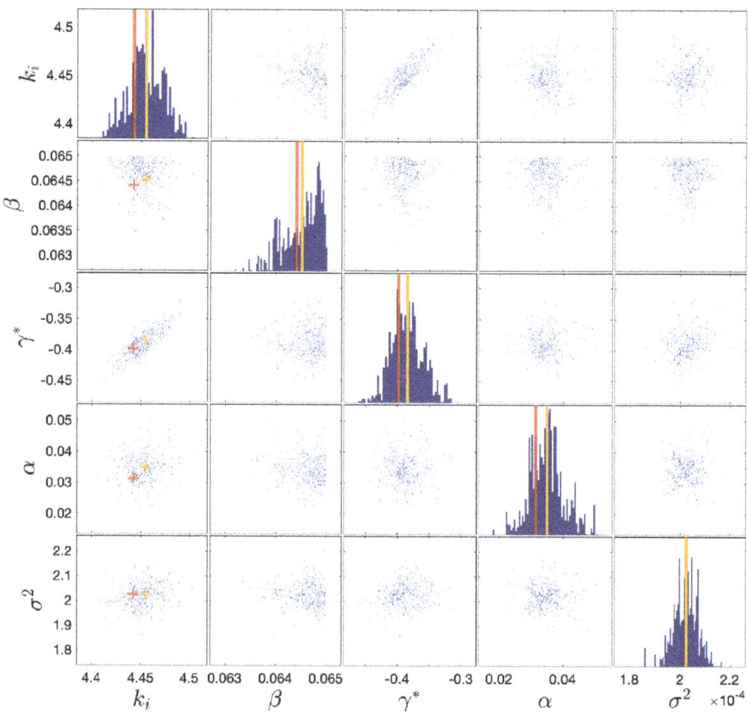

Figure 7. Posterior samples of the BWBN benchmark model parameters. The vertical lines denote the mean of the distribution (in yellow) and the maximum a posteriori estimate (MAP) (in red).

3.2. Demonstration on a Case Study

Having established that the UQ framework can significantly help to gain insight into the uncertainties intrinsic to calibration, thus furnishing a realistic level of confidence on model prediction, this can now be applied to a case study. The aim is to study how the UQ works with an identified hysteretic and degrading model that emulates the experimental response of a real monitored masonry building, the Town Hall of Pizzoli, subjected to its real earthquake recorded data. This is a three-story stone masonry building located northwest of the city of L'Aquila (Abruzzo), which is away from the city. It was built around 1920 and it formerly hosted a school. The structure presents a u-shaped regular plan mainly distributed along one direction, and its elevations are characterized by regular openings distributed along three levels above the ground (the raised ground floor, the first floor and the under-roof floor). The total area is about 770 m^2 while the volume is about 5000 m^3. Figure 8 reports the schematizations of the analyzed building. The Town Hall of Pizzoli belongs to the network of buildings monitored by OSS [26]. The OSS monitors the structure of Pizzoli thanks to a dynamic monitoring system composed of eight accelerometers installed on the building and one placed in the basement. More specifically, a tri-axial accelerometer to record the seismic input in the three space directions is fixed in the basement. On the raised ground floor, no accelerometers are present. On the first floor, three biaxial accelerometers and one monoaxial accelerometer (in X direction) are present to capture the acceleration responses of the first floor in the two horizontal directions. Finally, the same scheme of sensors installed on the first floor is present on the second floor. It is worth noting that the acquisition system and the sensors setup was designed by OSS and represents an input data for our analyses. In this regard, to obtain an optimal positioning of the sensors, a finite element model should be used. In particular, the optimal sensor placement is done recurring to a linear finite element model (using eigen-analysis, singular value decomposition, etc.). The model is often linear to perform this task because for nonlinear analysis the linear elastic component commonly brings the predominant amount of the output variance.

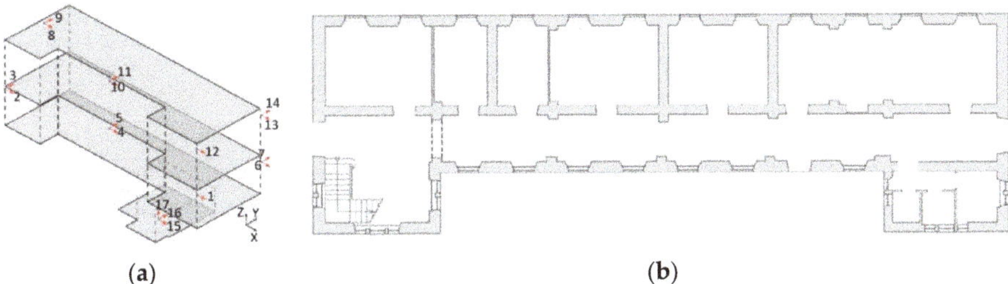

(a)   (b)

Figure 8. Building schematization: (**a**) floors schematization with sensors location; (**b**) plan.

The system sampled the signals at 250 Hz, for a total duration of about 50 s for the analyzed earthquake. The acquisition system recorded the sequence that struck central Italy in 2016. For this study, the earthquake acceleration responses of the building recorded on 30 October 2016 were used as reference signals. It is worth noting that accelerometric data are used in this work to define the likelihood function during the calibration. This is true as the use of inter-storey drift in the definition of the likelihood function would require its estimation. This estimate is often complicated because the filtering and integration operation on experimental data usually brings errors during the elaboration, especially with respect to the time alignment of two different recorded signals (quantities that are needed to estimate the inter-storey drift).

Since the scope of this study is to show how the discrepancy of models affects the calibration of model parameters value for seismic SHM, only the inter-storey response of the under-roof floor (2nd DoF) has been considered as reference output quantity. The acceleration of the first floor (1st DoF) is considered as the input signal of the reference model. The decision to use an inter-storey law is based on the fact that the results of the calibration will be compared with the results of calibration performed with discrepant models. In this sense, a single DoF model helps to make a direct comparison of the results, since the correspondence of parameters belonging to different models becomes less obvious for multi DoF models. The idea is to present a numerical case study derived from a real-world structure, with realistic values for hysteretic model parameters.

3.2.1. Reference Model

In this subsection, the model that came as the result of a nonlinear identification previously performed on the Pizzoli Town Hall structure is described for completeness. The nonlinear identification, like in [41,42], relied on a plane frame model in the direction of minor inertia of the building. Assuming the mass contribution of the raised ground floor in the dynamics of the structure is negligible, the building has been modelled with two lumped masses in correspondence to the first and under-roof floor.

The identification procedure consisted of using absolute acceleration data measured at the floors to approximate the response with a generalized linear model obtained by expanding the Bouc–Wen model of hysteresis.

The basic functions of the generalized linear model ended up being nonlinear in terms of the exponential parameters of the BW model. These parameters were therefore identified with specialized optimization algorithms, while for the remaining parameters a direct estimate in the joint time-frequency domain was performed. The identification provided also instantaneous values of the BW model parameters made explicit by the generalized linear model. The main findings of the identification can be summarized as follow:

- The adopted procedure allowed verification of the consistency of the assumed nonlinear model. This was done by checking the stability of the values of the model parameters over time.
- The resulting model satisfactorily reproduced the experimental response. This was validated in [43] by comparing the model results with the experimental findings obtained by the authors and by other researchers [44] independently,
- The procedure allowed for the collection of timely information on the health of the structure immediately after the occurrence of the earthquake.

In the identification process, a global box-like behavior was assumed, also based on the results of on-site inspections that led to the verification of the existence of good connections between walls and floor-walls before the occurrence of seismic events. In addition, no strength deterioration was accounted for (i.e., $v\left(\varepsilon^h\right) = 1$), while for the stiffness degradation, only the proportional component was maintained (i.e., $\eta\left(\varepsilon^h\right) = 1$ and $A\left(\varepsilon^h\right) = 1 - \delta_A \varepsilon^h(t)$). The parameter α was instead imposed to zero. These choices were made to make the reference model consistent with the model identified in [43], which was calibrated with experimental data.

The reference two DoFs model has been reconducted to a DoF by imposing the acceleration response of the first floor of the Town Hall (recorded by the monitoring system), $\ddot{u}_1(t)$, as the input signal. The output was obtained for the inter-story drift, $u(t)$, between the under-roof and the first floor. For further information on the identified model one can refer to [43].

Thus, the reference model used for the analysis from now on becomes:

$$m \cdot \ddot{u}(t) + f(t) = -m \cdot \ddot{u}_1(t); \tag{16}$$

$$\dot{f}_L(t) = k\left(\varepsilon^h(t), t\right) \cdot \dot{u}(t); \tag{17}$$

$$k\left(\varepsilon^h(t), t\right) = k_i \cdot \left(1 - \delta_A \cdot \varepsilon^h(t)\right) = k_i - k_i \delta_A \cdot \varepsilon^h(t); \tag{18}$$

$$\dot{f}_{NL}(t) = -\beta \dot{u}(t)\left|f(t)\right|^N sign\left[\dot{u}(t)f(t)\right] - \gamma \dot{u}\left|f(t)\right|^N. \tag{19}$$

The reference model is depicted in Figure 9 with the acceleration input $\ddot{u}_1(t)$, while the corresponding model parameters, identified in a previous study with the Pizzoli Town Hall building responses, are reported in Table 4. For the numerical simulations, a viscous damping term (with a damping ratio of 3% set as a deterministic value in the calibration process) has been added to the model of Equation (16).

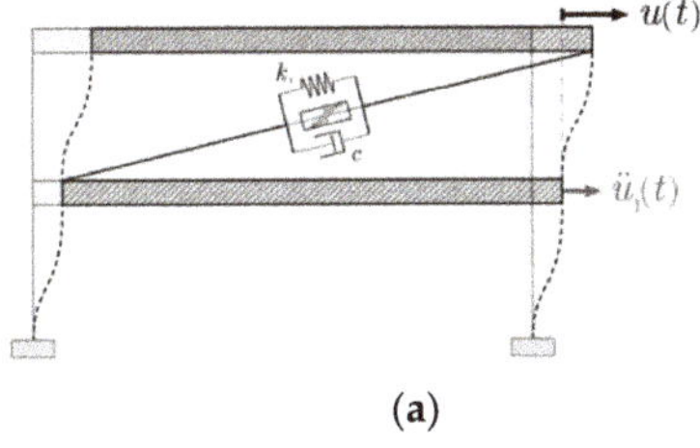
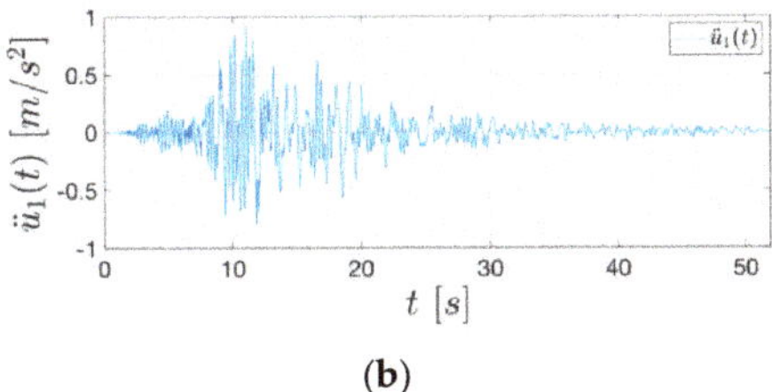

(a)

(b)

Figure 9. Reference models. (**a**) Structural model; (**b**) acceleration of the first floor.

Table 4. Reference model parameters.

Parameter	Value
$k_i\ (N/m)$	5.735×10^8
$\beta\ (1/m)$	35.01
γ	-16.68
N	1
$\delta_A\ (1/J)$	1.812×10^{-6}
$m\ (kg)$	$573{,}459$

3.2.2. Bayesian Calibration of the Reduced Single DoF Reference Model

The reduced single DoF BWBN model was used to validate the entire procedure in the first phase. Tables 5 and 6 report the prior information on model parameters and the results of the Bayesian inversion, respectively. Figure 10 depicts the posterior distributions of the parameters, while Figure 11 depicts the velocity response of the oscillator once the model has been calibrated. From the inspection of the latter, it can be concluded that uncertainties on the input parameters do not affect the response prediction.

Table 5. Uncertain parameters of the BWBN model for the case study.

Parameter	Distribution	Support	Mean	Std. Dev.
$k_i\ (N/m)$	Lognormal	$[0, \infty]$	6.2×10^2	0.5×10^2
$\beta\ (m^{-1})$	Uniform	$[20,\ 65]$	4.25	1.3
$\gamma*$	Uniform	$[-1,\ 1]$	0	0.57
N	Uniform	$[0.1,\ 2]$	1.05	5.5×10^{-1}
δ_A	Uniform	$\left[1 \times 10^{-8}, 1 \times 10^{-4}\right]$	5×10^{-5}	2.88×10^{-5}

Table 6. Marginal posterior distribution of the BWBN model parameters for the case study.

Parameter	Mean	Std. Dev.	(0.05–0.95) Quant.	MAP
$k_i\ (N/m)$	5.7×10^2	0.55×10^{-1}	$(5.7\text{–}5.7) \times 10^2$	5.7×10^2
$\beta\ (m^{-1})$	34	0.26	$(34\text{–}35)$	34
$\gamma*$	-0.47	0.007	$(-0.49\text{–}-0.46)$	-0.48
N	0.98	0.15	$(0.96\text{–}1)$	1
δ_A	4.2×10^{-5}	6.8×10^{-6}	$(2.8\text{–}5.3) \times 10^{-5}$	3.8×10^{-5}
σ^2	1.2×10^{-10}	3.6×10^{-11}	$(0.52\text{–}1.7) \times 10^{-10}$	3.5×10^{-11}

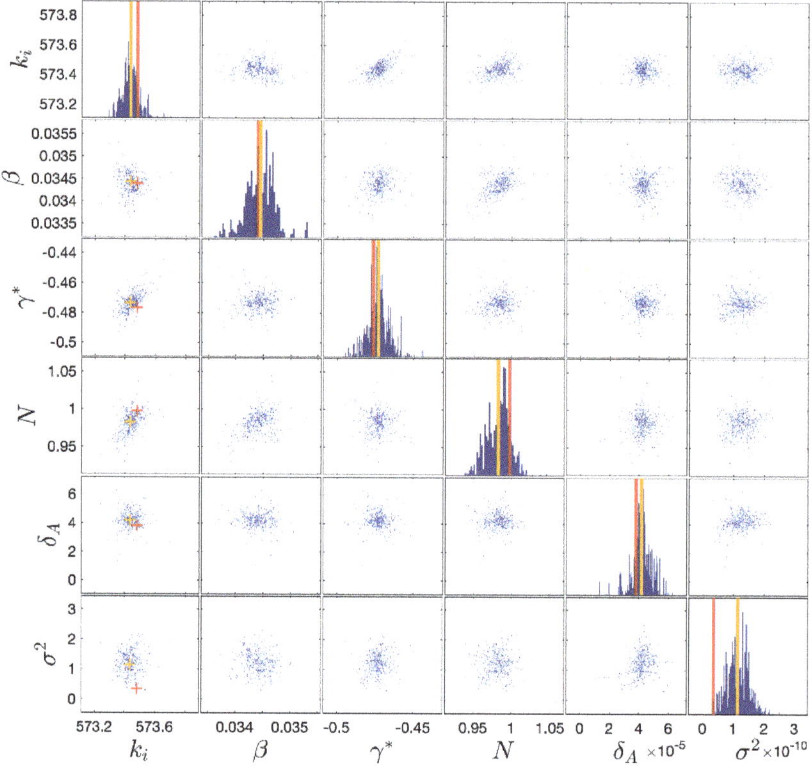

Figure 10. Posterior samples of the BWBN model parameters for the case study. The vertical lines denote the mean of the distribution (in yellow) and the maximum a posteriori estimate (MAP) (in red).

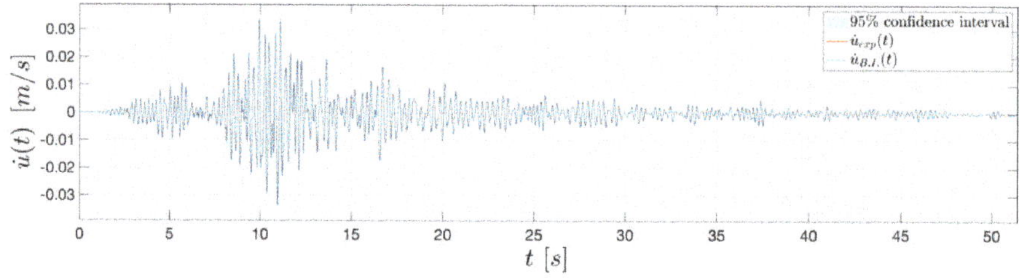

Figure 11. Velocity response prediction of the BWBN model for the case study.

Moreover, the value of the degrading term δ_A turns out not to influence the response prediction. This is due to the fact that during the earthquake shaking (PGA ≈ 1 m/s^2) the structure exhibited a low level of damage [44].

Both parametric and non-parametric degradation models have been used herein. Two models per each family are first mathematically introduced, then insights into the accuracy afforded by each model are provided.

3.2.3. Models for Stiffness Degradation

Besides the original BWBN degrading model, an additional parametric model is now introduced to investigate the capability of the Gaussian model discrepancy term to cover model uncertainty. An important part of the current literature on damage indices focuses on exponential function of the dissipated energy.

Consequently, an energy-based exponential function, according to [45], has been used in this work to replace the original BWBN stiffness degradation term $(1 - \delta_A \varepsilon(t))$ that multiplies the initial stiffness k_i:

$$D_E(t, \varepsilon, \beta) = \exp\left(-\beta \frac{\varepsilon(t)}{\int (-m\ddot{u} - ma_g)dt}\right),\tag{20}$$

where β is an additional parameter to infer jointly with k_i. In order to replicate the degrading effect on the initial stiffness k_i through non-parametric models, probability distribution functions have been used herein. First, according to the procedure used in [46], a Weibull distribution function was adopted:

$$D_W(t, R_W, p_1, p_2) = \left(1 - \left[R_W - R_W \, exp\left(-\frac{t}{p_1}\right)^{p_2}\right]\right),\tag{21}$$

where R, p_1, p_2 are the distribution parameters to infer jointly with k_i. Meaning to these parameters may be given as follows: p_1 represents the instant of time at which there is loss of stiffness; p_2 controls its rate; and R_W controls the amount of damage. Secondly, a logistic distribution function was adopted:

$$D_L(t, R_L, \mu, s) = \left(1 - \frac{R_L}{1 + exp\left(-\frac{t-\mu}{s}\right)}\right).\tag{22}$$

In this case, parameter μ is the instant of time in which the loss of stiffness occurs, while parameter s represents its rate, and R_L defines the amount of degradation.

3.2.4. Comparison of the Calibrated Models

It is useful to make a comparison among all the models introduced to get insights on their accuracy in predicting the response of the oscillator. For the sake of simplicity, from now *Model 1*, *Model 2*, *Model 3* and *Model 4* will refer to the models with the degrading law of the original BWBN model, the exponential law, the Weibull and the logistic distribution functions, respectively. It is also worth noticing that, in order to reduce the number of parameters to infer, only the initial stiffness k_i and the degrading term have been involved in the calibration process. This can be easily justified by the fact that, for practical earthquake engineering applications, the focus relies on the quantification of the structure's stiffness and its drop (especially for masonry structures which are prone to crack during a seismic event). Prior information on the possible value of the model parameters was obtained by visual inspection of the dissipated energy versus time curve. For both parametric and non-parametric models, the velocity response of the system obtained was almost identical to Figure 11 for a low-level of degradation. Figure 12 depicts the posterior distributions of the initial stiffness parameter k_i for each model considered. From this inspection, it can be seen that all models were able to predict the correct value of the initial stiffness. The choice of the MAP (or of the mean of the distributions) as a point estimate after the uncertainty propagation led to a slightly underrated initial stiffness regardless of the model adopted. However, this is neglectable from an engineering point of view. Moreover, Figure 12 depicts the distributions of the discrepancy variance term σ^2 as well. All the models share the same order of magnitude for σ^2 and, more or less, the same variance. This can be read as follows: the adopted Gaussian discrepancy term, with mean null and unknown variance, performs as a good function to embody all the model errors arising. Thus, at the end, it appears clear how the choice of the model used to infer the stiffness of the oscillator is not meaningful in the presence of a low-level of damage.

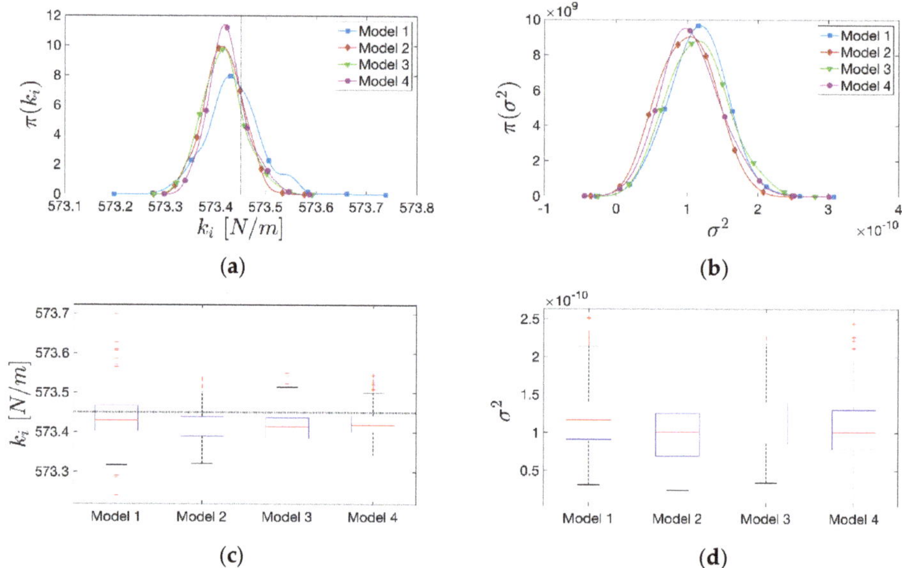

Figure 12. (**a**) Kernel density estimations (KDE) of the initial stiffness model parameter k_i (on the left) and; (**b**) of the discrepancy variance σ^2 for each model (low-level of degradation, PGA $\approx$ 1 m/s^2). (**c**) Boxplot representations (based on the five-number summary: minimum, first quartile, median, third quartile and maximum) of the posterior samples for the model parameter k_i and; (**d**) σ^2. The horizontal and vertical dotted lines in the plots represent the calibrated reference value of k_i.

3.3. Influence of the Degradation Level on Model Parameters Inference

Having established that the choice of the models used to infer the stiffness parameter k_i appears to not be relevant for a low-level of degradation, the aim is now to investigate what happens in possible further scenarios in which a higher level of damage is exhibited. This study is important when working in the field of model calibration because it explains the importance of linking the choice of the discrepancy model distribution—which is needed as input in the calibration process—to the degradation amount and the characteristics of the external force (e.g., amplitude and frequency).

To this aim, the input excitation was rescaled. Specifically, two cases are examined as follows: (i) one in which the input has been rescaled with a PGA of 6 m/s^2 (medium level of degradation); (ii) one in which the input has been rescaled with a PGA of 9 m/s^2 (high level of degradation).

3.3.1. Medium Level of Degradation

When increasing the level of excitation to a medium value (i.e., PGA of 6 m/s^2) and keeping the same reference damage, all the models were still able to predict the correct value of the initial stiffness k_i. However, an increase of the uncertainties in the model prediction was registered, this time due to the greater values of the discrepancy variance. The choice of the MAP (or of the mean of the distributions) as point estimate led to a slight overrating of the initial stiffness. As the estimation error was negligible, the adopted discrepancy term still performs as a good function to embody all the model errors. However, increasing the level of the input without accounting for a growth of the damage level appears to be unrealistic. This is because higher seismic excitations generally involve greater dissipation of energy. Therefore, a loss up to 40% of the initial stiffness k_i was taken into account as well, resulting in Model 1 (i.e., the one consistent with the BWBN degrading model, which is able to predict the correct value of the initial stiffness). On the contrary, all the models inconsistent with the BWBN model (i.e., Models 2, 3 and 4) overrate the initial stiffness value in a non-negligible proportion (Figure 13), leading to higher level of uncertainty in the model response prediction. It is also worth noticing how the estimation error was amplified for non-parametric models. This is a considerable indicator of the inadequacy of the discrepancy function adopted. In order to recover the response records, more accurate and well-suited discrepancy models should be investigated, but this is beyond the scope of this paper.

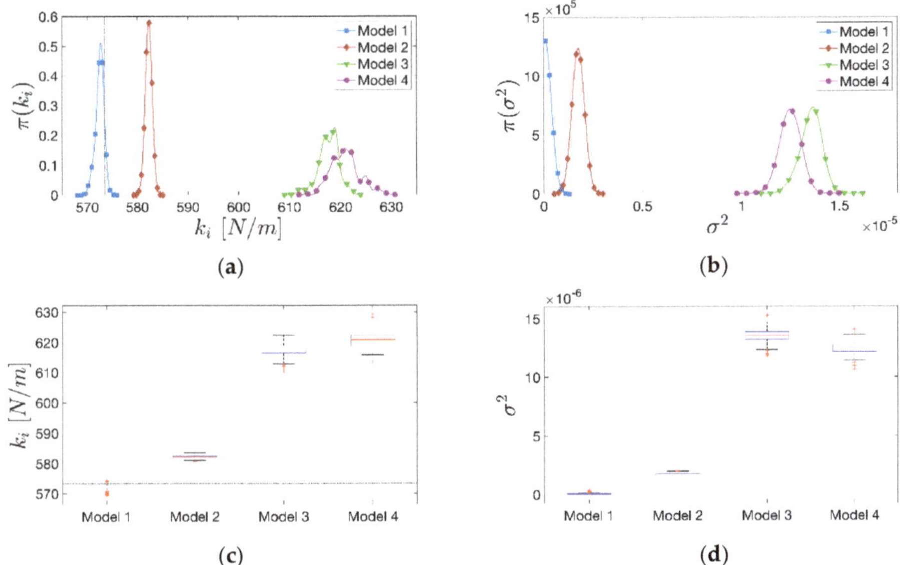

Figure 13. (a) Kernel density estimations (KDE) of the initial stiffness model parameter k_i and; (b) of the discrepancy variance σ^2 for each model (medium level of degradation: PGA $\approx$ 6 m/s^2, stiffness reduction up 40% of the initial one). (c) Boxplot representations (minimum, first quartile, median, third quartile and maximum) of the posterior samples for the model parameter k_i and; (d) σ^2. The horizontal and vertical dotted lines in the plots represent the calibrated reference value of the initial stiffness.

3.3.2. High Level of Degradation

The final step consisted of increasing the level of input to a high value (i.e., PGA of 9 m/s^2). Again, when keeping the same level of reference damage, all models were able to predict the correct value of the initial stiffness although with a visibly greater variance than before. When the level of damage increased, taking into account a stiffness loss of up to the 80% of the initial one, the same conclusions made in Section 3.3.1 were also confirmed, stressing the fact that level of uncertainty is now greater due to both greater excitation and a higher level of damage (Figure 14).

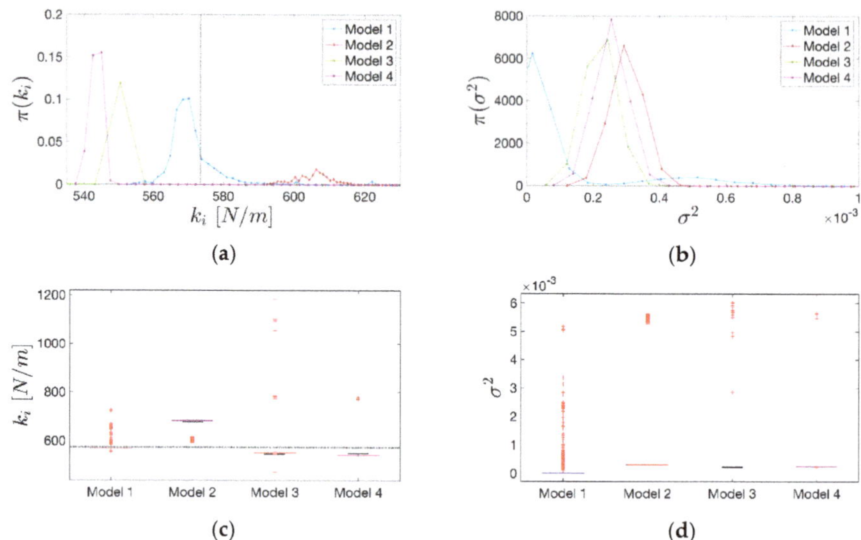

Figure 14. (**a**) Kernel density estimations (KDE) of the initial stiffness model parameter k_i and; (**b**) of the discrepancy variance σ^2 for each model (medium level of degradation: PGA $\approx$ 9 m/s^2, stiffness reduction up to 80% of the initial one). (**c**) Boxplot representations (minimum, first quartile, median, third quartile and maximum) of the posterior samples for the model parameter k_i and; (**d**) σ^2. The horizontal and vertical dotted lines in the plots represent the calibrated reference value of the initial stiffness.

4. Discussion

This realistic case study has shown that a Gaussian discrepancy term, with null mean and unknown variance, is an effective choice to tackle the model inaccuracy rising for low levels of degradation when the aim is inference of the system's initial stiffness. At the same time, it has been proven that this kind of discrepancy still works fine for low levels of damage even if the input is rescaled up to 1 g. Consequently, the discrepancy term adopted can be judged suitable to cover model errors with increasing values of the PGA. However, higher levels of PGA introduce higher uncertainties in the model prediction. It has been shown that increasing the amount of degradation jointly to the level of damage leads to inaccurate predictions for those models that are inconsistent with the original physical BWBN model and more uncertainty also arises in the latter. This can be appreciated, for example, in the velocity response prediction of Model 3 (Figure 15). This is the model that leads to the worst model prediction for high level of degradation, owing to the presence of a greater number of overestimated outliers.

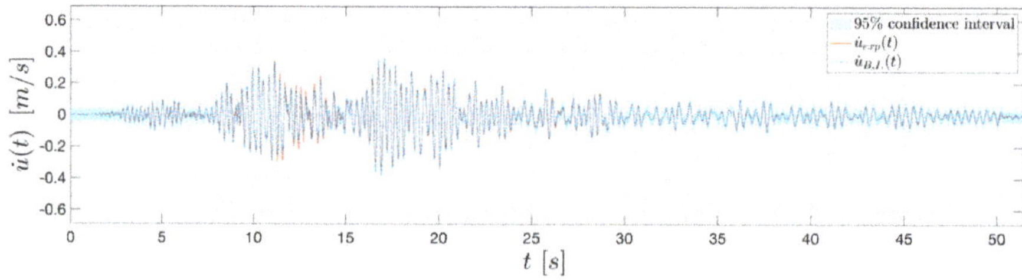

Figure 15. Velocity response prediction of the BWBN Model 3 for high level of degradation (PGA of 9 m/s^2, stiffness reduction up to 80% of the initial one).

What has been determined can be easily read from the graphs in Figure 16. The latter depicts the MSE error over the time of the velocity response prediction for each case. Specifically, Figure 16a–c display the variation of MSE computed for all the post-predictive model runs, for each model and for each level of degradation (low, medium and high). Instead, the comparison between the average MSE computed on all the post-predictive models runs versus the MSE of the MAP estimate prediction at different levels of degradation, is presented in Figure 16d. From the inspection of latter, we can conclude how the error in prediction increases as the level of degradation increases. This leads to more uncertain predictions.

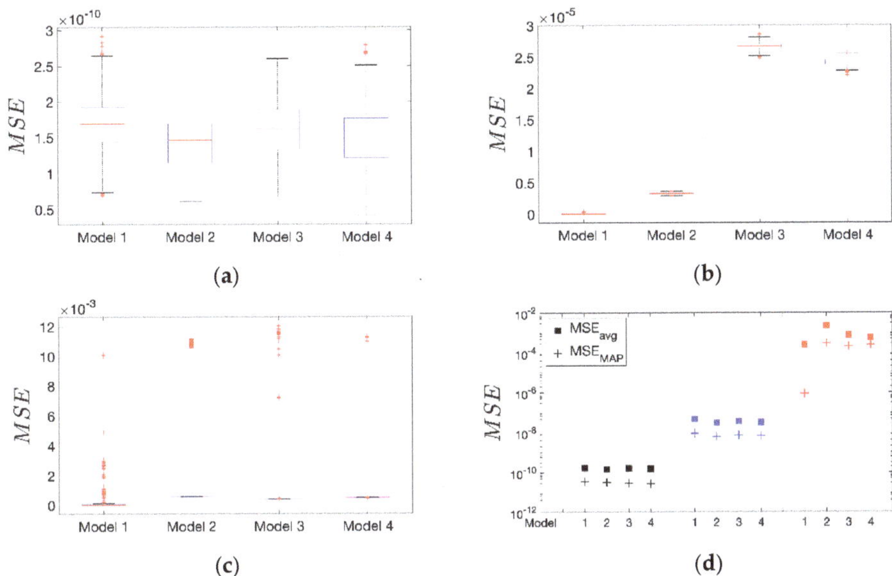

Figure 16. Boxplot representations (minimum, first quartile, median, third quartile and maximum) of the mean square error (MSE) computed on all the post-predictive models runs. Specifically: (**a**) boxplot of MSE for low level of degradation (PGA $\approx$ 1 m/s^2); (**b**) boxplot of MSE for medium level of degradation (PGA $\approx$ 6 m/s^2); (**c**) boxplot of MSE for high level of degradation (PGA $\approx$ 9 m/s^2). (**d**) Comparison between the average MSE computed over all the post-predictive model runs for each model (square markers), versus the MSE computed for the MAP estimate (cross markers). The comparison is differentiated into three colored subsets: (i) in black, MSEs computed for a low level of degradation (PGA $\approx$ 1 m/s^2); (ii) in blue, MSEs computed for a medium level of degradation (PGA $\approx$ 6 m/s^2); (iii) in red, MSEs computed for a high level of degradation (PGA $\approx$ 9 m/s^2).

At last, for those cases inconsistent with the physics model that presents from a medium to high level of damage, it can be concluded that the simple discrepancy function adopted is not sufficient to embody all the model errors. This means that the choice of the discrepancy function should be linked to the level of damage and, in actual practice applications, to the level of excitation.

Thus, to summarize the considerations on the use of Bayesian updating and the uncertainty quantification framework, one can state that the model errors and the observation errors can be taken into account by adding a discrepancy term in the model calibration process. This brings benefits in terms of calibration outcomes (i.e., in addition to good time-series fitting, the order of the parameters is close to the real value). However, for some cases, the Bayesian approach within the uncertainty quantification framework could bring trivial results in terms of model parameter values. This especially occurs if the discrepancy model distribution is not chosen properly. The findings of this paper show that a good way to choose the discrepancy model distribution could be to study potential correlations between it and the statistical nature of the external force (e.g., amplitude and frequency) and the statistical nature of the modal characteristics (e.g., natural frequencies) of a system, evaluated in operational conditions. For example, for external forces with frequency content close to the natural frequencies of the system, there is a high chance of high degradation to occur, and thus a discrepancy model distribution close to a Gaussian distribution could bring trivial results. This is clearly valid for a given model, since the model selection framework is another research topic that was not evaluated in this study and is complementary to the actual work.

This finding could stimulate further research on the statistical correlation of discrepancy models and the magnitude of excitation, helping to calibrate mathematical models that try to emulate real word systems and structures.

5. Conclusions

This article addressed the Bayesian probabilistic calibration of nonlinear hysteretic systems for model-driven seismic SHM purposes. The well-known Bayesian inversion framework was first illustrated through a numerical benchmark of a SDOF Bouc–Wen type oscillator, and then applied to a reference model derived from a calibrated 2-DOFs system of a real monitored masonry building hit by the 2016 Central Italy earthquake. The main novelty of this study is the application of this framework on models representing data measured directly on masonry structures during seismic events. In the examined cases, the model calibration procedures was carried out by incorporating the posterior uncertainty linked to model discrepancy in order to:

(i)   explicitly define the errors and uncertainties present in the model;
(ii)   provide the full multivariate distribution of the calibrated parameters;
(iii)   estimate the model discrepancy posterior distribution;
(iv)   provide insights on quantities of interest (e.g., maximum a posteriori estimates, time-history response prediction).

Furthermore, the study points to the importance of choosing a discrepancy model function correlated to the underlying seismic characteristics. This consideration can be drawn from the results obtained for the case study associated to Pizzoli Town Hall, for which several parametric and non-parametric degrading models were investigated to gain insight into how model inaccuracy could be embodied by the discrepancy function adopted. Specifically, with reference to non-parametric models, it has been determined that:

(i)   for low level of damage, and even for high levels of PGA, accurate predictions can be achieved adopting a Gaussian discrepancy term with null mean and unknown variance, which overcomes the low sensitivity of the term used to model the degradation in the response;
(ii)   for high levels of damage, on the contrary, the simple Gaussian discrepancy function is unable to tackle the model inaccuracy rising.

This means that, when the magnitude of the external excitation is very high (for the system considered), or the system is subjected to resonance, the system is likely to undergo damage (e.g., to high level of degradation) and thus simplified assumptions on the model discrepancy term are not reasonable. For these cases, the use of a fuller and more specific model for the mean of the discrepancy term should be used to get reliable estimates. These results suggest further studies to better relate discrepancy models to the statistical nature of the excitation and other factors that can influence probabilistic model calibration for use in seismic SHM.

Author Contributions: Methodology, A.F.; case study, A.F. and G.M.; review, R.C. and G.M., editing, A.F.; supervision, R.C. All authors have read and agreed to the published version of the manuscript.

Funding: This research received no external funding.

Acknowledgments: The data of the Town Hall of Pizzoli were supplied by the Seismic Observatory of Structures within the DPC-ReLUIS project. Computational resources were provided by HPC@POLITO (http://hpc.polito.it).

Conflicts of Interest: The authors declare no conflict of interest.

References

1. Biegler, L.; Biros, G.; Ghattas, O.; Heinkenschloss, M.; Keyes, D.; Mallick, B.; Tenorio, L.; van Bloemen Waanders, B.; Willcox, K.; Marzouk, Y. *Large-Scale Inverse Problems and Quantification of Uncertainty*; Wiley Online Library: Hoboken, NJ, USA, 2011; Volume 712.
2. Bernagozzi, G.; Mukhopadhyay, S.; Betti, R.; Landi, L.; Diotallevi, P.P. Output-only damage detection in buildings using proportional modal flexibility-based deflections in unknown mass scenarios. *Eng. Struct.* **2018**, *167*, 549–566. [CrossRef]
3. Liu, Y.; Mei, Z.; Wu, B.; Bursi, O.S.; Dai, K.-S.; Li, B. Seismic behaviour and failure-mode-prediction method of a reinforced-concrete rigid-frame bridge with thin-walled tall piers: Investigation by model-updating hybrid test. *Eng. Struct.* **2020**, *208*, 110302. [CrossRef]
4. Moon, F.; Catbas, N. *Structural Identification of Constructed Systems: Approaches, Methods, and Technologies for Effective Practice of St-Id*; American Society of Civil Engineers (ASCE): Reston, WV, USA, 2013. [CrossRef]
5. Beck, J.L.; Katafygiotis, L.S. Updating Models and Their Uncertainties. I: Bayesian Statistical Framework. *J. Eng. Mech.* **1998**, *124*, 455–461. [CrossRef]
6. Wagner, P.-R.; Nagel, J.; Marelli, S.; Sudret, B. *UQLab User Manual—Bayesian Inversion for Model Calibration and Validation*; Chair of Risk, Safety and Uncertainty Quantification: ETH Zurich, Switzerland, 2019.
7. Noel, J.; Kerschen, G. Nonlinear system identification in structural dynamics: 10 more years of progress. *Mech. Syst. Signal Process.* **2017**, *83*, 2–35. [CrossRef]
8. Ching, J.; Beck, J.L.; Porter, K.A. Bayesian state and parameter estimation of uncertain dynamical systems. *Probabilistic Eng. Mech.* **2006**, *21*, 81–96. [CrossRef]
9. Ortiz, G.A.; Alvarez, D.A.; Bedoya-Ruiz, D. Identification of Bouc–Wen type models using the Transitional Markov Chain Monte Carlo method. *Comput. Struct.* **2015**, *146*, 252–269. [CrossRef]
10. Chen, H.-P.; Mehrabani, M.B. Reliability analysis and optimum maintenance of coastal flood defences using probabilistic deterioration modelling. *Reliab. Eng. Syst. Saf.* **2019**, *185*, 163–174. [CrossRef]
11. Erazo, K.; Nagarajaiah, S. An offline approach for output-only Bayesian identification of stochastic nonlinear systems using unscented Kalman filtering. *J. Sound Vib.* **2017**, *397*, 222–240. [CrossRef]
12. Zhou, K.; Tang, J. Uncertainty quantification in structural dynamic analysis using two-level Gaussian processes and Bayesian inference. *J. Sound Vib.* **2018**, *412*, 95–115. [CrossRef]
13. Yan, G.; Sun, H. A non-negative Bayesian learning method for impact force reconstruction. *J. Sound Vib.* **2019**, *457*, 354–367. [CrossRef]
14. Yan, W.-J.; Chronopoulos, D.; Papadimitriou, C.; Cantero-Chinchilla, S.; Zhu, G.-S. Bayesian inference for damage identification based on analytical probabilistic model of scattering coefficient estimators and ultrafast wave scattering simulation scheme. *J. Sound Vib.* **2020**, *468*, 115083. [CrossRef]
15. Imholz, M.; Faes, M.; Vandepitte, D.; Moens, D. Robust uncertainty quantification in structural dynamics under scarse experimental modal data: A Bayesian-interval approach. *J. Sound Vib.* **2020**, *467*, 114983. [CrossRef]
16. Hizal, Ç.; Turan, G. A two-stage Bayesian algorithm for finite element model updating by using ambient response data from multiple measurement setups. *J. Sound Vib.* **2020**, *469*, 115139. [CrossRef]
17. Muto, M.; Beck, J.L. Bayesian Updating and Model Class Selection for Hysteretic Structural Models Using Stochastic Simulation. *J. Vib. Control.* **2008**, *14*, 7–34. [CrossRef]
18. Wu, M.; Smyth, A.W. Application of the unscented Kalman filter for real-time nonlinear structural system identification. *Struct. Control. Health Monit.* **2007**, *14*, 971–990. [CrossRef]
19. Astroza, R.; Ebrahimian, H.; Conte, J. Material Parameter Identification in Distributed Plasticity FE Models of Frame-Type Structures Using Nonlinear Stochastic Filtering. *J. Eng. Mech.* **2015**, *141*, 04014149. [CrossRef]
20. Conte, J.; Astroza, R.; Ebrahimian, H. Bayesian Methods for Nonlinear System Identification of Civil Structures. In Proceedings of the MATEC Web Conference, Zürich, Switzerland, 19–21 October 2015; Volume 24, p. 03002. [CrossRef]
21. Jalayer, F.; Iervolino, I.; Manfredi, G. Structural modeling uncertainties and their influence on seismic assessment of existing RC structures. *Struct. Saf.* **2010**, *32*, 220–228. [CrossRef]

22. Gardoni, P.; Der Kiureghian, A.; Mosalam, K.M. Probabilistic Capacity Models and Fragility Estimates for Reinforced Concrete Columns based on Experimental Observations. *J. Eng. Mech.* **2002**, *128*, 1024–1038. [CrossRef]

23. Bracchi, S.; Rota, M.; Magenes, G.; Penna, A. Seismic assessment of masonry buildings accounting for limited knowledgeon material by Bayesian updating. *Bull. Earthq. Eng.* **2016**, *14*, 2273–2297. [CrossRef]

24. Ramos, L.F.; Miranda, T.; Mishra, M.; Fernandes, F.; Manning, E. A Bayesian approach for NDT data fusion: The Saint Torcato church case study. *Eng. Struct.* **2015**, *84*, 120–129. [CrossRef]

25. Bartoli, G.; Betti, M.; Facchini, L.; Marra, A.; Monchetti, S. Bayesian model updating of historic masonry towers through dynamic experimental data. *Procedia Eng.* **2017**, *199*, 1258–1263. [CrossRef]

26. Dolce, M.; Nicoletti, M.; De Sortis, A.; Marchesini, S.; Spina, D.; Talanas, F.; De Sortis, A. Osservatorio sismico delle strutture: The Italian structural seismic monitoring network. *Bull. Earthq. Eng.* **2015**, *15*, 621–641. [CrossRef]

27. Marelli, S.; Sudret, B. UQLab: A Framework for Uncertainty Quantification in Matlab. *Vulnerability Uncertain. Risk* **2014**, 2554–2563. [CrossRef]

28. Goodman, J.; Weare, J. *Communications in Applied Mathematics and Computational Science*; Math. Sci. Publishing: Berkeley, CA, USA, 2010.

29. Kyprianou, A.; Worden, K.; Panet, M. Identification Of Hysteretic Systems Using The Differential Evolution Algorithm. *J. Sound Vib.* **2001**, *248*, 289–314. [CrossRef]

30. Worden, K.; Hensman, J. Parameter estimation and model selection for a class of hysteretic systems using Bayesian inference. *Mech. Syst. Signal Process.* **2012**, *32*, 153–169. [CrossRef]

31. Ben Abdessalem, A.; Dervilis, N.; Wagg, D.J.; Worden, K. Model selection and parameter estimation in structural dynamics using approximate Bayesian computation. *Mech. Syst. Signal Process.* **2018**, *99*, 306–325. [CrossRef]

32. Ceravolo, R.; Demarie, G.V.; Erlicher, S. Instantaneous Identification of Degrading Hysteretic Oscillators Under Earthquake Excitation. *Struct. Health Monit.* **2010**, *9*, 447–464. [CrossRef]

33. Ceravolo, R.; Demarie, G.V.; Erlicher, S. Instantaneous Identification of Bouc-Wen-Type Hysteretic Systems from Seismic Response Data. *Key Eng. Mater.* **2007**, *347*, 331–338. [CrossRef]

34. Ceravolo, R.; Demarie, G.V.; Erlicher, S. Identification of degrading hysteretic systems from seismic response data. In Proceedings of the 7th European Conference on Structural Dynamics, EURODYN 2008, Leuven, Belgium, 20–22 September 2008.

35. Bouc, R. Modèle mathématique d'hystérésis: Application aux systèmes à un degré de liberté. *J. Acust.* **1969**, *24*, 16–25.

36. Wen, Y.-K. Method for Random Vibration of Hysteretic Systems. *J. Eng. Mech. Div. ASCE* **1976**, *102*, 249–263.

37. Baber, T.T.; Noori, M. Random Vibration of Degrading, Pinching Systems. *J. Eng. Mech.* **1985**, *111*, 1010–1026. [CrossRef]

38. Ma, F.; Zhang, H.; Bockstedte, A.; Foliente, G.; Paevere, P. Parameter Analysis of the Differential Model of Hysteresis. *J. Appl. Mech.* **2004**, *71*, 342–349. [CrossRef]

39. Erlicher, S.; Bursi, O.S. Bouc–Wen-Type Models with Stiffness Degradation: Thermodynamic Analysis and Applications. *J. Eng. Mech.* **2008**, *134*, 843–855. [CrossRef]

40. Ikhouane, F.; Rodellar, J. *Systems with Hysteresis*; John Wiley & Sons, Ltd.: West Sussex, UK, 2007.

41. Bursi, O.S.; Ceravolo, R.; Erlicher, S.; Fragonara, L.Z. Identification of the hysteretic behaviour of a partial-strength steel-concrete moment-resisting frame structure subject to pseudodynamic tests. *Earthq. Eng. Struct. Dyn.* **2012**, *41*, 1883–1903. [CrossRef]

42. Ceravolo, R.; Erlicher, S.; Fragonara, L.Z. Comparison of restoring force models for the identification of structures with hysteresis and degradation. *J. Sound Vib.* **2013**, *332*, 6982–6999. [CrossRef]

43. Miraglia, G.; Lenticchia, E.; Surace, C.; Ceravolo, R. Seismic damage identification by fitting the nonlinear and hysteretic dynamic response of monitored buildings. *J. Civ. Struct. Health Monit.* **2020**, *10*, 457–469. [CrossRef]

44. Cattari, S.; Degli Abbati, S.; Ottonelli, D.; Marano, C.; Camata, G.; Spacone, E.; Da Porto, F.; Modena, C.; Lorenzoni, F.; Magenes, G.; et al. Discussion on data recorded by the italian structural seismic monitoring network on three masonry structures hit by the 2016-2017 central italy earthquake. *Proc. COMPDYN* **2019**, *1*, 1889–1906.

45. Colombo, A.; Negro, P. A damage index of generalised applicability. *Eng. Struct.* **2005**, *27*, 1164–1174. [CrossRef]
46. Deng, J.; Gu, D. On a statistical damage constitutive model for rock materials. *Comput. Geosci.* **2011**, *37*, 122–128. [CrossRef]

Article

Piezoelectric Electro-Mechanical Impedance (EMI) Based Structural Crack Monitoring

Tao Wang [1,*], Bohai Tan [1], Mingge Lu [1], Zheng Zhang [2] and Guangtao Lu [2]

[1] Key Laboratory of Metallurgical Equipment and Control Technology, Wuhan University of Science and Technology, Ministry of Education, Wuhan 430081, China; bohaitan@outlook.com (B.T.); lmg980129@gmail.com (M.L.)

[2] Hubei Key Laboratory of Mechanical Transmission and Manufacturing Engineering, Wuhan University of Science and Technology, Wuhan 430081, China; zhangzheng1532@gmail.com (Z.Z.); luguangtao@wust.edu.cn (G.L.)

* Correspondence: wangtao77@wust.edu.cn

Received: 19 May 2020; Accepted: 2 July 2020; Published: 5 July 2020

Abstract: To detect small cracks in plate like structures, the high frequency characteristics of local dynamics were studied with the piezoelectric electro-mechanical impedance (EMI) method, and damages were monitored by the changes of the EMI. The finite element simulation model of EMI was established, and numerical analysis was conducted. The simulation results indicated that the peak frequency of the piezoelectric admittance signal is a certain order resonance frequency of the structure, and the piezoelectric impedance method could effectively detect the dynamic characteristics of the structure. The piezoelectric admittance simulation and experimental study of aluminum beams with different crack sizes were performed. Simulation and experimental results revealed that the peak admittance frequency decreases with the increase of crack size, and the higher resonance frequency is more sensitive to the small-scale damage. The proposed method has good repeatability and strong signal-to-noise ratio to monitor the occurrence and development of small-scale crack damage, and it has an important application prospect.

Keywords: crack damage detection; piezoelectric impedance; piezoelectric admittance; peak frequency

1. Introduction

During the life cycle of some engineering structures, cracks will occur due to the effects of dynamic load [1,2], varying temperature, environmental corrosion [3–5] and other factors [6]. If the cracks are not detected and restrained after the initial occurrence, the crack growth will accelerate, leading to eventual structural failure with potentially significant losses. To avoid catastrophic consequences caused by cracks in structures, it is of particularly importance to detect initial cracks in these structures. At present, the techniques of structural crack detection include magnetic particle detection, ultrasonic detection and radiographic detection, among others [7,8]. Most of the above methods have their shortcomings, such as limited applicability, low detection efficiency and inability to detect crack growth in real time.

Compared with the traditional detection methods mentioned, structural health monitoring (SHM) [9–11] is advancing rapidly because of its real time capacity. In SHM, vibration-based structural damage detection is a widely used technology. Since for any structure, the structural dynamic characteristic is a function of its structural parameters such as stiffness, mass and damping, structural damage means the change of structural parameters, and the change of the structural parameters will inevitably lead to the change of structural dynamic characteristics, such as resonance frequency [12,13] and vibration mode shapes [14–16]. Vibration information of structures in different states can be collected by real-time monitoring with sensors, which are installed on the structure. Then, by analyzing the vibrational signal, the dynamic characteristic changes such as resonance frequency, vibration mode, modal curvature and modal stiffness can be extracted and taken as the damage feature in the structural system [17–19]. The vibration-based monitoring method has great advantages, such as its low cost and real-time capacity. However, vibration-based damage detection and identification technologies use the low-order dynamic information of the structure and ignore the higher-order dynamic characteristics which are more sensitive to small scaled cracks.

Therefore, with the traditional strain gauge, scholars have conducted research on the early damage detection based on strain mode. Lakshmi et al. [20] used the strain mode to monitor the cracks of beams, and used the change of resonance frequency to locate the damage position. Tondreau [21] utilized strain mode shapes and local mode filters, combined with sensor arrays and statistical control charts to identify damage on the beam automatically. Strain mode is a feasible method to detect the early damage; however, this method needs the sensors installed very close to the crack location, which limits its application.

Since the piezoelectric material has small size, the ability to act as both a sensor and an actuator simultaneously, and can be embedded into the structure to form intelligent structures, piezoelectric sensing technology has been applied in SHM. Piezoelectric sensing technology can be adopted as passive sensing methods such as acoustic emission technique [22], active sensing method such as vibro-acoustic modulation technique [23], and piezoelectric impedance method. Acoustic emission technology was successfully applied to detect rolling contact fatigue damage [24] and crack initiation and propagation of low-speed rotating shaft [25]. However, due to the limitation of environment and application conditions, acoustic emission technology is difficult to apply to real time monitoring of structures in a wide range.

The vibro-acoustic modulation method implements the vibro-modulation technique with an ultrasonic vibration field transmitted through a cracked specimen undergoing an additional low frequency structural vibration. If the specimen is damaged, the amplitude modulation appears; otherwise, the two vibration fields do not interact [26]. In combination with vibro-acoustic modulation, Lee et al. [27] proposed an algorithm that can comprehensively analyze the results by comparison with ambient noise, and cracks are detected correctly in damaged specimens. The method is reported as being sensitive to the initial state of opening and closing of the crack [28,29], and early bolt looseness [23]. Wang et al. [30] extended the original vibro-acoustic modulation method by using piezoceramic transducers and ascertained a quantitative correlation between vibro-acoustic nonlinear distortion and the degree of bolt loosening.

Piezoelectric active sensing technology is also widely used in bolt looseness detection [31,32], self-sensing CFRP (carbon fiber reinforced polymer) fabrics in reinforced concrete structures [33], bond slips between the reinforcing bars and concrete [34], and in damage shape and size detection of alloy materials [35–37] among other applications. Piezoceramic-based smart aggregate pairs, such as actuators and sensors for stress wave propagation within concrete structures, can be used for stress wave communication and remain effective under low signal-to-noise ratios [38–40]. Parvasi et al. [41] investigated the effect of piezoelectric coefficients on the characteristics of generated waves with actuating and sensing compressive, flexural, and shear wave modes in metallic plates, using compressive and shear piezoelectric transducers. The piezoelectric material is used to excite the ultrasonic signal and receive the ultrasonic signal after the crack and other defects. The crack defect is identified and located by techniques such as signal feature extraction and imaging technology. Many scholars performed extensive research on damage imaging of cracks in metal or composite plate structures, including tomography [42], stress wave based imaging [43] and multi-delay-and-sum imaging algorithms [44,45]. Taylor et al. used piezoelectric guided waves to detect the small damage of composite rotor blades of wind turbines [46]. Dumoulin et al. [47] used a piezoelectric smart aggregate to study the derivation and development of cracks in concrete based on piezoelectric active sensing. With the time reversal technique combined with guided wave, Sohn and Park [48] utilized the piezoelectric arrays to conduct bridge damage detection, and proposed a symmetric index (SYMI) to evaluate the damage degree.

The piezoelectric impedance method is an active sensing method that uses only one transducer and is applied in the bolt looseness detection [49], stress monitoring [50] and structural damage diagnosis [51]. It is sensitive to structural small-scale damage due to its high frequency detection band, and can realize on-line monitoring under certain conditions. Zagrai et al. [52] verified that the electro-mechanical impedance method can identify the presence of damage in a thin plate by theoretical and experimental study. Xu and Giurgiutiu [53] presented a low-cost impedance analyzer for active structural health monitoring, and succeeded detecting a disbonding on a spacecraft panel of aluminum. Bhalla et al. [54] and Yang et al. [55] interrogated the damage condition of the structure by extracting the changes of equivalent mass, stiffness and damping of the structure through the piezoelectric impedance signature before and after damage, establishing a damage interrogation method. Ai et al. [56] applied the piezoelectric impedance technology to monitor the corrosion process of metals, and the mechanical impedance of the host structure was extracted from the electromechanical coupling impedance by the united mechanical impedances (UMI) method. It found that the mechanical impedance damage index extracted by the proposed UMI method was better than that by the traditional method.

In this paper, the piezoelectric electro-mechanical impedance (EMI) was adopted for the detection of the small crack in a beam. By analyzing the piezoelectric admittance formula of the electromechanical coupling system and combining with the harmonic response analysis of the finite element method (FEM), it found that the local dynamic characteristics of the structure would change when cracks occur and propagate in the beam. This dynamic variation is represented by the shift of peak frequency in the imaginary part of the piezoelectric admittance spectrum. The method of piezoelectric admittance peak frequency shift was proposed for the detection of the small-scale crack detection with a relatively higher monitoring frequency band. An experimental device was constructed to measure the admittance signal of a piezoelectric patch installed on the structure by a precision impedance meter. The relationship between the peak frequency in the piezoelectric admittance spectrum and the change of the crack size was analyzed. The finite element simulation and experimental results show that the small crack in the beam can be detected by the proposed method.

2. Principle of Piezoelectric Electro-Mechanical Impedance Method

Because of its strong piezoelectric effect, a PZT (lead–zirconate–titanate) type piezoelectric transducer has been successfully used in SHM and is adopted in this research. For a PZT patch of length l_a, width b_a, and thickness h_a is bonded on a structure, based on the assumption of elastic connection (represented by a pair of stiffness springs $2K_{str}$) between piezoelectric material and host structure, as shown in Figure 1, Giurgiutiu and Zagrai [57] proposed an EMI model and the piezoelectric admittance expression is given as:

$$Y(\omega) = i\omega C\left[1 - \kappa_{31}^2\left(1 - \frac{1}{\varphi \cdot \cot(\varphi) + K_{str}/K_{PZT}}\right)\right] \tag{1}$$

where ω is the angular frequency of excitation signal, ρ is the density of the PZT patch, κ_{31}^2 is the electromechanical coupling factor, $\varphi = \gamma \cdot l_a/2$, while $\gamma = \omega/c$ is the wave number, $c = \sqrt{1/\rho s_{11}^E}$ is the wave speed, s_{11}^E is the mechanical compliance at zero field, $C = l_a b_a \widetilde{\varepsilon}_{33}^T h_a^{-1}$ is the conventional capacitance of the PZT patch, K_{str} is the stiffness of the host structure, $K_{PZT} = A_a/\left(s_{11}^E l_a\right)$ is the stiffness of the PZT patch, $A_a = h_a \times b_a$ is the cross-sectional area of PZT patch, and $\widetilde{\varepsilon}_{33}^T$ is dielectric constant at zero stress. The expressions in Equations (1) cover the complete frequency spectrum and encompass both structure and sensor dynamics, overcoming low-frequency structure-focused analysis of Liang's model [58].

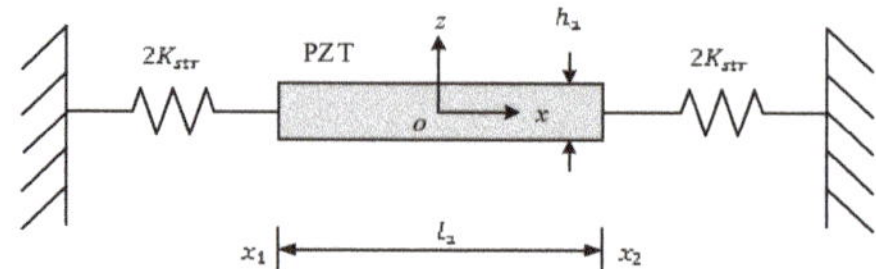

Figure 1. Simplified piezoelectric electro-mechanical impedance (EMI) Model.

Equation (1) reveals that the change of admittance Y of the coupled system is only related on the host structural stiffness K_{str} while the parameters of piezoelectric material and host structural material are determined. In order to find out the effect of the stiffness of the host structure on the admittance signal, Equation (1) is derived with ω simultaneously on both sides. Here, we expand the cotangent function by a Taylor series and substitute the first two terms in the Taylor series into Equation (1):

$$\frac{dY}{d\omega} = iC\left[1 - \kappa_{31}^2 + \kappa_{31}^2 \frac{1 + \frac{K_{str}}{K_{PZT}} + \frac{l_a^2}{12c^2}\omega^2}{\left(1 + \frac{K_{str}}{K_{PZT}} - \frac{l_a^2}{12c^2}\omega^2\right)^2}\right] \tag{2}$$

Some peak values in the admittance signature represent the resonant frequency of the host structure. In this case, the peak frequency is a certain order resonance frequency, ω_n, of the coupled system of the host structure and the piezoelectric patch. Thus, in the position of the resonance frequency, ω_n, the derivative of piezoelectric admittance signal equals zero, i.e., $dY/d\omega_n = 0$. In Equation (2), both c and K_{PZT} are complex.

Considering that the imaginary part of c and K_{PZT} depend on the material damping and the effect of material damping is small, the imaginary part is neglected in order to simplify the analysis process. Meanwhile, let $\kappa_{31}^2 = a + bi$ in Equation (2). By extracting the imaginary part of Equation (2) and making it equal to zero, the expressions about K_{str} and ω_n can be obtained as

$$\left(\frac{l_a^2}{12c^2}\right)^2 \omega_n^4 - \frac{l_a^2}{12c^2}(3 + K_{str}/K_{PZT})\omega_n^2 + (1 + K_{str}/K_{PZT}) \cdot K_{str}/K_{PZT} = 0 \tag{3}$$

Solving Equation (3) results in,

$$\omega_n^2 = 6c^2 \frac{3 + 2K_{str}/K_{PZT} - \sqrt{9 + 8K_{str}/K_{PZT}}}{l_a^2} \tag{4}$$

Equation (4) shows that the peak frequency ω_n in the piezoelectric admittance signature depends on the stiffness of the host structure. When a crack occurs and propagates in the host structure, the local stiffness of the host structure will reduce. While the local stiffness of the host structure decreases, the peak frequency in the piezoelectric admittance signature also decreases [59]. The piezoelectric patch bonded near the crack can transform this stiffness variation into peak frequency changes in the piezoelectric admittance signature. By analyzing the peak frequency changes in the piezoelectric admittance signature, the structural crack can be detected and thus it provides a theoretical basis for monitoring the crack status of a host structure with the peak frequency changes of the piezoelectric impedance.

3. Finite Element Analysis of Piezoelectric Impedance

To verify the above proposed method, the finite element analysis (FEA) is adopted to simulate the piezoelectric impedance and to find out the influence of crack and crack size on the peak frequency in the piezoelectric admittance signature, i.e., the local resonance frequency of the coupled system. The host structure is an aluminum alloy beam with a dimension of 250 × 30 × 5 mm. A PZT patch, with a size of 8 × 7 × 1 mm, is bonded on the center of the beam surface with epoxy resin adhesive. The adhesive layer thickness is kept 0.125 mm. The properties of the piezoelectric patch are listed in

Table 1 [60], and the material constants of the adhesive layer and the aluminum alloy beam are listed in Table 2.

Table 1. Properties of piezoelectric patch.

Parameter	Symbol	Value
Density	ρ	7850 kg/m^3
Dielectric constant	ε^T_{33}	3.01×10^{-8} F/m
Piezoelectric constant	d_{31}	-274×10^{-12} C/N

Table 2. Material coefficients of the beam and the adhesive layer.

Item	Density /kg·m^{-3}	Young's Modulus /GPa	Poisson Ratio
Beam	2700	70	0.3
Adhesive layer	2000	0.8	0.4

Cracks are simulated by setting rectangular grooves on a beam. The width of each rectangular groove is 0.5 and the depth is 1 mm. Only the length of groove is changed to simulate the different crack sizes. The length of the groove is set as 0, 6 and 12 mm, respectively. The location of the piezoelectric patch and the crack on the beam is shown in Figure 2. In the numerical analysis, free boundaries are set at both ends of the beam. The system damping ratio is set to 0.005.

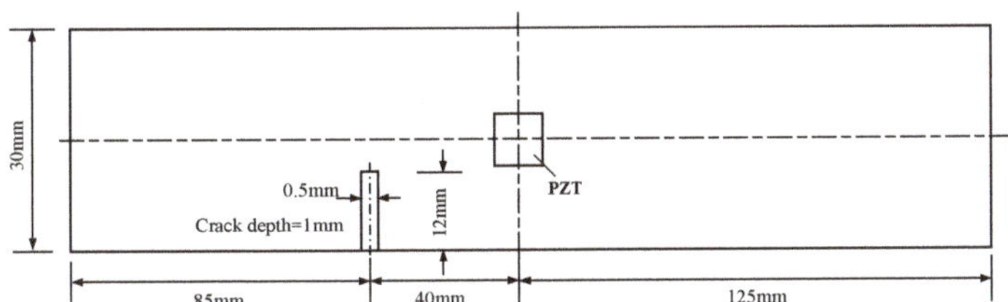

Figure 2. Schematic diagram of piezoelectric patch and crack location.

The finite element model of the cracked beam with piezoelectric patch is built, as shown in Figure 3a. The model is meshed by a mapping grid with an element size of 2 mm, except where the crack generates and the piezoelectric patch is bonded. For the region where the crack is located, the mapping mesh with the element size of 0.25 mm is used. In the vicinity of the crack, the free mesh is used to gradually transfer the specialized element size of 0.25 mm to the overall model element size of 2 mm. The mesh at the crack is detailed in Figure 3b. The piezoelectric patch is discretized with an element size of 0.5 mm by mapping, and the adhesive layer under the piezoelectric patch is discretized by free mesh with a 1 mm element size gradually changing to a 2 mm element size.

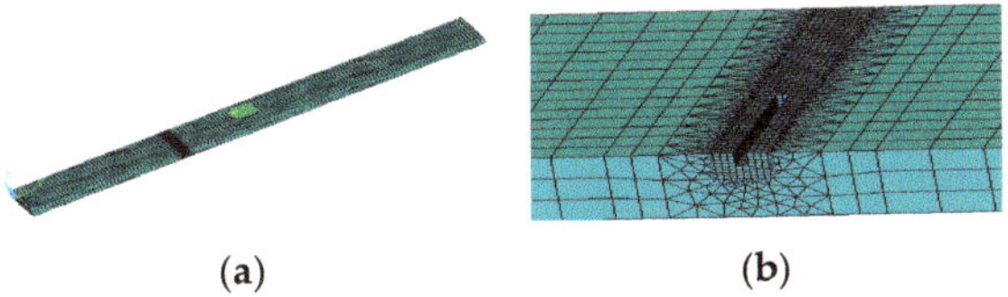

(a)   (b)

Figure 3. Overall (**a**) and local (**b**) finite element model of cracked beam and lead–zirconate–titanate (PZT) patch.

In the simulation, the modal analysis is firstly carried out to obtain the resonance frequency, and then the piezoelectric harmonic analysis is performed. In the piezoelectric harmonic analysis, the top and bottom surface piezoelectric degrees of the piezoelectric patch is coupled, and the voltage of

1 V and 0 V are applied on the top and bottom surfaces of the piezoelectric patch respectively to simulate 1 V AC voltage. The selected frequency range in the harmonic analysis is near the modal resonant frequencies obtained from the modal analysis. The time-varying charge on the piezoelectric surface is obtained, and then the piezoelectric impedance, piezoelectric admittance and other information are calculated. Comparing the results of the modal analysis with the piezoelectric admittance analysis, it is found that the modal resonant frequencies correspond approximately to the peak frequencies in the admittance signature, indicating that some of the peak admittance frequencies are the resonant frequencies of the coupled system. The peak frequencies in the admittance signature can reflect the local dynamic characteristics of the host structure with the piezoelectric patch.

After comparative analysis, the piezoelectric admittance around some resonant frequencies is simulated and analyzed with different crack sizes. The scanning frequency bands are selected around 19, 62 and 174 kHz, which is near the coupled structure resonance frequencies.

The detailed frequency bands of 19.2–19.3, 62.5–63 and 174.5–175 kHz, covering from relatively low to relatively high frequencies, are chosen as scanning frequency bands for the finite element harmonic response analysis. Then the piezoelectric impedance is computed. The piezoelectric admittance spectra and peak frequencies in the admittance spectrum with different cracks lengths (0 mm crack length means primitive state) on the beam are shown from Figures 4–6, the vertical coordinate is an imaginary part of the piezoelectric admittance spectrum, whose unit is siemens (S).

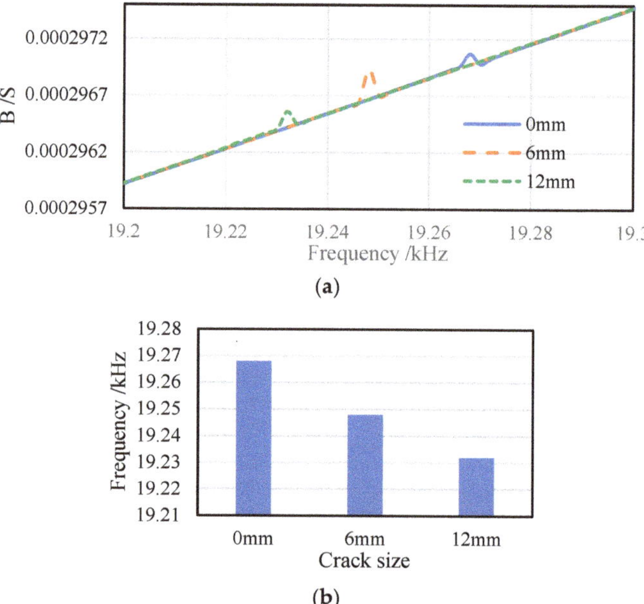

Figure 4. Numerical simulation results: (**a**) piezoelectric admittance spectrum and (**b**) peak frequency tendencies in 19.2 kHz19.3 kHz with different crack sizes.

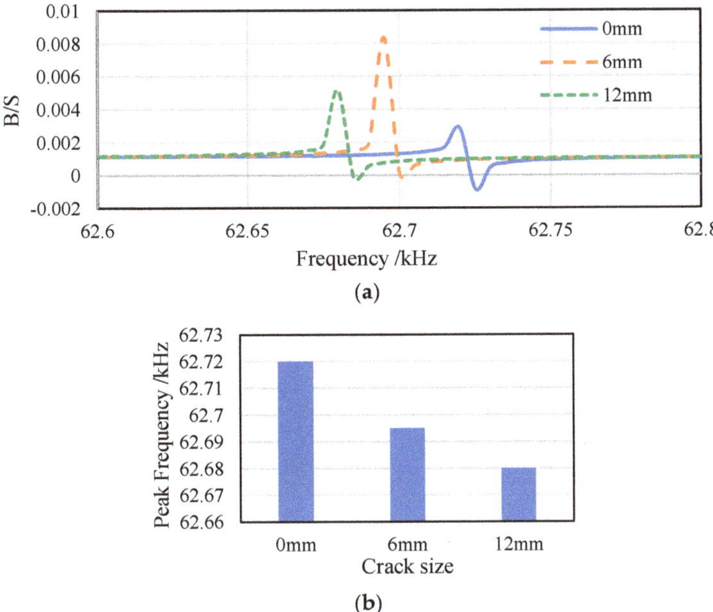

(a)

(b)

Figure 5. Numerical simulation results (**a**) piezoelectric admittance spectrum and (**b**) peak frequency tendencies in 62.6 kHz–62.8 kHz with different crack sizes.

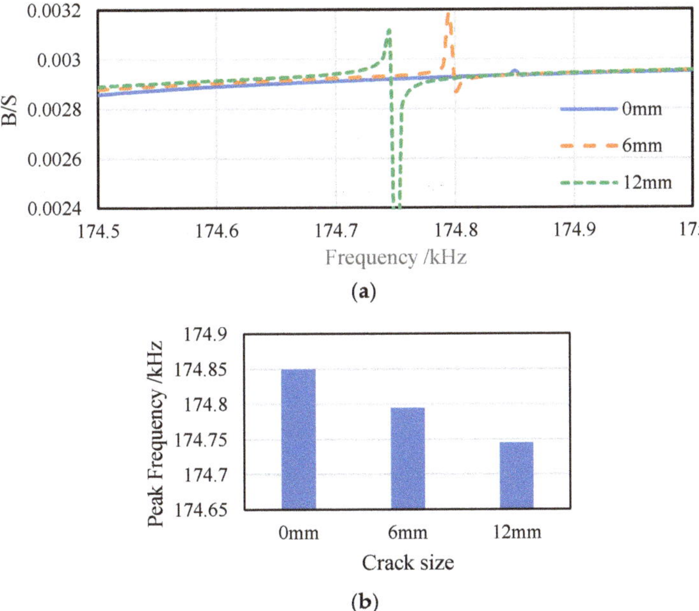

(a)

(b)

Figure 6. Numerical simulation results (**a**) piezoelectric admittance spectrum and (**b**) peak frequency tendencies in 174.5 kHz–175 kHz with different crack sizes.

From Figures 4a, 5a and 6a, it is clear that as the crack appears and grows, in both selected frequency bands, the peak frequencies in the piezoelectric admittance spectra shift to the left. It means that the peak frequency decreases when the crack grows, as shown in Figures 4b, 5b and 6b. Based on the above theoretical analysis, the local stiffness of the host structure around the crack decreases when the crack occurs and propagates. Once the local stiffness decreases, the resonant frequency of the host structure will decrease, shifting the peak frequency to left in the piezoelectric admittance spectrum. The numerical simulation results are consistent with the theoretical analysis.

Based on the above findings, the peak frequency in the piezoelectric admittance is chosen as the feature parameter to characterize the crack propagation. The peak frequency shift between crack-free and 6 mm crack is defined as Δf_1, while the peak frequency offset between crack-free and 12 mm crack is defined as Δf_2. The peak frequency shifts at different frequency bands with different crack sizes are extracted as shown in Table 3. It shows that the higher the scanning frequency band, the greater the peak frequency shift, thus the higher the frequency band, the higher the sensitivity of the proposed method.

Table 3. Peak frequency shifts at different frequency bands.

	19.2–19.3 kHz	62.6–62.8 kHz	174.5–175 kHz
Δf_1 (Hz)	−10	−25	−55
Δf_2 (Hz)	−18	−40	−105

4. Experiments and Discussion

To verify the above theoretical model and finite element analysis, an experimental device was set up, as shown in Figure 7, and experiments were conducted. The dimension of the test specimen was consistent with the simulation model and the piezoelectric patch was bonded on the center of the one surface of the beam, as shown in Figure 8. When the piezoelectric patch was bonded to the host structure with epoxy resin, two optical fibers of 0.125 mm diameter were placed parallel on the surface of the beam with the epoxy resin to ensure the consistent adhesive layer thickness for all test pieces. First, the PZT patch, bonded on the without crack beam, was performed the piezoelectric impedance test by the WK6530B impedance analyzer (Wayne Kerr, West Sussex, UK). Then, a crack with length of 6 mm was machined and the test was performed again. At last, the piezoelectric impedance test was carried out with crack length of 12 mm machined on the beam. The piezoelectric admittance spectra of the beam with different crack lengths (0, 6 and 12 mm) were measured and the peak frequencies in the admittance spectra are extracted, as shown in Figures 9–11. As the figures indicate, as the crack grows, the peak frequency, i.e., the resonant frequency of the coupled system shifts to the left. That means the frequency decreases as the crack grows. There may be a slight difference between the simulation and experimental results, which may be resulted from an error between the actual and the designed positions of the damage and PZT. The experimental results have the same trend with the theoretical analysis and finite element simulation results. The tested resonant frequencies are also consistent with the simulation results; therefore, the finite element simulation results can be used to guide the experiment to select an appropriate scanning frequency band of the piezoelectric impedance measurement.

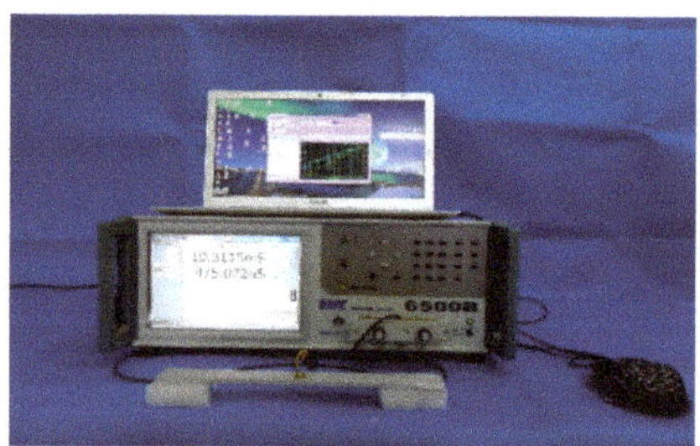

Figure 7. Testing device.

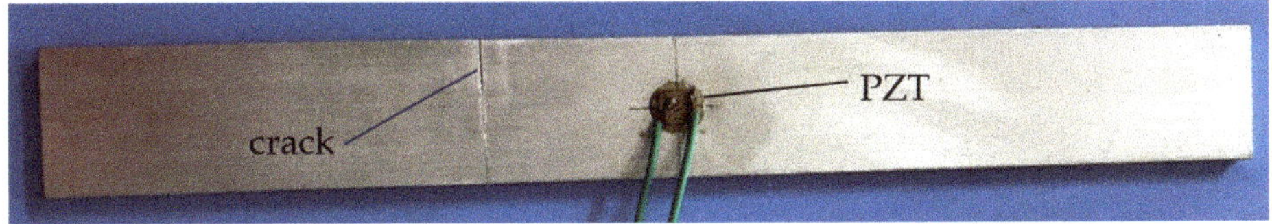

Figure 8. The aluminum plate sample with a crack.

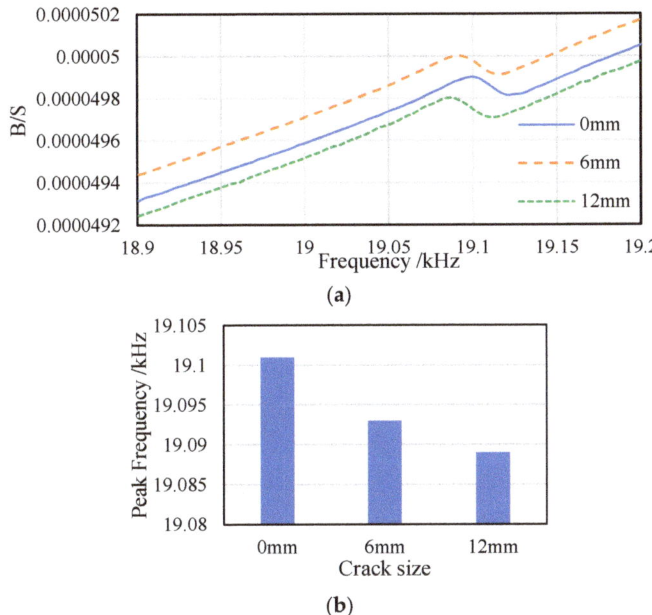

Figure 9. Experimental results (**a**) piezoelectric admittance spectrum and (**b**) peak frequency tendencies in 18.9 kHz–19.2 kHz with different crack sizes.

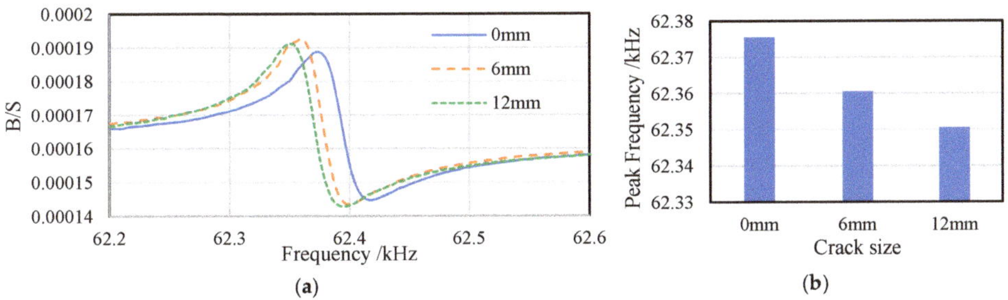

Figure 10. Experimental results (**a**) piezoelectric admittance spectrum and (**b**) peak frequency tendencies in 62.2 kHz–62.6 Hz with different crack sizes.

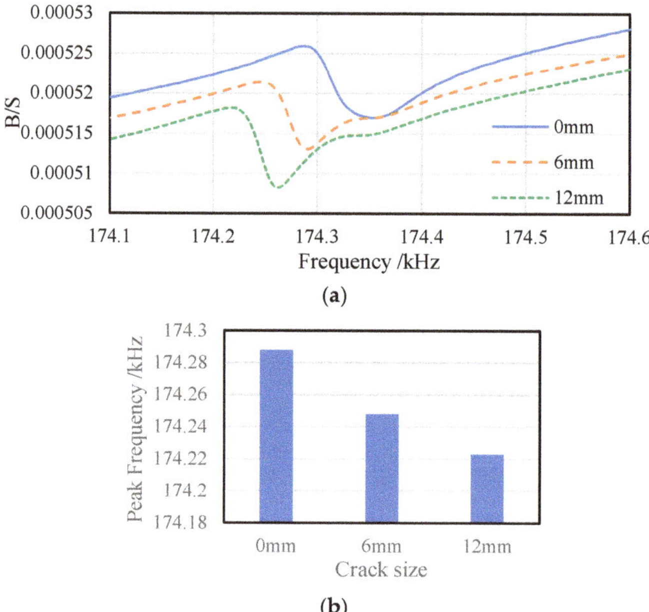

Figure 11. Experimental results (**a**) piezoelectric admittance spectrum and (**b**) peak frequency tendencies in 174.1 kHz–174.6 kHz with different crack sizes.

The peak frequency shifts extracted from the piezoelectric admittance spectra at different frequency bands are shown in Table 4. The peak frequencies in the piezoelectric admittance spectra decrease with the increase of crack length in each frequency band. Moreover, at the same crack length, the peak frequency shifts increase with the increase of the scanning frequency band, which means increasing the scanning frequency band can increase the sensitivity of the proposed method. Therefore, the crack occurrence and propagation can be interrogated by the peak frequency shift in the imaginary part of the piezoelectric admittance spectrum. With the increase of the scanning frequency, the higher the frequency band is, and the larger the shift of peak frequency is, i.e., the higher the frequency band is, the more sensitive is the method to damage.

Table 4. The frequency shifts at different frequencies bands.

	18.9–19.2 kHz	62.2–62.6 kHz	174.1–174.6 kHz
Δf_1 (Hz)	−8	−15	−40
Δf_2 (Hz)	−12	−25	−65

The experimental studies reveal that the peak frequency in the piezoelectric admittance spectrum, i.e., the resonant frequency of the coupled system of the beam and piezoelectric patch, decreases with the increase of crack length, which indicates that the peak frequency shift can well reflect the size of cracks. Therefore, the peak frequency can be used as a feature parameter to characterize the crack length. With the increase of the scanning frequency band, the shift of peak frequency tends to increase. In application, the detection sensitivity can be improved by properly increasing the monitoring frequency band, and small cracks can be detected and crack propagation can be monitored.

At the same time, the experimental results are basically consistent with the finite element numerical analysis results, which shows the accuracy and validity of the finite element model. That indicates the established finite element model can effectively help choosing the sensitive frequency band and the installation location of the piezoelectric patch in practical applications.

5. Conclusions

For small crack detection and crack propagation monitoring in structures, the method based on the peak frequency shift in the piezoelectric admittance spectrum is proposed. The feasibility of the method is analyzed theoretically. The FEA and experimental results verified the effectiveness of the method. The following conclusions are drawn:

(1) The peak frequency in the imaginary part of the piezoelectric admittance spectrum can sensitively detect a crack in a structure, and the peak frequency decreases with the increase of the crack length.

(2) Appropriately increasing the detection frequency band can increase the peak frequency shift at the same crack size and improve the detection sensitivity to the crack, therefore, making the detection of a small crack in a structure possible.

(3) The experimental results are consistent with the FEM simulation results, which show the accuracy and validity of the finite element model. In future research, the resonant frequencies of complex structures can be found by a finite element model firstly, then the scanning frequency bands of piezoelectric admittance can be set to include the resonant frequencies, and finally the damage degree of structures can be judged by the peak frequency shift in the piezoelectric admittance signature.

(4) Since the frequency is selected as a feature parameter for damage monitoring, the proposed method has a better repeatability and signal-to-noise ratio than the root mean square deviation based damage index.

Since the method proposed in this paper is based on the change of local dynamic characteristics to monitor the small crack in a structure, there still needs to be more in-depth study on the local dynamic modeling and analysis. Further research is still needed on how to extract the feature parameter to characterize the crack propagation quantitatively, determine the optimal detection frequency band, and select the sensitive mode frequency bands, etc. Moreover, the influence of the damage size including the width and depth, the damage position, different excitation positions, the structural load and the damage identification of composite materials will be investigated in-depth in our future work.

Author Contributions: T.W., B.T., and Z.Z. conceived and designed the experiments; B.T., Z.Z., M.L. and G.L. performed the experiments; B.T., Z.Z. and M.L. performed the simulation; T.W., Z.Z., B.T. and M.L. analyzed the data; T.W. and G.L. contributed reagents/materials/analysis tools; T.W., B.T., M.L., Z.Z. and G.L. wrote the paper. All authors have read and agreed to the published version of the manuscript.

Funding: This research was funded by the Natural Science Foundation of China under Grant No. 51375354 and 51808417, 61703215.

Acknowledgments: The authors would like to thank for the support by National Natural Science Foundation of China under the Grant 51375354, 51808417, 61703215.

References

1. Jiang, T.; Zhang, Y.; Wang, L.; Zhang, L.; Song, G. Monitoring fatigue damage of modular bridge expansion joints using piezoceramic transducers. *Sensors* **2018**, *18*, 3973. [CrossRef] [PubMed]
2. Li, N.; Wang, F.; Song, G. New entropy-based vibro-acoustic modulation method for metal fatigue crack detection: An exploratory study. *Measurement* **2020**, *150*, 107075. [CrossRef]
3. Peng, J.; Xiao, L.; Zhang, J.; Cai, C.S.; Wang, L. Flexural behavior of corroded HPS beams. *Eng. Struct.* **2019**, *195*, 274–287. [CrossRef]
4. Li, W.; Ho, S.C.M.; Song, G. Corrosion detection of steel reinforced concrete using combined carbon fiber and fiber Bragg grating active thermal probe. *Smart Mater. Struct.* **2016**, *25*, 045017. [CrossRef]
5. Linsheng, H.; Chuanbo, L.; Tianyong, J.; Hong-Nan, L. Feasibility study of steel bar corrosion monitoring using a piezoceramic transducer enabled time reversal method. *Appl. Sci.* **2018**, 2304. [CrossRef]
6. Ojanomare, C.; Cornetti, P.; Romagnoli, R.; Surace, C. Fatigue crack growth analysis of drill pipes during rotary drilling operations by the multiple reference state weight function approach. *Eng. Fail. Anal.* **2017**, *74*, 11–34. [CrossRef]

7.   Gholizadeh, S. A review of non-destructive testing methods of composite materials. *Procedia Struct. Integr.* **2016**, *1*, 50–57. [CrossRef]

8.   Drinkwater, B.W.; Wilcox, P.D. Ultrasonic arrays for non-destructive evaluation: A review. *NDT E Int.* **2006**, *39*, 525–541. [CrossRef]

9.   Liu, Y.; Zhang, M.; Yin, X.; Huang, Z.; Wang, L. Debonding detection of Reinforced Concrete (RC) beam with Near-Surface Mounted (NSM) Pre-stressed Carbon Fiber Reinforced Polymer (CFRP) Plates using embedded piezoceramic Smart Aggregates (SAs). *Appl. Sci.* **2019**, *10*, 50. [CrossRef]

10.   Kong, Q.; Robert, R.H.; Silva, P.; Mo, Y.L. Cyclic crack monitoring of a reinforced concrete column under simulated pseudo-dynamic loading using piezoceramic-based smart aggregates. *Appl. Sci.* **2016**, *6*, 341. [CrossRef]

11.   Lv, Y.; Yuan, R.; Wang, T.; Li, H.; Song, G. Health degradation monitoring and early fault diagnosis of a rolling bearing based on CEEMDAN and improved MMSE. *Materials* **2018**, *11*, 1009. [CrossRef] [PubMed]

12.   Ruotolo, R.; Surace, C.; Crespo, P.; Storer, D. Harmonic analysis of the vibrations of a cantilevered beam with a closing crack. *Comput. Struct.* **1996**, *61*, 1057–1074. [CrossRef]

13.   Pugno, N.; Surace, C.; Ruotolo, R. Evaluation of the non-linear dynamic response to harmonic excitation of a beam with several breathing cracks. *J. Sound Vib.* **2000**, *235*, 749–762. [CrossRef]

14.   Yan, Y.J.; Cheng, L.; Wu, Z.Y.; Yam, L.H. Development in vibration-based structural damage detection technique. *Mech. Syst. Signal. Process.* **2007**, *21*, 2198–2211. [CrossRef]

15.   Fan, W.; Qiao, P. Vibration-based damage identification methods: A review and comparative study. *Struct. Health Monit.* **2011**, *10*, 83–111. [CrossRef]

16.   Corrado, N.; Durrande, N.; Gherlone, M.; Hensman, J.; Mattone, M.; Surace, C. Single and multiple crack localization in beam-like structures using a Gaussian process regression approach. *J. Vib. Control* **2017**, *24*, 4160–4175. [CrossRef]

17.   Koto, Y.; Konishi, T.; Sekiya, H.; Miki, C. Monitoring local damage due to fatigue in plate girder bridge. *J. Sound Vib.* **2019**, *438*, 238–250. [CrossRef]

18.   Perera, R.; Huerta, C.; Orqui, J.M. Identification of damage in RC beams using indexes based on local modal stiffness. *Constr. Build. Mater.* **2008**, *22*, 1656–1667. [CrossRef]

19.   Bovsunovsky, A.P.; Surace, C. Considerations regarding superharmonic vibrations of a cracked beam and the variation in damping caused by the presence of the crack. *J. Sound Vib.* **2005**, *288*, 865–886. [CrossRef]

20.   Lakshmi, N.K.; Jebaraj, C. Sensitivity analysis of local/global modal parameters for identification of a crack in a beam. *J. Sound Vib.* **1999**, *228*, 977–994. [CrossRef]

21.   Tondreau, G.; Deraemaeker, A. Automated data-based damage localization under ambient vibration using local modal filters and dynamic strain measurements: Experimental applications. *J. Sound Vib.* **2014**, *333*, 736–7385. [CrossRef]

22.   Caesarendra, W.; Kosasih, B.; Tieu, A.K.; Zhu, H.; Moodie, C.A.S.; Zhu, Q. Acoustic emission-based condition monitoring methods: Review and application for low speed slew bearing. *Mech. Syst. Signal. Process.* **2016**, *72*, 134–159. [CrossRef]

23.   Wang, F.; Song, G. Bolt early looseness monitoring using modified vibro-acoustic modulation by time-reversal. *Mech. Syst. Signal. Process.* **2019**, *130*, 349–360. [CrossRef]

24.   Rahman, Z.; Ohba, H.; Yoshioka, T.; Yamamoto, T. Incipient damage detection and its propagation monitoring of rolling contact fatigue by acoustic emission. *Tribol. Int.* **2009**, *42*, 807–815. [CrossRef]

25.   Elforjani, M.; Mba, D. Detecting natural crack initiation and growth in slow speed shafts with the Acoustic Emission technology. *Eng. Fail. Anal.* **2009**, *16*, 2121–2129. [CrossRef]

26.   Klepka, A.; Staszewski, W.J.; Jenal, R.B.; Szwedo, M.; Iwaniec, J.; Uhl, T. Nonlinear acoustics for fatigue crack detection—Experimental investigations of vibro-acoustic wave modulations. *Struct. Health Monit.* **2011**, *11*, 197–211. [CrossRef]

27.   Lee, S.E.; Lim, H.J.; Jin, S.; Sohn, H.; Hong, J.-W. Micro-crack detection with nonlinear wave modulation technique and its application to loaded cracks. *NDT E Int.* **2019**, *107*, 102132. [CrossRef]

28.   Duffour, P.; Marco, M.; Peter, C. A study of the vibro-acoustic modulation technique for the detection of cracks in metals. *J. Acoust. Soc. Am.* **2006**, *119*, 1463–1475. [CrossRef]

29.   Bovsunovsky, A.; Surace, C. Non-linearities in the vibrations of elastic structures with a closing crack: A state of the art review. *Mech. Syst. Signal. Process.* **2015**, *62*, 129–148. [CrossRef]

30. Wang, T.; Liu, S.; Shao, J.; Li, Y. Health monitoring of bolted joints using the time reversal method and piezoelectric transducers. *Smart Mater. Struct.* **2016**, *25*, 025010. [CrossRef]

31. Yin, H.; Wang, T.; Yang, D.; Liu, S.; Shao, J.; Li, Y. A Smart Washer for Bolt Looseness Monitoring Based on Piezoelectric Active Sensing Method. *Appl. Sci.* **2016**, *6*, 320. [CrossRef]

32. Tashakori, S.; Baghalian, A.; Unal, M.; Fekrmandi, H.; Senyürek, V.Y.; McDaniel, D.; Tansel, I.N. Contact and non-contact approaches in load monitoring applications using surface response to excitation method. *Measurement* **2016**, *89*, 197–203. [CrossRef]

33. Feng, Q.; Ou, J. Self-sensing CFRP fabric for structural strengthening and damage detection of reinforced concrete structures. *Sensors* **2018**, *18*, 4137. [CrossRef]

34. Xu, K.; Ren, C.; Deng, Q.; Jin, Q.; Chen, X. Real-time monitoring of bond slip between GFRP bar and concrete structure using piezoceramic transducer-enabled active sensing. *Sensors* **2018**, *18*, 2653. [CrossRef] [PubMed]

35. Lu, G.; Wang, T.; Zhou, M.; Li, Y. Characterization of Ultrasonic Energy Diffusion in a Steel Alloy Sample with Tensile Force Using PZT Transducers. *Sensors* **2019**, *19*, 2185. [CrossRef] [PubMed]

36. Lu, G.; Li, Y.; Zhou, M.; Feng, Q.; Song, G. Detecting damage size and shape in a plate structure using PZT transducer array. *J. Aerosp. Eng.* **2018**, *31*, 04018075. [CrossRef]

37. Tashakori, S.; Baghalian, A.; Cuervo, J.; Senyurek, V.; Tansel, I.N.; Uragun, B. Inspection of the machined features created at the embedded sensor aluminum plates. In Proceedings of the 2017 8th International Conference on Recent Advances in Space Technologies, Istanbul, Turkey, 19–22 June 2017; Volume 6, pp. 517–522. [CrossRef]

38. Ji, Q.; Ho, M.; Zheng, R.; Ding, Z.; Song, G. An exploratory study of stress wave communication in concrete structures. *Smart Struct. Syst.* **2015**, *15*, 135–150. [CrossRef]

39. Siu, S.; Ji, Q.; Wu, W.; Song, G.; Zhi, D. Stress wave communication in concrete: I. Characterization of a smart aggregate based concrete channel. *Smart Mater. Struct.* **2014**, *23*, 125030. [CrossRef]

40. Siu, S.; Qing, J.; Wang, K.; Song, G.; Ding, Z. Stress wave communication in concrete: II. Evaluation of low voltage concrete stress wave communications utilizing spectrally efficient modulation schemes with PZT transducers. *Smart Mater. Struct.* **2014**, *23*, 125031. [CrossRef]

41. Parvasi, S.M.; Ji, Q.; Song, G. Structural health monitoring of Plate-Like structures using Compressive/Shear modes of piezoelectric transducers. *Earth Space 2016* **2016**, 1033–1044. [CrossRef]

42. Hay, T.R.; Royer, R.L.; Gao, H.; Zhao, X.; Rose, J.L. A comparison of embedded sensor Lamb wave ultrasonic tomography approaches for material loss detection. *Smart Mater. Struct.* **2006**, *15*, 946–951. [CrossRef]

43. Wang, C.H.; Rose, J.T.; Chang, F.-K. A synthetic time-reversal imaging method for structural health monitoring. *Smart Mater. Struct.* **2004**, *13*, 415–423. [CrossRef]

44. Lu, G.; Li, Y.; Wang, T.; Xiao, H.; Huo, L.; Song, G. A multi-delay-and-sum imaging algorithm for damage detection using piezoceramic transducers. *J. Intell. Mater. Syst. Struct.* **2017**, *28*, 1150–1159. [CrossRef]

45. Lu, G.; Li, Y.; Song, G. A delay-and-Boolean-ADD imaging algorithm for damage detection with a small number of piezoceramic transducers. *Smart Mater. Struct.* **2016**, *25*, 095030. [CrossRef]

46. Taylor, S.G.; Farinholt, K.; Choi, M.; Jeong, H.; Jang, J.; Park, G.; Lee, J.; Todd, M.D. Incipient crack detection in a composite wind turbine rotor blade. *J. Intell. Mater. Syst. Struct.* **2014**, *25*, 613–620. [CrossRef]

47. Dumoulin, C.; Karaiskos, G.; Deraemaeker, A. Monitoring of crack propagation in reinforced concrete beams using embedded piezoelectric transducers. *Acoust. Emiss. Relat. Non-Destr. Eval. Tech. Fract. Mech. Concr.* **2015**, 161–175. [CrossRef]

48. Park, H.W.; Kim, S.B.; Sohn, H. Understanding a time reversal process in Lamb wave propagation. *Wave Motion* **2009**, *46*, 451–467. [CrossRef]

49. Shao, J.; Wang, T.; Yin, H.; Yang, D.; Li, Y. Bolt Looseness Detection Based on Piezoelectric Impedance Frequency Shift. *Appl. Sci.* **2016**, *6*, 298. [CrossRef]

50. Wang, T.; Wei, D.; Shao, J.; Li, Y.; Song, G. Structural stress monitoring based on piezoelectric impedance frequency shift. *J. Aerosp. Eng.* **2018**, *31*, 04018092. [CrossRef]

51. Park, G.; Sohn, H.; Farrar, C.R.; Inman, D.J. Overview of piezoelectric impedance-based health monitoring and path forward. *Shock Vib. Dig.* **2003**, *35*, 451–464. [CrossRef]

52. Zagrai, A.N.; Giurgiutiu, V. Electro-Mechanical Impedance Method for Crack Detection in Thin Plates. *J. Intell. Mater. Syst. Struct.* **2016**, *12*, 709–718. [CrossRef]

53. Xu, B.; Giurgiutiu, V. A Low-Cost and Field Portable Electromechanical (E/M) Impedance Analyzer for Active Structural Health Monitoring. In Proceedings of the 5th International Workshop on Structural Health Monitoring, Stanford University, Stanford, CA, USA, 15–17 September 2005.

54. Dixit, A.; Bhalla, S. Prognosis of fatigue and impact induced damage in concrete using embedded piezo-transducers. *Sens. Actuators A Phys.* **2018**, *274*, 116–131. [CrossRef]

55. Yang, Y.; Hu, Y.; Lu, Y. Sensitivity of PZT impedance sensors for damage detection of concrete structures. *Sensors* **2008**, *8*, 327–346. [CrossRef] [PubMed]

56. Ai, D.; Zhu, H.; Luo, H.; Yang, J. An effective electromechanical impedance technique for steel structural health monitoring. *Constr. Build. Mater.* **2014**, *73*, 97–104. [CrossRef]

57. Giurgiutiu, V.; Zagrai, A.N. Characterization of piezoelectric wafer active sensors. *J. Intell. Mater. Syst. Struct.* **2000**, *11*, 959–976. [CrossRef]

58. Liang, C.; Sun, F.P.; Rogers, C.A. Coupled electro-mechanical analysis of adaptive material systems-determination of the actuator power consumption and system energy transfer. *J. Intell. Mater. Syst. Struct.* **1997**, *8*, 335–343. [CrossRef]

59. Ruotolo, R.; Surace, C. Natural frequencies of a bar with multiple cracks. *J. Sound Vib.* **2004**, *272*, 301–316. [CrossRef]

60. Wuhan Hitrusty Electronice, LTD. Available online: http://www.hi-trusty.com/ (accessed on 12 June 2020).

Article

Robust Structural Damage Detection Using Analysis of the CMSE Residual's Sensitivity to Damage

Mingqiang Xu [1], Shuqing Wang [1,*], Jian Guo [1] and Yingchao Li [2]

[1] Shandong Provincial Key Laboratory of Ocean Engineering, Ocean University of China, Qingdao 266100, China; xumingqiang@stu.ouc.edu.cn (M.X.); guojian@stu.ouc.edu.cn (J.G.)

[2] Department of Port and Waterway Engineering, College of Civil Engineering, Ludong University, Yantai 264000, China; yingchao.ouc@163.com

* Correspondence: shuqing@ouc.edu.cn

Received: 28 March 2020; Accepted: 16 April 2020; Published: 19 April 2020

Abstract: This paper presents a robust damage identification scheme in which damage is predicted by solving the cross-modal strain energy (CMSE) linear system of equations. This study aims to address the excessive equations issue faced in the assemblage of the CMSE system. A sensitivity index that, to some extent, measures how the actual damage level vector satisfies each CMSE equation, is derived by performing an analysis of the defined residual's sensitivity to damage. The index can be used to eliminate redundant equations and enhance the robustness of the CMSE system. Moreover, to circumvent a potentially ill-conditioned problem, a previously published iterative Tikhonov regularization method is adopted to solve the CMSE system. Some improvements to this method for determining the iterative regularization parameter and regularization operator are given. The numerical robustness of the proposed damage identification scheme against measurement noise is proved by analyzing a 2-D truss structure. The effects of location and extent of damage on the damage identification results are investigated. Furthermore, the feasibility of the proposed scheme for damage identification is experimentally validated on a beam structure.

Keywords: noise robustness; sensitivity analysis; cross-modal strain energy; damage detection

---

1. Introduction

Structural damage identification is a fundamental element of structural health monitoring (SHM) that has become a vital tool in maintaining the safety and integrity of structures [1–7]. Research on vibration-based damage identification has been rapidly expanding over recent decades. The basic idea behind this technology is that modal parameters (notably frequencies, mode shapes, and modal damping) are functions of the physical properties of the structure. Therefore, changes in the physical properties (such as stiffness reduction caused by damage) will cause detectable changes in the modal properties. Conversely, these changes can be used to reflect damage.

Vibration-based structural damage identification can be formulated as a linear inverse problem [8–12], which requires the determination of the unknown input (i.e., structural damage) to a linear system from the known output (i.e., extracted modal parameters from the vibration measurements of the structure). The discretization of the linear inverse problem [12,13] typically gives rise to the linear system of equations with a very ill-conditioned matrix $\mathbf{C}^*$ [14],

$$\mathbf{C}^*\boldsymbol{\alpha} = \boldsymbol{b}^*, \ \mathbf{C}^* \in \mathbb{R}^{N_q \times N_d}, \ \boldsymbol{\alpha} \in \mathbb{R}^{N_d}, \ \boldsymbol{b}^* \in \mathbb{R}^{N_q}, \tag{1}$$

where the coefficient matrix $\mathbf{C}^*$ and the right-hand side vector $\boldsymbol{b}^*$ are both assembled by the identified modal parameters and will be given in Equation (16); $\boldsymbol{\alpha}$ is the unknown damage level vector to be solved, whose definition will be given in Equation (13); and the superscript "*" indicates a damaged version of physical or modal parameters. In other words, these parameters are dependent on α; and N_q and N_d are the number of linear equations in the system and the potentially damaged elements, respectively. Usually, $N_q \geq N_d$ is required to ensure that the linear system of equations has a unique solution. In general, the computation of a meaningful approximate solution of the linear system requires that the system be replaced by a nearby system that is less sensitive to perturbations. This replacement is referred to as regularization. Tikhonov regularization is one of the most popular regularization techniques that can damp out the small measurement errors and single out a relatively accurate solution [15]. So far, it has been widely applied to damage detection [16–18] and other related research, such as model updating [19,20] and load identification [21]. Björck [22] gave a general form of Tikhonov regularization as follows:

$$\left(\mathbf{C}^{*T}\mathbf{C}^* + \xi^2 \mathbf{L}^T \mathbf{L}\right)\alpha = \mathbf{C}^{*T}b^* \tag{2}$$

where ξ^2 and $\mathbf{L}$ are the regularization parameter and the regularization operator, respectively. Then, the solution α of Equation (2) satisfies the minimization problem

$$\min_{\alpha \in \mathbb{R}^{N_d}} \{\|\mathbf{C}^*\alpha - b^*\|^2 + \xi^2 \|\mathbf{L}\alpha\|^2\} \tag{3}$$

Here and below, $\|\cdot\|$ denotes the Euclidean norm. The determination of a suitable value of ξ^2 and $\mathbf{L}$ is an important task. In many previously developed methods, ξ^2 is selected using an L-curve criterion [12,15] or the generalized cross-validation method [20,23], but both methods are time-consuming computational tasks. Besides, the regularization parameter $\mathbf{L}$ is usually selected as an identity operator and thus leads to over-smoothing of the solution, which goes against the sparse feature of isolated damage at the early stage of structural deterioration [18]. Wang et al. [24] proposed the iterative Tikhonov regularization (ITR) method specialized for the identification of the isolated damage to structures. The method has some advantages, including the incorporation of an adaptive strategy to simultaneously determine ξ^2 and $\mathbf{L}$. Additionally, the ITR method shows good convergence behavior.

Over the past decades, numerous techniques for damage identification have been reported. Carden and Fanning [25] gave a review of vibration-based condition monitoring that revealed numerous and diverse algorithms using data in the time, frequency, and modal domains. Yan et al. [26] presented a general summary of the state-of-art of intelligent algorithms, and their application prospects in structural damage identification were introduced. Among many reviews of the existing literature concerning damage identification methods, Fan and Qiao [27] dealt in particular with the subset of methods related to variations in basic modal properties. Changes or lack of smoothness in mode shape or mode shape curvature, analysis of dynamically measured flexibility, and updating of structural model parameters provide different examples of this family of methods. Dessi and Camerlengo [28] also focused on technique processing information about mode shape curvature or strain modes with or without knowledge of baseline data. The general modal strain energy (MSE), as an extension of mode shape analysis, is widely appreciated because of its excellent damage-sensitive features [29,30]. It is formed by the product of the stiffness matrix and the second power of mode shape:

$$P^*_{j,n} = \left(\phi^*_j\right)^T \mathbf{K}^*_n \phi^*_j \left(j = 1, 2, \cdots, N_j\right) \tag{4}$$

where N_j is the number of measured modes; ϕ^*_j and $\mathbf{K}^*_n$ are the j-th mode shape and the n-th stiffness submatrix of the damaged structure, respectively; and $P^*_{j,n}$ is the MSE of the n-th damaged structural element associated with the j-th measured mode. When damage occurs, the distribution of strain energy originally stored in the structure will change in a more pronounced manner in the detected areas. Therefore, changes in the modal strain energy distributions of the healthy and damaged structures can be used to detect the existence, location, and extent of the damage. A review of MSE-based methods was given by Wang and Xu [29], and will not be discussed here.

In this study, a special MSE method named the CMSE method, proposed by Hu et al. [10] and later improved by Wang et al. [31], is used for damage detection. The adjective "cross" here indicates that MSE-like terms are product terms extending over the baseline finite element model (FEM) of the healthy structure and the measured damaged structure, also extending over various modes. The CMSE is defined as

$$C^*_{m,n} = (\phi_i)^T \mathbf{K}_n \phi^*_j \left(i = 1, 2, \cdots, N_i, \ j = 1, 2, \cdots, N_j \right) \tag{5}$$

where ϕ_i and $\mathbf{K}_n$ are the i-th analytical mode shape and the n-th stiffness submatrix of the baseline FEM, respectively, and N_i is the number of analytical modes. In practice, it is easy to obtain the analytical modes of the baseline FEM, but difficult or expensive to extract the measured modes of the damaged structure [10]; therefore, one may select a much larger N_i than N_j, i.e., $N_i \gg N_j$.

It is clear that the CMSE does not require matching modes between the healthy and damaged structures, and thus has more available mode combinations (MCs) than the general MSE, i.e., $N_i \times N_j \gg N_j$. Like many other damage identification techniques, the CMSE method can establish a linear system of the same number of equations as the available measured mode or MCs. Therefore, a distinct advantage of this method is that it breaks through the constraint that the number of available measured modes is smaller than that of actual damages, i.e., $N_j < N_d$. It further raises the possibility that one can assume all N_e ($N_e \geq N_d$) structural elements are damaged in case of omitting any actual damage only if sufficient equations $\left(N_i \times N_j \geq N_e \right)$ are constructed. Nevertheless, the disadvantages of constructing excessive equations are obvious. On the one hand, the damage identification result by solving a linear system assembled by different equations varies, leading to complicated decisions surrounding the actual damage state of the structure. On the other hand, the redundant equations do not contribute to damage detection but complicate the system, and exacerbate the damage identification result.

The main purpose of this study is to address the excessive equations issue faced in the assemblage of the CMSE linear system of equations. From the standpoint of solving a linear system of equations, optimizing the assemblage of the system by eliminating redundant equations serves as an effective tool to improve identification of damage, which has been rarely investigated in previous research. The ITR method is adopted as a tool for solving the linear system of equations given by Equation (1), with some improvements to this method also presented. This study contributes towards supplying a guideline for the elimination of the redundant equations to enhance the robustness of the CMSE system. A sensitivity index that, to some extent, measures how the actual damage level vector satisfies each CMSE equation is derived by performing an analysis of the defined residual's sensitivity to damage. Finally, the numerical and experimental robustness of the proposed damage identification scheme against measurement noise is investigated.

2. Theoretical Background

2.1. CMSE Method

If $\mathbf{M}$, $\mathbf{K}$ and $\mathbf{M}^*$, $\mathbf{K}^*$ are used to denote the mass and stiffness matrices for baseline FEM of the healthy structure and the measured damaged structure, respectively, the following can be written:

$$\mathbf{K}\phi_i = \lambda_i \mathbf{M}\phi_i \tag{6}$$

$$\mathbf{K}^*\phi^*_j = \lambda^*_j \mathbf{M}^*\phi^*_j \tag{7}$$

where λ_i and λ^*_j are the i-th and the j-th eigenvalue of the baseline FEM of the healthy structure and the measured damaged structure, respectively.

The CMSE method is developed under the assumption that the mass distributions of the healthy and damaged structures are not known, but do not change, that is $\mathbf{M}^* = \mathbf{M}$. In the following derivation, as λ_i, ϕ_i, $\mathbf{K}$, λ^*_j, and ϕ^*_j are presumably known, the unknowns are $\mathbf{K}^*$ and $\mathbf{M}$.

Pre-multiplying Equation (6) by $(\phi_j^*)^T$ and Equation (7) by $(\phi_i)^T$ yields

$$(\phi_j^*)^T \mathbf{K} \phi_i = \lambda_i (\phi_j^*)^T \mathbf{M} \phi_i \tag{8}$$

$$(\phi_i)^T \mathbf{K}^* \phi_j^* = \lambda_j^* (\phi_i)^T \mathbf{M} \phi_j^* \tag{9}$$

Since $\mathbf{M}$ and $\mathbf{K}$ are both symmetric matrices, also noting the transpose of a scalar equals to itself, one thus has

$$(\phi_j^*)^T \mathbf{M} \phi_i = (\phi_i)^T \mathbf{M} \phi_j^* \tag{10}$$

$$(\phi_j^*)^T \mathbf{K} \phi_i = (\phi_i)^T \mathbf{K} \phi_j^* \tag{11}$$

Assuming that all structural elements are damaged, the stiffness matrix of the damaged structure can be written as

$$\mathbf{K}^* = \mathbf{K} - \sum_{n=1}^{N_e} \alpha_n \mathbf{K}_n \tag{12}$$

where α_n $(0 \le \alpha_n \le 1)$ is the extent of damage of the n-th element. It is difficult to model the damage in sufficient detail for an unknown type of damage. Here, it is assumed that the damage to a structure can be represented by a decrease in the modulus of elasticity of each structural element:

$$E_n^* = (1 - \alpha_n) E_n \tag{13}$$

where E_n and E_n^* are the moduli of elasticity of the n-th element of the baseline FEM and the damaged structure, respectively. Herein, E_n is regarded as a known constant but E_n^* is regarded as an unknown variable. The objective herein is to evaluate α_n corresponding to each structural element.

Dividing Equation (9) by Equation (8), and using the scalar identities of Equations (10) and (11) yields

$$\frac{(\phi_i)^T \mathbf{K}^* \phi_j^*}{(\phi_j^*)^T \mathbf{K} \phi_i} = \frac{\lambda_j^*}{\lambda_i} \tag{14}$$

Finally, substituting Equation (12) into Equation (14), the following is obtained:

$$\sum_{n=1}^{N_e} \alpha_n C_{m,n}^* = b_m^* \tag{15}$$

where $C_{m,n}^* = (\phi_i)^T \mathbf{K}_n \phi_j^*$, as given by Equation (5), is the CMSE of the n-th structural element associated with the m-th MC, and the right-hand side $b_m^* = (1 - \lambda_j^*/\lambda_i) C_m^*$, where $C_m^* = \sum_{n=1}^{N_e} C_{m,n}^* = (\phi_i)^T \mathbf{K} \phi_j^*$ is the CMSE of the overall structural system. If the first N_i analytical modes and N_j measured modes are obtained, then $N_q = N_i \times N_j$ MCs are available and N_q linear equations can be constructed, which can be assembled in a matrix of the form as given by Equation (1), where

$$\mathbf{C}^* = \begin{bmatrix} C_1^* \\ \vdots \\ C_m^* \\ \vdots \\ C_{N_q}^* \end{bmatrix} = \begin{bmatrix} C_{1,1}^* & \cdots & C_{1,n}^* & \cdots & C_{1,N_e}^* \\ \vdots & \ddots & \vdots & \ddots & \vdots \\ C_{m,1}^* & \cdots & C_{m,n}^* & \cdots & C_{m,N_e}^* \\ \vdots & \ddots & \vdots & \ddots & \vdots \\ C_{N_q,1}^* & \cdots & C_{N_q,n}^* & \cdots & C_{N_q,N_e}^* \end{bmatrix} \tag{16}$$

$$\boldsymbol{\alpha} = \begin{bmatrix} \alpha_1 & \cdots & \alpha_n & \cdots & \alpha_{N_e} \end{bmatrix}^T$$

$$\boldsymbol{b}^* = \begin{bmatrix} b_1^* & \cdots & b_m^* & \cdots & b_{N_q}^* \end{bmatrix}^T$$

If N_q is greater than N_e, more equations are available than unknowns. Hence, it would expected that the least-squares solution for α can be taken as

$$\hat{\alpha} = \left(\mathbf{C}^{*T}\mathbf{C}^*\right)^{-1}\mathbf{C}^{*T}b^*$$

2.2. ITR Method

The CMSE-based structural damage identification can be treated as a linear inverse problem, whose discretization gives rise to the linear system as given by Equation (1) with a very ill-conditioned matrix $\mathbf{C}^*$. This means that a small perturbation in measurements can lead to unrealistically large perturbations in the predicted damage level vector. For this case, the least-squares solution is not sufficient. To address this problem, the ITR method that was tested and that outperformed the generalized least-squares method to provide a proper sparse solution [24] is introduced with the following solution:

$$\hat{\alpha}^{(k)} = \left\{\mathbf{C}^{*T}\mathbf{C}^* + \left[\xi^{(k)}\right]^2\left[\mathbf{L}^{(k)}\right]^T\mathbf{L}^{(k)}\right\}^{-1}\mathbf{C}^{*T}b^* \tag{17}$$

where $\xi^{(k)}$ and $\mathbf{L}^{(k)}$ are the iterative regularization parameter and regularization operator, respectively, in iteration k. In the context of the identification of isolated damage, the actual damage level vector bears sparse features. Proper sparsity of the Tikhonov solution depends on both the regularization parameter and regularization operator that have to be optimally selected. In the ITR method, an adaptive strategy is presented to determine them. The regularization parameter is selected as the intermediate singular value of interval $[\sigma_b, \sigma_a]$ as follows:

$$\xi^{(k)} = \sigma_p, \ p = \left[\frac{a+b}{2}\right], 1 \leq a \leq b \leq rank(\mathbf{C}^*) \tag{18}$$

where σ_a, σ_b, and σ_p are the a-th, b-th and p-th non-zero singular values of the coefficient matrix $\mathbf{C}^*$, respectively, and the square brackets can round the inside value to the nearest integer less than or equal to itself. Moreover, the regularization operator is selected as

$$\mathbf{L}^{(k)} = diag\left\{\left[\widetilde{\alpha}_1^{(k-1)} + \varepsilon\right]^{-1}, \cdots, \left[\widetilde{\alpha}_n^{(k-1)} + \varepsilon\right]^{-1}, \cdots, \left[\widetilde{\alpha}_{N_e}^{(k-1)} + \varepsilon\right]^{-1}\right\} \tag{19}$$

where $\widetilde{\alpha}_n^{(k-1)}$ is the predicted and adjusted extent of damage of n-th structural element in iteration $k-1$; and ε is a given small parameter in the case that the diagonal entries of $\mathbf{L}^{(k)}$ approach infinity when $\widetilde{\alpha}_n^{(k-1)}$ approaches 0. A detailed process of the ITR method can be found in the paper by Wang et al. [24]

In this paper, two modifications to the adaptive strategy of determining the iterative regularization parameter and regularization operator are introduced, although this is not the main aim of this study.

First, it can be noticed that the regularization parameter is selected as the intermediate singular value of interval $[\sigma_b, \sigma_a]$. When $a = b$, the calibration of the regularization parameter is terminated although the iteration continues. In order to continuously calibrate the regularization parameter and accelerate the convergence of the iteration, a dichotomy method in logarithmic space is introduced

$$\xi^{(k)} = \exp\left\{\frac{\left[\ln\left(\eta_a^{(k)}\right) + \ln\left(\eta_b^{(k)}\right)\right]}{2}\right\} \tag{20}$$

where $\left[\eta_b^{(k)}, \eta_a^{(k)}\right]$ is the range for the regularization parameter selection in iteration k, and in the initial iteration $\eta_b^{(0)} = \sigma_r$ and $\eta_a^{(0)} = \sigma_1$, where σ_1 and σ_r are the first and last non-zero singular values of matrix $\mathbf{C}^*$, respectively. Here the reason that the logarithmic space is used is that the singular values of matrix $\mathbf{C}^*$ usually belong to a log-linear distribution.

Second, to give a more proper regularization, the regularization operator is modified as

$$\mathbf{L}^{(k)} = \mathrm{diag}\left\{\left[\sqrt{\tilde{\alpha}_1^{(k-1)}} + \varepsilon\right]^{-1}, \cdots, \left[\sqrt{\tilde{\alpha}_n^{(k-1)}} + \varepsilon\right]^{-1}, \cdots, \left[\sqrt{\tilde{\alpha}_{N_e}^{(k-1)}} + \varepsilon\right]^{-1}\right\} \tag{21}$$

according to considerable computational experience.

3. Robust Damage Identification Scheme

From the standpoint of solving a linear system of equations, optimizing the assemblage of the system by eliminating redundant equations serves as an effective tool to improve identification of damage. However, this approach was rarely investigated in previous research. Therefore, this section approaches discrimination and elimination of redundant equations, thereby enhancing the robustness of the CMSE system against noise.

If the actual and predicted damage level vectors of the measured damaged structure are denoted as α_0 and $\hat{\alpha}$, respectively, then $\hat{\alpha} = \alpha_0 + \Delta\alpha$ is due to measurement errors, in which $\Delta\alpha$ is a small perturbation in $\hat{\alpha}$. In the context of the least-squares problem, $\hat{\alpha}$ instead of α_0 is the optimal solution that can minimize the Euclidean norm $\|\mathbf{C}^*\alpha - b^*\|_2$. One would also expect that α_0 could nearly minimize $\|\mathbf{C}^*\alpha - b^*\|_2$ so that α_0 can be obtained by solving $\mathbf{C}^*\alpha = b^*$ via a specific tool, such as the ITR method. In view of this, narrowing the difference between $\|\mathbf{C}^*\hat{\alpha} - b^*\|_2$ and $\|\mathbf{C}^*\alpha_0 - b^*\|_2$ may raise the possibility of obtaining α_0. Hence, the following CMSE residual function is defined:

$$\mathbf{R}(\alpha) = \mathbf{C}^*\alpha - b^* \tag{22}$$

where $\mathbf{R}(\alpha)$ collects N_q residual sub-functions $R_m(\alpha) = C_m^*\alpha - b_m^*$.

Sensitivity analyses are widely used in engineering to evaluate the effect of changes of one variable on another variable. Herein, a sensitivity analysis of $\mathbf{R}(\alpha)$ with respect to α is performed to measure the difference between $\|\mathbf{R}(\hat{\alpha})\|_2$ and $\|\mathbf{R}(\alpha_0)\|_2$ to some degree. If $\mathbf{R}(\alpha)$ has a large sensitivity at $\hat{\alpha}$, a small perturbation $\Delta\alpha$ tends to cause a big change of $\|\mathbf{R}(\alpha)\|_2$, or a great difference between $\|\mathbf{R}(\hat{\alpha})\|_2$ and $\|\mathbf{R}(\alpha_0)\|_2$. In other words, α_0 will not satisfy the currently assembled CMSE system. Hence, a linear system that corresponds to a small residual sensitivity to damage, i.e., $\|\mathbf{R}'(\hat{\alpha})\|_2$ is expected.

To further illustrate this point, one considers two special residual sub-functions $Y_1 = R_1(\alpha)$ and $Y_2 = R_2(\alpha)$ in a CMSE linear system. It should be noted that all symbols are presented by a scalar, indicating the measured damaged structure only has one assumed damage location. The curves of the Euclidean norm of these two residual functions, i.e., $l_1 : y_1 = \|R_1(\alpha)\|_2$ and $l_2 : y_2 = \|R_2(\alpha)\|_2$, are plotted against α, respectively, as shown in Figure 1. Because $\hat{\alpha}$ is the least-squares solution of the CMSE linear system, one assumes that both y_1 and y_2 have their near-minimum value R_{min} when $\alpha = \hat{\alpha}$. However, their right derivations, i.e., $\|R_1'(\hat{\alpha})\|_2$ and $\|R_2'(\hat{\alpha})\|_2$, are different, more specifically, $\|R_2'(\hat{\alpha})\|_2 > \|R_1'(\hat{\alpha})\|_2$. It can be observed by comparing curves l_1 and l_2 that, at the actual solution α_0, $\|R_1(\alpha_0)\|_2$ is closer to the minimum residual R_{min} than $\|R_2(\alpha_0)\|_2$. This means that α_0 can satisfy the CMSE equation corresponding to residual sub-function Y_1 better than that corresponding to Y_2. It can be further envisaged that α_0 has a larger chance of being predicted when the linear system is completely composed of CMSE equations corresponding to Y_1 like residual sub-functions.

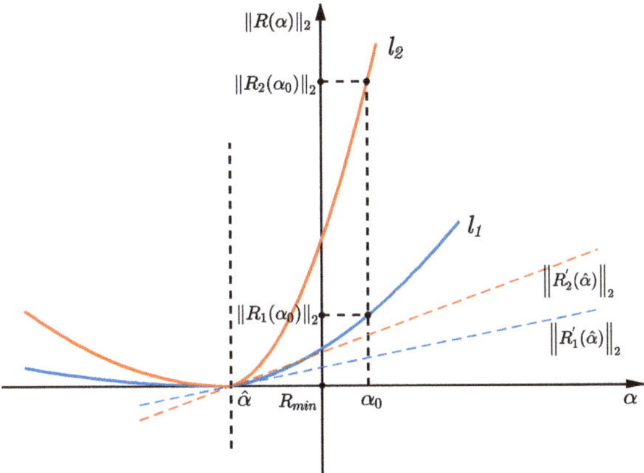

Figure 1. A comparison of two residual sub-functions with different residual-damage sensitivity.

The main task hereafter is discriminating and eliminating the redundant equations that correspond to relatively large $\|R'_m(\hat{\alpha})\|_2$, which can be obtained by calculating the residual sensitivity of the CMSE to damage:

$$R'_m(\alpha) = \frac{\partial C^*_m}{\partial \alpha^T}\alpha + C^*_m\frac{\partial \alpha}{\partial \alpha^T} - \frac{\partial b^*_m}{\partial \alpha^T} \tag{23}$$

It is assumed that only a few elements are slightly damaged; hence, $\alpha \approx 0$. Then Equation (23) is reduced to

$$R'_m(\alpha) \approx C^*_m - \frac{\partial b^*_m}{\partial \alpha^T} \tag{24}$$

By considering $\partial b^*_m/\partial \alpha^T = [\partial b^*_m/\partial \alpha_1,\cdots,\partial b^*_m/\partial \alpha_n,\cdots,\partial b^*_m/\partial \alpha_{Ne}]$, $\|R'_m(\hat{\alpha})\|_2$ can be calculated and defined as a sensitivity index to ascertain whether the m-th equation is redundant:

$$S_m = \|R'_m(\hat{\alpha})\|_2 = \mathrm{sqrt}\left(\sum_{n=1}^{Ne}\left|C^*_{m,n} - \partial b^*_m/\partial \alpha_n\right|^2\right) \tag{25}$$

Substituting $C^*_{m,n} = (\phi_i)^T K_n \phi^*_j$ and $b^*_m = (1 - \lambda^*_j/\lambda_i)(\phi_i)^T K\phi^*_j$ into Equation (26), it can be shown that

$$S_m = \mathrm{sqrt}\left\{\sum_{n=1}^{Ne}\left|(\phi_i)^T K_n \phi^*_j - \partial\left[(1 - \lambda^*_j/\lambda_i)(\phi_i)^T K\phi^*_j\right]/\partial \alpha_n\right|^2\right\} \tag{26}$$

Substituting Equation (13) into Equation (27) yields

$$S_m = \mathrm{sqrt}\left\{\sum_{n=1}^{Ne}\left|(\phi_i)^T K_n \phi^*_j + \partial\left[(1 - \lambda^*_j/\lambda_i)(\phi_i)^T K\phi^*_j\right]E_n/\partial E^*_n\right|^2\right\} \tag{27}$$

In order to compute Equation (28), one first considers computing $\partial \lambda^*_j/\partial E^*_n$. The derivative of Equation (7) with respect to E^*_n can be expressed as

$$(K^* - \lambda^*_j M)\frac{\partial \phi^*_j}{\partial E^*_n} = -\left(\frac{\partial K^*}{\partial E^*_n} - \lambda^*_j\frac{\partial M}{\partial E^*_n}\right)\phi^*_j + \frac{\partial \lambda^*_j}{\partial E^*_n}M\phi^*_j \tag{28}$$

Assuming that the eigenvector ϕ^*_j is mass-normalized yields

$$(\phi^*_j)^T M\phi^*_j = 1 \tag{29}$$

Pre-multiplying Equation (29) by $(\phi_j^*)^T$ and substituting Equations (7) and (30) into Equation (29) yields

$$\frac{\partial \lambda_j^*}{\partial E_n^*} = (\phi_j^*)^T \left(\frac{\partial \mathbf{K}^*}{\partial E_n^*} - \lambda_j^* \frac{\partial \mathbf{M}}{\partial E_n^*} \right) \phi_j^* \tag{30}$$

Here, it is always $\partial \mathbf{M}/\partial E_n^* = 0$ because the mass distributions of the structure are independent of the modulus of elasticity. Further, defining a purely geometric stiffness submatrix $\mathbf{K}_n^0 = \mathbf{K}_n^*/E_n^*$, also noting $\mathbf{K}^* = \sum_{t=1}^{Ne} \mathbf{K}_t^*$ and $\partial \mathbf{K}_t^*/\partial E_n^* = 0$ when $t \neq n$, Equation (31) can be simplified as

$$\partial \lambda_j^* / \partial E_n^* = P_{j,n}^{*0} \tag{31}$$

where $P_{j,n}^{*0} = (\phi_j^*)^T \mathbf{K}_n^0 \phi_j^*$ is the purely geometric MSE of the n-th structural element associated with the j-th mode of the damaged structure.

Moreover, as for $\partial \left[(\phi_i)^T \mathbf{K} \phi_j^* \right]/\partial E_n^*$, a compact matrix form can be written [32]:

$$\frac{\partial \left[(\phi_i)^T \mathbf{K} \phi_j^* \right]}{\partial E_n^*} = \left[\begin{array}{cc} (\phi_i)^T \mathbf{K} & 0 \end{array} \right] \left\{ \begin{array}{c} \frac{\partial \phi_j^*}{\partial E_n^*} \\ \frac{\partial \lambda_j^*}{\partial E_n^*} \end{array} \right\} \tag{32}$$

According to Lee and Jung [33], the eigenpairs derivations for the damaged structure with respect to E_n^* can be obtained by

$$\left\{ \begin{array}{c} \frac{\partial \phi_j^*}{\partial E_n^*} \\ \frac{\partial \lambda_j^*}{\partial E_n^*} \end{array} \right\} = \left[\begin{array}{cc} \mathbf{K}^* - \lambda_j^* \mathbf{M} & -\mathbf{M}\phi_j^* \\ -\left(\phi_j^*\right)^T \mathbf{M} & 0 \end{array} \right]^{-1} \left\{ \begin{array}{c} -\left(\frac{\partial \mathbf{K}^*}{\partial E_n^*} - \lambda_j^* \frac{\partial \mathbf{M}}{\partial E_n^*} \right) \phi_j^* \\ \frac{1}{2} \left(\phi_j^*\right)^T \frac{\partial \mathbf{M}}{\partial E_n^*} \phi_j^* \end{array} \right\} \tag{33}$$

As it is still assumed that only a few elements are slightly damaged, the overall stiffness matrix of the damaged structure can be replaced by the healthy one. Equation (34) can thus be reduced to

$$\left\{ \begin{array}{c} \frac{\partial \phi_j^*}{\partial E_n^*} \\ \frac{\partial \lambda_j^*}{\partial E_n^*} \end{array} \right\} \approx - \left[\begin{array}{cc} \mathbf{K} - \lambda_j^* \mathbf{M} & -\mathbf{M}\phi_j^* \\ -\left(\phi_j^*\right)^T \mathbf{M} & 0 \end{array} \right]^{-1} \left\{ \begin{array}{c} \mathbf{K}_n^0 \phi_j^* \\ 0 \end{array} \right\} \tag{34}$$

Substituting Equation (35) into Equation (33), the derivative of $(\phi_i)^T \mathbf{K} \phi_j^*$ with respect to E_n^* can be obtained:

$$\frac{\partial \left[(\phi_i)^T \mathbf{K} \phi_j^* \right]}{\partial E_n^*} = - \left[\begin{array}{cc} (\phi_i)^T \mathbf{K} & 0 \end{array} \right] \left[\begin{array}{cc} \mathbf{K} - \lambda_j^* \mathbf{M} & -\mathbf{M}\phi_j^* \\ -\left(\phi_j^*\right)^T \mathbf{M} & 0 \end{array} \right]^{-1} \left\{ \begin{array}{c} \mathbf{K}_n^0 \phi_j^* \\ 0 \end{array} \right\} \tag{35}$$

Then, substituting Equations (32) and (36) into Equation (21) and simplifying yields the final expression of the sensitivity index as follows:

$$S_m^* = \text{sqrt} \left\{ \sum_{n=1}^{Ne} \left| C_{m,n}^* - \frac{\widetilde{P}_{j,n}^* C_m^*}{\lambda_i} - \left(1 - \frac{\lambda_j^*}{\lambda_i} \right) \left[\begin{array}{cc} (\phi_i)^T \mathbf{K} & 0 \end{array} \right] \left[\begin{array}{cc} \mathbf{K} - \lambda_j^* \mathbf{M} & -\mathbf{M}\phi_j^* \\ -\left(\phi_j^*\right)^T \mathbf{M} & 0 \end{array} \right]^{-1} \left\{ \begin{array}{c} \mathbf{K}_n \phi_j^* \\ 0 \end{array} \right\} \right|^2 \right\} \tag{36}$$

where $\widetilde{P}_{j,n}^* = P_{j,n}^{*0} E_n = (\phi_j^*)^T \mathbf{K}_n \phi_j^*$ is an approximation for $P_{j,n}^*$.

Good robustness of the CMSE system against noise necessitates finding $N_m \left(N_e \leq N_m \leq N_q \right)$ CMSE equations corresponding to the first N_m lowest sensitivity indices. Since the sensitivity index as given by Equation (37) is computable, these N_m CMSE equations can be selected to reassemble the CMSE system and identify damage. An interesting question then arises as to what guidelines one can follow to determine N_m. Minimizing the level of correlation between the measured and predicted natural frequencies and mode shapes provides a simple but effective way to indicate whether the selected N_m is optimal for this case. In this regard, the optimal N_m is the one within the range $\left[N_e, N_q\right]$ that maximizes the following fitness function:

$$f(N_m) = -\lg\left[\sum_{j=1}^{N_j} \frac{\left|\omega_j^* - \widetilde{\omega}_j^*(N_m)\right|}{\omega_j^*} + \sum_{j=1}^{N_j} \frac{\left\|\phi_{F,j}^* - \widetilde{\phi}_{F,j}^*(N_m)\right\|_2}{\left\|\phi_{F,j}^*\right\|_2}\right] \tag{37}$$

where ω_j^* and $\phi_{F,j}^*$ are the j-th natural frequency and mode shape partitioned to F measured degrees of freedom (DoFs) of the damaged structure, respectively; likewise, $\widetilde{\omega}_j^*(N_m)$ and $\widetilde{\phi}_{F,j}^*(N_m)$ are the correspondingly predicted values produced by the selected N_m CMSE equations. Note that compared with directly using all N_q equations, a larger amount of computation is required in order to ascertain the optimal subset of equations. However, there is only a small increase in computational costs but a clear improvement in damage detection performance.

It is also noted from Equation (37) that the sensitivity index S_m^* is determined by the eigen-parameters associated with a unique MC, i.e., the i-th mode of the healthy structure and the j-th mode of the damaged structure. Therefore, it can also serve as a pre-test tool to select suitable modes for the CMSE-based methods and, in turn, lead to sensor placement requirements in a built-in health monitoring system.

4. Numerical Simulation

In this section, the numerical robustness of the proposed damage identification scheme against measurement noise is demonstrated by considering a truss structure. A comparative study is conducted between the CMSE method with and without the sensitivity analysis. For convenience, the classical CMSE method without sensitivity analysis is still called the CMSE method, but the one with sensitivity analysis is called the robust CMSE (RCMSE) method.

4.1. Description of the Truss Structure

The example adopted in this numerical study is a 2-D clamped–clamped truss structure. As shown in Figure 2, the examined structure consists of 29 elements comprising 14 horizontal brace elements (HBs), 7 vertical brace elements (VBs), and 8 diagonal brace elements (DBs). In addition, the sectional area of each member is 1.5×10^{-3} m^2 and the moment of inertia is 3.13×10^{-7} m^4. The structure is made of plexiglass, with a Young's modulus of 3 GPa, linear mass density of 1300 kg/m^3, and Poisson's ratio of 0.3. Although only a truss structure modeled by beam elements is used as an illustrative example to validate the proposed method, it can be applied to any more general structure, such as a 3-D structure that needs to be modeled with solid elements.

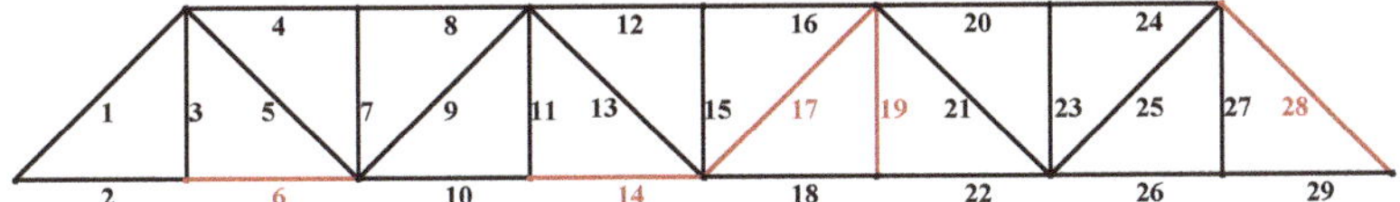

Figure 2. Sketch of the truss structure.

4.2. Damage Cases

As listed in Table 1, six damage cases were simulated involving three types of elements, namely, HB elements Nos. 6 and 14, VB element No. 19, and DB elements Nos. 17 and 28, as well as three levels of damage. Damage cases A to D were used to investigate the performance of the proposed scheme with respect to damage location, and damage cases C, E, and F were used to investigate the performance with respect to damage level. For the simulated damage cases, the corresponding natural frequencies obtained by eigenvalue analysis are listed in Table 1.

Table 1. The simulated damage cases of the truss structure.

Damage Case	Location	Extent	Natural Frequencies (Hz)		
			1st	2nd	3rd
Baseline	N/A	N/A	13.878	22.257	23.832
A	6	30%	13.845	22.074	23.634
B	19	30%	13.877	21.996	23.740
C	28	30%	13.701	22.166	23.601
D	14, 17	30%, 30%	13.837	21.946	23.761
E	28	20%	13.774	22.202	23.686
F	28	10%	13.832	22.232	23.763

In practice, it is easy to obtain the analytical modes of the healthy structure, but difficult or expensive to extract the measured modes of the damaged structure. Therefore, one may choose a much larger N_i than N_j [10]. Herein, the first 20 analytical modes and the first 3 measured modes are used for tests, i.e., $N_i = 20$ and $N_j = 3$, respectively. Hence, there are 60 MCs or equations in total available for determining the damage level vector.

4.3. Robustness Performance Investigation

A parametric study is conducted to investigate the effects of location and extent of damage on the robustness performance of the proposed damage identification scheme against noise. In order to consider noise interference, the measurements of the i-th polluted frequency and mode displacement at the v-th DOF of the damaged structure, denoted by $\hat{\omega}_j^*$ and $\hat{\phi}_{v,j}^*$, respectively, are simulated by adding a Gaussian random error to the corresponding true values:

$$\hat{\omega}_j^* = \left(1 + n_\omega \varsigma_\omega\right)\omega_j^* \tag{38}$$

$$\hat{\phi}_{v,j}^* = \left(1 + n_\phi \varsigma_\phi\right)\phi_{v,j}^* \tag{39}$$

where ς_ω and ς_ϕ are two Gaussian random numbers both with zero mean and unit standard deviation, and n_ω and n_ϕ are the modal noise levels for natural frequency and mode shape, respectively. Previous studies [34] suggest that mode shape estimates have error levels as much as 20 times greater than those in the natural frequency estimates. Therefore, in this study, the modal noise levels were set as $n_\phi = 20n_\omega$. In the following, the mentioned noise level refers to n_ϕ.

4.3.1. Effects of Damage Location

In this section, damage detection analysis is performed by considering different damage locations. It starts from damage case A ($\alpha_6 = 30\%$) at a 3% level of noise. The least-squares method is firstly used to solve the CMSE system of linear equations. Figure 3 shows a comparison of the resulting location and extent of damage by using the classical CMSE and RCMSE methods.

It is observed from Figure 3a that the localization indicator ambiguously locates the actually damaged element at a 3% level of noise, and both beam elements Nos. 2 and 25 seem to be candidates, as they are affected by the damage in the same way. Moreover, the estimated damage level of element No. 6 is far smaller than the expected value of 30%, indicating that the CMSE method cannot correctly detect stiffness reduction by using all 60 equations together.

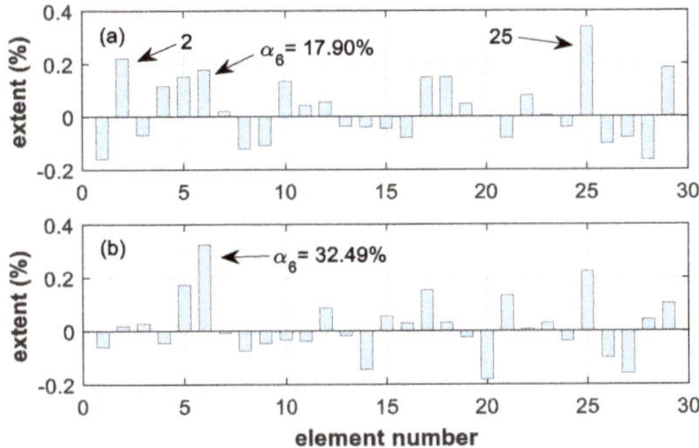

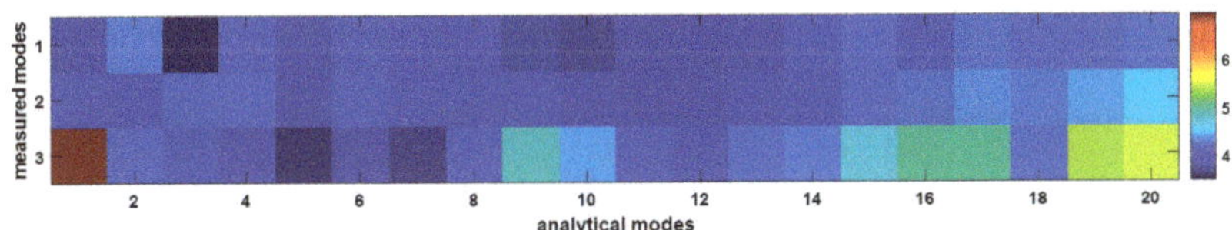

Figure 3. The damage identification results of damage case A at a 3% level of noise, with the least-squares solution: (**a**) CMSE method; (**b**) RCMSE method.

For the RCMSE method, a sensitivity analysis of the CMSE residual function with respect to damage is first performed to discriminate and eliminate redundant CMSE equations. Figure 4 shows the corresponding sensitivity indices of all 60 CMSE equations corresponding to 60 MCs. It is an expected result that the CMSE equations associated with the third mode of the damaged structure have larger sensitivity overall. The fitness function given in Figure 5 suggests that 43 equations are an optimum number for identifying the damage, so the CMSE system is reconstructed by eliminating 17 equations that correspond to the first 17 largest sensitivity indices, and damage identification analysis is performed again. It is shown in Figure 3b that the actual damage to element No. 6 is located accurately by the RCMSE method, whereas the estimated damage level is a little larger than the expected value, indicating that RCMSE outperforms the classical CMSE method both in damage localization and quantification.

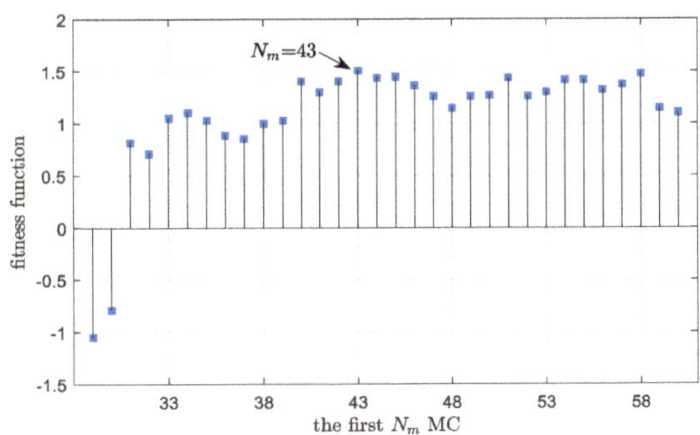

Figure 4. The sensitivity index of damage case A versus MCs $N_i = 20$ and $N_j = 3$, at a 3% level of noise.

Figure 5. Fitness function of damage case A at a 3% level of noise for the least-squares method.

Note that in Figure 5, only the fitness functions corresponding to 32 tested N_m $\left(N_e \leq N_m \leq N_q\right)$ are presented. This is because there are $N_e = 29$ elements simulated for the 2-D truss structure, which also means at least 29 equations are required to ensure that the reduced CMSE system has a unique solution.

It is seen from Figure 3 that the damage identification results obtained by the least-squares method are not sparse, i.e., massive false-positive alarms of damage caused by noise contamination are produced, masking the actually damage-induced alarm. Here, an iterative Tikhonov solution can be tried to address this problem. Figure 6 shows a comparison between the identified results of the classical CMSE and RCMSE methods. By comparing Figures 3 and 6, the obtained Tikhonov solutions are clearly sparse and the estimated extents of the actual damage are both closer to the expected value. This comparison illustrates that the ITR method can eliminate false-positive alarms of damage and is more suitable for detecting isolated damage to the structure. It is thus exclusively used to solve the CMSE system of linear equations hereafter.

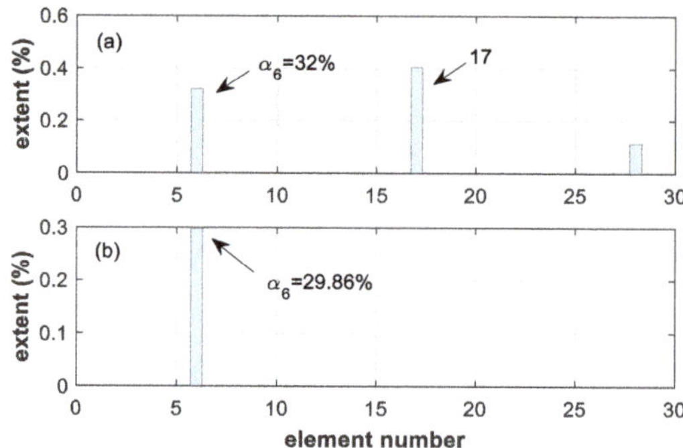

Figure 6. Damage identification results of damage case A at a 3% level of noise, with an iterative Tikhonov solution: (**a**) CMSE method; (**b**) RCMSE method.

It is also noted from Figure 6 that the RCMSE method remains more accurate than the classical CMSE method both in damage localization and quantification, even in terms of an iterative Tikhonov solution. This improvement strongly benefits from the elimination of 25 redundant equations according to the fitness function given by Figure 7. It can also be seen from Figure 7 that the fitness value obtained by using the corresponding first 35 equations is much larger than that obtained using all 60 equations, which illustrates that the predicted damage state of the structure agrees better with the real state when some redundant equations are eliminated.

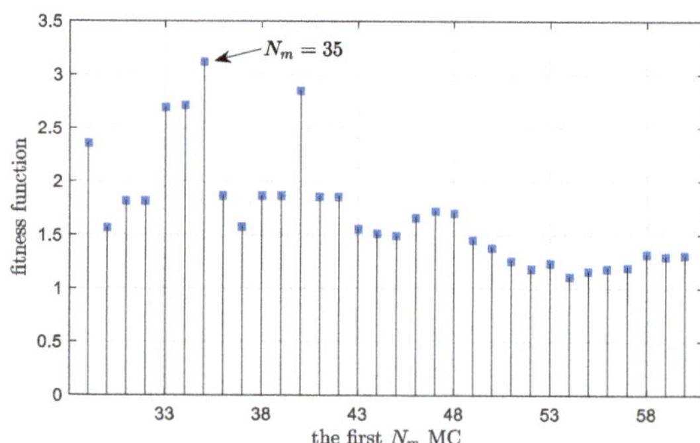

Figure 7. Fitness function of damage case A at a 3% level of noise, for the ITR method.

For damage case B, Figure 8 shows a comparison between the damage identification results of the CMSE and RCMSE methods. The top panel of Figure 8 illustrates that the damage of element No. 19 can be located by using all equations, but its extent is overestimated. In comparison, the RCMSE method, as shown in the bottom panel of Figure 8, exhibits enhancement of damage quantification.

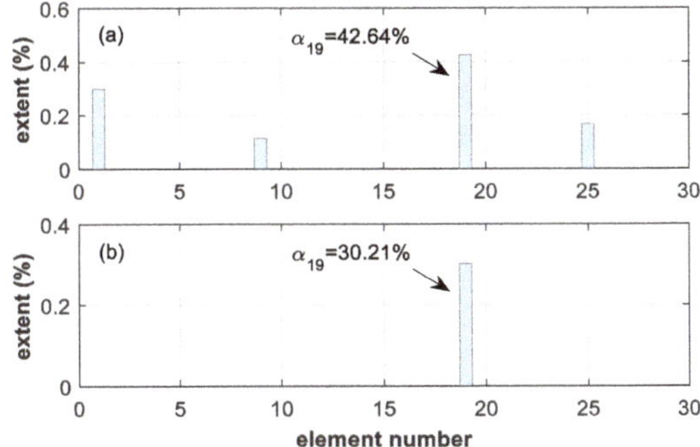

Figure 8. Damage identification results of damage case B at a 3% level of noise, with an iterative Tikhonov solution: (**a**) CMSE method; (**b**) RCMSE method.

It should be noted that in this case only 40 equations are used to predict the preset damage location and extent of damage (see Figures 9 and 10), confirming that not all equations contribute to damage detection but complicate the CMSE system. Furthermore, by performing the analysis of CMSE equations' sensitivity to damage, numerous redundant equations are eliminated, thereby simplifying the CMSE system.

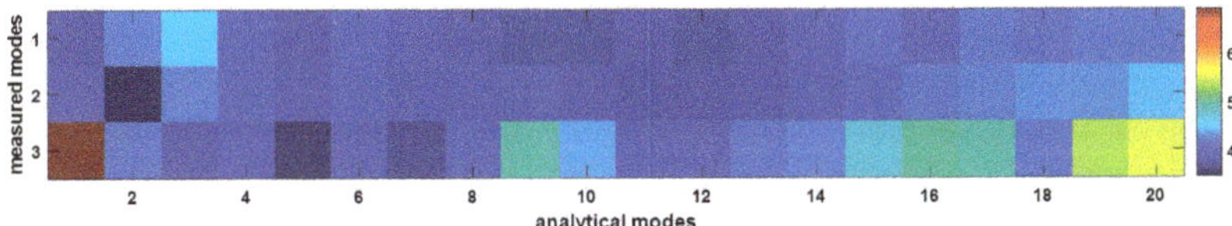

Figure 9. Sensitivity index of damage case B versus MCs $N_i = 20$ and $N_j = 3$, at a 3% level of noise.

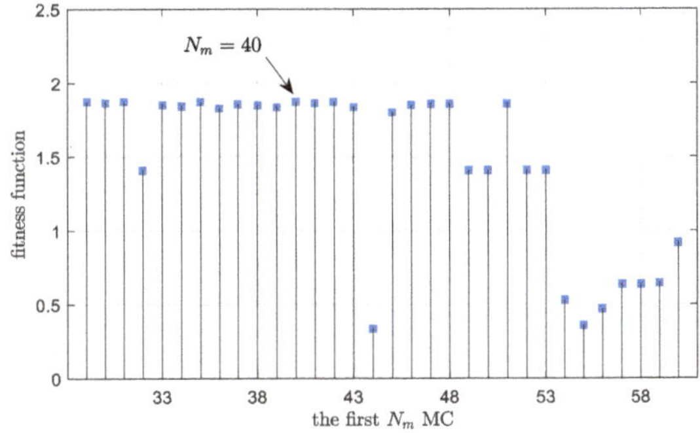

Figure 10. Fitness function of damage case B at a 3% level of noise, for the ITR method.

For damage case C, the analysis of CMSE equations' sensitivity to damage indicates that 54 equations are most appropriate for solving the preset location and extent of damage (see Figures 11 and 12).

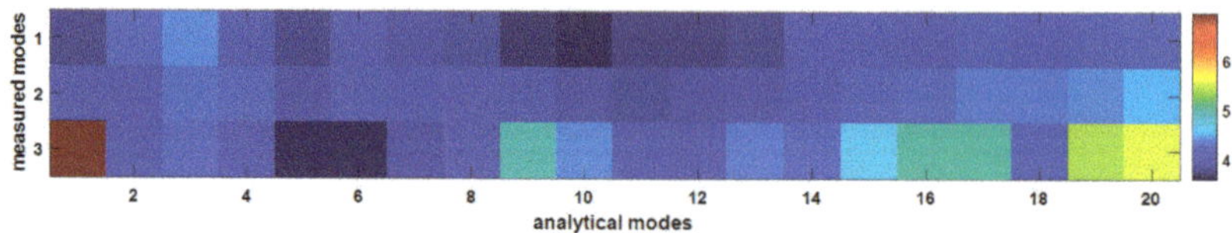

Figure 11. Sensitivity index of damage case C versus MCs $N_i = 20$ and $N_j = 3$, at a 3% level of noise.

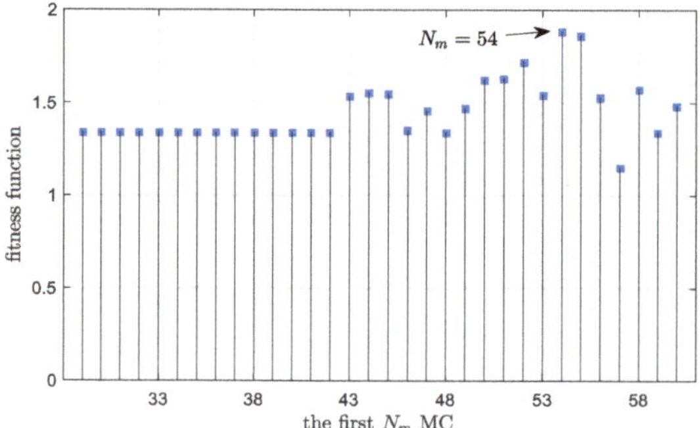

Figure 12. Fitness function of damage case C at a 3% level of noise, for the ITR method.

Following this, the CMSE method employing 60 equations and the RCMSE method employing 54 equations are used to identify the damage, respectively. Figure 13 shows that both methods can correctly locate the actual damage. However, two false-positive alarms of damage to elements Nos. 5 and 17 are produced when using the CMSE method. At the same time, both methods overestimate the preset extent of the damage. This reveals that the robustness of the CMSE method against noise is slightly improved by eliminating six redundant equations.

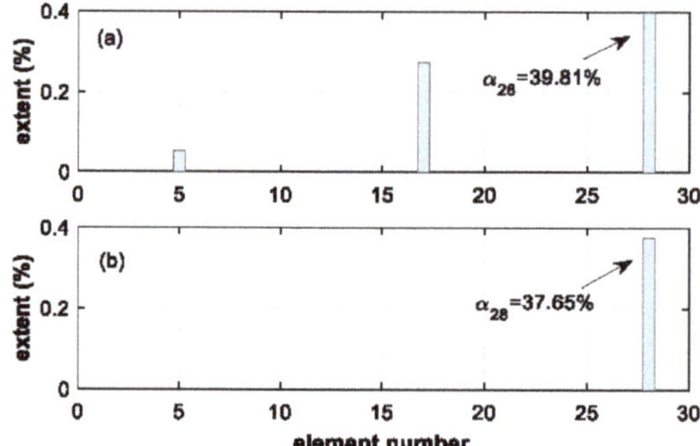

Figure 13. Damage identification results of damage case C at a 3% level of noise, with an iterative Tikhonov solution: (**a**) CMSE method; (**b**) RCMSE method.

Damage case D is a double-damage case, in which elements Nos. 14 and 17 both exhibit a 30% stiffness loss. For this case, 45 equations are suggested for identifying damage according to the sensitivity analysis (see Table 2). The sensitivity index and fitness function are not shown here because of space limitations.

Table 2. The number of equations (N_m) and the corresponding condition number (κ) of the original and reduced CMSE systems for damage cases A to D.

Damage Case	Original System		Reduced System	
	N_m	κ	N_m	κ
A	60	126.36	35	95.92
B	60	126.79	40	28.23
C	60	128.45	54	22.31
D	60	123.86	45	55.15

Figure 14 provides a comparison between the damage identification results of the CMSE and RCMSE methods. It is observed that the CMSE method omits the actually damaged element No. 17 and the extent of damage of element No. 14 is clearly overestimated. The RCMSE method demonstrates good damage localization and quantification performances, while it slightly overestimates the extent of the preset damage.

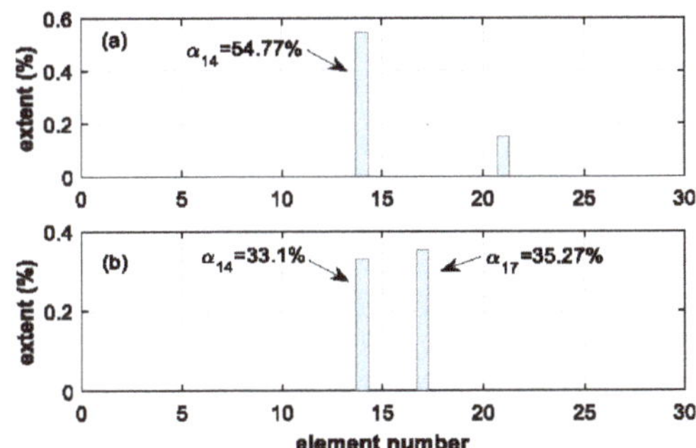

Figure 14. Damage identification results of damage case D at a 3% level of noise, with an iterative Tikhonov solution: (**a**) CMSE method; (**b**) RCMSE method.

All results shown above demonstrate an improvement of damage identification by considering the sensitivity analysis scheme. It is worth noting that when the equation subsets are eliminated from the original CMSE system, the corresponding coefficient matrix of the reduced CMSE system, determining the accuracy of solution to some extent, is simplified. Hence, one would expect to explain this improvement from an ill-conditioned problem viewpoint. The condition number of the coefficient matrix $\mathbf{C}$, defined as

$$\kappa = \|\mathbf{C}^{*+}\| \cdot \|\mathbf{C}^*\| \tag{40}$$

measures how ill-conditioned a CMSE system is, where $\mathbf{C}^{*+}$ is the Moore–Penrose inverse of $\mathbf{C}^*$ because it is usually not a square matrix.

Table 2 shows the number of equations and the corresponding condition number of the original and reduced CMSE systems. It can be seen that in all simulated cases, the condition number is reduced as some redundant equations are eliminated. Taking damage case A as an example, the condition number is reduced from 126.36 to 95.92 as 25 equations are eliminated. This might confirm that a reasonable reduction of the dimension of the CMSE system yields a better-conditioned system and a better damage identification.

To further investigate the noise robustness of the proposed scheme, 1000 Monte Carlo simulations are performed by varying the level of noise from 0.5% to 5%. The damage identification probability [24] is used to measure the prediction accuracy, which is defined as

$$p_d = \frac{n_d}{n_s} \times 100\% \tag{41}$$

where $n_s = 1000$ is the total number of Monte Carlo simulations for a given level of noise and n_d is the number of realizations in which the actual damage is detected.

Figure 15 shows the obtained damage identification probability of damage case A to damage case C, in which at least two findings are observable.

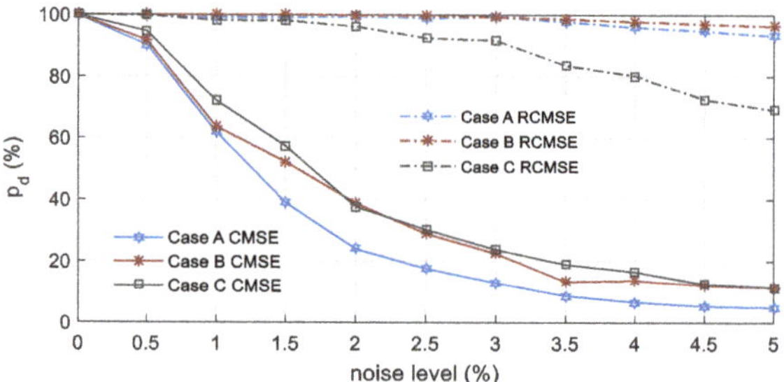

Figure 15. The identification probability of damage cases A to C against the level of noise.

First, it is expected that the damage localization performance of the RCMSE method is better than that of the CMSE method at every level of noise, thereby confirming the effectiveness of eliminating redundant equations from the classical CMSE method.

Second, it is observed that for the CMSE method, the damage identification probability of DB element No. 28 and HB element No. 6 is always greater than that of HB element No. 19 in the examined range of noise level. When using a vibration-based method, the detectability of damage depends highly upon the modal parameter changes that the damage caused. It is observed by recalling Table 1 that the overall change in the first three natural frequencies due to the damage of element No. 19 is the smallest among these three damage cases, which may account for the difference in the damage localization performance of the CMSE method for different damage locations. However, for the RCMSE method, the damage identification probability of DB element No. 28 comes last. It is realized by comparing Figures 4, 9 and 11 that the MCs associated with the third mode of the damaged structure, on the whole, have larger sensitivity of the residual function to damage. In other words, the CMSE equations constructed by the third mode of the damaged structure are likely classified as redundant and eliminated from damage detection analysis. It can also be observed from Table 1 that the damage of element No. 28 mainly changes the third mode rather than the others. The discrepancy of the employed MCs may explain the different performances of these two methods.

4.3.2. Effects of Damage Level

In this section, damage detection analysis is performed to consider how varying the damage level would affect the damage identification performance of the proposed scheme. The results of damage cases C ($\alpha_{28} = 30\%$), E ($\alpha_{28} = 20\%$), and F ($\alpha_{28} = 10\%$), are shown in Figures 13, 16 and 17, respectively.

Upon comparing the top panels of these three figures, as expected, the damage identification results of the classical CMSE method gradually worsen as the damage level decreases. Especially for damage case F, the actually damaged element No. 28 is classified as undamaged at a 3% level of noise. This demonstrates that the classical CMSE method has weak robustness against noise and its performance of the CMSE method is highly relevant to the damage level of the structure.

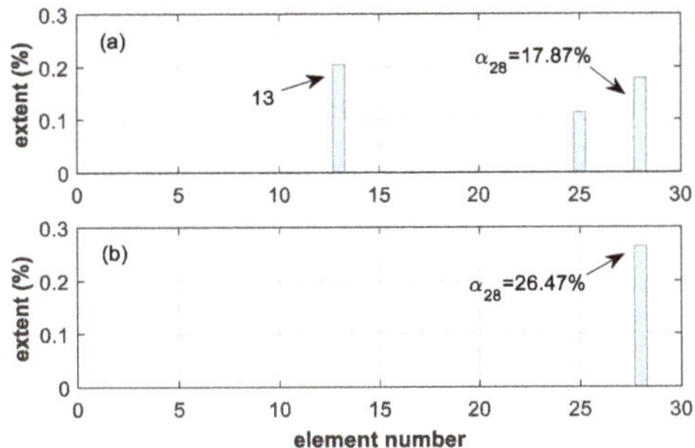

Figure 16. Damage identification results of damage case E at a 3% level of noise, with an iterative Tikhonov solution: (a) CMSE method; (b) RCMSE method.

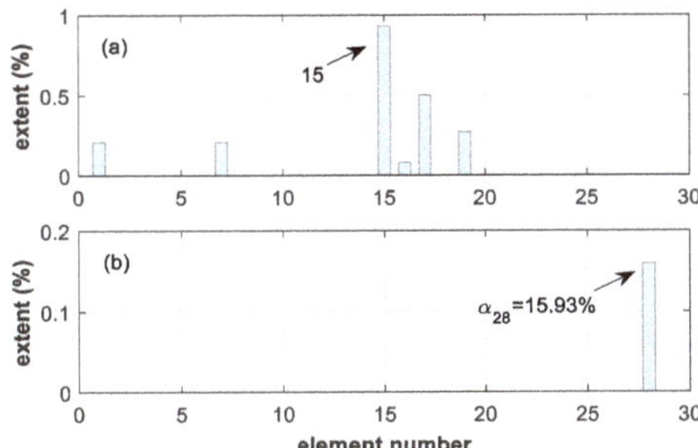

Figure 17. Damage identification results of damage case F at a 3% level of noise, with an iterative Tikhonov solution: (a) CMSE method; (b) RCMSE method.

For the RCMSE method, it can be seen from the bottom panels of Figures 13, 16 and 17 that all damage cases are located, indicating the RCMSE method outperforms the classical CMSE method with regard to damage localization. Moreover, it can be seen that the RCMSE method always over- or under-estimates the extent of the damage. This is because, as mentioned earlier, the damage of element No. 28 mainly changes the third modal parameters of the structure; these parameters, however, contribute little to the damage detection analysis.

Table 3 shows the number of equations and the corresponding condition number of the original and reduced CMSE systems for damage cases E and F. It is also observed for each damage case that the condition number is reduced as some redundant equations are eliminated. This, again, confirms that reasonable reduction of the dimension of the CMSE system yields a better-conditioned system and better damage identification.

Table 3. The number of equations (N_m) and the corresponding condition number (κ) of the original and reduced CMSE systems for damage cases E and F.

Damage Case	Original System		Reduced System	
	N_m	κ	N_m	κ
E	60	126.30	56	36.61
F	60	124.03	40	25.71

Then, 1000 Monte Carlo simulations are performed to obtain the identification probability of these three damage cases, and the results are displayed in Figure 18. Two findings are observable. First, by means of eliminating redundant equations and optimizing the construction of CMSE systems, the damage localization performance of the classical CMSE method is significantly improved. In addition, there is always a lower damage identification probability as the damage level decreases because severe damage always causes a larger change in the structural modal parameter. This trend also reveals that the CMSE and RCMSE methods are both sensitive to the damage level of the structure.

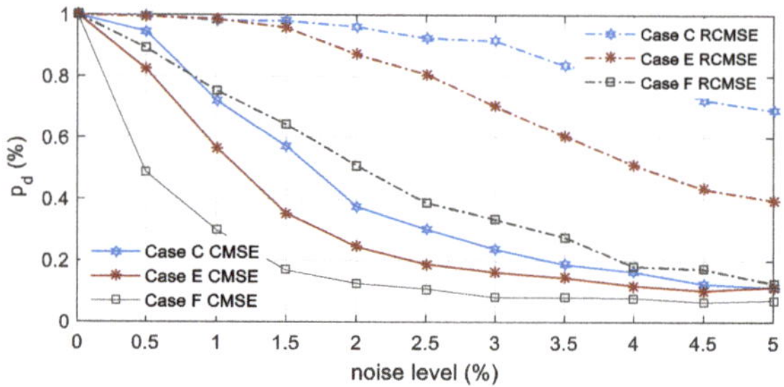

Figure 18. Identification probability of damage cases C, E, and F against the level of noise.

5. Experimental Validation

5.1. Description of the Beam Structure

Experimental data from a cantilever beam structure are used to evaluate the effectiveness of the proposed damage identification scheme. The cantilever beam, as displayed in Figure 19, has a length, width, and thickness of 200, 5.0, and 2.8 cm, respectively, and is simulated by 20 equal Euler–Bernoulli beam elements.

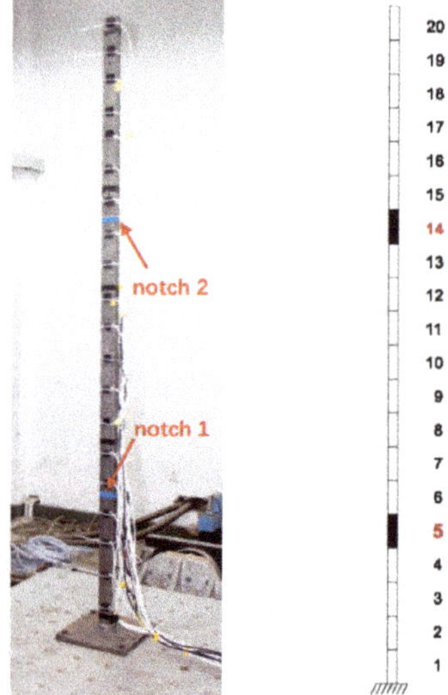

Figure 19. Beam structure used in the experiment, physical (**left**) and finite element (**right**) models.

Twenty accelerometers were vertically installed at every 10 cm on the beam to collect its acceleration time histories. A shock excitation on the beam was generated by means of an impulse hammer. To generate a maximum amplitude of the vibration signal, the impact location was selected at the free end of the beam. The excitation force was not measured. For output-only modal identification, the acceleration signal was processed by the eigensystem realization algorithm [35] to obtain the incomplete mode shapes, and Guyan's method [36] was subsequently used for modal expansion.

The experiment started by measuring the dynamic responses of the undamaged beam, and a baseline FE model was constructed to represent the dynamic characteristics of the undamaged beam. Subsequently, two notches were generated by a saw cut on the beam in the width direction. Again, the dynamic responses of the damaged beam were measured for modal identification and damage detection. Three damage cases including two single- and one double-damage cases were considered by producing two notches. As shown in Figure 19, the location of the first notch was 44.2 cm away from the clamped end of the beam, approximately at the middle of element No. 5, and the second notch was located approximately at the middle of element No. 14. Two extents of damage are considered, including the notch depths of about 1/4 and 1/2 of the beam thickness. The length of the notch along the beam axis was about 1 mm. In the experiment, the first three modes of the beam were always identified from the measurement data, i.e., $N_j = 3$. The simulated damage cases and the measured natural frequencies are listed in Table 4 for clarity.

Table 4. Simulated damage cases and measured natural frequencies in the experiment.

Case	Location	Extent	Natural Frequencies (Hz)		
			1st	2nd	3rd
Undamaged	N/A	N/A	5.524	34.711	97.200
I	5	1/4 thickness	5.473	34.768	96.874
II	5	1/2 thickness	5.299	34.755	95.089
III	5 and 14	1/2 and 1/2 thickness	5.289	33.837	91.485

The essential geometric and material properties of the baseline FE model are as follows. The linear mass density is 7920 kg/m^3 by considering the mass of beam, sensors, and wires; and the Young's modulus of the elements at the non-clamped position is uniformly 1.91×10^{11} N/m^2, whereas that of the element at the clamped edge is reduced to 1.80×10^{11} N/m^2 to consider the possibly imperfect connection.

To examine the accuracy of the baseline FE model, an eigenvalue analysis was performed to obtain its modal characteristics, i.e., the analytical modal parameters. The measured and analytical natural frequencies and mode shape correlations represented by modal assurance criterions (MACs) are summarized in Table 5 for clarity. It is evident that the analytical modal parameters match the corresponding measured values very well, thereby confirming the accuracy of the model updating process.

Table 5. Measured and analytical natural frequencies and MACs of the undamaged beam.

Data Type	Natural Frequencies (Hz)			MACs		
	1st	2nd	3rd	1st	2nd	3rd
Measured	5.524	34.711	97.200	0.999	0.997	0.998
Analytical	5.523	34.644	97.079			

5.2. Results and Discussions

In this experiment, the first 15 modes of the undamaged beam calculated from the updated baseline FE model as well as the first 3 modes of the damaged beam identified from the measured data are used for damage identification. That is to say, a total of 45 CMSE equations can be formed. Therefore, for the classical CMSE method, 45 equations are always used to identify the damage, whereas for the RCMSE method, an analysis of the residual function's sensitivity to damage is performed first to discriminate redundant equations.

For damage case I (1/4-thickness notch), Figures 20 and 21 give the 42 equations that are most appropriate to solve the actual location and extent of the damage. Then, these equations are selected to construct the CMSE system.

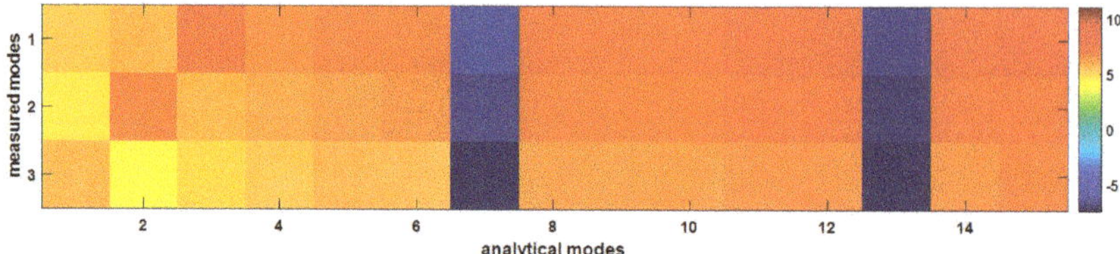

Figure 20. Sensitivity index of damage case I versus MCs $N_i = 15$ and $N_j = 3$.

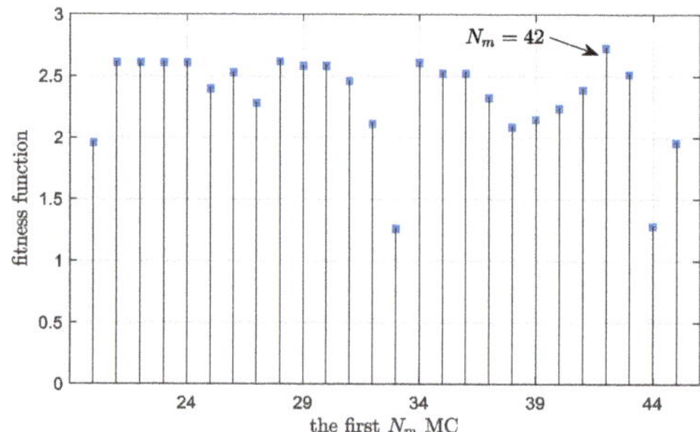

Figure 21. Fitness function of damage case I, for the ITR method.

Figure 22 shows a comparison of the resulting location and extent of damage by using the classical CMSE and the RCMSE methods. It is evident that the damage of element No. 5 can be located by the RCMSE method but cannot be located by the CMSE method, indicating an enhancement of damage identification by eliminating three redundant equations.

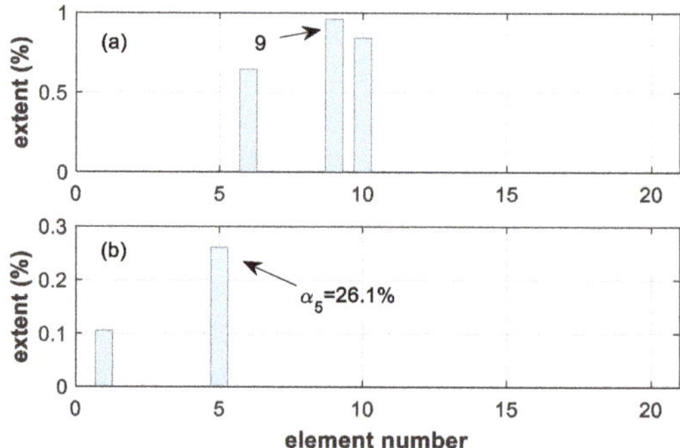

Figure 22. Damage identification results of damage case I, with an iterative Tikhonov solution: (**a**) CMSE method; (**b**) RCMSE method.

It is worth mentioning that the performance of the CMSE and RCMSE methods with regard to damage severity estimation is not discussed here. This is because the actual stiffness reduction of the beam is unknown.

For damage case II (1/2-thickness notch), 43 equations are used to identify damage for the RCMSE method. One can observe from Figure 23 that damage at element No. 5 also can be located clearly by the RCMSE method. However, for the classical CMSE method, although the actually damaged element can be identified, a false-alarm of damage at element No. 15 is also produced. The difference between the results, again, illustrates the effectiveness of the elimination of redundant equations.

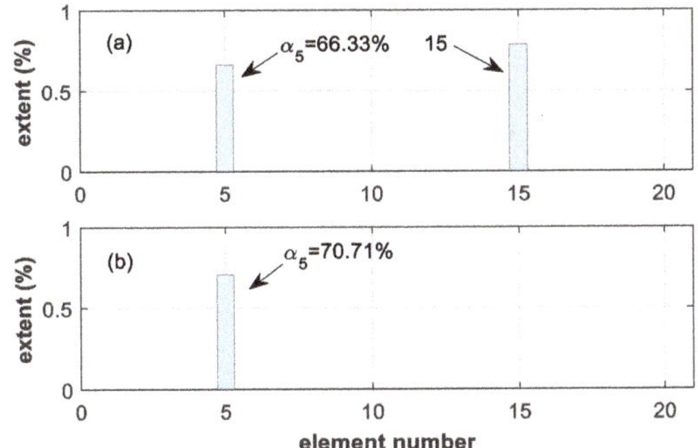

Figure 23. Damage identification results of damage case II, with an iterative Tikhonov solution: (a) CMSE method; (b) RCMSE method.

Furthermore, by comparing Figures 22b and 23b, it can be seen that RCMSE also produces a false-positive damage at element No. 1 for damage case I, hinting that smaller damage always results in a smaller change in the structural modal parameter (see Table 4) and leads to lower damage detectability.

For damage case III (1/2-thickness notches), 23 equations are used to identify damage for the RCMSE method. Figure 24 shows that both methods cannot yield a sparse solution. In other words, both methods produce several false-positives of damage except for the correct indications of damage at elements Nos. 5 and 14. This illustrates that double-damage is difficult to identify for both methods. However, it is also shown that the estimated extents of damage at the actually damaged locations by the RCMSE method are greater than others. In other words, the actually damaged elements are not masked by false-positives. In addition, the estimated extents of these two damaged elements, i.e., for both 1/2-thickness notches, by the RCMSE method are almost the same, implicating an improvement in damage identification performance as compared to the CMSE method.

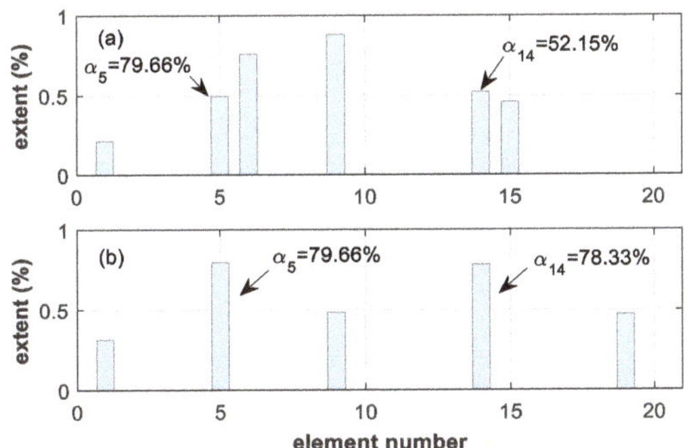

Figure 24. Damage identification results of damage case III, with an iterative Tikhonov solution: (a) CMSE method; (b) RCMSE method.

6. Conclusions

This paper presented a CMSE-based damage identification scheme. An analysis of the defined residual function's sensitivity to damage was performed to measure how the actual damage level vector satisfied each CMSE equation. A sensitivity index was formulated to discriminate and eliminate the redundant equations that do not contribute to damage detection but complicate the system. The numerical and experiment robustness of the proposed damage identification scheme against measurement noise was investigated. Two main aspects were emphasized by the results.

First, the damage identification performance of the classical CMSE method was clearly improved by eliminating redundant equations from the original CMSE system, confirming the validity of the sensitivity analysis process. The improvement of damage identification might inherently benefit from an improvement of the originally ill-conditioned problem by a reasonable reduction of the dimension of the CMSE system.

Second, for the proposed CMSE-based damage identification scheme, the detectability of damage depended highly upon the modal parameter changes that the damage caused. For different locations of damage even with the same damage level, the resulting modal parameter change in different modes varied, and the employed modes were also different. This is the reason for the discrepancy of the damage localization performance for different damage locations. Besides, the damage identification probability gradually decreased as the damage level decreased. The phenomenon is due to the fact that smaller damage always results in a smaller change in structural modal parameters and leads to lower damage detectability.

Overall, this study provides a contribution towards a clear and simple guideline for eliminating redundant equations in order to enhance the robustness of the CMSE system, which is solved by the ITR method to obtain spare solutions. Since a broad range of regression techniques [37,38] that can solve the CMSE system are now available, it is extremely necessary to compare the performance of the ITR method and these state-of-art techniques.

Author Contributions: Funding acquisition, S.W. and Y.L.; Investigation, M.X., J.G. and Y.L.; Methodology, M.X. and S.W.; Software, M.X., J.G. and S.W.; Supervision, S.W.; Validation, S.W. and Y.L.; Experiment, S.W.; Writing—Original draft, M.X.; Writing—Review and Editing, M.X., Y.L. and S.W. All authors have read and agreed to the published version of the manuscript.

Funding: This work was supported by the National Science Fund for Distinguished Young Scholars (51625902), the National Key Research and Development Program of China (2019YFC0312404), the Major Scientific and Technological Innovation Project of Shandong Province (2019JZZY010820), the National Natural Science Foundation of China (51809134), the Natural Science Foundation of Shandong Province (ZR2017MEE007), and the Taishan Scholars Program of Shandong Province (TS201511016).

Conflicts of Interest: The authors declare no potential conflicts of interest with respect to the research, authorship, and/or publication of this article.

References

1. Flynn, E.B.; Todd, M.D. A Bayesian approach to optimal sensor placement for structural health monitoring with application to active sensing. *Mech. Syst. Signal Process.* **2010**, *24*, 891–903. [CrossRef]
2. Pérez, J.E.R.; Rodríguez, R.; Vázquez-Hernández, A.O. Damage detection in offshore jacket platforms with limited modal information using the damage submatrices method. *Mar. Struct.* **2017**, *55*, 78–103. [CrossRef]
3. Oliveira, G.; Magalhães, F.; Cunha, Á.; Caetano, E. Vibration-based damage detection in a wind turbine using 1 year of data. *Struct. Control Health Monit.* **2018**, *25*, e2238. [CrossRef]
4. Soman, R.; Mieloszyk, M.; Ostachowicz, W. A two-step damage assessment method based on frequency spectrum change in a scaled wind turbine tripod with strain rosettes. *Mar. Struct.* **2018**, *61*, 419–433. [CrossRef]
5. Chaabane, M.; Mansouri, M.; Ben Hamida, A.; Nounou, H.; Nounou, M. Multivariate statistical process control-based hypothesis testing for damage detection in structural health monitoring systems. *Struct. Control Health Monit.* **2019**, *26*, e2287. [CrossRef]
6. Ding, Z.H.; Li, J.; Hao, H. Structural damage identification using improved Jaya algorithm based on sparse regularization and Bayesian inference. *Mech. Syst. Signal Process.* **2019**, *132*, 211–231. [CrossRef]
7. Wang, S.Q.; Wang, H.Y.; Xu, M.Q.; Guo, J. Identifying the presence of structural damage: A statistical hypothesis testing approach combined with residual strain energy. *Mech. Syst. Signal Process.* **2020**, *140*, 106655. [CrossRef]
8. Shi, Z.Y.; Law, S.S.; Zhang, L.M. Structural damage localization from modal strain energy change. *J. Eng. Mech.* **2000**, *218*, 1216–1223. [CrossRef]
9. Zhu, H.P.; Xu, Y.L. Damage detection of mono-coupled periodic structures based on sensitivity analysis of modal parameters. *J. Sound Vib.* **2005**, *285*, 365–390. [CrossRef]

10. Hu, S.L.J.; Wang, S.Q.; Li, H.J. Cross-Modal Strain Energy Method for Estimating Damage Severity. *J. Eng. Mech.* **2006**, *132*, 429–437.

11. Zhan, J.W.; Xia, H.; Chen, S.Y.; Roeck, G.D. Structural damage identification for railway bridges based on train-induced bridge responses and sensitivity analysis. *J. Sound Vib.* **2011**, *330*, 757–770. [CrossRef]

12. Zheng, Z.D.; Lu, Z.R.; Chen, W.H.; Liu, J.K. Structural damage identification based on power spectral density sensitivity analysis of dynamic responses. *Comput. Struct.* **2015**, *146*, 176–184. [CrossRef]

13. Golub, G.; Hansen, P.; O'Leary, D. Tikhonov regularization and total least squares. *J. Matrix Anal. Appl.* **1999**, *21*, 185–194. [CrossRef]

14. Calvetti, D.; Morigi, S.; Reichel, L.; Sgallari, F. Tikhonov regularization and the L-curve for large discrete ill-posed problems. *J. Comput. Appl. Math.* **2000**, *123*, 423–446. [CrossRef]

15. Hansen, P.C. Analysis of Discrete Ill-Posed Problems by Means of the L-Curve. *SIAM Rev.* **1992**, *34*, 561–580. [CrossRef]

16. Zhang, C.D.; Xu, Y.L. Comparative studies on damage identification with Tikhonov regularization and sparse regularization. *Struct. Control Health Monit.* **2016**, *23*, 560–579. [CrossRef]

17. Entezami, A.; Shariatmadar, H.; Sarmadi, H. Structural damage detection by a new iterative regularization method and an improved sensitivity function. *J. Sound Vib.* **2017**, *399*, 285–307. [CrossRef]

18. Fan, X.; Li, J.; Hao, H.; Ma, S. Identification of Minor Structural Damage Based on Electromechanical Impedance Sensitivity and Sparse Regularization. *J. Aerosp. Eng.* **2018**, *31*, 04018061. [CrossRef]

19. Weber, B.; Paultre, P.; Proulx, J. Structural damage detection using nonlinear parameter identification with Tikhonov regularization. *Struct. Control Health Monit.* **2007**, *14*, 406–427. [CrossRef]

20. Rezaiee-Pajand, M.; Entezami, A.; Sarmadi, H. A sensitivity-based finite element model updating based on unconstrained optimization problem and regularized solution methods. *Struct. Control Health Monit.* **2020**, *27*, e2481. [CrossRef]

21. Yan, G.; Sun, H.; Büyüköztürk, O. Impact load identification for composite structures using Bayesian regularization and unscented Kalman filter. *Struct. Control Health Monit.* **2017**, *24*, e1910. [CrossRef]

22. Björck, Å. A bidiagonalization algorithm for solving large and sparse ill-posed systems of linear equations. *BIT Numer. Math.* **1988**, *28*, 659–670. [CrossRef]

23. Golub, G.H.; Heath, M.; Wahba, G. Generalized Cross-Validation as a Method for Choosing a Good Ridge Parameter. *Technometrics* **1979**, *21*, 215–223. [CrossRef]

24. Wang, S.Q.; Xu, M.Q.; Xia, Z.P.; Li, Y.C. A novel Tikhonov regularization-based iterative method for structural damage identification of offshore platforms. *J. Mar. Sci. Technol.* **2019**, *24*, 575–592. [CrossRef]

25. Carden, E.P.; Fanning, P. Vibration Based Condition Monitoring: A Review. *Struct. Health Monit.* **2004**, *3*, 355–377. [CrossRef]

26. Yan, Y.J.; Cheng, L.; Wu, Z.Y.; Yam, L.H. Development in Vibration-Based Structural Damage Detection Technique. *Mech. Syst. Signal Process.* **2007**, *21*, 2198–2211. [CrossRef]

27. Fan, W.; Qiao, P.Z. Vibration-based Damage Identification Methods: A Review and Comparative Study. *Struct. Health Monit.* **2011**, *10*, 83–111. [CrossRef]

28. Dessi, D.; Camerlengo, G. Damage identification techniques via modal curvature analysis: Overview and comparison. *Mech. Syst. Signal Process.* **2015**, *52*, 181–205. [CrossRef]

29. Wang, S.Q.; Xu, M.Q. Modal Strain Energy-based Structural Damage Identification: A Review and Comparative Study. *Struct. Eng. Int.* **2019**, *29*, 234–248. [CrossRef]

30. Xu, M.Q.; Wang, S.Q.; Jiang, Y.F. Iterative two-stage approach for identifying structural damage by combining the modal strain energy decomposition method with the multiobjective particle swarm optimization algorithm. *Struct. Control Health Monit.* **2019**, *26*, e2301. [CrossRef]

31. Wang, S.Q.; Li, H.J.; Hu, S.L.J. Cross Modal Strain Energy Method for Damage Localization and Severity Estimation. In Proceedings of the ASME 2007 26th International Conference on Offshore Mechanics and Arctic Engineering, San Diego, CA, USA, 10–15 June 2007; pp. 245–249.

32. Yan, W.J.; Ren, W.X. A direct algebraic method to calculate the sensitivity of element modal strain energy. *Int. J. Numer. Methods Biomed. Eng.* **2011**, *27*, 694–710. [CrossRef]

33. Lee, I.W.; Jung, G.H. An efficient algebraic method for the computation of natural frequency and mode shape sensitivities—Part I. Distinct natural frequencies. *Comput. Struct.* **1997**, *62*, 429–435. [CrossRef]

34. Messina, A.; Williams, E.J.; Contursi, T. Structural damage detection by a sensitivity and statistical-based method. *J. Sound Vib.* **1998**, *216*, 791–808. [CrossRef]

Appl. Sci. **2020**, *10*, 2826

35. Juang, J.N.; Pappa, R.S. An eigensystem realization algorithm for modal parameter identification and model reduction. *J. Guid. Control Dyn.* **1985**, *8*, 620–627. [CrossRef]
36. Guyan, R.J. Reduction of stiffness and mass matrices. *AIAA J.* **1965**, *3*, 380. [CrossRef]
37. Bertsimas, D.; King, A.; Mazumder, R. Best Subset Selection via a Modern Optimization Lens. *Mathematics* **2016**, *44*, 813–852. [CrossRef]
38. Hastie, T.; Tibshirani, R.; Tibshirani, R.J. Extended Comparisons of Best Subset Selection, Forward Stepwise Selection, and the Lasso. *arXiv* **2017**, arXiv:1707.08692.

Article

A Novel Dense Full-Field Displacement Monitoring Method Based on Image Sequences and Optical Flow Algorithm

Guojun Deng [1,2], Zhixiang Zhou [1,2,*], Shuai Shao [1,2], Xi Chu [1] and Chuanyi Jian [1]

[1] School of Civil Engineering, Chongqing Jiaotong University, Chongqing 400074, China; guojunforsea@gmail.com (G.D.); shuai.shaos@foxmail.com (S.S.); chuxi1986@163.com (X.C.); chuanyi_jian_cqjtu@163.com (C.J.)

[2] College of Civil and Transportation Engineering, Shenzhen University, Shenzhen 518061, China

* Correspondence: zhixiangzhou@cqjtu.edu.cn

Received: 17 February 2020; Accepted: 15 March 2020; Published: 20 March 2020

Featured Application: This method can be applied to health monitoring of large-scale bridge structures, and the deformation of bridge structures can be monitored regularly and nondestructively using camera as a noncontact sensor. In order to improve measurement accuracy, a uniaxial automatic cruise acquisition device was designed to obtain the deformation of bridge elevation. The measurement points using the proposed method are denser than those of the traditional sensor measurement method. It can also detect abnormal deformation caused by the damage, and it is more efficient and easier to use.

Abstract: This paper aims to achieve a large bridge structural health monitoring (SHM) efficiently, economically, credibly, and holographically through noncontact remote sensing (NRS). For these purposes, the author proposes a NRS method for collecting the holographic geometric deformation of test bridge, using static image sequences. Specifically, a uniaxial automatic cruise acquisition device was designed to collect static images on bridge elevation under different damage conditions. Considering the strong spatiotemporal correlations of the sequence data, the relationships between six fixed fields of view were identified through the SIFT algorithm. On this basis, the deformation of the bridge structure was obtained by tracking a virtual target using the optical flow algorithm. Finally, the global holographic deformation of the test bridge was derived. The research results show that: The output data of our NRS method are basically consistent with the finite-element prediction (maximum error: 11.11%) and dial gauge measurement (maximum error: 12.12%); the NRS method is highly sensitive to the actual deformation of the bridge structure under different damage conditions, and can capture the deformation in a continuous and accurate manner. The research findings lay a solid basis for structure state interpretation and intelligent damage identification.

Keywords: noncontact remote sensing (NRS); optical flow algorithm; structural health monitoring (SHM); uniaxial automatic cruise acquisition device

---

1. Introduction

With the elapse of time, it is inevitable for a bridge to face structural degradation under the long-term effects of natural factors (e.g. climate and environment). In extreme cases, the bridge structure will suffer from catastrophic damages with the continuous increase in traffic volume and heavy vehicles, due to the booming economy [1]. The traditional approach of structural management, which is mainly manual periodic inspection, can no longer satisfy the demand of modern transport facilities. The traditional approach is inefficient, uncertain, and highly subjective, lacking in scientific or quantitative bases [2–5].

Structural health monitoring (SHM) aims to monitor, analyze, and identify all kinds of loads and structural responses during the service life of the target structure, to realize the evaluation of its structural performance and safety status, and provide support for the proprietor in structural management and making maintenance decisions [6–8]. In order to achieve this goal, a new sensor technology must be developed in combination with interdisciplinary theories and methods to provide advanced monitoring methods and reliable data sources for SHM [9–11]. Displacement is an important index of structural state and performance evaluations [7]. The static and dynamic characteristics of a structure, such as bearing capacity [12], deflection [13], deformation [14], load distribution [15], load input [16], influence line [17], influence surface [18], and modal parameters [19,20], can be calculated by structural displacement, to convert them further into physical indicators of response for structural safety assessment.

Since the 1990s, SHM systems have been set up on important large-span bridges across the globe. The main functions are to monitor the state and behavior of the bridge structure, while tracking and recording environmental conditions. On the upside, these systems have high local accuracy, run on an intelligent system, and support long-term continuous observation. On the downside, these systems are too costly to construct, the sensors cannot be calibrated periodically, and the layout of monitoring points is limited by a local terrain and the structure type. The geometric deformation of the bridge structure can only be collected by a few discrete monitoring points, making it difficult to characterize the local or global holographic geometry of bridge safety [21–25].

With the continuous development of machine vision technology and image acquisition equipment, structural displacement monitoring methods based on computer vision continues to emerge and have been verified in practical engineering applications [26–30]. Given its long-distance, noncontact, high-precision, time-saving, labor-saving, multi-point detection, and many other advantages, as well as increasing attention has been received from scientific researchers and engineers [31]. This method is mainly used to track the target of the measured structure video, which is captured by the camera, to obtain the moving track of the measuring point in the image, and then determine the displacement information of the structure through the set relationship between the image and the real world. The camera is mounted on a fixed point which is far from the structure to be tested, eliminating the requirement to install a fixed support point on the structure for the contact displacement detection method. In addition, low-cost multi-point measurement it is easy to achieve because the camera field of view can cover multiple measurement areas of the structure. Structural displacement monitoring based on computer vision has been applied in many tasks of bridge health monitoring, such as bridge deflection measurement [32], bridge alignment [33], bearing capacity evaluation [34], finite element model calibration [35], modal analysis [36], damage identification [18], cable force detection [29], and dynamic weighing assistance [12]. The images can also improve the accuracy of the estimation part through several means, such as deblurring, denoising and image enhancement, and even satellite images are gradually applied in structural monitoring [37,38].

Although the application of machine vision technology in structural displacement monitoring has many advantages, there are still some problems to be solved. In the recent research, part of it is aimed at small-scale structure, which can capture all structure displacements with a camera's field of view. However, for large-scale structure monitoring, to achieve sufficient accuracy, only the structural displacement for a specific area of the structure can be obtained, ignoring the overall deformation state of the structure from the macro view as a whole. Meanwhile, as a kind of holographic data, machine vision can obtain the displacement information of every pixel of an image in theory, but is mainly focused on obtaining the displacement information of key points in practice.

2. Purpose and Concept

On this basis, this paper proposes a method of obtaining the displacement information of an whole bridge structure from different views. According to the field of view of a camera, the bridge structure is divided into several areas, and image data are collected under different working conditions to form the image database of the bridge facade under the time–space sequence. The entire field displacement information of the bridge is obtained by establishing the connection between the time and space series. For this purpose, noncontact remote sensing (NRS) is designed to obtain the time–space sequence image data of the test bridge under different test conditions.

In order to obtain the full field displacement deformation of the whole structure, it is no longer used to track the artificial marked and corner points as the target points, but to track the displacement information by extracting every pixel of the lower edge contour line of the main beam as the virtual marked points. Modelling correction and damage identification is more conducive than measuring of finite points. Abundant data can be accumulated to provide more details data for machine learning and life-cycle maintenance.

The paper is organized as follows. Section 3 covers the theoretical background of the intelligent NRS system and the layout of lab. Section 4 proposes the algorithm to connect the fields of view and track the deformation, presents the analysis with the theoretical model, and discusses the validation of the proposed sensor and algorithms in full-field noncontact displacement and vibration measurement. The results are summarized in Section 5.

3. Test Overview

3.1. Intelligent NRS System

This paper designs an intelligent NRS system for the holographic monitoring of bridge structure based on virtual pixel sensors and several cutting-edge techniques (i.e., modern panoramic vision sensing, pattern recognition, and computer technology). As shown in Figure 1a, this intelligent NRS system mainly consists of an active image acquisition device, an automatic cruise remote control platform, an environmental monitoring unit, a signal transmission unit, and a data storage and analysis unit.

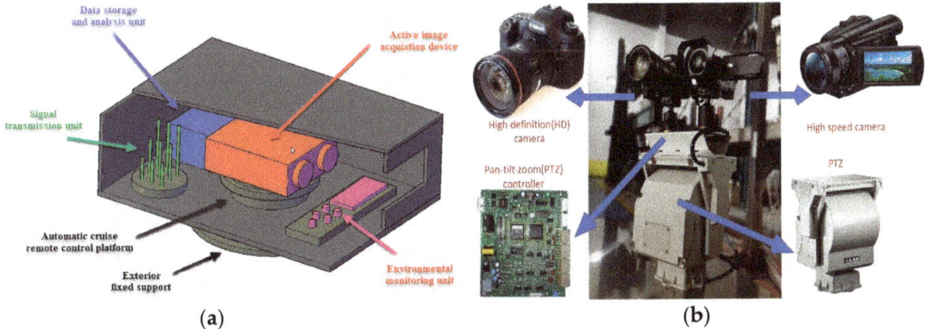

Figure 1. (a)The architecture of the intelligent noncontact remote sensing (NRS) system; (b) the intelligent NRS system for our load tests.

To monitor the holographic geometry of the bridge structure, the automatic cruise parameters (preset position, watch position, cruise time, and sampling time) are configured by a computer to remotely control the active image acquisition device and the environmental monitoring unit. In this way, the dynamic and static images of the bridge structure can be captured in the current field of view. Figure 1b is a photo of the intelligent NRS system for our load tests on the reduced-scale model of a super long span self-anchored suspension bridge. The workflow of the intelligent NRS system is explained in Figure 2.

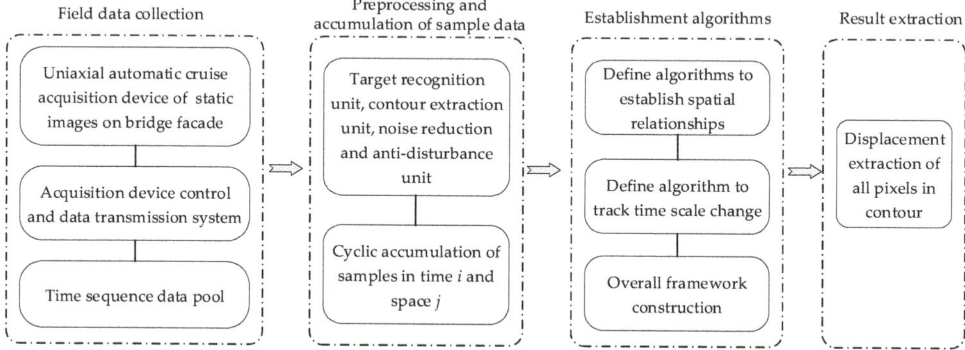

Figure 2. The workflow of the intelligent NRS system.

3.2. Object and data Collection

According to the previous research of our research team [39–41], a 1:30 model was constructed for Taohuayu Yellow River Bridge. A total of 52 C30 concrete deck slabs (1.16 × 0.45 × 0.2 m) were prepared and laid on the steel box girder to simulate vehicles on the bridge and serve as the counterweight.

The main cable is composed of 37 steel wire ropes with a diameter of 2mm, an elastic modulus of 195 GPa and a f_{tk} of 1860 MPa. The suspender is composed of steel wire ropes with a diameter of 4mm, an elastic modulus of 195 GPa, and a f_{tk} of 1860 MPa. The main tower adopts a thin-walled box made of steel plate with a thickness of 6 mm and Q345D material.

The standard section of the main beam is shown in Figure 3 in which the top, web, and bottom plates are made of 2 mm thick steel plates. Owing to the strong axial force of the main beam, considering the stability of the box girder, the top and bottom of the girder have four solid steel stiffeners (Φ 6 rebar), which are connected with the top and bottom plates of the box girder by spot welding.

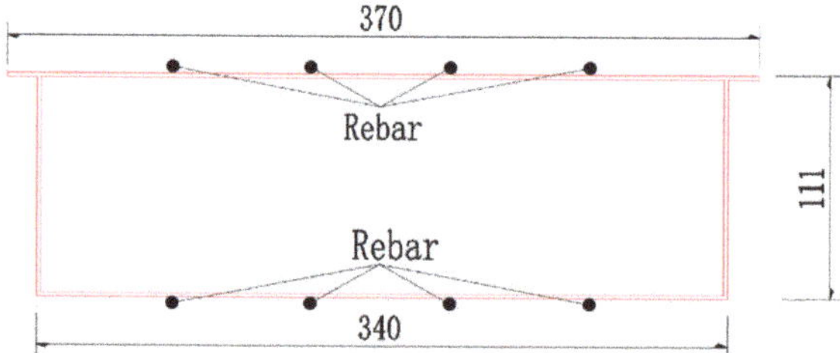

Figure 3. The standard section diagram of a girder model (unit: Mm).

In order to fully simulate the geometric similarity between the suspender and the main beam, the rigid arm is extended from both sides of the main beam at the lifting point, and the anchor plate is set on the rigid arm so that the suspender can be connected with the stiffener. Steel is selected for the rigid arm. To ensure the local stability of the main beam, one diaphragm is set every 450 mm (i.e., the section of the lifting point) of the main beam model, and the diaphragm and the main beam can be connected with the steel box girder by welding. The diaphragm plate is made of Q345D steel with a thickness of 2mm. As shown in Figures 4 and 5, respectively.

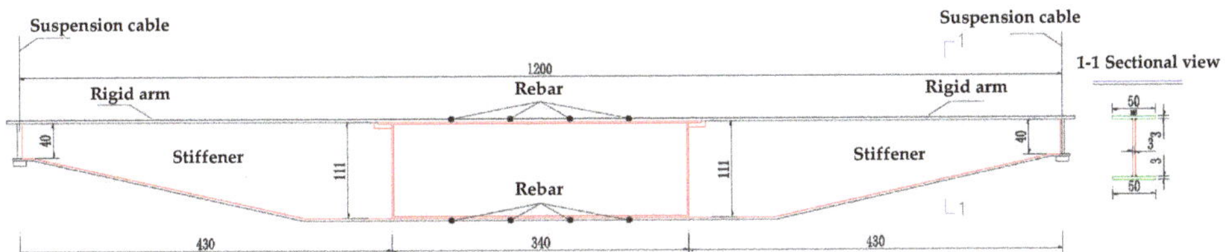

Figure 4. Section at lifting point of main beam (unit: Mm).

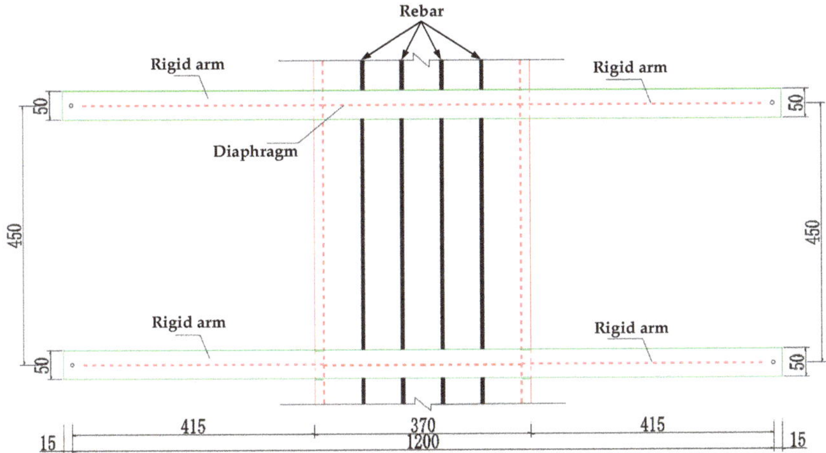

Figure 5. Schematic plan view of the main beam (unit: Mm).

The rigid arm has 5 cm wide and 5 mm thick steel plates, which are arranged through the cross-section of the main beam. The rigid arm is connected to the top plate of the steel box girder through welding; the stiffener adopts a 3 mm thick steel plate, which is connected to the main beam and the rigid arm through welding. The main beam is processed in the factory. The total length of the main beam is 24.2 m, which is divided into 24 sections: 1.3 + 6 × 3.6 + 1.3 m. Figure 6 is a photo of the reduced-scale model.

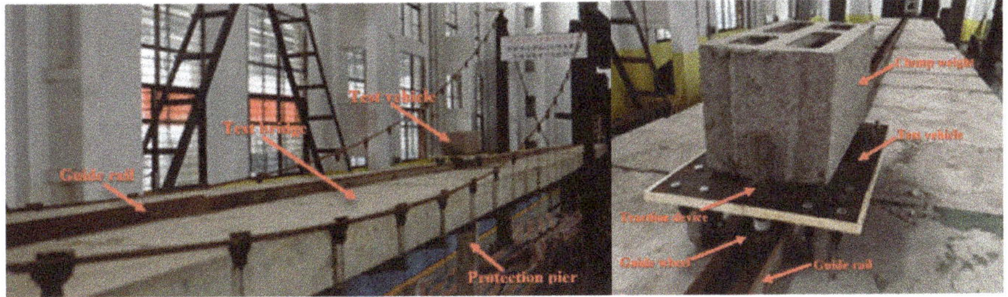

Figure 6. The reduced-scale model.

The intelligent NRS system was set up 5m away from the bridge façade. Then, a computer-controlled camera rotated at fixed angles to collect the images of specific sections of the bridge from fixed positions. The layout of the lab and the principle of image collection are shown in Figures 7 and 8, respectively.

Figure 7. The layout of the lab.

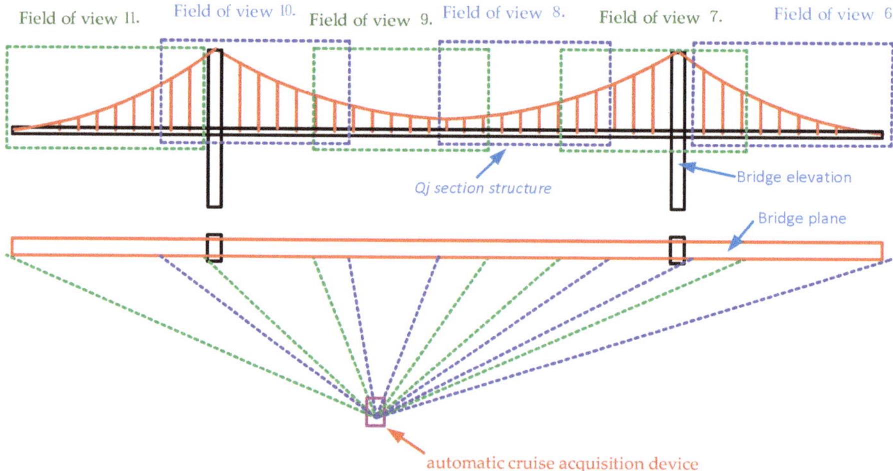

Figure 8. The principle of image collection.

To verify the feasibility of the image collection method, 11 dial gauges were arranged along the axis of the bridge to capture the shape change while the camera took photos of the bridge, DH5902N is adopted for data acquisition equipment. The arrangement of the dial gauges is displayed in Figure 9 below.

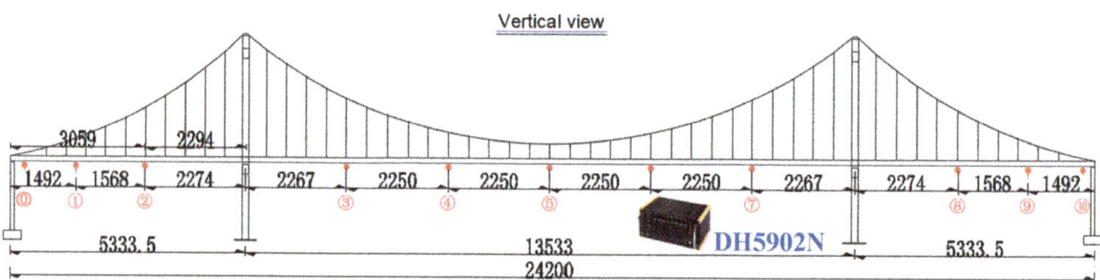

Figure 9. The arrangement of dial gauges.

3.3. Test Contents

The structural deformation data of the bridge were collected under two scenarios to obtain the deformation of the test bridge with the intelligent NRS system and provide more samples for the tracking algorithm. In the first scenario, the bridge had no damage, the test load (50kg) was placed on the middle of the test bridge, and image data on the structural change were collected. In the second scenario, different suspension cables were damaged to simulate varied degrees of bridge damages at different positions, the test load (50 kg) was placed on the same place as the first scenario, and image data on the structural change were collected. Table 1 lists the positions and numbers of damaged suspension cables. The serial numbers of suspension cables are provided in Figure 10.

Table 1. The positions and numbers of damaged suspension cables.

Serial Number	Damage Conditions		Data Collection Method	
	Position	Number	Traditional Method	Visual Method
1	0	0	Dial gauges	Intelligent NRS system
2	24	2	Dial gauges	Intelligent NRS system
3	23, 24	4	Dial gauges	Intelligent NRS system
4	22, 23, 24	6	Dial gauges	Intelligent NRS system
5	21, 22, 23, 24	8	Dial gauges	Intelligent NRS system
6	20, 21, 22, 23, 24	10	Dial gauges	Intelligent NRS system

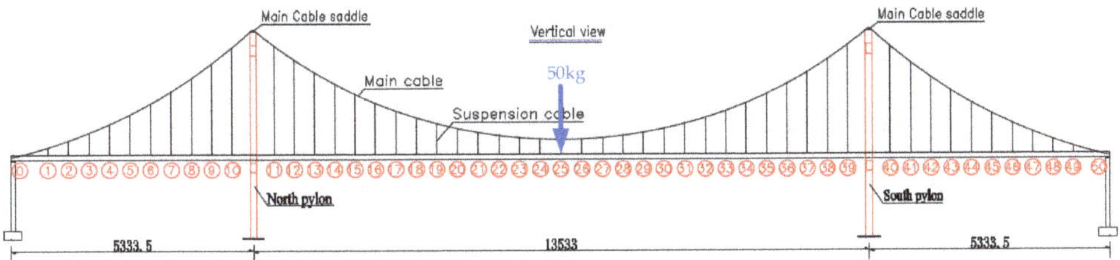

Figure 10. Serial number of suspension cables.

3.4. Finite Element Model

The finite element model is established by Midas Civil [39–41]. The ratio of side to span of the self-anchored suspension test bridge is 1:2.5, and the ratio of rise to span is 1:5.8. The structure is a spatial bar model. The main tower, main beam and cross beam are all simulated by a beam element. The main beam is simulated by fishbone type, and the main cable and suspender are simulated by cable element. The whole bridge model consists of 388 elements and 293 nodes. A rigid connection is adopted between the end of main cable and main beam. The main cable and the suspender, the main cable and tower, and the suspender and the main beam share the same nodes, and no connection is set, as shown in Figure 11. Main material parameters are shown in Table 2.

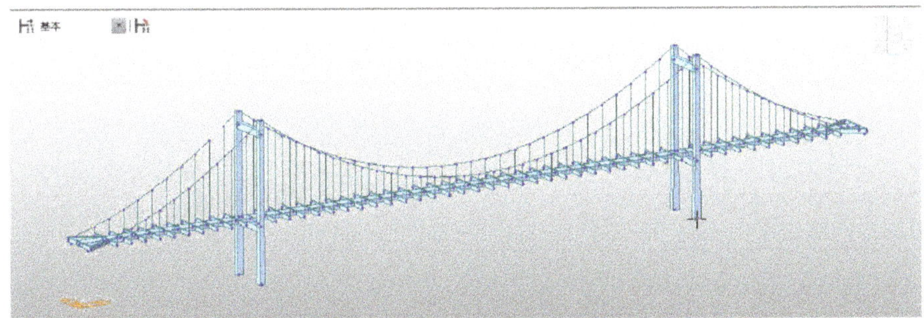

Figure 11. Finite element model of test bridge.

Table 2. Main material parameters.

Serial Number	Item	Section Shape	E (GPa)	f_{tk} (MPa)	σ_s (MPa)	Poisson's Ratio
1	Main cable	○	195	1860	/	0.3
2	suspender	○	195	1860	/	0.3
3	Main beam	⊐	206	/	345	0.3
4	main tower	□	206	/	345	0.3

Main modeling steps: (1) According to the overall design of the suspension bridge, select the corresponding material and section characteristics of each component to initially generate the linearity of the main cable and the initial internal force of the main cable. (2) establish a complete bridge calculation model. (3) define the update node group and vertical analysis function to accurately calculate the structure, and obtain the internal force data of the balance unit node to obtain the initial balance of the suspension bridge state. (4) adjust the cable force of the suspender to a reasonable completion state until the bending moment of the main beam meets the design requirements. The distribution of the suspender force is shown in Figure 12.

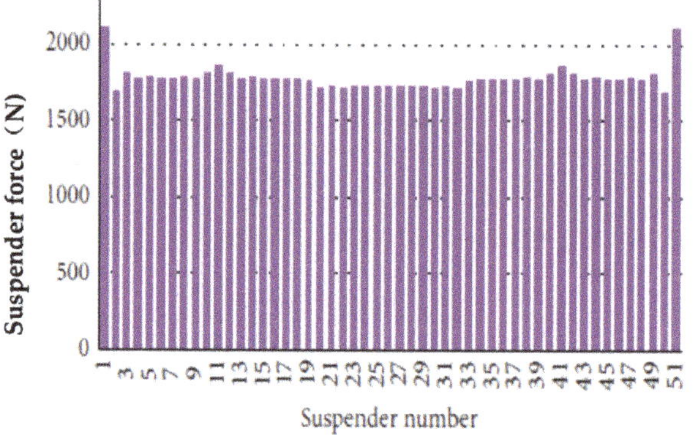

Figure 12. The distribution of the suspender force.

4. Design of Multipoint Displacement Monitoring Algorithm for Bridge Structure

This experiment simulates two problems faced by the noncontact measurement of a large-scale bridge structure. One is the inability to obtain the monitoring image data of the whole bridge through one field of view, which obviously reduces the accuracy of the captured structural changes. Another is the accurate transformation of the displacement of the concerned part in the time series image data. In view of these two problems, this paper proposes a method of acquiring the structural deformation of a large-scale bridge structures by using a uniaxial automatic cruise acquisition device to collect data in different fields of view and establish the relationship between data images in time and space.

4.1. Location and Extraction Method of Bridge Structure Contour

Many studies on camera calibration and perspective transformation have been conducted, and the corresponding theory and application are relatively mature. In this paper, the calibration method of Zhang Zhengyou [42,43] and the perspective transformation method of Jack Mezirow [44] were used directly and are not explained in detail.

The images collected by the intelligent NRS system contain the time sequences in a fixed field of view. Hence, the grayscales and contours were extracted from six images by MATLAB edge function [45], as shown in Figure 13.

The Canny edge detector was adopted for the extraction process. This operator finds the edge points in four steps: smoothing the images with a Gaussian filter, computing the gradient amplitude and direction through finite-difference computing with first-order derivative, applying non-maximum suppression to the gradient amplitude, and using a double threshold to detect and connect the edges.

The Canny edge detector could effectively extract the contours of the bridge structure from the static images collected by the intelligent NRS system. The extracted contours were further processed by a graphics processing software to decontextualize the contours of the useless parts, leaving only the lower edge contour of the deck slabs to reflect the variation in structural shape.

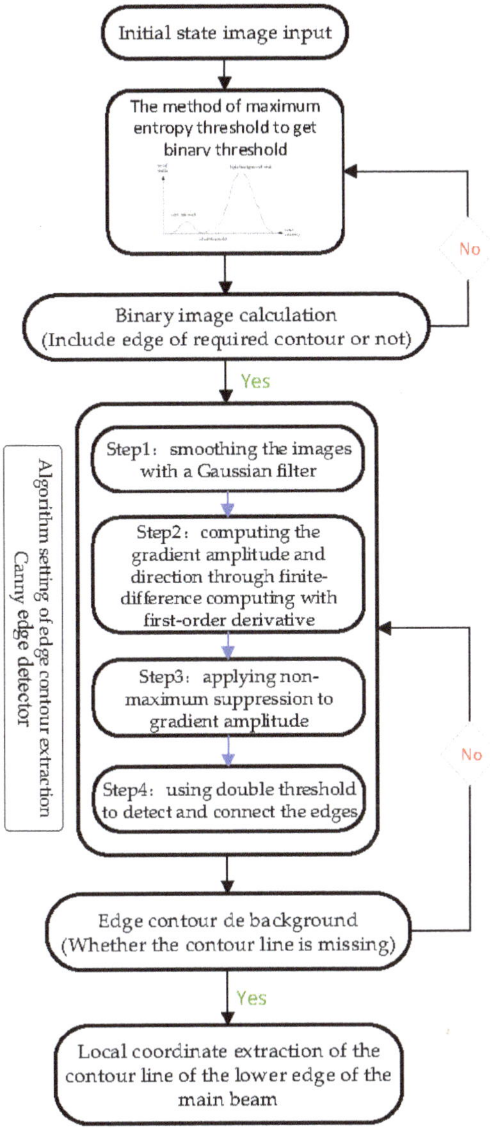

Figure 13. Flowchart of the MATLAB algorithm.

Since the fields of view in the six images are fixed, the contours of the bridge structure were located by the following method. The six images containing the initial boundary of the bridge structure were taken as the original images. The coordinates of each pixel in the boundary were extracted from the six images. Based on these coordinates, each pixel was marked in the original images, revealing the position of the initial boundary. The manual marking helps to suppress the noises in the images. In the subsequent deep learning, the contours could be automatically tracked based on the marked pixels, revealing the change features of the bridge structure. The specific flow of denoising and marking is shown in Figure 14.

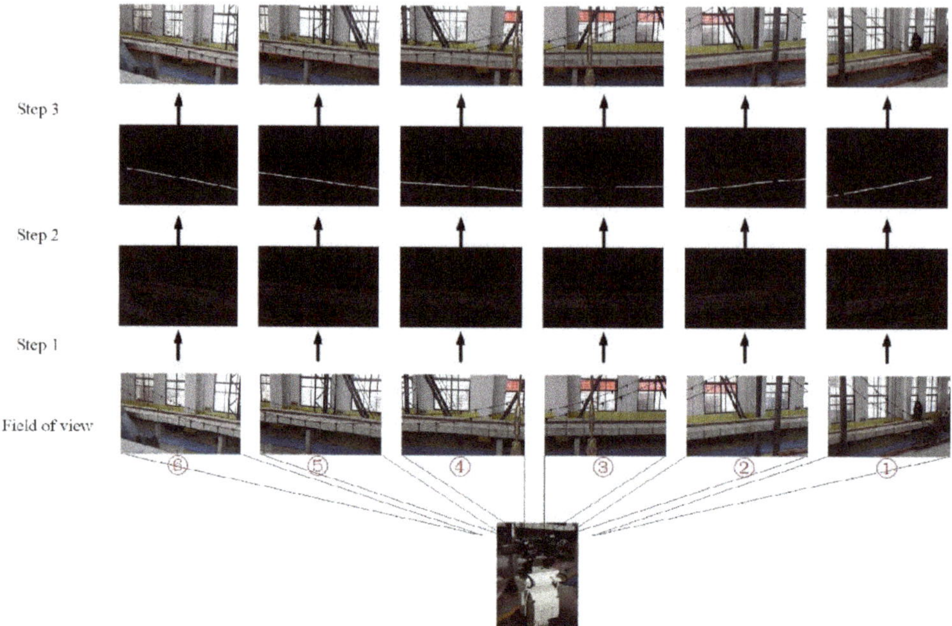

Figure 14. The workflow of denoising and marking of bridge contours. Step 1: Edge detection of static images; step 2: Acquisition of bridge contours through decontextualization of useless contours; step 3: Marking the original images based on the coordinates of boundary pixels.

4.2. The Method of Establishing the Space-Time Relationship of Image Sequence Data

4.2.1. Dataset Construction Based on Spatiotemporal Static Image Sequences

To realize the holographic monitoring of the bridge structure with the uniaxial automatic cruise acquisition device, the key lies in setting up the global and local holographic data based on the dynamic and static image sequences, which were captured at different times from multiple angles and fields of view.

The data in the static image sequences have four main features: Multi-time, multi-field of view, multi-angle, and a strong correlation between time and space. First, the holographic data collected in different fields of view differed in time history. Second, based on technical and economic considerations, the local details of the bridge structure were monitored with a few devices in different fields of view, yielding the local holographic data in each field of view. Third, the data were collected by the automatic cruise device at different watch positions, and the resulting angle difference should be adaptively equalized in the data processing. Fourth, the spatiotemporal features of the original data were determined by the random impact of the entire bridge at the current moment or period, and the structural response in local field of view reflects the overall state of the whole structure to different degrees.

Meanwhile, the cameras responsible for six fields of view (overlap ratio: 20%–30%) each cruised seven times under each damage condition. For the stability of the collected data, seven sets of images were collected on the load in the same field of view under each damage condition.

During data acquisition, the time and the space pointers were constructed based on the features of the intelligent NRS system and the image sequences. The former (time dimension) indicates the current damage condition and the field of view, and the latter (spatial dimension) reflects the position of the current local area relative to the global structure. The spatiotemporal features of the data sequences in the static images are presented in Figure 15 below.

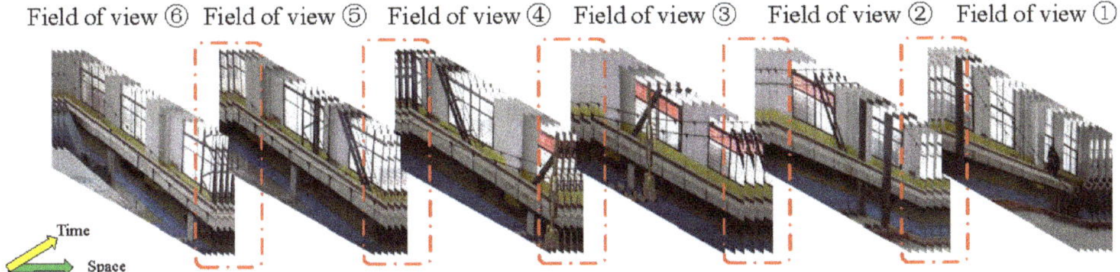

Figure 15. The spatiotemporal features of static image sequences.

In view of the data features, temporal information and spatial information were added into the dataset as labels before deep learning. The temporal information indicates the variation in damage condition and the order of images in the same field of view, and the spatial information reflects the correlation between a local structure with the global structure in a field of view. On this basis, the temporal, spatial and angular data were constructed for the original data, and then integrated with the environmental data (i.e. temperature, humidity, and illumination). The labels can be expressed as:

$$labels\{i,j\}(m,n) = np.arange(Time\_lable, Space\_lable, Angle\_lable, Env.\_lable)$$

where i is the serial number of damage conditions of the test bridge (i = 1–6); j is the label position under different damage conditions (1 for Time, 2 for Space, 3 for Angle, ...); m is the invocation parameter of the data on labels Time, Space, Angle and Environment in a local field of view; n is the serial number of measurements under the same damage condition, i.e. the time history of the same damage condition in the same field of view; *Time_label*, *Space_label*, *Angle_label* and *Env._label* are the matrices of labels Time, Space, Angle, and Environment, respectively.

After tagging the photos in the spatiotemporal sequence through the above steps, the photos are connected in the spatial sequence by using the overlapping part between field of view n and field of view $n + 1$ (n = 1–5). The rich feature points and the SIFT feature points of the test bridge structure must be matched [46,47] in the image data. As shown in Figure 16, the yellow line is the corresponding relationship between the local feature points of the structure in different fields of view and the overall feature points of the structure. The matching results of the feature points of the structure itself and the feature points of the structure SIFT are good. The red line in Figure 11 is the most similar line between the field of view n of the test bridge and the feature points of the whole bridge. The line after matching according to the similarity in the calculation process of the algorithm is used as the constraint condition of edge registration theory and the basis of the displacement measurement information (the red line is the need of explanation, but the actual line should be the yellow line). However, many mismatches in the calculation of the algorithm remain, such as the connection of several characteristic points on the bridge tower and the reaction frame. Therefore, the greedy algorithm is used in this study to re-express the matching set of feature points, and the matching similarity rate is calculated by traversing the proximal and the sub proximal points in the process of filtering. The error matching points in the feature matching set are eliminated by optimizing the selection of each calculation in the process of traversing.

$$Euclidean\ Distance:\ d_{i,j} = \sqrt{\sum_{i=1}^{n} (S_i - S_j)^2}$$

$$HMF:\ Obj(x) = \sum_{i=1}^{m} \sum_{j=1}^{n} \left| \frac{H_{ij}(x,y,z) - H'_{ij}(x,y,z)}{H_{ij}(x,y,z)} \right| (S_i - S_j)^2$$

According to the Euclidean distance calculation, the correlation degree between the spatial feature points of the test bridge structure is used for the optimal matching of the feature point set, where $d_{i,j}$ is the Euclidean distance between feature points, S_i and S_j are the spatial feature point set. *HMF* is the control equation of superposition analysis of displacement calculation in the space–time domain, where m and n are the serial numbers of spatial feature points; x, y, and z are the spatial coordinates of feature points; $H_{ij}(x, y, z)$ is the structural holographic morphological response measurement of a certain time in a certain field of view; and $H'_{ij}(x, y, z)$ is the reference state of the structural holographic morphological response measurement.

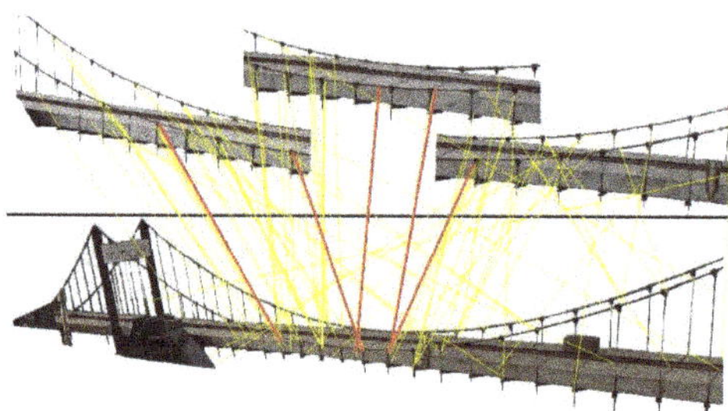

Figure 16. Description of text bridge feature points.

4.2.2. Target Tracking and Displacement Calculation

In the previous research, the static image is used to extract the contour of the bridge structure at different times and under different load-damage conditions, and carry out stacking to obtain the deformation of the bridge structure. Given that this method has many manual interventions, the optical flow optical flow algorithm which is widely used in computer vision, was adopted for target tracking and displacement calculation in the static image sequence data collection [48,49].

The algorithm using optical flow must satisfy two hypotheses: (1) constant brightness: The brightness of the same point does not change with time; (2) small motion: The position will not change drastically with over time, such as finding the derivative of grayscale relative to the position. In our research, both hypotheses were satisfied by the data collected by the NRS. The brightness of each point in the collected images remained constant because of the small data interval of the seven time sequences; the small motion was also fulfilled, and the bridge structure had a limited deformation for the monitoring object.

Basic constraint equation. Consider the light intensity of a pixel $f(x, y, t)$ on the first picture (where t represents its time dimension). It moves the distance (dx, dy) to the next picture because it is the same pixel point, and according to the first assumption above, the light intensity of the pixel before and after the motion is constant.

$$f(x, y, t) = f(x + dx, y + dy, t + dt) \tag{1}$$

The Taylor expansion on the right end of Formula (1) is as follows:

$$f(x + dx, y + dy, t + dt) = f(x, y, t) + \frac{\partial f}{\partial x}dx + \frac{\partial f}{\partial y}dy + \frac{\partial f}{\partial t}dt + \varepsilon \tag{2}$$

where ε represents the second order infinitesimal term, which could be ignored; then (2) is substituted into (1) and divided by dt:

$$\frac{\partial f}{\partial x}\frac{dx}{dt} + \frac{\partial f}{\partial y}\frac{dy}{dt} + \frac{\partial f}{\partial t} = 0 \tag{3}$$

Let $f_x = \frac{\partial f}{\partial x}$, $f_y = \frac{\partial f}{\partial y}$, $f_t = \frac{\partial f}{\partial t}$ represent the partial derivatives of the gray level of the pixels in the image along the x, y, and t directions. Summing up,

$$f_x u + f_y v + f_t = 0 \tag{4}$$

where, f_x, f_y, and f_t can be obtained from the image data, and (u, v) is the optical flow vector that must be solved.

At this time, there is only one constraint equation but two unknowns. In this case, the exact values of u and v cannot be obtained. Constraints need to be introduced from another perspective. The introduction of constraints from different angles leads to different methods of optical flow field calculation. According to the difference between the theoretical basis and the mathematical method, they are divided into four kinds: Gradient-based method, matching-based method, energy-based method, phase-based method, and neurodynamic method. In addition to distinguishing the optical flow method according to different principles, the optical flow method can also be divided into dense optical flow and sparse optical flow according to the density of a two-dimensional vector in the optical flow field.

Dense optical flow is a kind of image registration method that matches the image or a specific area point by point. It calculates the offset of all the points on the image to form a dense optical flow field. Through this dense optical flow field, pixel level image registration can be performed. In contrast to dense optical flow, sparse optical flow does not calculate every pixel of the image point by point. It usually needs to specify a group of points for tracking, which is better to have some obvious characteristics, such as the Harris corner, for a relatively stable and reliable tracking. The computation cost of sparse tracking is much less than that of dense tracking.

Since the collected image has less data than the video and does not pursue timeliness and according to the object extraction method described in Section 3.1, the displacement of all points on the image need not be calculated, but the offset of each pixel on the extracted contour line for pixel level image registration must be calculated. Thus, the Horn-Schunck algorithm [50] was selected. The Horn-Schunck algorithm belongs to the dense optical flow with the best accuracy. The objective function of the Horn-Schunck algorithm is as follows:

$$\min_{u,v} E(u, v) = \iint [(T(x, y) - f(x + u, y + v))^2 + \lambda \cdot (u_x^2 + u_y^2 + v_x^2 + v_y^2)] dx dy \tag{5}$$

In order to facilitate the calculation, the approximation method in the sparse optical flow is used to approximate the data items linearly to obtain a new objective function:

$$\min_{u,v} E(u, v) = \iint [(f_x u + f_y v + f_t)^2 + \lambda \cdot (u_x^2 + u_y^2 + v_x^2 + v_y^2)] dx dy \tag{6}$$

The definite integral of the above formula can be rewritten into discrete form. Then, the corresponding Jacobian iterative formula can be obtained by the super relaxation iterative method (SOR):

$$u = \bar{u} - f_x \frac{f_x \bar{u} + f_y \bar{v} + f_t}{a + f_x^2 + f_y^2} \tag{7}$$

$$v = \bar{v} - f_y \frac{f_x \bar{u} + f_y \bar{v} + f_t}{a + f_x^2 + f_y^2} \tag{8}$$

According to Equations (11), (14), and (15), optical flow vectors t_i to t_{i+1} can be calculated, and all time u_i and v_i can be summed to calculate the displacement of the pixel point.

5. Extraction Deformation and Discussion

The MATLAB edge function and the Canny edge detector were adopted to extract the grayscale and contour from each image in the static image sequences of the test bridge under different damage conditions. Based on the extracted feature, the contours were marked on the original images. Then, the marked images and the subsequently taken images were compiled into a dataset. After that, the SIFT algorithm was applied to establish the spatial relationship between the fields of view, and the optical flow algorithm was used to track the displacement information of the lower edge contour of the main beam, based on the images collected by the NRS method (proposed in this paper). Since there are too many working conditions and the generated data to show well, the bridge tower consolidation point was used as the reference point to compare the nondestructive working condition (con. 1) obtained by the two methods and the working condition (con. 6) that shows the most obvious damage and deformation. In the figure, "- 1" represents the data obtained from finite element calculation, and "- 2" represents the data obtained from the method proposed in this paper, as shown in Figure 17.

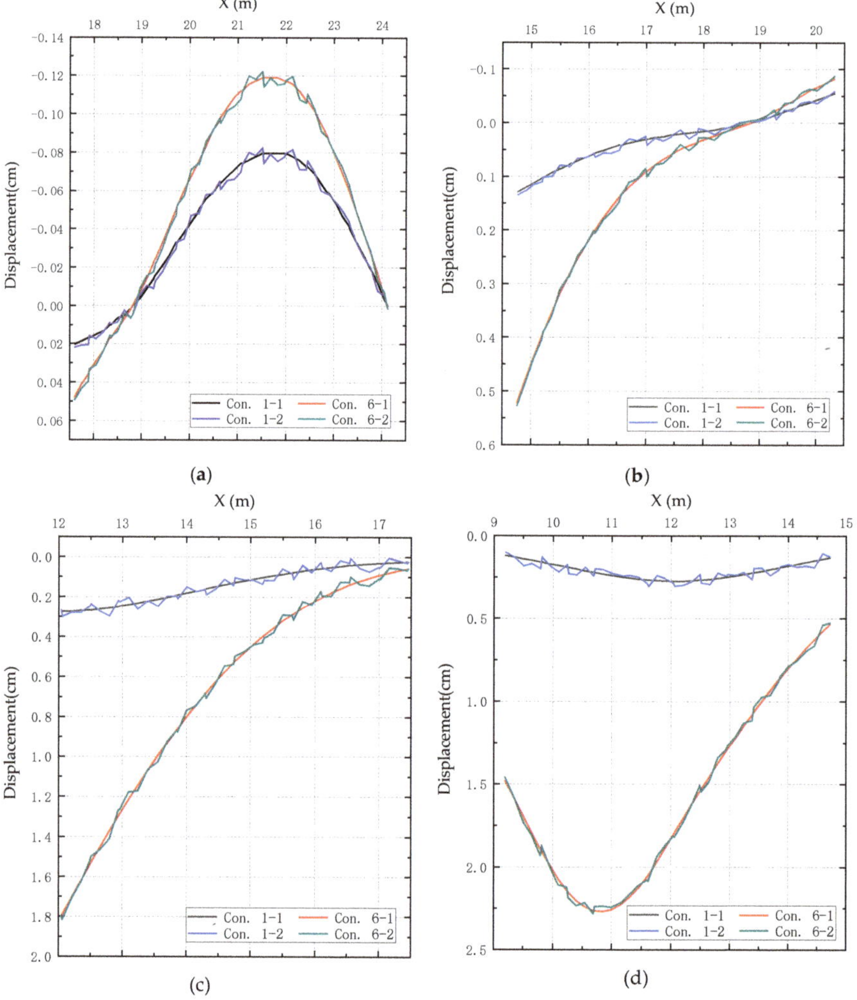

Figure 17. *Cont.*

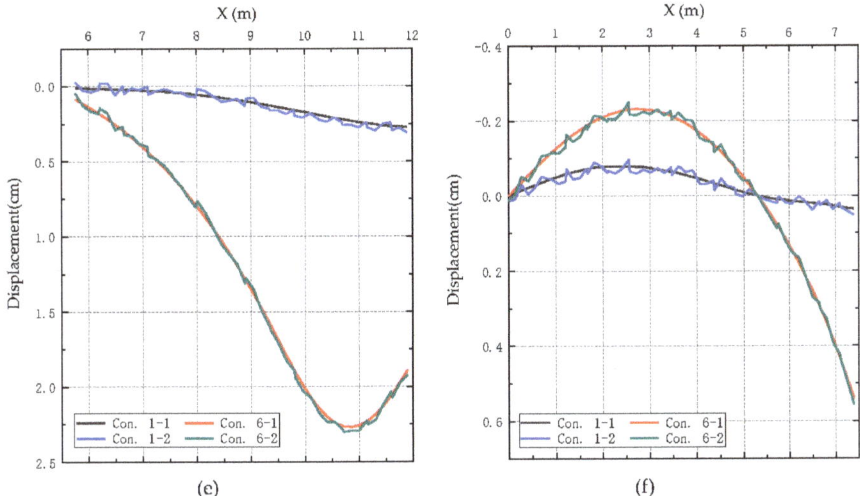

(e)   (f)

Figure 17. Comparison of two methods for obtaining deformation value for different fields of views: (**a**) Comparison for field of view 1; (**b**) comparison for field of view 2; (**c**) comparison for field of view 3; (**d**) comparison for field of view 4; (**e**) comparison for field of view 5; (**f**) comparison for field of view 6.

As shown in Figure 17, the deformation curves of our method are less smooth than those of the finite-element method. There are two possible reasons for the lack of smoothness. Firstly, the lower edge of the deck slabs marked in the original images, which was considered as the contour of the bridge structure, is not smooth and even discrete in some places. Secondly, the positions of the marked pixels changed greatly after the bridge deformed, and were not captured accurately through deep learning.

The first problem was solved through the contour stacking analysis in structural deformation monitoring, which is a method previously developed by our research team. This method treats the initial contours as known white noises of the system, and subtracts them from the contours acquired under different damage conditions. The second problem calls for improvement of the capture algorithm. Here, the improvement is realized through manual intervention. In this way, the bridge deformation data in the six fields of view were integrated into the global holographic deformation of the test bridge (Figure 18).

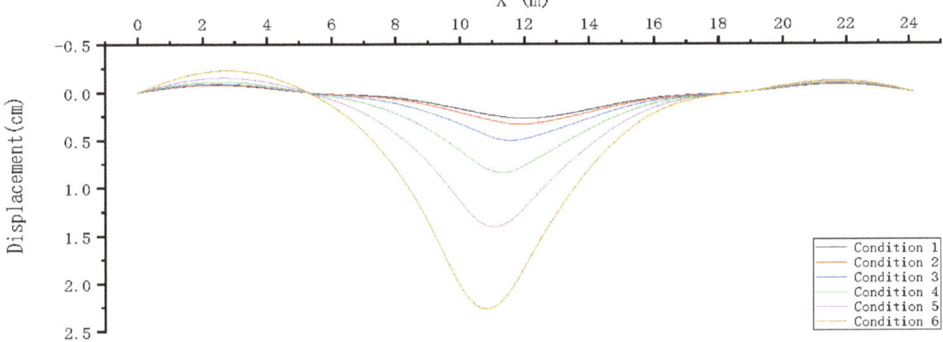

Figure 18. The global holographic deformation under different damage conditions.

The deformation map of the test bridge based on the 11 dial gauges is not presented here. Even if fitted, the data collected by these gauges were discrete to demonstrate the global deformation features of the test bridge. Moreover, the initial state of the test bridge was not measured at the completion. Consequently, the actual stress state of the bridge structure at that moment is impossible to determine. However, the relative deformation of the test bridge in the monitoring period can be obtained from Figure 18. The obtained results were compared with the relative deformation recorded by the dial gauges to verify the accuracy of our method.

Out of the many damage conditions, the greatest difference lies between damage conditions 1 (no damage) and damage condition 6 (suspension cables 20–24 are damaged). Thus, these two damage conditions were subjected to stacking analysis and compared in detail (Table 3).

Table 3. Comparison results.

No.	Deformation of Stacking Analysis (mm)			Measured Deviation %	Relative Error %
	Dial gauge Measurement R1	Finite-Element Method R2	Noncontact Remote Sensing R3	|R3–R1|/R1	|R3–R2|/R2
1	0.11	0	0.1	9.09%	/
2	0.99	1	1.08	9.09%	8.00%
3	1.56	1.55	1.68	7.69%	8.39%
4	5.42	5.55	5.87	8.30%	5.77%
5	17.46	17.32	18.75	7.39%	8.26%
6	15.16	15.3	16.43	8.38%	7.39%
7	5.18	5.24	5.67	9.46%	8.21%
8	0.93	0.96	1.02	9.68%	6.25%
9	0.37	0.38	0.41	10.81%	7.89%
10	0.35	0.33	0.37	5.71%	12.12%
11	0.09	0	0.08	11.11%	/

Table 3 shows that our NRS method accurately derives the deformation features of the bridge structure from those collected in local fields of view. Compared with the dial gauge measurement and finite-element results, the maximum errors of our method were 11.11% on the 11th measuring point and 12.12% on the 12th measuring point, indicating that the global holographic deformation curves obtained through the decking analysis of contours are accurate enough for engineering practices.

In order to more clearly reflect the advantages of dense full-field displacement, the deformation of the whole bridge girder obtained from NRS under condition 6 is compared with the deformation of the whole bridge girder obtained from nine dial indicators, as shown in Figure 19. The comparison of Figure 19 and Table 3, it can be derived that the position with the maximum deformation in Table 3 is the measuring point 4, but the position with the maximum deformation of the actual test bridge is the position with the length of 10.71m between the measuring points 5 and 4; the comparison of the six working conditions, it can be derived that the position with the maximum deformation of the girder changes under different damage conditions, as shown in Figure 20. If the contact sensor, such as the dial gauge is used for measurement, a reasonable arrangement way of obtaining the deformation characteristics of the bridge structure from the finite points is difficult to determine. One arrangement approach cannot meet the deformation characteristics of the bridge structure caused by the change of structural stiffness at different positions. The noncontact remote measurement method used in this paper can capture the pixel level change of any position of the bridge structure, the accuracy can meet the requirements of engineering application, and the advantages are obvious. Meanwhile, model updating according to more accurate line shape can make the model closer to the real bridge, increasing the authenticity and credibility of the finite element calculation results, and exerting the same positive significance on the application of digital twin. The dense full-field displacement monitoring provides a larger amount of real data, which is the basis of machine learning. Using this method for regular or long-term monitoring can obtain a larger amount of real data than traditional ways, laying the foundation for using machine learning for structural health monitoring.

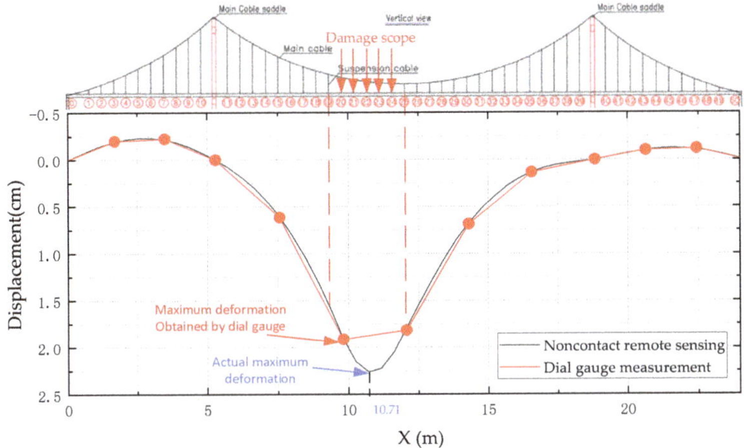

Figure 19. Comparison of girder alignment obtained by two measure method under condition 6.

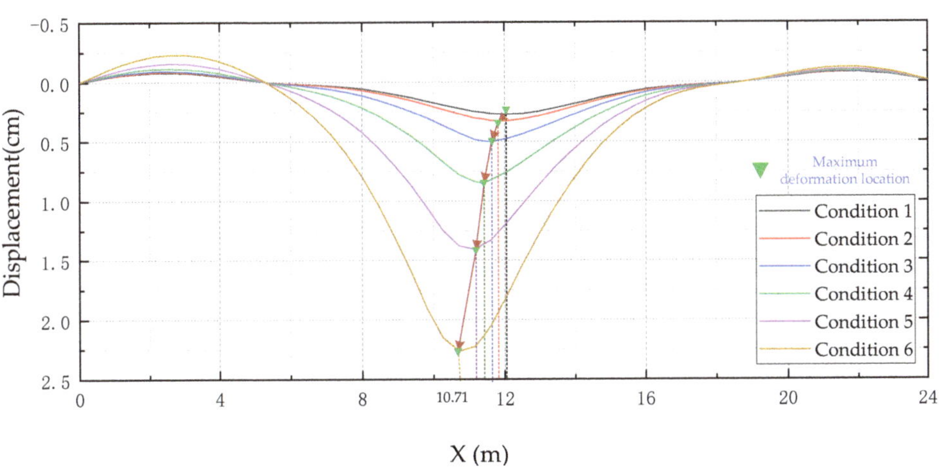

Figure 20. Deformation trend chart.

6. Conclusions

In this study, the dense full-field deformation of a reduced-scale model for a 24m-span self-anchored suspension bridge under multiple damage conditions was captured with the noncontact remote measurement method in a multi visual field. The spatiotemporal sequence static image data under different work conditions were collected to establish the relationship between spatial and temporal. Furthermore, the dense full-field displacement monitoring data of the girder of the test bridge were obtained, and compared with the finite element calculation results. The measurement results of the dial gauge displacement meter were compared. The main conclusions are as follows:

(1)   A fixed point uniaxial automatic cruise acquisition device was designed to collect the static images of the bridge façade under different damage conditions. Then, the spatiotemporal sequences of static images were processed by the edge detection method, the edge pixel virtual marker point, the SIFT algorithm and the optical flow algorithm to obtain a dense full-field displacement of the whole test bridge girder, which can be used as the data base to make the structural health monitoring technology more economical, efficient and direct. Compared with other monitoring

methods, the girder dense points displacement information provides a data-base for more accurate model updating and damage identification. Meanwhile, the technology proposed in this paper is low-cost and can be used as a long-term regular monitoring method to accumulate massive real structural displacement information and provide big data set for the subsequent study of machine learning for damage identification.

(2) The optical flow algorithm, which is widely used in video analysis, was used in the static image data set to track the target and calculate the displacement, overcoming the shortcomings of many manual interventions in the early stage of research group. Meanwhile, the number of monitoring points remains the same (i.e. the displacement of each pixel of the lower edge contour line of the girder). The output data are basically consistent with the finite-element prediction and dial gauge measurement. The global holographic deformation curves of the test bridge exhibit similar trends under different damage conditions, with an error of less than 12%. This means that the proposed method in this paper satisfies the engineering requirement on measurement accuracy.

(3) A new method of making a virtual target was used. The coordinates of the required lower edge contour of the girder were extracted and then used it to make the pixels of the initial image of the lower edge of the girder as a virtual target back, and then track and calculate the displacement information of all pixels of the contour through the optical flow algorithm. Although this method needs a certain amount of manual intervention in the early stage, it can locate accurately and obtain more measuring point displacement simultaneously.

(4) The information obtained from the combination of several points does not really reflect the structural deformation characteristics of the bridge under different damage conditions, and the abnormal local deformation information caused by the damage will be lost. Thus, the dense full-field displacement information is more sensitive to the structural stiffness change.

(5) The characteristics of the linear change of the test bridge under different damage conditions indicate a strong correlation between the damage location and degree and the linear change. The relationship between the three can be established, and the method of amplifying the damage and deformation characteristics and carrying out the quantification requires further study.

(6) This work is only the first exploration of the dense full-field displacement monitoring of the whole bridge girder using NRS. It involves less in the optimization of parameters in the experiment, less in the improvement of the algorithm and the accuracy of the algorithm, which needs to be further studied in the future. Meanwhile, it only shows that the dense full-field displacement is more sensitive to the damage identification, but the damage identification is not involved.

Author Contributions: Conceptualization, Z.Z.; formal analysis, G.D.; investigation, X.C.; methodology, G.D.; resources, G.D., S.S. and C.J.; writing – original draft, G.D.; writing – review and editing, G.D. and S.S. All authors have read and agreed to the published version of the manuscript.

Funding: This research was funded by the National Natural Science Foundation of China, grant number 51778094, the National Science Foundation for Distinguished Young Scholars of China, grant number 51608080, the National Science Foundation for Distinguished Young Scholars of China, grant number 51708068, and the Science and Technology Innovation Project of Chongqing Jiaotong University, grant number 2019S0141.

Acknowledgments: Special thanks to J.L. Heng at the Shenzhen University, and Y.M. Gao at the State Key Laboratory of Mountain Bridge and Tunnel Engineering, and Chongqing Jiaotong University.

Conflicts of Interest: The authors declare no conflict of interest.

References

1. Editorial Department of China Journal of Highway and Transport. Review on China's Bridge Engineering Research: 2014. *China J. Highw. Transp.* **2014**, *27*, 1–96.
2. He, S.-H.; Zhao, X.-M.; Ma, J.; Zhao, Y.; Song, H.-S.; Song, H.-X.; Cheng, L.; Yuan, Z.-Y.; Huang, F.-W.; Zhang, J.; et al. Review of Highway Bridge Inspection and Condition Assessment. *China J. Highw. Transp.* **2017**, *30*, 63–80.

3. Nhat-Duc, H. Detection of Surface Crack in Building Structures Using Image Processing Technique with an Improved Otsu Method for Image Thresholding. *Adv. Civ. Eng.* **2018**, *2018*, 3924120.

4. Li, H.; Bao, Y.-Q.; Li, S.-L. Data Science and Engineering Structural Health Monitoring. *J. Eng. Mech.* **2015**, *32*, 1–7.

5. Bao, Y.-Q.; James, L.B.; Li, H. Compressive Sampling for Accelerometer Signals in Structural Health Monitoring. *Struct. Health Monit.* **2011**, *10*, 235–246.

6. Gul, M.; Dumlupinar, T.; Hattori, H.; Catbas, N. Structural monitoring of movable bridge mechanical components for maintenance decision-making. *Struct. Monit. Maint.* **2014**, *1*, 249–271. [CrossRef]

7. Gul, M.; Catbas, F.N.; Hattori, H. Image-based monitoring of open gears of movable bridges for condition assessment and maintenance decision making. *J. Comput. Civ. Eng.* **2015**, *29*, 04014034. [CrossRef]

8. Garcia-Palencia, A.; Santini-Bell, E.; Gul, M.; Çatbaş, N. A FRF-based algorithm for damage detection using experimentally collected data. *Struct. Monit. Maint.* **2015**, *24*, 399–418. [CrossRef]

9. Spencer, B.F., Jr.; Hoskere, V.; Narazaki, Y. Advances in computer vision-based civil infrastructure inspection and monitoring. *Engineering* **2019**, *5*, 199–222. [CrossRef]

10. Yang, Y.; Jung, H.K.; Dorn, C.; Park, G.; Farrar, C.; Mascareñas, D. Estimation of full-field dynamic strains from digital video measurements of output-only beam structures by video motion processing and modal superposition. *Struct. Control Health Monit.* **2019**, *26*, e2408. [CrossRef]

11. Kim, H.; Shin, S. Reliability verification of a vision-based dynamic displacement measurement for system identification. *J. Wind Eng. Ind. Aerod.* **2019**, *191*, 22–31. [CrossRef]

12. Ojio, T.; Carey, C.H.; Obrien, E.J.; Doherty, C.; Taylor, S.E. Contactless bridge weigh-in-motion. *J. Bridge Eng.* **2016**, *217*, 04016032. [CrossRef]

13. Moreu, F.; Li, J.; Jo, H.; Kim, R.E. Reference-free displacements for condition assessment of timber railroad bridges. *J. Bridge Eng.* **2016**, *21*, 04015052. [CrossRef]

14. Xu, Y.; Brownjohn, J.; Kong, D. A non-contact vision-based system for multipoint displacement monitoring in a cable-stayed footbridge. *Struct. Control Health Monit.* **2018**, *25*, e2155. [CrossRef]

15. Hester, D.; Brownjohn, J.; Bocian, M.; Xu, Y. Low cost bridge load test: Calculating bridge displacement from acceleration for load assessment calculations. *Eng. Struct.* **2017**, *143*, 358–374. [CrossRef]

16. Celik, O.; Dong, C.Z.; Catbas, F.N. A computer vision approach for the load time history estimation of lively individuals and crowds. *Comput. Struct.* **2018**, *200*, 32–52. [CrossRef]

17. Catbas, F.N.; Zaurin, R.; Gul, M.; Gokce, H.B. Sensor networks, computer imaging, and unit influence lines for structural health monitoring: Case study for bridge load rating. *J. Bridge Eng.* **2012**, *17*, 662–670. [CrossRef]

18. Khuc, T.; Catbas, F.N. Structural identification using computer vision–based bridge health monitoring. *J. Struct. Eng.* **2018**, *144*, 04017202. [CrossRef]

19. Dong, C.Z.; Celik, O.; Catbas, F.N. Marker-free monitoring of the grandstand structures and modal identification using computer vision methods. *Struct. Health Monit.* **2019**, *18*, 1491–1509. [CrossRef]

20. Yang, Y.; Dorn, C.; Mancini, T.; Talkend, Z.; Kenyone, G.; Farrara, C.; Mascareñasa, D. Blind identification of full-field vibration modes from video measurements with phase-based video motion magnification. *Mech. Syst. Signal Process.* **2017**, *85*, 567–590. [CrossRef]

21. Bao, Y.-Q.; Bao, Y.-Q.; Ou, J.; Li, Hui. Emerging Data Technology in Structural Health Monitoring: Compressive Sensing Technology. *J. Civ. Struct. Health Monit.* **2014**, *4*, 77–90. [CrossRef]

22. Bao, Y.-Q.; Li, H.; Sun, X.-D.; Ou, J.-P. A Data Loss Recovery Approach for Wireless Sensor Networks Using a Compressive Sampling Technique. *Struct. Health Monit.* **2013**, *12*, 78–95. [CrossRef]

23. Bao, Y.-Q.; Zou, Z.-L.; Li, H. *Compressive Sensing Based Wireless Sensor for Structural Health Monitoring*; 90611W-1-10; SPIE Smart Structures/NDE: San Diego, CA, USA, 2014.

24. Bao, Y.-Q.; Yan, Y.; Li, H.; Mao, X.; Jiao, W.; Zou, Z.; Ou, J. Compressive Sensing Based Lost Data Recovery of Fast-moving Wireless Sensing for Structural Health Monitoring. *Struct. Control Health Monit.* **2014**, *22*, 433–448. [CrossRef]

25. Guzman-Acevedo, G.M.; Becerra, G.E.V.; Millan-Almaraz, J.R.; Rodríguez-Lozoya, H.E.; Reyes-Salazar, A.; Gaxiola-Camacho, J.R.; Martinez-Felix, C.A. GPS, Accelerometer, and Smartphone Fused Smart Sensor for SHM on Real-Scale Bridges. *Adv. Civ. Eng.* **2019**, *2019*, 6429430. [CrossRef]

26. Xu, Y.; Brownjohn, J.M.W. Review of machine-vision based methodologies for displacement measurement in civil structures. *J. Civ. Struct. Health Monit.* **2018**, *8*, 91–110. [CrossRef]

27. Feng, D.; Feng, M.Q.; Ozer, E.; Fukuda, Y. A vision-based sensor for noncontact structural displacement measurement. *Sensors* **2015**, *15*, 16557–16575. [CrossRef] [PubMed]
28. Feng, D.; Feng, M.-Q. Identification of structural stiffness and excitation forces in time domain using noncontact vision-based displacement measurement. *J. Sound Vib.* **2017**, *406*, 15–28. [CrossRef]
29. Feng, D.; Scarangello, T.; Feng, M.-Q.; Ye, Q. Cable tension force estimate using novel noncontact vision-based sensor. *Measurement* **2017**, *99*, 44–52. [CrossRef]
30. Dong, C.Z.; Ye, X.W.; Jin, T. Identification of structural dynamic characteristics based on machine vision technology. *Measurement* **2018**, *126*, 405–416. [CrossRef]
31. Ye, X.-W.; Dong, C.-Z.; Liu, T. A review of machine vision-based structural health monitoring: Methodologies and applications. *J. Sens.* **2016**, *2016*, 7103039. [CrossRef]
32. Khuc, T.; Catbas, F.N. Computer vision-based displacement and vibration monitoring without using physical target on structures. *Struct. Infrastruct. Eng.* **2017**, *13*, 505–516. [CrossRef]
33. Tian, L.; Pan, B. Remote bridge deflection measurement using an advanced video deflectometer and actively illuminated LED targets. *Sensors* **2016**, *16*, 1344. [CrossRef] [PubMed]
34. Lee, J.J.; Cho, S.; Shinozuka, M.; Yun, C.; Lee, C.; Lee, W. Evaluation of bridge load carrying capacity based on dynamic displacement measurement using real-time image processing techniques. *Int. J. Steel Struct.* **2006**, *6*, 377–385.
35. Feng, D.; Feng, M.-Q. Model updating of railway bridge using in situ dynamic displacement measurement under trainloads. *J. Bridge Eng.* **2015**, *20*, 04015019. [CrossRef]
36. Chen, J.-G.; Adams, T.M.; Sun, H.; Bell, E.S. Camera-based vibration measurement of the world war I memorial bridge in portsmouth, New Hampshire. *J Struct. Eng.* **2018**, *144*, 04018207. [CrossRef]
37. Abraham, L.; Sasikumar, M. Analysis of satellite images for the extraction of structural features. *IETE Tech. Rev.* **2014**, *31*, 118–127. [CrossRef]
38. Milillo, P.; Perissin, D.; Salzer, J.-T.; Lundgren, P.; Lacava, G.; Milillo, G.; Serio, C. Monitoring dam structural health from space: Insights from novel InSAR techniques and multi-parametric modeling applied to the Pertusillo dam Basilicata, Italy. *Int. J Appl. Earth Obs. Geoinf.* **2016**, *52*, 221–229. [CrossRef]
39. Wang, S.-R.; Zhou, Z.-X.; Gao, Y.-M.; Xu, J. Newton-Raphson Algorithm for Pre-offsetting of Cable Saddle on Suspension Bridge. *China J. Highw. Transp.* **2016**, *29*, 82–88.
40. Wang, S.-R.; Zhou, Z.-X.; Wu, H.-J. Experimental Study on the Mechanical Performance of Super Long-Span Self-Anchored Suspension Bridge in Construction Process. *China Civ. Eng. J.* **2014**, *47*, 70–77.
41. Wang, S.-R.; Zhou, Z.-X.; Wen, D.; Huang, Y. New Method for Calculating the Pre-Offsetting Value of the Saddle on Suspension Bridges Considering the Influence of More Parameters. *J. Bridge Eng.* **2016**, *2016*, 06016010. [CrossRef]
42. Zhang, Z.-Y. A flexible new technique for camera calibration. *IEEE Trans. Pattern Anal. Mach. Intell.* **2000**, *22*, 1330–1334. [CrossRef]
43. Zhang, Z.-Y. Camera calibration with one-dimensional objects. *IEEE Trans. Pattern Anal. Mach. Intell.* **2004**, *26*, 892–899. [CrossRef] [PubMed]
44. Mezirow, J. Perspective Transformation. *Adult Educ. Q.* **2014**, *28*, 100–110. [CrossRef]
45. Deng, G.-J.; Zhou, Z.-X.; Chu, X.; Lei, Y.-K.; Xiang, X.-J. Method of bridge deflection deformation based on holographic image contour stacking analysis. *Sci. Technol. Eng.* **2018**, *18*, 246–253.
46. Grabner, M.; Grabner, H.; Bischof, H. Fast approximated SIFT. *ACCV* **2006**, *3851*, 918–927.
47. Liu, Y.; Liu, S.-P.; Wang, Z.-F. Multi-focus image fusion with dense SIFT. *Inf. Fusion* **2015**, *23*, 139–155. [CrossRef]
48. Lucena, M.J.; Fuertes, J.M.; Gomez, J.I.; de la Blanca, N.P.; Garrido, A. Optical flow-based probabilistic tracking. In *Seventh International Symposium on Signal Processing and Its Applications*; IEEE: New York, NY, USA, 2003.
49. Roth, S.; Black, M.J. On the Spatial Statistics of Optical Flow. *Int. J. Comput. Vis.* **2007**, *74*, 33–50. [CrossRef]
50. Horn, B.K.P.; Schunck, B.G. Determining optical flow. *Artif. Intell.* **1981**, *17*, 185–203. [CrossRef]

 *applied sciences*

Article

Railway Wheel Flat Recognition and Precise Positioning Method Based on Multisensor Arrays

Chenyi Zhou, Liang Gao *, Hong Xiao and Bowen Hou

School of Civil Engineering, Beijing Jiaotong University, Beijing100044, China; 16121225@bjtu.edu.cn (C.Z.); mgubao@163.com (H.X.); bwhou@bjtu.edu.cn (B.H.)
* Correspondence: lgao@bjtu.edu.cn

Received: 15 January 2020; Accepted: 12 February 2020; Published: 14 February 2020

Featured Application: Wheel flats are one of the most threatening defects during the service lifespan of railway vehicle wheels. This study proposes a new long-term monitoring method for wheel flats based on multisensor arrays. The dynamic strain responses of rails are captured by sensor arrays mounted on the rail web, ensuring that all the wheels are assessed during the train passage. Through data fusion among multiple sensors, the method locates the specific position of wheel defects. This study provides potential guidance for the maintenance of vehicles soon after the occurrences of defects.

Abstract: Wheel flats have become a major problem affecting the long-term service of railway systems. Wheels with flats create intermittent impact loads to trains and rails. This not only accelerates the deterioration of vehicle and track components but also leads to abnormal wheel-rail contact conditions. An effective method for detecting wheel conditions is urgently needed to ensure the operation of the railway and provide guidance for the repair of wheels. However, most previous researches have used qualitative detection methods, and hence have been unable to achieve accurate positioning of the wheel flats. In addition, the theoretical basis for the layout scheme for wheel flat detection sensors is lacking, making it impossible to meet the needs of field applications. In this study, we simulated the spatial distribution characteristics of rail strain, under different wheel flat conditions, and based on this, a layout scheme of multisensor arrays was proposed which more effectively captured the responses of the wheel flats. A wheel flat recognition and precise positioning method based on multisensor fusion was designed. The algorithm was validated through the combination of experimental and simulation methods. The result shows that the algorithm can ideally detect and locate the wheel flats under complex conditions.

Keywords: wheel flat; real-time monitoring; strain distribution characteristics; multisensor array; precise positioning

---

1. Introduction

In the field of railway transportation, the health state of wheels has a crucial impact on vehicle operation safety. Therefore, it is of great significance to monitor the generation and development of wheel defects in real time to enhance the safety and reliability of railway operation. Wheel flats are the most typical form of defect during long-term service of train wheels, and this defect induces the failure of both the vehicles and the infrastructures. Wheel flats are mainly the result of the following two aspects: (1) The anomalous wear of the wheel tread due to sudden braking of the moving vehicle [1] and (2) the local reduction of wheel-rail adhesion force, which is often caused by rail surface foreign matters, leading to complete sliding between the wheel and rail [2].

Wheels with flats create intermittent impact loads on trains and rails while moving, which strongly increases the dynamic responses of the vehicle-track coupling system [3]. In light of previous studies, the wheel-rail impact loads caused by wheel flats are far higher than the wheel-rail contact force under normal circumstances [4]. These high impact loads accelerate the aging of the vehicle and track components, inducing defects such as axle abnormal vibration, rail abrasion, and fastener fracture. In some extreme situations, the wheel-rail contact conditions change dramatically because of wheel flats, which directly threatens the operation safety.

To solve this problem, railway operation departments generally take measures with respect to both precautions and renovation. In terms of precautions, at present, most passenger trains are equipped with advanced anti-sliding systems, which, to some extent, alleviate the frequency of wheel-rail sliding [5,6]. Nevertheless, with increasing operating speed and axle load, wheel flats cannot be completely avoided. In addition, with regards to those freight trains without an anti-sliding system, the condition of the wheels is usually even worse, which has a significant impact on the long-term service of the trains and infrastructures.

Concerning the repair of wheel flats, nowadays, wheel reprofiling is the most effective, and therefore the most popular approach for repairing wheel defects [7]. However, under the premise that the health state of wheels is difficult to monitor in real time, how to develop a reasonable wheel tread reprofiling strategy becomes an open issue. Excessive repairs not only cause a surge in management costs but also shorten the life of healthy wheels, since all the wheels are subject to grinding during a reprofiling procedure [8]. Therefore, it is of great importance to reprofile the wheelsets according to their real conditions.

In recent years, researchers have proposed many different methods for in-service measuring of wheel defects, which have, generally, been divided into two types, on-board and wayside measurements [9]. Most on-board techniques are based on vibration, acoustic, image detection, and ultrasonic technologies. Bosso, Gugliotta et al. [10] proposed a diagnosis method based on the time domain analysis of the train axial accelerations and introduced a new index to judge whether the wheel flats exist. Gao, Shang et al. [11] put forward an acoustic emission (AE) signal-based detection approach. By installing an acoustic sensor near the wheel, anomalous wheel-rail noises excited by wheel flats are collected and, then, analyzed. Verkhoglyad, Kuropyatnik et al. [12] used an infrared camera to capture alterations of temperature extension on the wheel tread to detect the surface cracks. A limitation for on-board methods is that this method is quite time-consuming since the detection is only implemented after the rail is heated. Cavuto, Martarelli et al. [13] proposed an air-coupled ultrasonic method. They installed the sensors on the bogie approaching the wheels to measure the radial and circumferential defect of wheel. However, the high cost of equipment makes it difficult to promote this approach. It is worthwhile to note that all of the above methods only detect the health state of the wheels near the location where the sensors are deployed. In order to achieve comprehensive detection and management of all wheels, the only solution is to install sensors on every wheel. Hence, on-board detection methods are usually used to evaluate track structure rather than long-term monitoring of wheel conditions [14].

On the contrary, currently, wayside measurement methods are an ideal solution for identification of wheel defects, since the condition of all wheels is assessed during the train passage. Stratman, Liu et al. [15] proposed a wheel condition estimation method based on wheel impact load detectors. From the statistical characteristics of the major trend of the sensor output signals, two criteria of wheel removal were put forward. Liu, Ni et al. [16] installed more than 20 FBG sensors at the rail foot along the longitudinal direction to measure the bending moments of rail under a defected wheel. Gao, He et al. [17] designed a wheel flat detection device based on a parallelogram mechanism, on the assumption that all the rigid displacements of the rail are completely transmitted to the sensing unit. Filograno, Corredera et al. [18,19] mounted a total of 22 FBG sensors at various locations on a segment of rail to measure the temperature and strain changes of the rails, and two of the sensors were used to diagnose the state of the wheels. AE and laser technology has also been used in the wayside measurement [20]. Brizuela, Jose et al. [21,22] calculated the roundtrip time-of-flight (RTOF) of wheel flat impact echo to detect the wheel defect. The biggest limitation of this method is that high accuracy is only achieved when the train moves at a relatively low speed (5 to 15 km/h). Salzburger, Schuppmann et al. [23] developed a wheel roundness inspection system based on an ultrasonic technology, namely AUROPA III. Electromagnetic probes are mounted on the rail to generate magnetic field and a special sensor is used to detect the Rayleigh waves caused by the arrival of vehicles. Bollas, Papasalouros et al. [24] tried to improve the speed limitation of the AE measurement method. They mounted sensors on the rail to monitor the AE signals in real time and extract features to diagnose wheel flats. This method is useful when the train speed is 40 km/h. Amini, Entezami et al. [25] made an attempt to apply the AE method to detect the inner defect of wheels.

However, when formulating sensor layout schemes, most previous studies have been based on engineering experiences rather than sufficient theoretical analysis. As a result, it is difficult to ensure that all the mounted sensors are of high sensitivity to wheel flats and immune to interferences from other no-defective factors. Another limitation is that the existing detection methods have only focused on qualitative testing and do not accurately locate the specific location of the wheel flats, and therefore have been unable to meet the needs of scientific maintenance strategies based on actual conditions.

Therefore, through numerical inspections, this paper has established a vehicle-track system coupling dynamic model and analyzed the sensitivity and reliability of different sensor layout schemes with different wheel flat conditions. In view of this, a more effective scheme based on multisensor arrays has been proposed, which better captures the abnormal responses caused by wheel flats. Using simulated sensor output signals, a wheel flat recognition and precise positioning method has been developed. By jointly analyzing multisensor signals, this algorithm accurately identifies the specific moments when wheel flat impacts occur, as well as traces their source, which could provide railway operation departments with important basic information for the maintenance and health management of vehicles and track system.

2. Multisensor Array Layout Scheme Based on the Spatial Distribution Characteristics of Rail Strain

2.1. Establishment of Simulation Model

2.1.1. Wheel-Rail Force Simulation Model

Multibody dynamics (MBD) is the science that studies the laws of motion for a multibody system, which is generally composed of several flexible or rigid objects connected to each other. MBD simulation has been widely used in the design, homologation test, and research of railway transportation [26]. The MBD software, Simpack, was used to calculate the wheel-rail forces under different wheel flat conditions. As is shown in Figure 1, the vehicle model had 38 degrees of freedom and consisted of a vehicle body, two bogies, and four wheelsets. The modeling parameters are shown in Table 1.

In actual conditions, the rails are not perfect, which also affects the wheel-rail forces. Therefore, we used a sample of vertical track irregularity measured on a subway in Nanjing, China as the initial excitation of the wheel-rail contact. As shown in Figure 2, the total length of the sample is 700 m and the interval is 0.25 m.

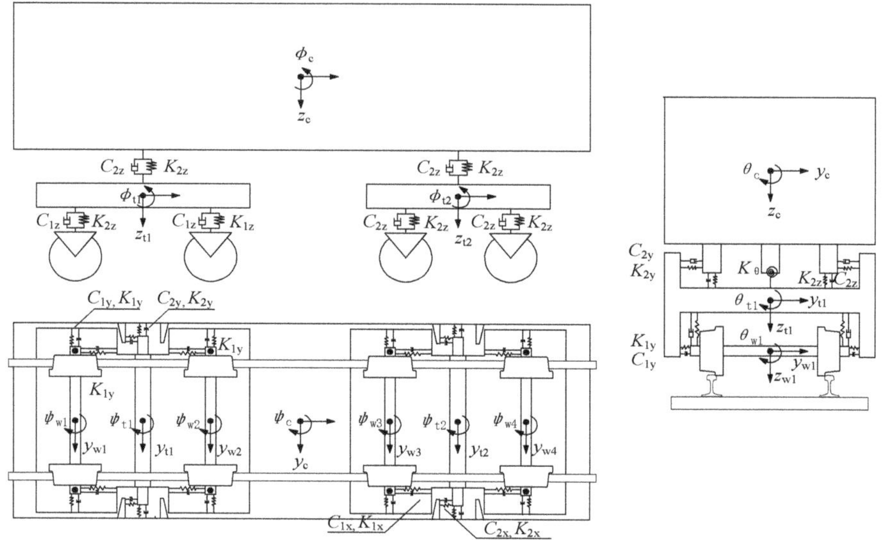

Figure 1. Vehicle-track system dynamic model.

Table 1. Parameters of the vehicle model.

Type	Parameter	Value	Unit
Mass	Vehicle body	33,593	kg
	Bogie	3730	kg
	Wheel	1730	kg
Moment of inertia	Vehicle body (around x/y/z axis)	50,000/1,101,000/1,099,000	kg·m^2
	Bogie (around x/y/z axis)	1194/876/2099	kg·m^2
	Wheel (around x/y/z axis)	1042/137/1042	kg·m^2
Stiffness	Primary suspension (longitudinal/lateral/vertical)	$17.62 \times 10^6 / 9.62 \times 10^6 / 5.35 \times 10^6$	N/m
	Secondary suspension (longitudinal/lateral/vertical)	$0.16 \times 10^6 / 0.16 \times 10^6 / 0.35 \times 10^6$	N/m
Damping	Secondary suspension (lateral)	5000	N·s/m
	Secondary suspension (vertical)	5000	N·s/m

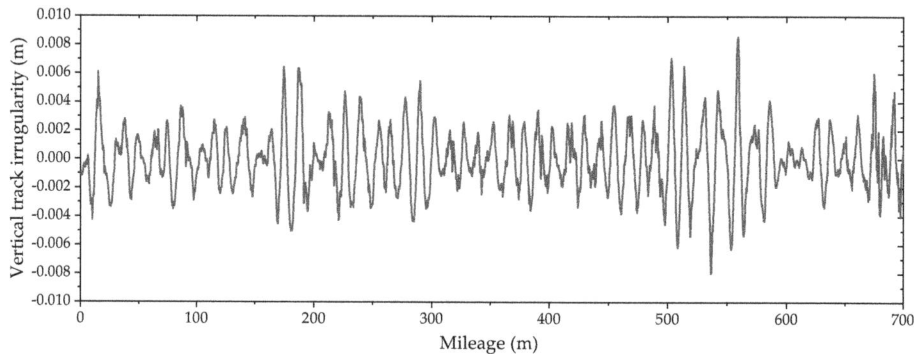

Figure 2. Track irregularity sample.

The wheel flats are simulated by defining the deviation of wheel radius. As shown in Figure 3, the equation of the undamaged wheel (plotted in blue) is expressed as follows:

$$z^2 + y^2 = R_0{}^2 \tag{1}$$

where z and y are the longitudinal and vertical position of a certain point relative to the center of the wheel and R_0 is the nominal radius of a perfect wheel.

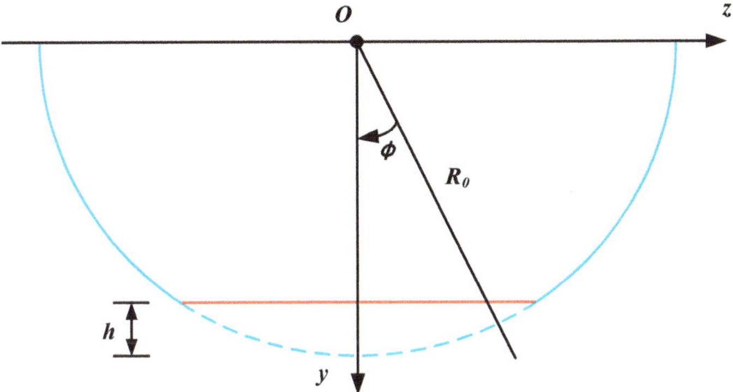

Figure 3. Geometric form of wheel flat.

According to the generation mechanism, a wheel with flats is approximately equivalent to cutting a plane from a certain position on the wheel tread [27]. Therefore, the equation of the flat (plotted in red) can be expressed as follows:

$$y = R_0 - h \tag{2}$$

where h is the depth of the wheel flat. Convert the above equations into polar coordinates, thus the deviation of wheel radius of a damaged wheel can be expressed as follows:

$$\Delta R(\varphi) = \begin{cases} 0, & \varphi \geq \frac{\alpha_R}{2} \text{ or } \varphi \leq -\frac{\alpha_R}{2} \\ R_0 - \left| \frac{R_0 - h}{\cos(\varphi)} \right|, & -\frac{\alpha_R}{2} < \varphi < \frac{\alpha_R}{2} \end{cases} \tag{3}$$

where φ is the angle between the radial line and the vertical line through the center of the wheel, α_R is central angle corresponding to the wheel flat region and equals $2 \times \cos^{-1}\left[(R_0 - h)/R_0\right]$.

By defining the radius deviation of wheels according to Equation (3), the geometric information of the wheel flats is input into the Simpack model. In this section, we have set nine damage conditions of different flat depths from 0 to 0.5 mm. Only one of the eight wheels has wheel flat and the vehicle runs at a speed of 20 m/s with a sampling frequency of 4 kHz. The calculated wheel-rail forces of different damage conditions are shown in Figure 4. The occurrence of wheel flats induces intermittent peaks to the load history and the amplitude increases as the depth grows.

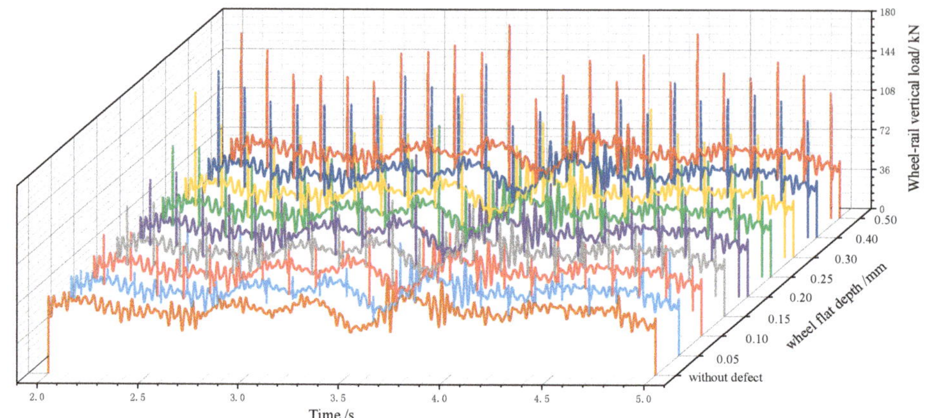

Figure 4. Wheel-rail force curves with wheel flats of different depths.

2.1.2. Refined Rail Response Simulation Model

A refined numerical model (shown in Figure 5) is developed with finite element method (FEM) software Abaqus to investigate the strain responses of rails under the previous calculated wheel-rail forces. The rails, sleepers, and slab track are modeled with solid elements according to their respective nominal geometry and material properties. The fastener system is modeled with multiple discrete linear springs and viscous damping to simulate the effect of rail pads and clamps.

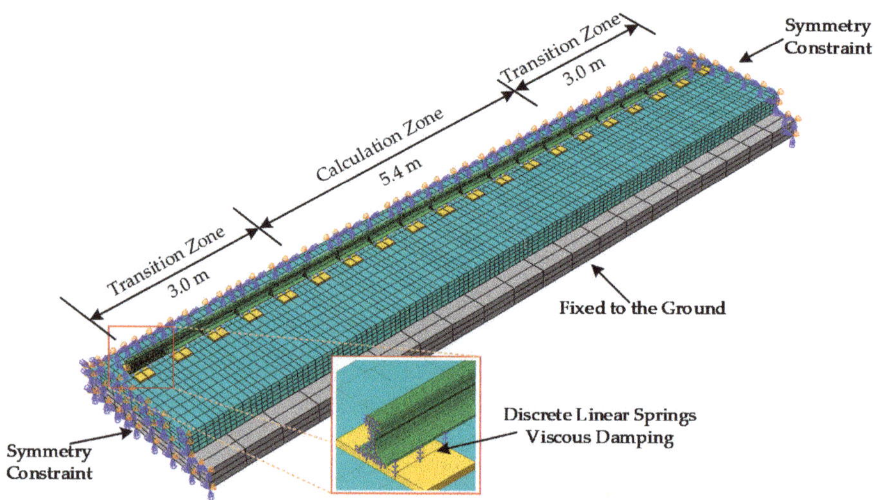

Figure 5. Refined rail responses simulation model.

The boundary settings are shown in the picture. To avoid the influence of boundary conditions on the calculation results, the model is divided into a calculation zone and two transition zones. The rail in the calculation zone is more precisely meshed. The mesh size in the longitudinal direction is 5 mm because the per unit time movement of the load on the rail is 5 mm at the given sampling frequency of 4 kHz. Since the sensors are usually mounted at the rail web or near the rail foot, due to the limitation of installation condition, the meshes at these regions are refined in the cross-section.

To transmit the wheel-rail forces to the FEM model, the user subroutine VDLOAD of Abaqus is deployed. The role of this subroutine is to define nonuniformly distributed loads as a function of position, time, and velocity, etc. According to the Hertz contact theory [28], the elastic deformation of the steel of the wheel and the rail creates an elliptic contact area. Assuming that the contact elliptic does not change during the movement, the effect of the vehicle can be regarded as a moving surface load applied to the tread of the rail. In the VDLOAD, the contact elliptic is simplified to a rectangular area (15 mm × 5 mm). The velocity of the moving load is 20 m/s and the load history are specified by the former section. The steps of the VDLOAD, at each point in time, are as follows: (1) Determine the center position of the load based on the speed of the vehicle and the time of the movement, (2) determine the contact area according to the predefined contact elliptic size, and (3) traverse all the nodes on the rail tread and calculate the stresses exerting on the nodes within the contact area in accordance with the load history.

2.2. Layout Scheme of Multisensor Arrays in the Rail Cross-Section

2.2.1. Sensitivity of Different Layout Schemes to Wheel Flats

Considering that the vertical rail responses surge intensively when wheel flats exist [29], measurement methods for vertical wheel-rail forces are feasible solutions to monitor the wheel defects. Conventional vertical wheel-rail forces measurement methods include: (1) sheer force method [30], (2) rail base bending moment method [31], (3) rail web bending moment method [32], and (4) rail web compression method [33]. Although they have different layout schemes (shown in Figure 6), they all achieve the purpose by establishing a relationship between the sensor outputs and the vertical loads.

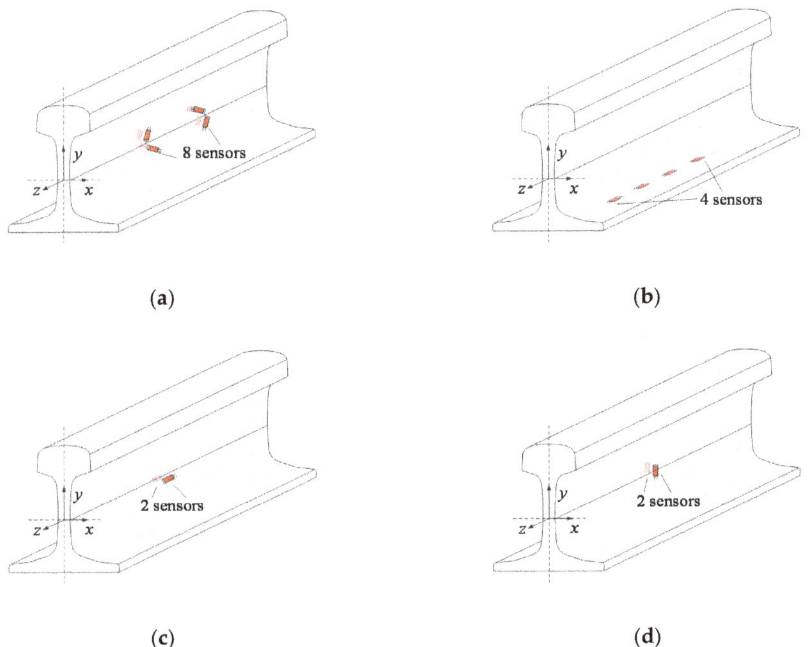

Figure 6. Commonly used wheel-rail force measurement methods: (**a**) Sheer force method; (**b**) rail base bending moment method; (**c**) rail web bending moment method; (**d**) rail web compression method.

Although the sheer force method is widely used in field tests for wheel-rail force, it is not a suitable choice for the long-term monitoring of wheel flats because this method uses eight strain gauges to form a bridge which is vulnerable to water and electromagnetic environment. In addition, the baseline drift is inevitable for strain gauges during long-term service, making it impossible to stay consistent for a long period of time. The rail base bending moment method is not an ideal choice since all the sensors are required to be attached to the bottom of the rail along the longitudinal direction which makes it difficult to ensure the installation accuracy without replacing the rail.

In contrast, the latter two methods, rail web bending moment and rail web compression, seem to have more potential for the long-term monitoring of wheel flats. For these methods, only two sensors need to be installed at the rail web. It is worth mentioning that the aim of this study is the recognition of wheel defects. Therefore, the sensors mounted on rails should capture the abnormal impact signals excited by wheel flats as effective as possible, rather than measure the exact dynamic loads. In view of this, we analyzed the sensitivity of these two methods under different defect conditions using the established refined rail response model. The predefined defect depths varied from 0 to 0.5 mm.

As illustrated in Figure 7, the output signals of these two methods have similar characteristics. Each series of signals consists of a major trend reflecting the wheel passage, as well as the fluctuation induced by a wheel flat. As the depth of the wheel flat increases, the fluctuation becomes more intense.

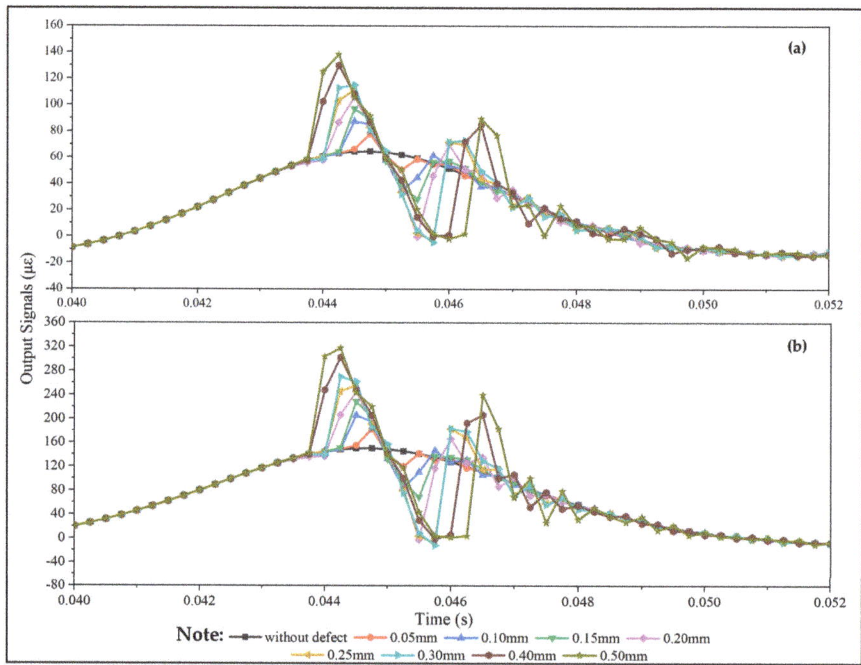

Figure 7. Sensitivity of sensor layout schemes to wheel flats of different defect conditions: (**a**) Rail web bending moment method; (**b**) rail web compression method.

It is obvious that signals collected by the rail web compression method are much stronger than those of its counterpart. Therefore, the rail web compression method is considered to be more sensitive to wheel flat impact because the signals collected by the rail web bending moment method are essentially the longitudinal dynamic bending strains caused by the wheel loads. Since the sensors are mounted near the neutral axis, the strains are relatively small. By adjusting the sensor positions away from the neutral axis, the test results are further improved. It is, nevertheless, inevitable that the installation of sensors is hindered by track equipment such as fasteners and sleepers.

2.2.2. Influences of Wheel-Rail Contact States on Different Layout Schemes

In addition to the vertical wheel loads, the rail is also subjected to bending moments and torques caused by the lateral wheel-rail forces and wheel eccentricity which makes the strain distribution in the rail cross-section much more complicated. Therefore, the selected method is supposed to be immune to these interferences, and therefore is better able to capture the vertical impacts caused by wheel flats.

To determine the influences of lateral forces, the model established in Section 2.1 was used to simulate the outputs of different sensor layout schemes during an intact wheel passage. Different levels of lateral forces (0, 10, 20, 30, and 40 kN) were exerted to the model during the process. Figure 8 shows the effects of the different lateral forces on the outputs of the rail web bending moment method and the rail web compression method. We concluded that the existence of lateral forces plays an important role in the measurement of the rail web bending moment method, while on the contrary, it has little effects on the test results of the rail web compression method.

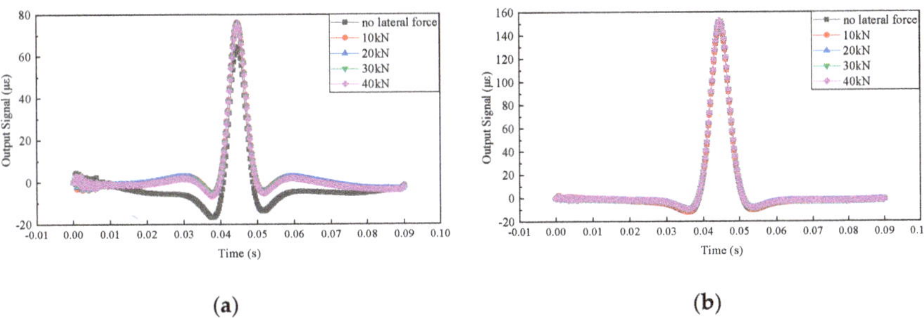

(**a**) (**b**)

Figure 8. Comparison of influences of lateral forces on different layout schemes: (**a**) Rail web bending moment method; (**b**) rail web compression method.

Similarly, we analyzed the influences of wheel eccentricity on different layout schemes. In the analysis, wheel eccentricity is defined as the distance between the wheel-rail contact center and the symmetry axis of the rail. Figure 9 shows a comparison of the output signals of these two methods under different wheel eccentricity conditions (0, 5, 10, and 15 mm). Apparently, wheel eccentricity has a greater impact on the rail web bending moment method.

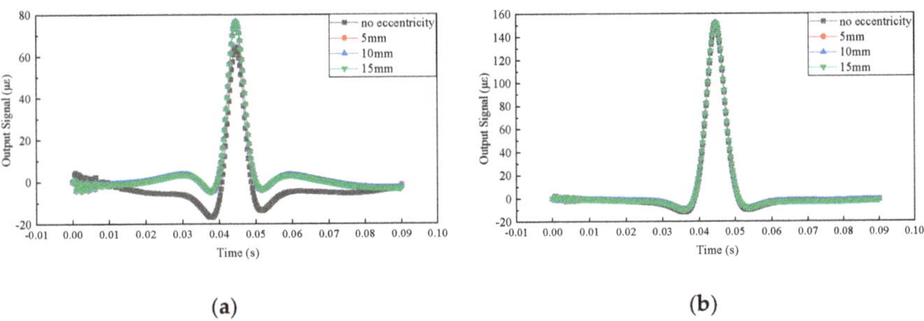

(**a**) (**b**)

Figure 9. Comparison of influences of wheel eccentricity on different layout schemes: (**a**) Rail web bending moment method; (**b**) rail web compression method.

Figure 10 concludes the influences of different wheel-rail contact states on these two layout schemes. The Y-axis represents the errors under different interferences. The result shows that the rail web compression method is the optimal solution for the sensor layout in the rail cross-section.

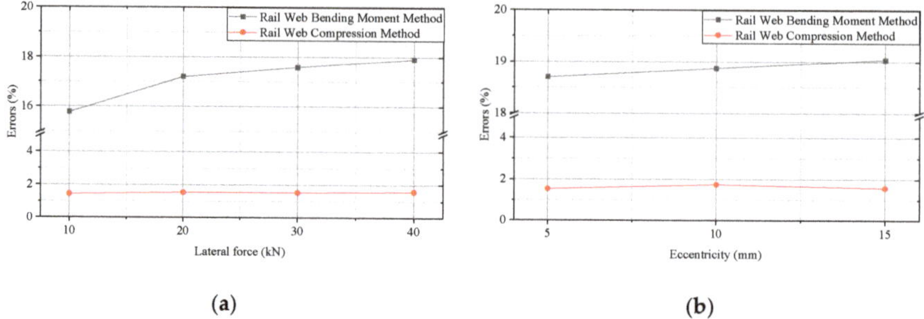

(a)　　　　　　　　　　　　(b)

Figure 10. Influences of wheel-rail contact states on different layout schemes: (**a**) Influences of lateral force; (**b**) influences of lateral eccentricity.

2.3. Layout Scheme of Multisensor Arrays along the Longitudinal Direction of Rail

2.3.1. Design of Sensor Arrangement Spacing in the Longitudinal Direction

In order to ensure that the multisensor arrays effectively capture the impact signals caused by wheel defects and avoids omissions, it is necessary to find out a reasonable spacing arrangement of sensors. Considering the limited spaces for railway field monitoring, three layout schemes for multisensor arrays in the longitudinal direction are proposed, as shown in Figure 11. Although other positions also have potential for the sensor arrangement, they are not discussed in this section because they are not easy to locate during installation.

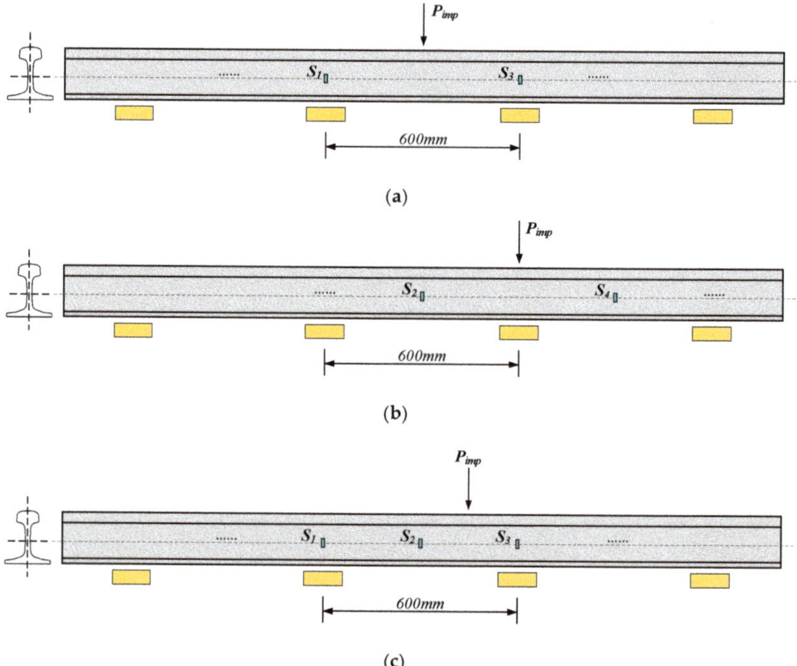

Figure 11. Plans of sensor arrangement: (**a**) Plan I, sensors are located above sleepers and intervals are 600 mm; (**b**) Plan II, sensors are located in the middle of the sleeper spacing and intervals are 600 mm; and (**c**) Plan III, sensors are evenly located along the rail and intervals are 300 mm.

2.3.2. Comparison of Different Sensor Arrangement Plans

The "most unfavorable impact position" is introduced to describe the impact position which is most difficult for adjacent sensors to recognize. The most unfavorable impact position is generally the middle position between every two sensors. When the wheel flat hits exactly this position, the impact signals collected by sensors are weakest.

The most unfavorable impact positions of the above plans are marked with P_{imp}. By adjusting the initial time of the simulation, the wheel flat impacts are ensured to occur just at those most unfavorable positions. Figure 12 shows the output signals of Plan I, where wheel flat impact occurs in the middle of the sleeper spacing. The wheel flat causes a sudden change in the signals of both adjacent sensors. Even when the wheel defect is very slight, for example, 0.1 mm, the sensors still completely capture the abnormal responses. As the depth increases, the sudden change becomes more intense.

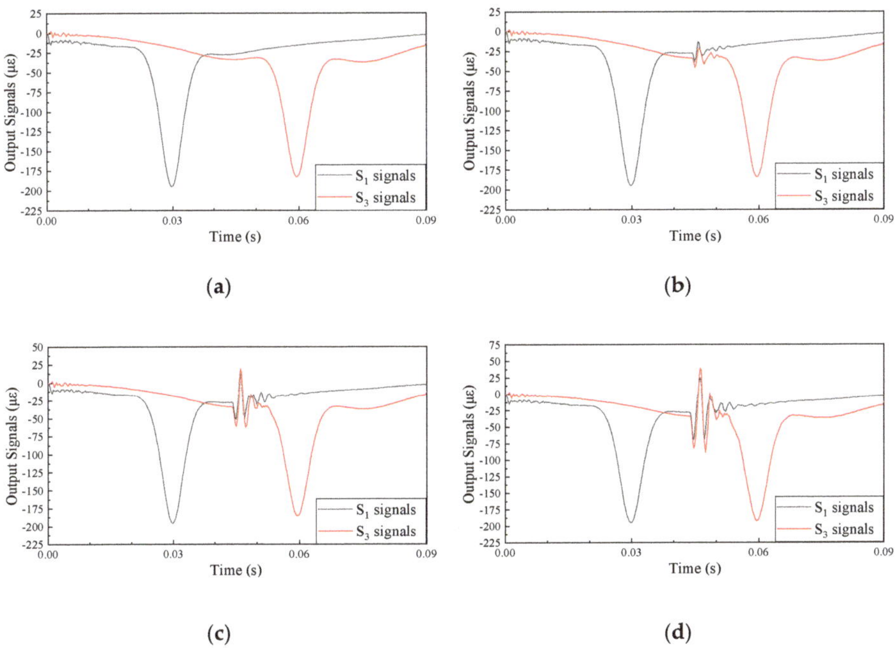

Figure 12. Output signals of Plan I under its most unfavorable condition: (**a**) Without wheel flat; (**b**) wheel flat depth of 0.1 mm; (**c**) wheel flat depth of 0.3 mm; and (**d**) wheel flat depth of 0.5 mm.

Figure 13 shows the output signals of Plan II where the wheel flat impact occurs right above the sleeper. Compared to Plan I, the signals of Plan II are much weaker, even if the wheel flat depth reaches 0.5 mm. When applied to the field test, the impact signals are likely to be submerged in system noises.

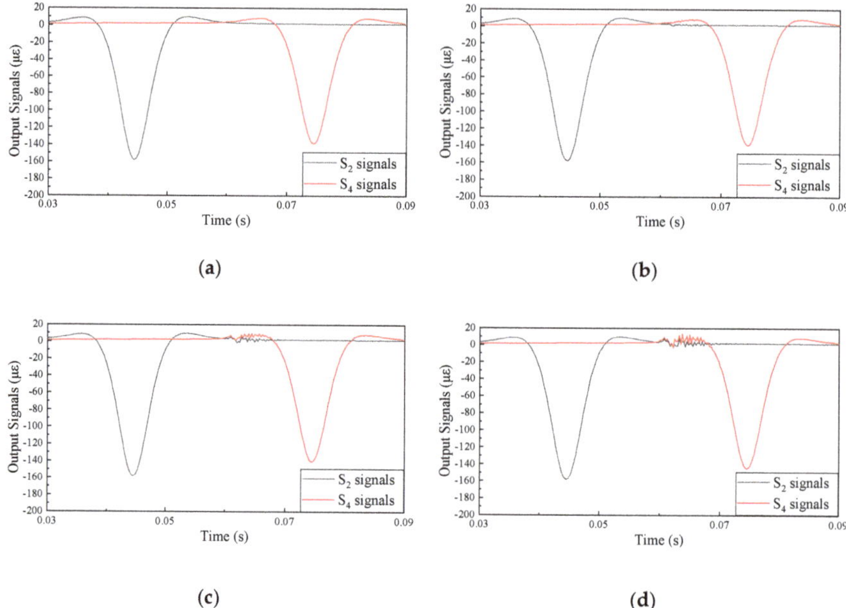

Figure 13. Output signals of Plan II under its most unfavorable condition: (**a**) Without wheel flat; (**b**) wheel flat depth of 0.1 mm; (**c**) wheel flat depth of 0.3 mm; and (**d**) wheel flat depth of 0.5 mm.

Plan III is essentially a combination of Plan I and Plan II. As shown in Figure 14, although the distance between sensors is halved, the test effects are not necessarily improved. Contrarily, the output signals are very dissimilar because the sensors are mounted in two different situations. In the middle of the sleeper spacing, the rail is relatively free in the vertical direction. Most of the wheel flat impact energy is dissipated through the vertical rigid body motion of the rail, leading to the smaller compression strain at the rail web. Therefore, the responses of sensors installed in these locations, like S_2, are much weaker. Whereas for the rail above the sleepers, it cannot freely vibrate because of the restraints of sleepers. The impact energy is mostly converted into elastic potential energy of the rail. As a result, these sensors, like S_3, have stronger output signals.

However, the fusion of multiple sensors is reliable only when the sensors are mounted in identical situations. Otherwise, the adjacent sensors have different transfer functions to the same excitation. This means that signals sampled by sensors are based on different baselines. Therefore, Plan III is not suitable for the sensor arrangement. By comparison, Plan I is considered as the best option.

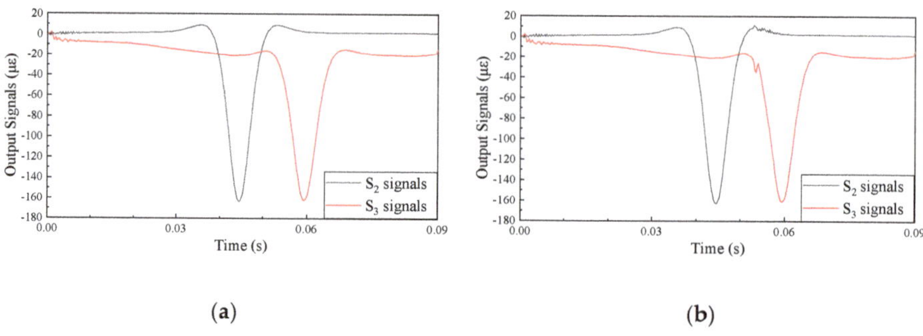

Figure 14. *Cont.*

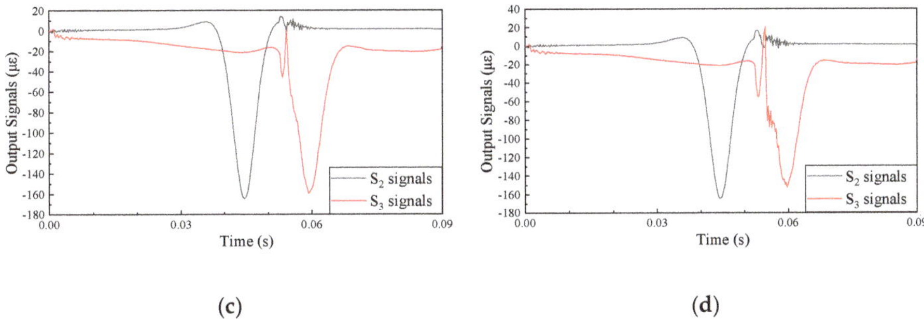

Figure 14. Output signals of Plan III under its most unfavorable condition: (**a**) Without wheel flat; (**b**) wheel flat depth of 0.1 mm; (**c**) wheel flat depth of 0.3 mm; and (**d**) wheel flat depth of 0.5 mm.

3. Wheel Flat Recognition and Positioning Algorithm Based on Multisensor Arrays

3.1. Design of the Algorithm

Using the established models, we simulated the output signals of the multisensor arrays during a bogie passage. Figure 15a shows the movement of the defected bogie and the configuration of the sensors. There is a wheel flat with a depth of 0.2 mm at the front wheel, whereas the rear wheel is intact. Ten pairs of sensors are mounted on the rail web, marked from S_1 to S_{10}. The length of a multisensor array is 5.4 m, which is longer than the circumference of the wheel. Thus, every wheel revolves two times during its passage of the measurement area. This ensures that a flat spot on the wheel hits at the rail tread within this segment no less than two times. Figure 15a shows the output signals of S_4. The successive two troughs represent the respective moments when the front and rear wheels arrive at the location of S_4. Similarly, Figure 15b,c displays the outputs of S_5 and S_6. We found that the outputs of different sensors have almost the identical main trend, except for the time delay due to the arrangement of sensors.

Apparently, the defected front wheel significantly influences the outputs of sensors. There are two obvious impact processes sensed by all the sensors. At 0.21 s, the defected wheel exactly hits the rail above S_5. Therefore, the fluctuation amplitude of S_5 at this moment is much greater than those of S_4 and S_6. With the rotation of the wheels, the flat spot on the front wheel hits the rail again between S_9 and S_{10}, at 0.34 s. However, this time the fluctuation caused by the impact process occurs just around the second trough of S_6 that reflects the arrival of the rear wheel. The reason is that the rear wheel is above S_6 at this moment. If we only observe the outputs of S_6, it is easy to draw the wrong conclusion that the rear wheel is also damaged.

Since the in situ condition is much more complex than this case (the only imperfectness is a flat spot on the front wheel), separate analysis can hardly acquire the accurate information. Even if we observe the responses of multiple sensors on the same wheel, it is still difficult to have a comprehensive understanding of the wheel states. Therefore, data fusion of multiple sensors is critical to the better diagnosis of wheel defects. On the basis of the correlations between sensors, we designed a wheel flat recognition and positioning algorithm. As shown in Figure 16, by solving the three key problems of "when", "where", and "which wheel", successively, it accurately accesses the damage states of the wheels.

3.2. Step I: When the Abnormal Impacts Occurred

The purpose of this step is to determine the specific moments when abnormal impacts occur. The fast Fourier transform (FFT) filter is applied to eliminate the major trends of output signals. With a cut-off frequency of 200 Hz, the high-frequency components caused by wheel flats are extracted, as shown in Figure 17. After the high-pass filtering, the defect signals become much more obvious.

All 10 sensors have sensed two anomalous responses caused by the wheel flat on the front wheel at 0.21 s and 0.34 s.

Additionally, an interesting fact is that the response amplitudes of sensors to a certain wheel flat seem to obey some special distribution. As the red dashed lines show, the responses to the first impact are almost symmetrically distributed around the location of S_6. As for the second impact, the amplitude of response increases along with an increase of the sensor's sequence number. This characteristic provides a hint as to how to locate where the impact happens, which will be later discussed in the Section 3.3.

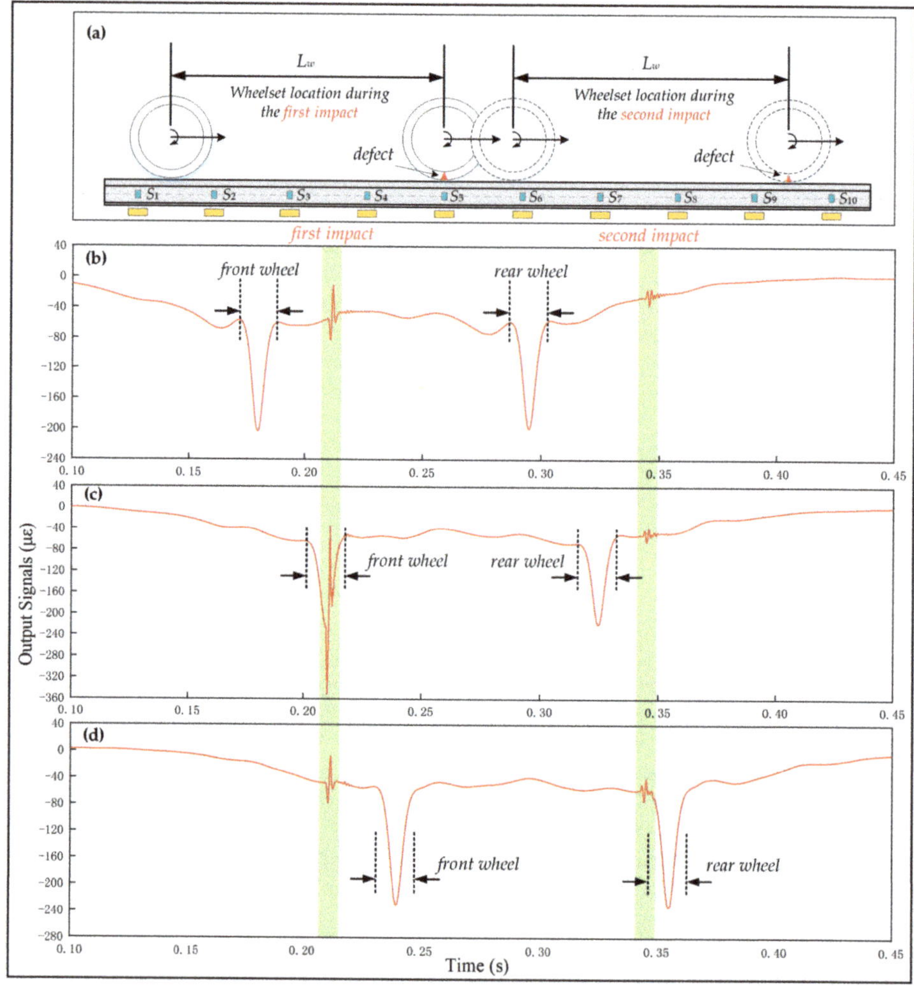

Figure 15. (a) The movement of the wheelset and the layout of the multisensor array, the output signals of (b) sensor S_4; (c) sensor S_5; and (d) sensor S_6.

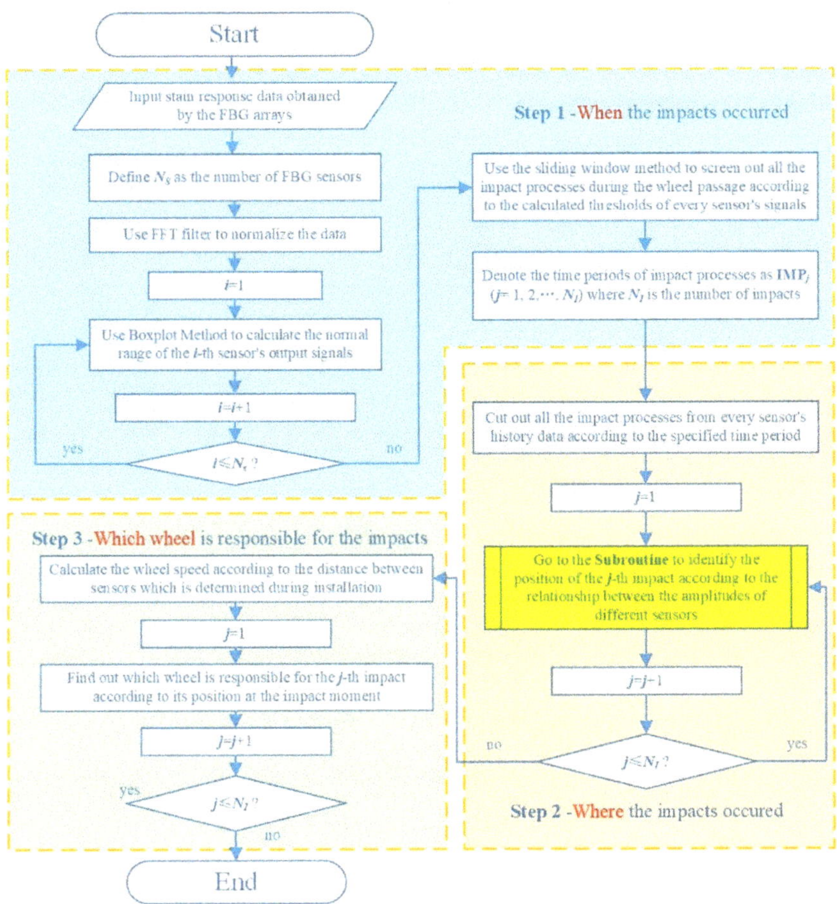

Figure 16. Main steps of the wheel flat recognition and positioning algorithm.

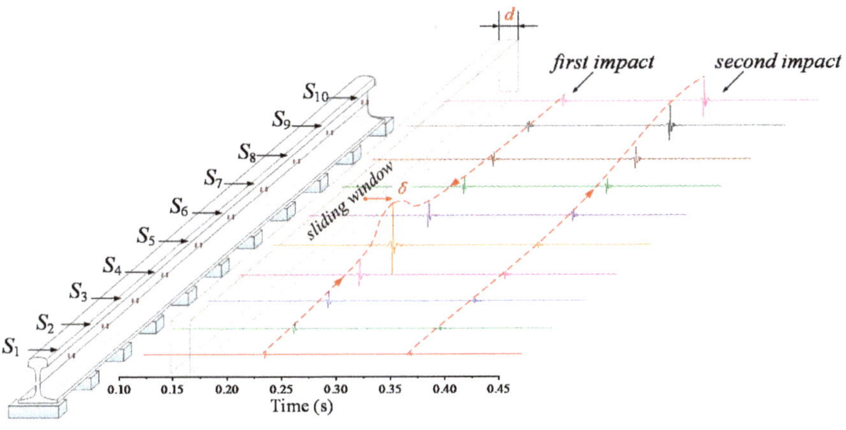

Figure 17. Processed signals after FFT high-pass filtering.

In order to screen out all the impact processes, the sliding window method is introduced in this step. As is shown in Figure 17, the sliding window, with a size of d, moves forward progressively by the step length of δ. The window is a matrix containing the signals of multiple sensors in a certain time range, which can be expressed as follows:

$$W_k = \begin{pmatrix} S_1, N_k & S_1, N_k+1 & \cdots & S_1, N_k+M-2 & S_1, N_k+M-1 \\ S_2, N_k & S_2, N_k+1 & \cdots & S_2, N_k+M-2 & S_2, N_k+M-1 \\ \vdots & \vdots & \ddots & \vdots & \vdots \\ S_{N_s-1}, N_k & S_{N_s-1}, N_k+1 & \cdots & S_{N_s-1}, N_k+M-2 & S_{N_s-1}, N_k+M-1 \\ S_{N_s}, N_k & S_{N_s}, N_k+1 & \cdots & S_{N_s}, N_k+M-2 & S_{N_s}, N_k+M-1 \end{pmatrix} \tag{4}$$

where N_s is the number of sensors. In this case, $N_s = 10$. M is the number of samples of each step determined by the window size d and the sampling frequency f. N_k represents the starting number of the sample in the k-th step. These two parameters can be calculated using the following equations:

$$M = d \times f \tag{5}$$

$$N_k = 1 + (k-1) \times \delta \times f \tag{6}$$

Outlier tests are performed during every step. Since the filtered signals are not normally distributed, the boxplot method is deployed to distinguish the outliers. This method uses quantiles to determine the thresholds of the normal data, which can effectively eliminate the influence of extreme values on the overall distribution. In light of this, the boxplot method has strong robustness in the outlier test for non-normally distributed data.

The upper and lower thresholds of the normal signals of i-th sensor specified by the boxplot method can be expressed in Equations (7) and (8), respectively [34].

$$Y_u = Q_3 + \alpha \times IQR \tag{7}$$

$$Y_l = Q_1 - \alpha \times IQR \tag{8}$$

where Q_3 and Q_1 are the upper and lower quartiles of the sample, IQR is difference between Q_3 and Q_1, and α is an adjustable coefficient which is often set as 1.5. Outliers are defined as values outside the range between Y_u and Y_l.

By comparing the maxima and minima of every row of W_k with the calculated thresholds, we discover how many sensors have sensed some abnormal fluctuations in the corresponding time range. Figure 18a illustrates the maxima and minima of the sensor S_1 during every sliding step. The black dots represent the maximums and the red ones stand for the minimums. Compared with Y_u and Y_l (black and red solid lines), outliers of S_1 during the measurement can be screened out. Similarly, the outliers of other sensors are shown in Figure 18b,c.

The outliers collected from the sliding window method are drawn in Figure 18d. It can be seen that all the outliers are clustered at two moments. To eliminate the interferences of sensor noises, the frequency of outliers in every step is calculated and shown with the red curve. If there are more than five sensors that have outliers in a certain step, the corresponding period is regarded as an impact process. These periods are denoted as time series IMP_j ($j = 1, 2, \ldots, N_I$) where N_I is the number of identified impact processes. In this case, two impact processes have been screened out through the sliding window method, i.e., IMP_1 (0.1875~0.225 s) and IMP_2 (0.325~0.35 s).

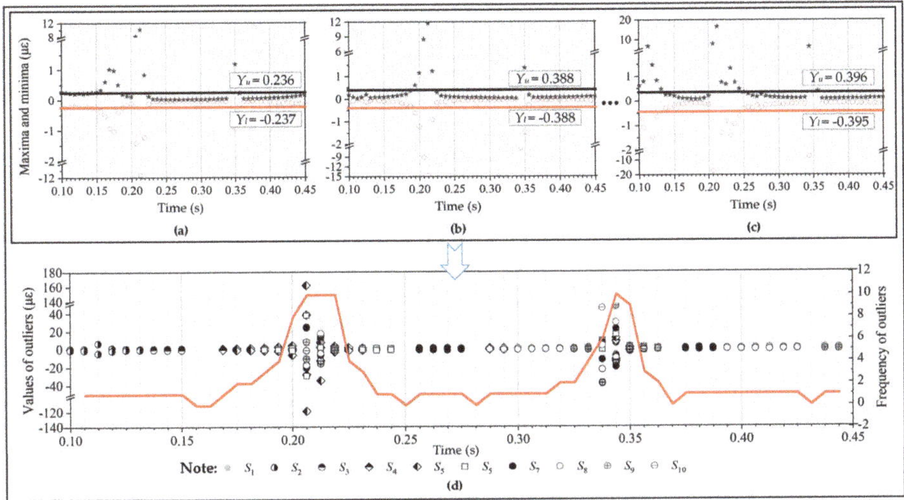

Figure 18. Identified impact processes using sliding window method: (**a**) Thresholds and outliers of S_1; (**b**) thresholds and outliers of S_2; (**c**) thresholds and outliers of S_{10}; and (**d**) the frequency of outliers during every step of sliding.

3.3. Step II: Where the Abnormal Impacts Occurred

The purpose of this step is to position the specific location of each abnormal impact. For every impact process identified in Section 3.2, the filtered signals during the corresponding period are extracted and arranged in order, according to the positional relationship of sensors. This gives us a different perspective, i.e., as if observing Figure 17 from the right side.

We found that there is an obvious correlation between the amplitudes and the impact location. Figure 19b illustrates the filtered signals of all sensors during IMP_1. During this process, the impact location is just over S_5, leading to the much higher output of S_5 than its counterparts. A similar conclusion can also be drawn according to Figure 19c.

Inspired by this, a data sequence containing the maximum values of every sensor's outputs during IMP_j is acquired as follows:

$$Ex_j = \left(Max_{j,1}, Max_{j,2}, \cdots, Max_{j,N_s}\right)^T \tag{9}$$

where $Max_{j,i}$ ($i = 1, 2, \dots, N_s$) is the maxima of the sensor S_i's outputs during IMP_j.

To better describe, a parameter, namely "flag", is introduced to represent the impact location. As is shown in Figure 19a, the value of flag is set as 1.5 in the beginning. All the elements of Ex_j are compared successively to determine the impact location. If $Max_{j,i+1}$ is greater than $Max_{j,i}$, then, the value of flag is increased by one, meaning that the impact is supposed to occur somewhere behind S_i. On the contrary, if $Max_{j,i+1}$ is less than $Max_{j,i}$, then the value of flag stays unchanged. After the calculation, the nearest two integers to the parameter flag suggest the location of a certain impact process. For instance, the final value of flag for IMP_2 is 9.5, which means that the most probable location of the second impact process is between S_9 and S_{10}.

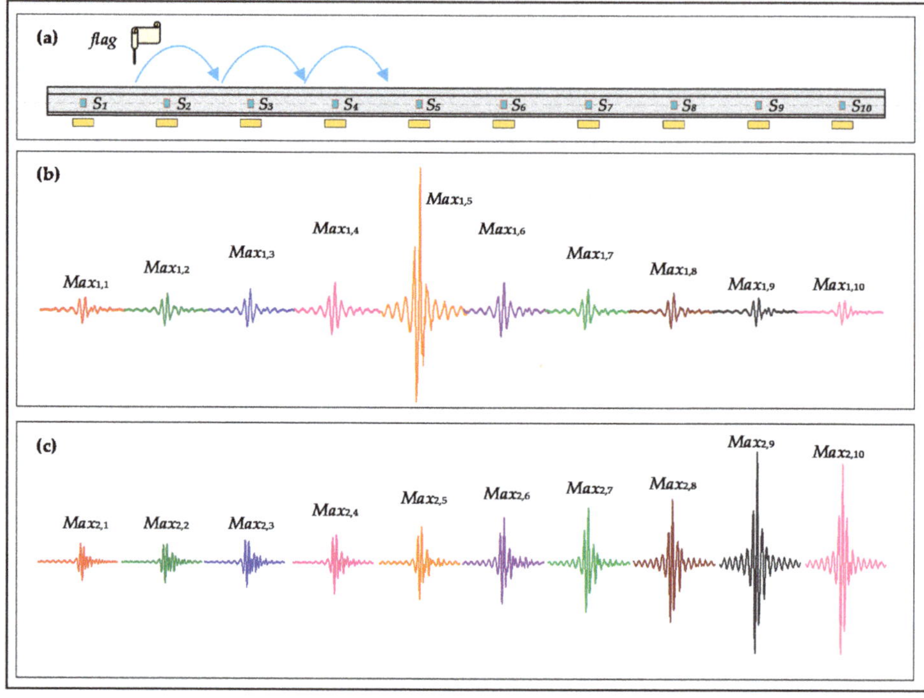

Figure 19. The relationship of response amplitudes and sensor location: (**a**) Movement rule of the parameter flag; (**b**) filtered signals during IMP_1 (0.1875~0.225 s); and (**c**) filtered signals during IMP_2 (0.325~0.35 s).

3.4. Step III: Which Wheel is Responsible for the Abnormal Impact

The first target of this step is to calculate the average speed of the wheels. Since the distance between the sensors is fixed once installed, the average speed of the wheels can be obtained by calculating the sensor spacing divided by the interval of troughs. The average speed of the wheels is denoted as v_w.

Indeed, there can only be one wheel at a specific time and position. Therefore, as long as the positions of the wheels are determined, it is possible to find out the source of impacts. The position of S_1 is regarded as the origin of the coordinate. The time when the front and rear wheels arrive at the origin are denoted as $t_{ini,\ front}$ and $t_{ini,\ rear}$. For the j-th ($j = 1, 2, \ldots, N_I$) impact process, the locations of the front and rear wheels can be described by the following equations, respectively:

$$S_{j,\ front} = v_w \times \left(t_{j,\ imp} - t_{ini,\ front}\right) \tag{10}$$

$$S_{j,\ rear} = v_w \times \left(t_{j,\ imp} - t_{ini,\ rear}\right) \tag{11}$$

where $t_{j,\ imp}$ is the middle time of the j-th impact process IMP_j.

According to the above method, the following result is obtained: During IMP_1 (0.1878~0.225 s), the front wheel is between S_5 and S_6, while the rear wheel is between S_2 and S_3. During IMP_2 (0.325~0.35 s), the front wheel is between S_{10} and S_{11}, while the rear wheel is between S_6 and S_7. According to the impact positions detected in Section 3.3, all the impact processes are identified as being generated by the front wheel, which is consistent with the predefined condition.

4. Validation of the Algorithm

Validation of the recognition and positioning algorithm was conducted via the experimental method. The offline wheel condition inspection was conducted to measure the radius deviation of the wheels. After several repeats of the experiments, two wheels with relatively severe wheel flats were chosen for the algorithm validation. Figure 20 shows the defect conditions of these two wheels. More specifically, there are two obvious wheel flats with approximate depths of 0.06 mm and 0.04 mm on the front wheel, and one wheel flat with an approximate depth of 0.07 mm on the rear wheel. The measured wheel profiles were input into the simulation model as the front and rear wheels, respectively. The sensor layout is consistent with Section 3.

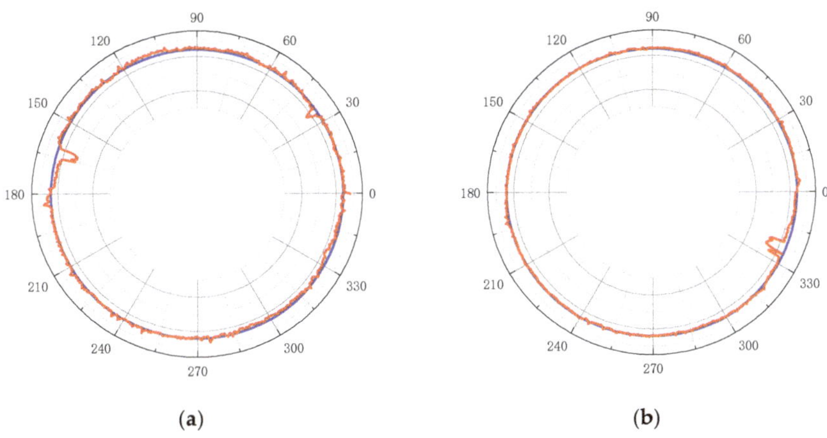

(a) (b)

Figure 20. The defect conditions of the chosen wheels: (**a**) Front wheel; (**b**) rear wheel.

To consider the influence of the monitoring system error, a 5% (SNR = 13 dB) Gaussian noise is added to the calculated output signals. Figure 21 shows the results of the simulation. For easier observation, the baselines of signals are shifted to different degrees. The abnormal data caused by multiple wheel defects mingles together, making it impossible to assess the real conditions of the wheels directly.

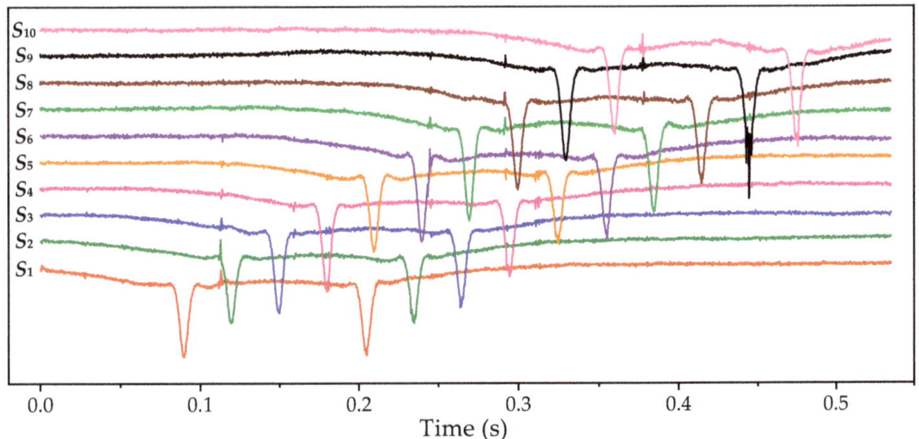

Figure 21. Raw outputs of sensors during the passage of wheelset under actual service state.

The algorithm proposed in Section 3 was used to identify the defect conditions of the wheels. The curves in Figure 22 illustrate the filtered signals and identified impact processes, while the histogram represents the frequency of outliers during every time step. The results show that there are 8 periods of time identified as potential impact processes. They are denoted as IMP_1 (0.1125~0.11875 s), IMP_2 (0.15625~0.1625 s), IMP_3 (0.18125~0.1875 s), IMP_4 (0.24375~0.25 s), IMP_5 (0.2875~0.29375 s), IMP_6 (0.30625~0.3125 s), IMP_7 (0.38125~0.3875 s), and IMP_8 (0.44375~0.45 s).

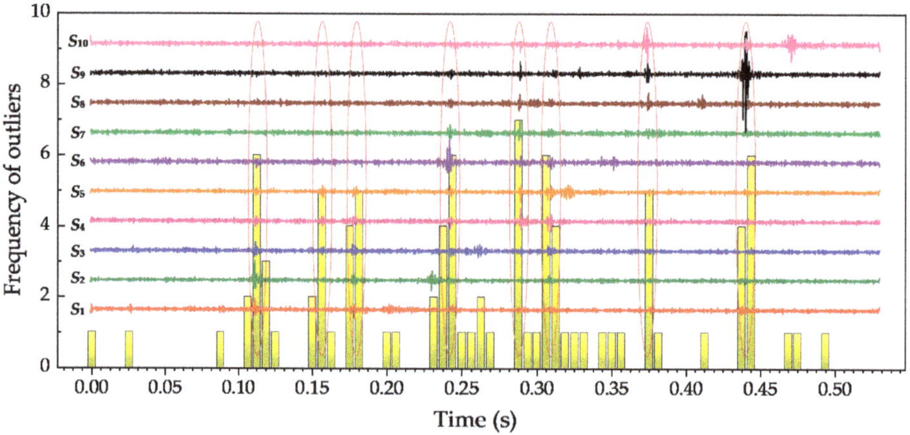

Figure 22. The filtered outputs and identified impact processes.

Furthermore, the position of every impact process, as well as the locations of the wheels, are determined. The results are concluded in Table 2. The initial time for calculating the locations of the front wheel is 0.0595 s. As for the rear wheel, the initial time is 0.17425 s. The reason for the negative values of the rear wheel position is that the rear wheel has not entered the test area at this moment. It is obvious that five abnormal impact processes are generated by the front wheel, and the others are due to the defects of the rear wheel.

Table 2. Results of the recognition and positioning algorithm.

Impact Processes	Impact Position	Middle Time/s	Front Wheel Position/m	Rear Wheel Position/m	Responsible Wheel
IMP_1	0.6~1.2 m	0.109375	1.1225	−1.1725	Front wheel
IMP_2	1.8~2.4 m	0.153125	1.9975	−0.2975	Front wheel
IMP_3	0~0.6 m	0.178125	2.4975	0.2025	Rear wheel
IMP_4	3.6~4.2 m	0.240625	3.7475	1.4525	Front wheel
IMP_5	4.2~4.8 m	0.284375	4.6225	2.3275	Front wheel
IMP_6	2.4~3 m	0.303125	4.9975	2.7025	Rear wheel
IMP_7	6.4~7 m	0.371875	6.4975	4.2025	Front wheel
IMP_8	5.4~6 m	0.440625	7.7475	5.4525	Rear wheel

To better describe the defect conditions of the wheels, the abnormal signals are converted into the polar coordinate system, where the responses represent the deviation of the radius. In this way, the wheel conditions can be deduced and illustrated in Figure 23. The impact signals collected by different sensors converge to the same center angle, reflecting that the wheel flat is most likely to be at this position. From the results of the algorithm, it can be deduced that there are two flat spots on the front wheel and one flat spot on the rear wheel, which matches well with the offline inspection results. Therefore, the algorithm is proven capable of effectively capturing the existences of wheel flats.

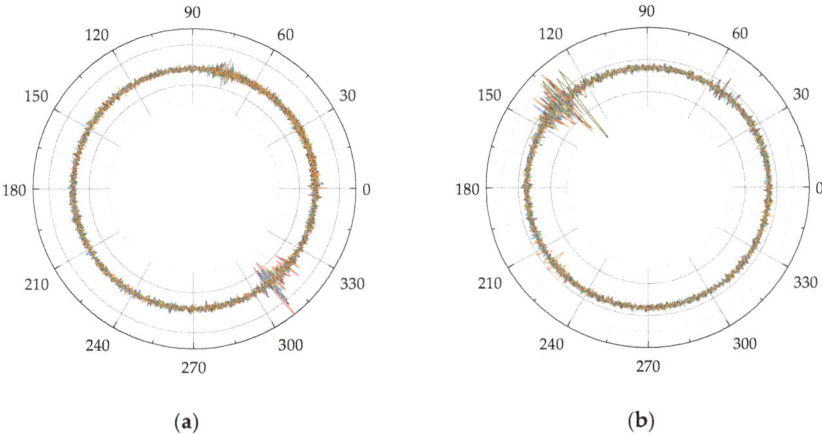

Figure 23. The deduced defect conditions of wheels: (**a**) Front wheel; (**b**) rear wheel.

5. Conclusions

In this paper, the strain distribution characteristics of the rails under different wheel flat conditions were analyzed using the numerical method. On the basis of this, we designed a comprehensive layout scheme based on multisensor arrays for the real-time detection of wheel conditions. To achieve accurate recognition and positioning of wheel flats, an algorithm based on multisource data fusion was proposed. To validate the algorithm, the offline inspected wheel profiles were input into the numerical model to simulate the output data of the multisensor arrays. The algorithm was, then, conducted to identify the defect conditions of wheels. The results match well with the offline inspection results, showing that the algorithm can accurately detect and locate the wheel flats during the wheel passage. The proposed sensor layout scheme, as well as the algorithm, can be a practical and effective solution for real-time monitoring and long-term health and management of railway vehicle wheels.

Author Contributions: Conceptualization, L.G. and H.X.; investigation, C.Z.; project administration, L.G.; software, C.Z. and B.H.; supervision, H.X.; validation, B.H.; writing—original draft, C.Z.; writing—review and editing, L.G. All authors have read and agreed to the published version of the manuscript.

Funding: This research was funded by the National Key R & D Program of China (2016YFB1200402) and the Key Program of National Natural Science Foundation of China (U1734206).

Conflicts of Interest: The authors declare no conflict of interest.

References

1. Jing, L. Wheel-rail impact by a wheel flat. In *Modern Railway Engineering*; Hessami, A., Ed.; IntechOpen: London, UK, 2018; pp. 31–49.
2. Wu, B.; An, B.; Wen, Z.; Wang, W.; Wu, T. Wheel–rail low adhesion issues and its effect on wheel–rail material damage at high speed under different interfacial contaminations. *Proc. Inst. Mech. Eng. Part C J. Mech. Eng. Sci.* **2019**, *233*, 5477–5490. [CrossRef]
3. Esveld, C. *Modern Railway Track*; MRT-Productions: Zaltbommel, The Netherlands, 2014; pp. 71–74.
4. Bian, J.; Gu, Y.; Murray, M.H. A dynamic wheel–rail impact analysis of railway track under wheel flat by finite element analysis. *Veh. Syst. Dyn.* **2013**, *51*, 784–797. [CrossRef]
5. Zuo, J.; Wu, M. Research on anti-sliding control of railway brake system based on adhesion-creep theory. In Proceedings of the IEEE International Conference on Mechatronics & Automation, Xi'an, China, 4–7 Auguest 2010.

6. Luo, R. Anti-sliding control simulation of railway vehicle braking. *Chin. J. Mech. Eng.* **2008**, *44*, 35–40. [CrossRef]
7. Zhu, W.; Yang, D.; Huang, J. A hybrid optimization strategy for the maintenance of the wheels of metro vehicles: Vehicle turning, wheel re-profiling, and multi-template use. *Proc. Inst. Mech. Eng. Part F J. Rail Rapid Transit* **2018**, *232*, 832–841. [CrossRef]
8. Mok, H.; Chiu, W.K.; Peng, D.; Sowden, M.; Jones, R. Rail wheel removal and its implication on track life: A fracture mechanics approach. *Theor. Appl. Fract. Mech.* **2007**, *48*, 21–31. [CrossRef]
9. Alemi, A.; Corman, F.; Lodewijks, G. Condition monitoring approaches for the detection of railway wheel defects. *Proc. Inst. Mech. Eng. Part F J. Rail Rapid Transit* **2017**, *231*, 961–981. [CrossRef]
10. Bosso, N.; Gugliotta, A.; Zampieri, N. Wheel flat detection algorithm for onboard diagnostic. *Measurement* **2018**, *123*, 193–202. [CrossRef]
11. Ruipeng, G.; Chunyang, S.; Hang, J. A fault detection strategy for wheel flat scars with wavelet neural network and genetic algorithm. *J. Xi'an Jiaotong Univ.* **2013**, *47*, 88–91.
12. Verkhoglyad, A.G.; Kuropyatnik, I.N.; Bazovkin, V.M.; Kuryshev, G.L. Infrared diagnostics of cracks in railway carriage wheels. *Russ. J. Nondestruct. Test.* **2008**, *44*, 664–668. [CrossRef]
13. Cavuto, A.; Martarelli, M.; Pandarese, G.; Revel, G.M.; Tomasini, E.P. Train wheel diagnostics by laser ultrasonics. *Measurement* **2016**, *80*, 99–107. [CrossRef]
14. Tsunashima, H. Condition monitoring of railway tracks from car-body vibration using a machine learning technique. *Appl. Sci.* **2019**, *9*, 2734. [CrossRef]
15. Stratman, B.; Liu, Y.; Mahadevan, S. Structural health monitoring of railroad wheels using wheel impact load detectors. *J. Fail. Anal. Prev.* **2007**, *7*, 218–225.
16. Liu, X.Z.; Ni, Y.Q. Wheel tread defect detection for high-speed trains using FBG-based online monitoring techniques. *Smart Struct. Syst.* **2018**, *21*, 687–694.
17. Gao, R.; He, Q.; Feng, Q. Railway wheel flat detection system based on a parallelogram mechanism. *Sensors* **2019**, *19*, 3614. [CrossRef]
18. Filograno, M.L.; Corredera, P.; Gonzalez-Herraez, M.; Rodriguez-Plaza, M.; Andres-Alguacil, A. Wheel flat detection in high-speed railway systems using fiber bragg gratings. *IEEE Sens. J.* **2013**, *13*, 4808–4816. [CrossRef]
19. Filograno, M.L.; Rodriguez-Barrios, A.; Corredera, P.; Martin-Lopez, S.; Rodriguez-Plaza, M.; Andres-Alguacil, A.; Gonzalez-Herraez, M. Real time monitoring of railway traffic using Fiber Bragg Gratings. *IEEE Sens. J.* **2011**, *12*, 85–92. [CrossRef]
20. Kundu, P.; Darpe, A.K.; Singh, S.P.; Gupta, K. A Review on Condition Monitoring Technologies for Railway Rolling Stock. In Proceedings of the European Conference of The Prognostics and Health Management Society 2018, Philadelphia, PA, USA, 24 September 2018.
21. Brizuela, J.; Fritsch, C.; Ibáñez, A. Railway wheel-flat detection and measurement by ultrasound. *Trans. Res. Part C Emerg. Technol.* **2011**, *19*, 975–984. [CrossRef]
22. Brizuela, J.; Ibañez, A.; Nevado, P.; Fritsch, C. Railway wheels flat detector using Doppler effect. *Phys. Procedia* **2010**, *3*, 811–817. [CrossRef]
23. Salzburger, H.; Schuppmann, M.; Li, W.; Xiaorong, G. In-motion ultrasonic testing of the tread of high-speed railway wheels using the inspection system AUROPA III. *Insight Non Destr. Test. Cond. Monit.* **2009**, *51*, 370–372. [CrossRef]
24. Bollas, K.; Papasalouros, D.; Kourousis, D.; Anastasopoulos, A. Acoustic emission inspection of rail wheels. *J. Acoust. Emiss.* **2010**, *28*.
25. Amini, A.; Entezami, M.; Kerkyras, S.; Papaelias, M. Condition monitoring of railway wheelsets using acoustic emission. In Proceedings of the 10th International Conference on Condition Monitoring and Machinery Failure Prevention Technologies, Krakow, Poland, 7 June 2013.
26. Weidemann, C. State-of-the-art railway vehicle design with multi-body simulation. *J. Mech. Syst. Transp. Logist.* **2010**, *3*, 12–26. [CrossRef]
27. Li, Y.; Zuo, M.J.; Lin, J.; Liu, J. Fault detection method for railway wheel flat using an adaptive multiscale morphological filter. *Mech. Syst. Signal. Process.* **2017**, *84*, 642–658. [CrossRef]
28. Yan, W.; Fischer, F.D. Applicability of the Hertz contact theory to rail-wheel contact problems. *Arch. Appl. Mech.* **2000**, *70*, 255–268. [CrossRef]

29. Wu, T.X.; Thompson, D.J. A hybrid model for the noise generation due to railway wheel flats. *J. Sound Vibr.* **2002**, *251*, 115–139.

30. Milković, D.; Simić, G.; Jakovljević, Ž.; Tanasković, J.; Lučanin, V. Wayside system for wheel–rail contact forces measurements. *Measurement* **2013**, *46*, 3308–3318. [CrossRef]

31. Zeng, S. *Railway Track Dynamic Measurement Techniques*; China Railway Publishing House: Beijing, China, 1988; pp. 46–61.

32. Ying, S.; Yanliang, D.; Baochen, S. Application of piezoelectric sensing technology in real-time monitoring of wheel/rail interaction. *J. Vibr. Shock* **2010**, *29*, 228–232.

33. Yuqing, Z.; Geming, Z.; Yan, Z.; Weidong, Y. Linear state method for continuous measurement of wheel/rail vertical force on ground. *China Rail. Sci.* **2015**, *36*, 111–119.

34. Li, A.; Feng, M.; Li, Y.; Liu, Z. Application of outlier mining in insider identification based on Boxplot method. *Procedia Comput. Sci.* **2016**, *91*, 245–251. [CrossRef]

applied
sciences

Article

Monitoring and Analysis of Dynamic Characteristics of Super High-rise Buildings using GB-RAR: A Case Study of the WGC under Construction, China

Lv Zhou [1,2], Jiming Guo [3], Xuelin Wen [1,*], Jun Ma [4], Fei Yang [3], Cheng Wang [5] and Di Zhang [3]

[1] College of Geomatics and Geoinformation, Guilin University of Technology, Guilin 541004, China; zhoulv@glut.edu.cn
[2] Key Laboratory of Geospace Environment and Geodesy, Ministry of Education, Wuhan University, Wuhan 430079, China
[3] School of Geodesy and Geomatics, Wuhan University, Wuhan 430079, China; jmguo@sgg.whu.edu.cn (J.G.); coffeeyang@whu.edu.cn (F.Y.); zhangdi@whu.edu.cn (D.Z.)
[4] China Railway Siyuan Survey and Design Group Co., LTD, Wuhan 430063, China; yangzhiqou.student@sina.com
[5] Guangxi Institute of Natural Resources Survey and Monitoring, Nanning 530023, China; wangcheng1509@163.com
[*] Correspondence: wenxuelin@glut.edu.cn

Received: 30 November 2019; Accepted: 21 January 2020; Published: 23 January 2020

Abstract: Accurate dynamic characteristics of super high-rise buildings serve as a guide in their construction and operation. Ground-based real aperture radar (GB-RAR) techniques have been applied in monitoring and analyzing the dynamic characteristics of different buildings, but only few studies have utilized them to derive the dynamic characteristics of super high-rise buildings, especially those higher than 400 m and under construction. In this study, we proposed a set of technical methods for monitoring and analyzing the dynamic characteristics of super high-rise buildings based on GB-RAR and wavelet analysis. A case study was conducted on the monitoring and analysis of the dynamic characteristics of the Wuhan Greenland Center (WGC) under construction (5–7 July 2017) with a 636 m design height. Displacement time series was accurately derived through GB-RAR and wavelet analysis, and the accuracy reached the submillimeter level. The maximum horizontal displacement amplitudes at the top of the building in the north–south and east–west directions were 18.84 and 15.94 mm, respectively. The roof displacement trajectory of the WGC was clearly identified. A certain negative correlation between the temperature and displacement changes at the roof of the building was identified. Study results demonstrate that the proposed method is effective for the dynamic monitoring and analysis of super high-rise buildings with noninvasive and nondestructive characteristics.

Keywords: dynamic characteristic; GB-RAR; super high-rise building; displacement

1. Introduction

The population and public infrastructures in large cities have intensively increased with limited land resources in recent years. Numerous super high-rise buildings have been built in large cities, and some of them exceeding 400 m in height have been constructed as new landmarks to improve the land resource utilization rate and demonstrate the prosperity and development of a city. At the end of 2017, China had 14 out of the top 20 completed super high-rise buildings in the world, in which all of them are more than 420 m in height. Super high-rise buildings will produce sway motions under the influences of sunlight, wind load, construction vibration, temperature, and other factors [1–3]. The structure of a super high-rise building will be destroyed when its horizontal displacement exceeds a certain limit value. The dynamic characteristics of a building directly affect the construction measurements that require a stable reference. Therefore, the dynamic characteristics of buildings should be accurately obtained and the performance of structures should be appropriately monitored and diagnosed for the safe construction and healthy operation of super high-rise buildings.

The main methods used to monitor and analyze the dynamic behavior of super high-rise buildings include total station [4], accelerometer sensor [5–7], digital vertical meter [8], and Global Navigation Satellite System (GNSS) technology [9–13]. Among the above methods, the total station method needs to install a prism on some feature points of the building, and the data sampling frequency of automatic tracking total station is low, making it difficult to achieve high-frequency data acquisition. The accelerometer sensor method obtains the vibration acceleration of a building by mounting a sensor and acquires the displacement changes by processing the acceleration data from the accelerometer [14,15]. Mounting of accelerometers are typically time-consuming tasks associated with the test. Furthermore, the installation of sensors for some special buildings (such as super tall towers) can be difficult, and accelerometers cannot directly provide accurate structural displacement responses. The digital vertical meter method uses the reserved holes in the floors of super high-rise buildings as the installation channel of the digital vertical meter system to obtain the diurnal oscillation law of the tower body. However, the vertical line is easily affected by the construction factors, wind load, and tower crane operation during the dynamic monitoring period. The total station, accelerometer sensor, and digital vertical meter methods cannot efficiently realize automatic real-time continuous dynamic monitoring of super high-rise buildings. With the improvement of GNSS satellite timing accuracy and the development of the corresponding data processing software, GNSS technology can realize high-frequency data acquisition in the dynamic monitoring of super high-rise buildings. GNSS has all-weather and all-day characteristics, making it widely used in automatic real-time continuous dynamic monitoring of super high-rise buildings, such as La Costanera Tower [4], Canton Tower [7,16], and Tianjin 117 Tower [17]. However, tower crane operation will seriously block the GNSS signal, the steel structure on the roof can cause serious multipath errors, and the vibration and wind load during construction will exert huge noise influence on the GNSS signal for super high-rise buildings under construction. On this basis, a large noise component is found in the GNSS-based deformation signals, making it difficult to accurately extract the high-precision dynamic characteristics of super high-rise buildings. Although the aforementioned methods are accurate and reliable, they only evaluate construction safety by retrieving and analyzing the dynamic characteristics of several feature points on the structure. Thus, effective monitoring and analysis of the overall structural characteristics are difficult. These methods are limited when they are used to monitor high-risk structures because physical contact is required with the monitored structures.

Interferometric Synthetic Aperture Radar (InSAR) has been gradually used in building deformation monitoring because of its noncontact, high monitoring accuracy (centimeter-to-millimeter level), and high spatial resolution. Kui et al. [18] adopted InSAR to monitor the deformation of Bohai Building in Tianjin, China using TerraSAR-X images. The InSAR-derived deformation was verified by leveling at an accuracy at the millimeter level. Wu et al. [19] used a Persistent Scatterer InSAR (PSInSAR) to measure the deformation of urban high-rise buildings and analyzed the temporal–spatial characteristics of building deformation combined with a Google Earth 3D model. InSAR can overcome the drawbacks of conventional monitoring methods (i.e., invasiveness, high cost, and low spatial resolution) and can retrieve the overall structural characteristics. However, this technique cannot be used for dynamic deformation monitoring of structures with high accuracy [20,21] because of the limitations related to its satellite platform (e.g., low revisiting time, geometrical distortion, and atmospheric delay) [22,23]. Ground-based interferometric radar technique has been proposed [24–26] to overcome the shortcomings of satellite-based InSAR. This technique can achieve real-time acquisition rates of 200 Hz and can perform high-resolution structural monitoring that reaches less than 1 m [27–29]. Additionally, Pieraccini et al. [30] proposed a high-speed CW step-frequency coherent radar that is portable and can be rapidly installed and operated. Ground-based interferometric radar can simultaneously obtain the responses over numerous points, and the radar sensor can be installed by selecting the most favorable orientation for the monitored structure on the basis of monitoring requirement [31]. The significant superiorities of ground-based interferometric radar are related to high-accuracy measurements, limited influence of atmospheric delay on measurement performance, and noninvasive monitoring [32]. This technique has been extensively used in building safety monitoring and analysis since its proposal. Luzi et al. [33] monitored the vibration of buildings (e.g., the Collserola Tower of Barcelona and multistory high-rise residential buildings) using a ground-based real aperture radar (GB-RAR) interferometer. Their results showed that the deformations ranging from micrometers to centimeters can be derived using GB-RAR under good monitoring conditions. Atzeni et al. [34] adopted GB-RAR to acquire the modal shapes of the Leaning Tower of Pisa vibrations and identify its resonance frequency under the effect of natural excitation. Gentile [35] used a ground-based microwave interferometer to measure the dynamic response of the cables of two different cable-stayed bridges, and the results were validated by a piezoelectric accelerometer, demonstrating a good consistency.

Montuori et al. [36] combined GB-RAR, ground-based synthetic aperture radar (GB-SAR), and satellite-based InSAR. The proposed method enabled the estimation of the dynamic characteristics of buildings and deformation monitoring over the surrounding areas at different spatial and temporal scales. Castagnetti et al. [37] discussed the performance of GB-RAR on the structural monitoring of ancient masonry towers. This technique was tested on the Saint Prospero bell tower in Northern Italy, and the results showed a high consistency with an accelerometer-based acquisition system. Previous studies have focused on many different types of buildings, but few studies have concentrated on super high-rise buildings, especially those more than 400 m in height. Therefore, this study proposed a method based on GB-RAR for the monitoring and analysis of dynamic characteristics of super high-rise buildings.

In this study, we used the Wuhan Greenland Center (WGC) which is under construction with a 636 m designed height as the research object. First, GB-RAR was utilized to derive the dynamic deformation information of the WGC from the north–south and east–west directions. Wavelet analysis was used to eliminate the noise in dynamic deformation signals and the accurate dynamic characteristics of the building, such as horizontal displacement, oscillation amplitude, and displacement trajectory, were extracted. Finally, the correlation between the building displacement and temperature was discussed and analyzed. A set of technical methods for monitoring and analyzing the dynamic characteristics of super high-rise buildings using GB-RAR was studied and established.

2. Methodology

The radar monitoring data from the north–south and east–west directions of the WGC were processed using the time series InSAR analysis to derive the dynamic time series (e.g., displacement and oscillation amplitude) of the building. First, the reference time of data processing was selected. Subsequently, the radar data collected during the dynamic monitoring period were processed through differential interference based on the reference time and the deformation changes between the corresponding time intervals were calculated. Finally, the deformation time series of target points of the building was acquired based on the deformation changes obtained in the previous step. Other parameters of dynamic characteristics of the building were obtained. The main data processing and analysis steps are described as follows.

2.1. Reference Time Selection

The start time of radar data acquisition was selected as the reference time for data processing. Assuming that the deformation of the building at this time was zero, the deformation calculated based on the radar data at other times were all related to the reference time.

2.2. Window Processing

The radar data acquired by the IBIS-S system (i.e., the ground-based interferometric radar) were the frequency domain sampling data of the radar signal echo. The frequency domain data must be transformed into the spatial domain through discrete inverse Fourier transform, that is, focusing to extract the deformation information of each resolution range [38]. Since many steel frame structures are found around the monitored building. The echo signal of these resolution units is generally strong, and their side lobes can interfere with the signal of adjacent resolution units, which may distort their deformation signals. Therefore, the Hann window function was applied to process the radar signal before focusing the radar monitoring data in the range direction to eliminate the sidelobe effect.

2.3. Differential Interferometric Processing

GB-RAR is a zero-baseline observation compared with satellite-based SAR. Thus, flat-earth and topographic phases have no influence on the interferometric phase model. Therefore, the differential interferometric phase model obtained by comparing the phase information difference of the target points acquired at different times can be expressed as follows [39]:

$$\phi_{diff} = \frac{4\pi d_{defo}}{\lambda} + \phi_{atm} + \phi_{noise},\tag{1}$$

where ϕ_{diff} denotes the interferometric phase, ϕ_{atm} represents the phase generated by the atmospheric disturbance between the radar and the target, and ϕ_{noise} is the random noise phase. Given that high-frequency (20 Hz) dynamic monitoring was adopted in the monitoring of the WGC, the phase produced by the atmospheric effect is basically negligible after differential interferometric processing. After removing the noise phase, the interferometric phase was unwrapped in one dimension. The monitoring point of deformation d_{defo} in the line-of-sight (LOS) direction can be obtained using Equation (1). The deformation time series of the target points in the range direction can be derived based on the monitoring time span.

2.4. Gross Error Detection

Considering that the WGC is in the construction state during the monitoring period, the actual monitoring environment is complex, and the radar signal is simultaneously affected by various factors, such as construction vibration, wind load, temperature, and sunlight. Gross errors may be found in the deformation time series obtained based on the radar signal. Therefore, the gross error in the deformation time series of the monitoring target point should be detected. The deformation corresponding to this moment is removed when the gross error is detected in the deformation time series. The deformation at this moment is interpolated based on the deformation data before and after this moment. Gross error detection was performed on the deformation time series until no gross error was found in the time series.

2.5. Wavelet Analysis Denoising

After gross error detection, the deformation time series of the target point may still have a certain amount of noise. To reduce the influence of noise and improve the signal-to-noise ratio, a wavelet threshold denoising method was used to process the deformation time series of the target point. During the processing, the wavelet function "sym4" was selected, the deformation time series was decomposed into seven layers by wavelet multi-scale analysis, the threshold value was estimated by Heuresure criterion, and the threshold value was quantified by soft threshold method [40–42]. On this basis, denoising analysis of the deformation time series of the target point was carried out.

3. Experiment Description

3.1. Monitored Object

The monitored object is the WGC in Wuhan, China, which is a super high-rise main building on plot A01 of the Wuhan Greenland International Financial City (Figure 1a). The building is located in the core area of the Wuchang Bin Jiang Business District, which is approximately 250 m away from the Yangtze River flood dyke. The main building of the WGC has five basement floors, with a building area of 70,171 m^2. The building has 125 floors above ground, with a building height of 636 m and building area of 302,399 m^2, and the average building area of each floor is 2419 m^2. The total steel consumption of the building is approximately 80,000 tons.

Figure 1. Wuhan Greenland Center (WGC): (**a**) architectural rendering; (**b**) architectural structure drawing.

The structural system of the WGC is a "steel frame concrete core tube" structural system, with a concrete core tube inside and steel frames on the outside and on top (Figure 1b). The main tower has four sets of wind grooves from top to bottom. Wind can pass through the hall to reduce the harm of strong wind to the building. The steel structure of the main tower is composed of 12 mega stiffening columns in the outer frame, 18 outer frame gravity columns, reinforced concrete core tubes with shear wall steel ribs, floor steel beams, 10 ring trusses, 4 outrigger trusses, and a 60 m high tower crown and canopy. The partial floor adopts profiled steel plate and cast-in-place concrete composite floor, and the pressed steel plate adopts galvanized steel plate with color coating. The total height of the concrete core tube is 552 m, the core tube is adducted twice at the 72nd and 88th floors, and the floor area is reduced accordingly. During the construction of the WGC, the construction progress of the external steel structure lags behind that of the concrete core tube. The concrete core tube is constructed in the top formwork, which adopts a steel platform structure climbing system to facilitate the construction and measurement of the workers.

3.2. Monitoring Scheme Design

The WGC under construction underwent dynamic deformation because of solar radiation, construction vibration, wind load, temperature, and other factors. Traditional monitoring methods (e.g., GNSS, total station, and sensors) need to be in contact with the building, and the monitoring points are difficult to arrange when monitoring the dynamic characteristics of the building. The large tower crane on top of the building during construction caused serious effects (such as signal blocking and multipath effect) on GNSS measurement, making it difficult to effectively obtain the dynamic characteristics of the building using GNSS. Satellite-based SAR cannot obtain the high-accuracy horizontal displacement dynamic information of the building because of the influences of the revisit period, imaging angle, and monitoring distance. GB-RAR can set reasonable monitoring base points based on the monitoring requirements, adjust the most favorable radar imaging angle to achieve the noncontact high-accuracy continuous deformation monitoring of the building, and can simultaneously monitor multiple target points in the radar field of view. Therefore, the IBIS-S system was adopted to conduct high-accuracy continuous monitoring of the WGC under construction, and the dynamic characteristics of the building were extracted and analyzed based on the acquired radar signal data.

In this study, two IBIS-S systems were simultaneously used to conduct high-accuracy continuous deformation monitoring of the WGC. Working base points S1 and S2 were established in the north–south and east–west directions of the building, and IBIS-S systems were installed in S1 and S2, as shown in Figure 2. The IBIS-S systems were used to conduct radar scanning of the WGC along the north–south and east–west directions for acquiring high spatial–temporal resolution radar images of the building and continuously obtain multiscene radar images. The displacement changes of the building along the monitoring direction can be obtained through interferometry, and the precise oscillation amplitude of the building can be calculated based on the GB-RAR-derived displacement. Radar monitoring adopted the high-frequency data acquisition mode (sampling frequency was 20 Hz). The horizontal displacement variation of the building and the displacement trajectory of the roof were computed through straightforward geometric projections based on the radar incidence angle, because the IBIS-S systems can only obtain the deformation of the radar LOS direction.

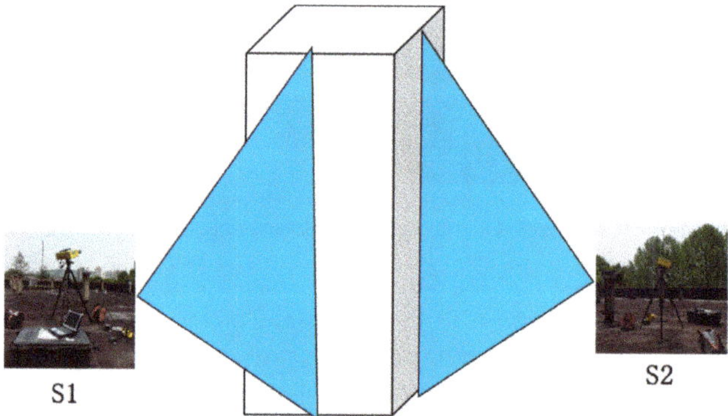

Figure 2. Schematic of ground-based real aperture radar (GB-RAR) working base points.

3.3. Measurement Campaign

We used GPS positioning to determine the field positions of working base points S1 and S2 in the two monitoring directions on the basis of field investigation, the intervisibility condition between radar and the monitored object, and the main monitoring building of the WGC within the main lobe of radar scanning. Working base points S1 and S2 were located in the north and west directions of the building, respectively, as shown in Figure 3.

We installed the IBIS-S-1 system on working base point S1, and the distance between the radar installation location and the WGC was 246 m. S1 was located at the roof of a two-story building of a farmer, as shown in Figure 4. The IBIS-S-2 system was rigidly fixed on working base point S2, which was located in the Wuchang River Beach Park with a distance of 241 m from the monitored building, as shown in Figure 5.

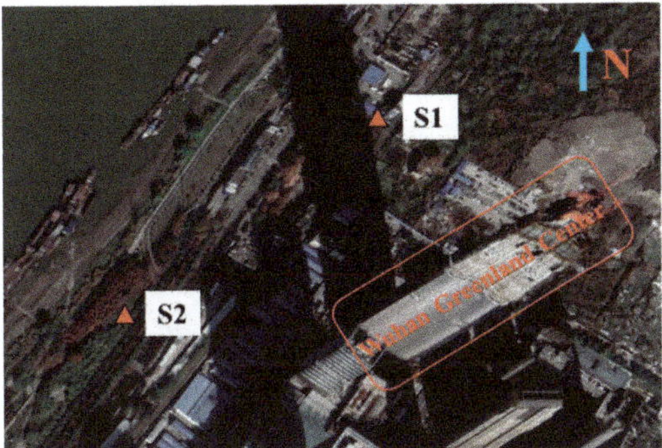

Figure 3. Distribution map of the two IBIS-S system installation locations (S1 and S2). The red triangle denotes the location of the IBIS-S system, and the monitored object (i.e., the WGC) is outlined by a red rectangle.

Figure 4. Location of the IBIS-S-1 system in the north–south direction. (**a**) View of the IBIS-S-1 interferometric radar on S1 (located in a farmhouse roof). (**b**) Weather station installed on the roof of the monitored building. (**c**) Ground view of the WGC from the location of the IBIS-S-1 interferometric radar.

Continuous dynamic monitoring of the WGC was conducted from 12:21 on 5 July 2017 to 12:14 on 7 July 2017, with a total duration of approximately 48 h. The times of starting and ending of the monitoring of the two IBIS-S systems are the same. Continuous and uninterrupted measurement was adopted for both IBIS-S systems. The two IBIS-S systems were configured to monitor the building up to a distance of 800 m with a sampling frequency of 20 Hz. The two systems used the same resolution (0.75 m) in range during the monitoring period. The configuration parameters of the two IBIS-S systems are the same, and the main configuration parameters of the measurement campaign are listed in Table 1. A weather station was fixed on the roof of the building (as shown in Figure 4b) to monitor the dynamic changes of meteorological parameters (such as temperature, atmospheric humidity, etc.) on the roof during the dynamic monitoring. The data collection time of the weather station ranged from 12:40 on 5 July 2017 to 09:40 on 7 July 2017, with a data sampling interval of 10 min.

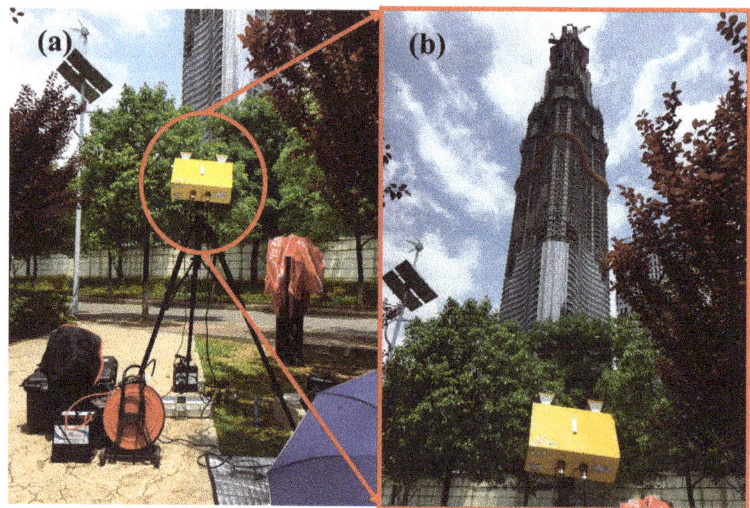

Figure 5. Location of the IBIS-S-2 system in the east–west direction. (**a**) View of the IBIS-S-2 interferometric radar on S2 (located in the Wuchang River Beach Park). (**b**) Ground view of the WGC from the location of the IBIS-S-2 interferometric radar.

Table 1. Configuration of IBIS-S systems for the measurement campaign.

Parameter	Value
Vertical tilt	50°
Antenna type	Type3 (Azimuth 17°, Vertical 15°)
Maximum range	800 m
Resolution in range	0.75 m
Sampling	20 Hz
Start time	12:21 5 July 2017
End time	12:14 7 July 2017

4. Results and Discussion

4.1. Horizontal Displacement and Amplitude Extraction Analysis

The power profiles obtained from working base points S1 and S2 are shown in Figures 6 and 7, respectively. Many peaks are visible in Figures 6 and 7. These peaks correspond to the positions of the monitored points characterized with good electromagnetic reflectivity. To analyze the displacement of the WGC, we selected the monitored points with good electromagnetic reflectivity located on the roof of building (i.e., the location of point A) and the middle position of the main structure of the building (i.e., the location of point B) to analyze its horizontal displacement during the monitoring period. Point A in the two figures is located on the roof of the WGC. The height of the roof relative to working base point S2 was 447 m, whereas Point B was 300 m high from the building relative to working base point S2 when we used GB-RAR to monitor the building. The heights of points A and B in the analysis are relative to working base point S2. As shown in Figures 6 and 7, the thermal signal-to-noise ratio (SNR) of the radar signal is basically greater than 30 dB within the range of 100 m to 447 m relative to the height of the building. The thermal SNRs of points A and B in the north–south monitoring direction are 37.4 and 60.8 dB, respectively, and the thermal SNRs of points A and B in the east–west monitoring direction are 39.0 and 62.2 dB, respectively. The above analysis demonstrates that the data collected by the radars at S1 and S2 are of good quality.

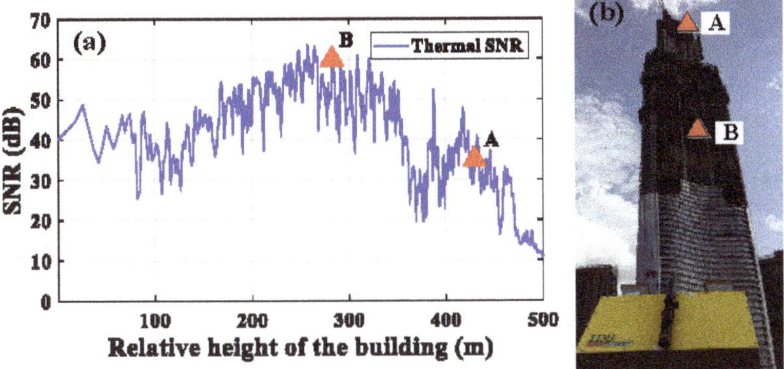

Figure 6. (a) Power profile of the WGC obtained from working base point S1 and selection of monitoring points. The red triangle denotes the monitoring point. The relative height of the building in the left figure is relative to working base point S1. Point A in the right figure is at the top of the building, with a height of 447 m relative to working base point S2. Point B is 300 m high relative to working base point S2. (b) Ground view of the WGC from the location of the IBIS-S-1 system.

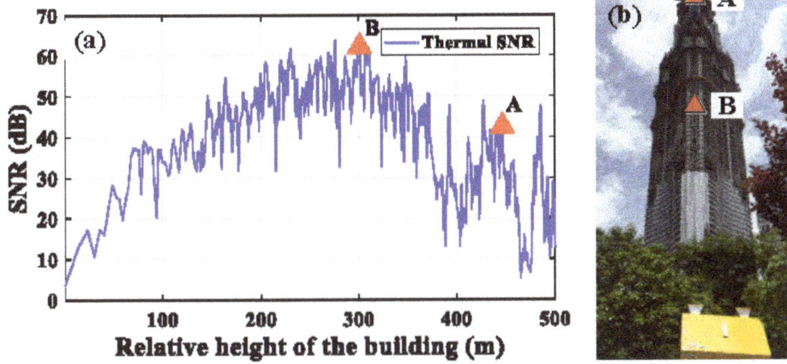

Figure 7. (a) Power profile of the WGC obtained from working base point S1 and selection of monitoring points. The red triangle denotes the monitoring point. The relative height of the building in the left figure is relative to working base point S1. Point A in the right figure is at the top of the building, with a height of 447 m relative to working base point S2. Point B is 300 m high relative to working base point S2. (b) Ground view of the WGC from the location of the IBIS-S-2 system.

We selected and analyzed some feature points of the WGC to analyze its horizontal displacement during monitoring. The feature points were distributed at points A and B of the building. The LOS displacement time series of feature points was extracted using the method described in Section 2, and the horizontal displacement time series of feature points was calculated through geometric projection based on the geometric relationship between the feature points and the radar position. Figure 8 shows the horizontal displacement time series in the north–south and east–west directions at the top of the WGC during the monitoring period, and the red and black curves represent the original and denoised horizontal displacement time series, respectively. The negative values in Figure 8a indicate that the building is moving westward, whereas the positive values indicate that the building is moving eastward. As shown in Figure 8b, the negative values indicate that the building is moving northward, whereas the positive value denotes that the building is moving southward. As shown in Figure 8b, the maximum negative horizontal displacement of the roof in the north–south direction is 7.99 mm, and the maximum positive horizontal displacement is 10.85 mm. Therefore, the maximum displacement amplitude of the roof of the WGC in the north–south direction during the monitoring period is 18.84 mm. For the east–west direction, the maximum negative and positive horizontal displacements of the roof are 7.10 and 8.84 mm, respectively, and the maximum east–west direction displacement amplitude of the roof is 15.94 mm. The displacement monitoring accuracies in the north–south and east–west directions at the top of the building are 0.15 and 0.17 mm, respectively.

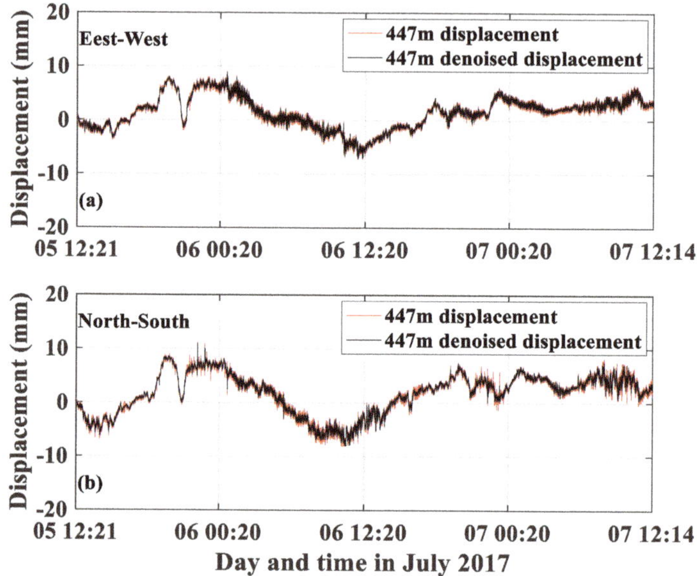

Figure 8. IBIS-S-derived horizontal displacement time series in east–west (**a**) and north–south (**b**) directions at 447 m height of the WGC. The red line denotes the displacement time series without denoising, and the black line represents the displacement time series after denoising.

Figure 9 shows the horizontal displacement time series in the north–south and east–west directions at the building height of 300 m during the monitoring period. The maximum negative and positive horizontal displacements in the north–south direction at the building height of 300 m are 6.99 and 7.74 mm, respectively, and the maximum horizontal displacement amplitude is 14.73 mm. For the east–west direction, the maximum negative and positive horizontal displacements are 4.94 and 7.30 mm respectively, and the maximum horizontal displacement amplitude reaches 12.24 mm. At the building height of 300 m, the monitoring accuracies of displacement in the north–south and east–west directions are 0.09 and 0.10 mm, respectively. The maximum horizontal displacement amplitudes at the building height of 300 m in the north–south and east–west directions are 4.11 and 3.70 mm less than that compared with the roof of the building, respectively.

The comparative analysis of Figures 6 and 7 shows that the displacement change of the building is relatively large before 7 July 2017, whereas the subsequent displacement steadily changes. The original displacement curve obtained based on radar data immensely fluctuated and was accompanied by many burr points. After denoising the original displacement curve through wavelet analysis [41,42], the burr phenomenon in the original displacement curve was basically eliminated, and the displacement curve became smooth, as shown in Figures 8 and 9 (black line). However, the displacement curve after denoising still exhibited a certain fluctuation, and the displacement curve at the building height of 447 m was more obvious than that at the building height of 300 m. This phenomenon may be attributed to the serious effect of noise caused by construction vibration, instrument system error, external environment, and other factors on the radar signal during the monitoring period. Although most of the noises were removed through wavelet analysis, a slight vibration still existed in the building itself when it swung because of the influences of construction vibration and other factors. Therefore, the displacement curve after wavelet denoising still had a small fluctuation, and this part of fluctuation may be caused by the slight vibration of the building itself when it swung.

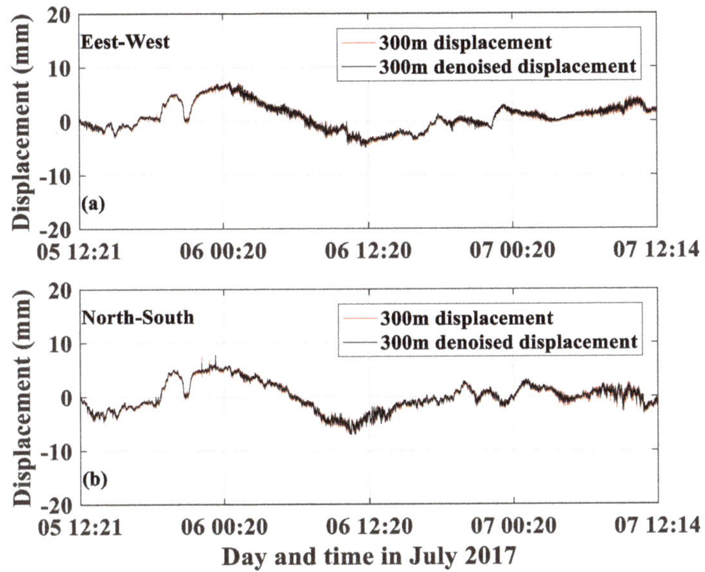

Figure 9. IBIS-S-derived horizontal displacement time series in east–west **(a)** and north–south **(b)** directions at 300 m height of the WGC. The red line indicates the displacement time series without denoising, and the black line denotes the displacement time series after denoising.

GB-RAR can realize high-accuracy continuous dynamic monitoring of super high-rise buildings, and the monitoring accuracy can reach submillimeter level. However, the monitoring accuracy gradually decreased with the increase in floors.

4.2. Roof Displacement Trajectory Analysis

To investigate the roof motion of the WGC, the displacement trajectory of the roof during the monitoring period is presented in this section on the basis of the horizontal displacement time series of the roof in the north–south and east–west directions obtained in the previous section, and the results are shown in Figure 10. As shown in Figure 10, the color bar on the right represents the time change, the negative value on the vertical axis denotes the westward movement, and the negative value on the horizontal axis indicates the northward movement. As illustrated in Figure 10, during the monitoring period, the general direction of roof movement was first toward the northwest for approximately 3 h, then moved toward the southeast to the maximum value, and subsequently moved along the northwest to the maximum value, and finally moved along the southeast to approximate the initial position (approximately a few millimeters deviation with respect to the initial position). The oscillation magnitude of the roof is basically less than 20 mm. To analyze the oscillation rule of the roof for a whole day, Figure 11 shows the displacement trajectory of the roof from 00:00 to 24:00 on 6 July 2017. Figure 11 intuitively reflects the diurnal oscillation rule of the roof, and the displacement trajectory is accompanied by slight fluctuation when it moves with time, which may be caused by the combined influences of construction vibration, wind load, and other factors.

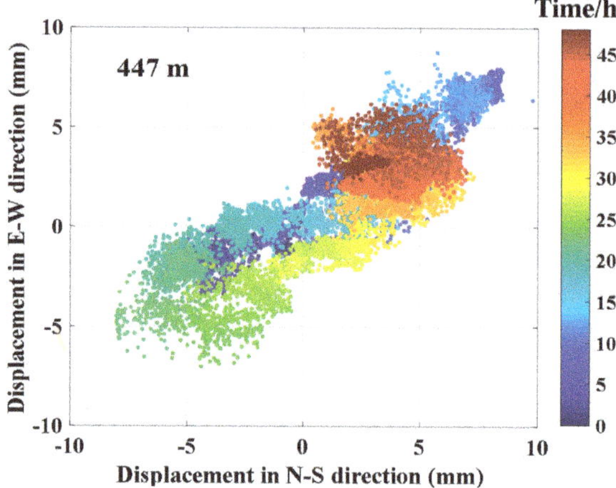

Figure 10. Displacement trajectory of the roof (from 5 July 2017 to 7 July 2017).

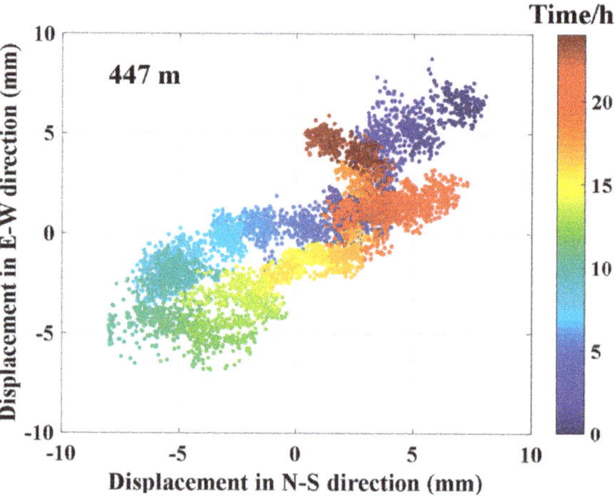

Figure 11. Displacement trajectory of the roof (from 00:00 to 24:00 on 6 July 2017).

4.3. Correlation between Roof Displacement and Temperature Changes

The roof displacement time series in the north–south and east–west directions and the roof temperature changes observed by the weather station were selected to compare and analyze their time series changes for determining the correlation between the displacement and temperature changes at the roof of the WGC. Given that the temperature data were collected from 12:40 on 5 July 2017 to 09:40 on 7 July 2017, the overlapping period data of the displacement time series derived by GB-RAR and temperature data were selected for comparative analysis. Figure 12 shows the comparison results between the temperature and displacement changes in the north–south and east–west directions during the study period.

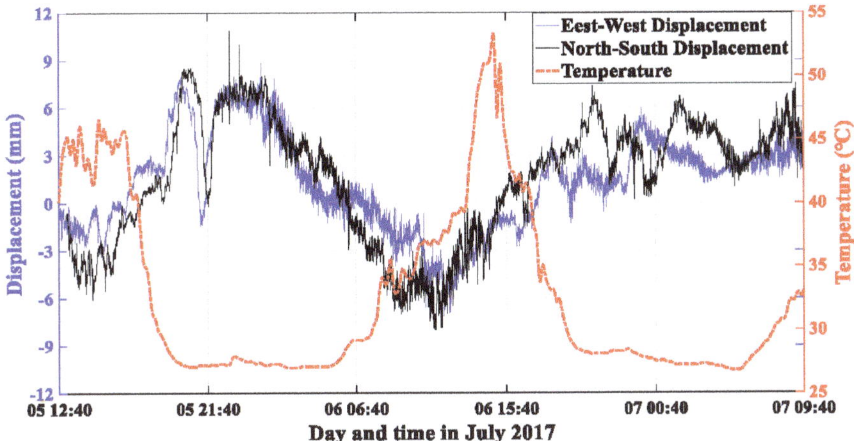

Figure 12. Displacement time series in east–west (**blue line**) and north–south (**black line**) directions versus temperature changes (**red dotted line**) at the roof of the WGC.

As illustrated in Figure 12, the temperature ranged from 26.8 °C to 53.4 °C during the study period, and the temperature difference reached 26.6 °C. The temperature was relatively stable from 21:00 to 6:00 at night and reached the maximum value of the day around 15:00 every afternoon. The maximum displacement amplitudes of the roof monitored using GB-RAR were 18.84 and 15.94 mm in the north–south and east–west directions, respectively. We conducted correlation analysis to analyze the time series between displacement and temperature changes for determining their correlation. The correlation coefficient between the displacement changes in the north–south direction and the temperature changes is −0.56, and that between the displacement changes in the east–west direction and the temperature changes is −0.65. A certain negative correlation is found between the temperature and displacement changes in the two directions, and the correlation between the displacement changes in the east–west direction and the temperature changes is strong. Additionally, the correlation coefficient between the displacement changes in the north–south and east–west directions is 0.84, indicating the displacement changes between the north–south and east–west directions show a relatively strong correlation.

Although the temperature changes have a certain influence on the displacement changes at the top of the super high-rise building, the construction vibration of the large tower crane on the top of the building, wind load, and other factors also have a great influence on the displacement changes at the top of the building because the WGC is under the construction stage. The displacement changes at the top of the building are caused by the joint action of temperature, construction vibration, wind load, and other factors. Therefore, a certain correlation is found between the temperature and displacement changes.

5. Conclusions

In this study, we utilized the GB-RAR to monitor and analyze the dynamic characteristics of the WGC under construction. We proposed and investigated a set of technical methods for monitoring and analyzing the dynamic characteristics of super high-rise buildings using GB-RAR. The GB-RAR was used to monitor the dynamic displacement information of the building in the north–south and east–west directions. The accurate dynamic characteristic information of the building, such as horizontal displacement, oscillation amplitude, and displacement trajectory, was extracted through wavelet analysis. The main conclusions are summarized as follows:

(1) During the monitoring of the WGC, the GB-RAR effectively extracted the dynamic characteristics (e.g., horizontal displacement and oscillation amplitude) of the building. The maximum horizontal displacement amplitudes at the top of the building in the north–south and east–west directions were 18.84 and 15.94 mm respectively, and the corresponding accuracies were 0.15 and 0.17 mm, respectively, suggesting that the accuracy of GB-RAR monitoring was high. However, the monitoring accuracy gradually decreased with the increase in floors. The roof displacement trajectory of the WGC was clearly identified. The displacement trajectory was accompanied by slight fluctuation, which may be caused by the combined influences of construction vibration, wind load, and other factors. The results demonstrate that the GB-RAR is effective for dynamic monitoring of super high-rise buildings.

(2) A certain negative correlation was identified between the temperature and displacement changes in the north–south and east–west directions at the roof of the building, and the correlation between the displacement changes in the east–west direction and the temperature changes was strong.

Author Contributions: L.Z. and J.G. conceived and designed the experiments. L.Z., C.W., J.M. and F.Y. carried out the data acquisition. L.Z. and X.W. performed data processing and analyses, and L.Z. contributed to the manuscript of the paper. J.G. and D.Z. discussed and analyzed the experimental results. All authors have read and approved the published version of the manuscript.

Funding: This work was supported by the Guangxi Science and Technology Plan Project (Grant No. GUIKE AD19110107); the Key Laboratory of Geospace Environment and Geodesy, Ministry of Education, Wuhan University (Grant No. 18-01-01); the Natural Science Foundation of Guangxi (Grant No. 2018GXNSFBA050006); the Wuhan Science and Technology Plan Project (Grant No. 2019010702011314); the National Natural Science Foundation of China (Grant No. 41604019); and the Foundation of Guilin University of Technology (Grant No. GUTQDJJ2018036).

Acknowledgments: We would like to thank the administrative staff of the Wuchang River Beach Park and the owner of the two-story building for providing the IBIS-S system installation sites and power supply.

Conflicts of Interest: The authors declare no conflict of interest.

References

1. Chen, Q.J.; Yuan, W.Z.; Li, Y.C.; Cao, L.Y. Dynamic response characteristics of super high-rise buildings subjected to long-period ground motions. *J. Cent. S. Univ.* **2013**, *20*, 1341–1353. [CrossRef]
2. Adachi, F.; Fujita, K.; Tsuji, M.; Takewaki, I. Importance of interstory velocity on optimal along-height allocation of viscous oil dampers in super high-rise buildings. *Eng. Struct.* **2013**, *56*, 489–500. [CrossRef]
3. Li, Q.S.; He, Y.C.; He, Y.H.; Zhou, K.; Han, X.L. Monitoring wind effects of a landfall typhoon on a 600 m high skyscraper. *Struct. Infrastruct. Eng.* **2019**, *15*, 54–71. [CrossRef]
4. Silva, I.D.; Ibañez, W.; Poleszuk, G. Experience of Using Total Station and GNSS Technologies for Tall Building Construction Monitoring. In Proceedings of the International Congress and Exhibition "Sustainable Civil Infrastructures: Innovative Infrastructure Geotechnology, Sharm El Sheikh, Egypt, 15–19 July 2017.
5. Xu, J.; Jo, H. Development of High-Sensitivity and Low-Cost Electroluminescent Strain Sensor for Structural Health Monitoring. *IEEE Sens. J.* **2016**, *16*, 1962–1968. [CrossRef]
6. Li, Q.S.; Xiao, Y.Q.; Wong, C.K. Full-scale monitoring of typhoon effects on super tall buildings. *J. Fluids Struct.* **2005**, *20*, 697–717. [CrossRef]
7. Xia, Y.; Zhang, P.; Ni, Y.Q.; Zhu, H.P. Deformation monitoring of a super-tall structure using real-time strain data. *Eng. Struct.* **2014**, *67*, 29–38. [CrossRef]
8. Wang, T.Y.; Xu, Y.M. Research on Dynamic Characteristic Monitoring Methods for Super High-rise Building. *Bull. Surv. Map.* **2017**, *4*, 89–92.
9. Zhang, X.; Zhang, Y.Y.; Li, B.F.; Qiu, G.X. GNSS-Based Verticality Monitoring of Super-Tall Buildings. *Appl. Sci.* **2018**, *8*, 991. [CrossRef]
10. Li, Q.S.; Zhi, L.H.; Yi, J.; To, A.; Xie, J.M. Monitoring of typhoon effects on a super-tall building in Hong Kong. *Struct. Control Health* **2014**, *21*, 926–949. [CrossRef]
11. Li, Q.S.; Yi, J. Monitoring of dynamic behaviour of super-tall buildings during typhoons. *Struct. Infrastruct. Eng.* **2016**, *12*, 289–311. [CrossRef]

12. Hwang, J.; Yun, H.; Park, S.K.; Lee, D.; Hong, S. Optimal Methods of RTK-GPS/Accelerometer Integration to Monitor the Displacement of Structures. *Sensors* **2012**, *12*, 1014–1034. [CrossRef] [PubMed]

13. Casciati, F.; Fuggini, C. Monitoring a steel building using GPS sensors. *Smart Struct. Syst.* **2011**, *7*, 349–363. [CrossRef]

14. Gentile, C.; Bernardini, G. An interferometric radar for non-contact measurement of deflections on civil engineering structures: Laboratory and full-scale tests. *Struct. Infrastruct. Eng.* **2010**, *6*, 521–534. [CrossRef]

15. Santana, J.; Hoven, R.V.D.; Liempd, C.V.; Colin, M.; Saillen, N.; Zonta, D.; Trapani, D.; Torfs, T.; Hoof, C.V. A 3-axis accelerometer and strain sensor system for building integrity monitoring. *Sens. Actuators A Phys.* **2012**, *188*, 141–147. [CrossRef]

16. Guo, Y.L.; Kareem, A.; Ni, Y.Q.; Liao, W.Y. Performance evaluation of Canton Tower under winds based on full-scale data. *J. Wind Eng. Ind. Aerodyn.* **2012**, *104*, 116–128. [CrossRef]

17. Xiong, C.; Niu, Y. Investigation of the Dynamic Behavior of a Super High-rise Structure using RTK-GNSS Technique. *KSCE J. Civ. Eng.* **2019**, *23*, 654–665. [CrossRef]

18. Yang, K.; Yan, L.; Huang, G.; Chen, C.; Wu, Z.P. Monitoring Building Deformation with InSAR: Experiments and Validation. *Sensors* **2016**, *16*, 2182. [CrossRef]

19. Wu, W.Q.; Cui, H.T.; Hu, J.; Yao, L.N. Detection and 3D Visualization of Deformations for High-Rise Buildings in Shenzhen, China from High-Resolution TerraSAR-X Datase. *Appl. Sci.* **2019**, *9*, 3818. [CrossRef]

20. Bardi, F.; Raspini, F.; Ciampalini, A.; Kristensen, L.; Rouyet, L.; Lauknes, T.R.; Frauenfelder, R.; Casagli, N. Space Borne and Ground Based InSAR Data Integration: The Åknes Test Site. *Remote Sens.* **2016**, *8*, 237. [CrossRef]

21. Bardi, F.; Frodella, W.; Ciampalini, A.; Bianchini, S.; Ventisette, C.D. Integration between ground based and satellite SAR data in landslide mapping: The San Fratello case study. *Geomorphology* **2014**, *223*, 45–60. [CrossRef]

22. Guo, J.M.; Zhou, L.; Yao, C.L.; Hu, J.Y. Surface Subsidence Analysis by Multi-Temporal InSAR and GRACE: A Case Study in Beijing. *Sensors* **2016**, *16*, 1495. [CrossRef] [PubMed]

23. Ji, L.; Zhang, Y.; Wang, Q.; Wan, Y.; Li, J. Detecting land uplift associated with enhanced oil recovery using InSAR in the Karamay oil field, Xinjiang, China. *Int. J. Remote Sens.* **2016**, *37*, 1527–1540. [CrossRef]

24. Farrar, C.R.; Darling, T.W.; Migliori, A.; Baker, W.E. Microwave interferometer for non-contact vibration measurements on large structures. *Mech. Syst. Signal Process.* **1999**, *13*, 241–253. [CrossRef]

25. Tapete, D.; Casagli, N.; Luzi, G.; Fanti, R.; Gigli, G.; Leva, D. Integrating radar and laser-based remote sensing techniques for monitoring structural deformation of archaeological monuments. *J. Archaeol. Sci.* **2013**, *40*, 176–189. [CrossRef]

26. Monserrat, O.; Crosetto, M.; Luzi, G. A review of ground-based SAR interferometry for deformation measurement. *ISPRS J. Photogramm.* **2014**, *93*, 40–48. [CrossRef]

27. Marchisio, M.; Piroddi, L.; Ranieri, G.; Sergio, V.C.; Paolo, F. Comparison of natural and artificial forcing to study the dynamic behaviour of bell towers in low wind context by means of ground-based radar interferometry: The case of the Leaning Tower in Pisa. *J. Geophys. Eng.* **2014**, *11*, 5004–5013. [CrossRef]

28. Roy, A.K.; Gangal, S.A.; Bhattacharya, C. Calibration of High Resolution Synthetic Aperture Radar SAR Image Features by Ground Based Experiments. *IETE J. Res.* **2017**, *63*, 381–391. [CrossRef]

29. Pieraccini, M.; Miccinesi, L.; Rojhani, N. A GBSAR Operating in Monostatic and Bistatic Modalities for Retrieving the Displacement Vector. *IEEE Geosci. Remote Sens.* **2017**, *14*, 1494–1498. [CrossRef]

30. Pieraccini, M.; Fratini, M.; Parrini, F.; Macaluso, G.; Atzeni, C. High speed CW step frequency coherent radar for dynamic monitoring of civil engineering structures. *Electron. Lett.* **2004**, *14*, 907–908. [CrossRef]

31. Zhou, L.; Guo, J.M.; Hu, J.Y.; Ma, J.; Wei, F.H.; Xue, X.Y. Subsidence analysis of ELH Bridge through ground-based interferometric radar during the crossing of a subway shield tunnel underneath the bridge. *Int. J. Remote Sens.* **2018**, *39*, 1911–1928. [CrossRef]

32. Stabile, T.A.; Perrone, A.; Gallipoli, M.R.; Ditommaso, R.; Ponzo, F.C. Dynamic Survey of the Musmeci Bridge by Joint Application of Ground-Based Microwave Radar Interferometry and Ambient Noise Standard Spectral Ratio Techniques. *IEEE Geosci. Remote Sens.* **2013**, *10*, 870–874. [CrossRef]

33. Luzi, G.; Monserrat, O.; Crosetto, M. The Potential of Coherent Radar to Support the Monitoring of the Health State of Buildings. *Res. Nondestruct. Eval.* **2012**, *23*, 125–145. [CrossRef]

34. Atzeni, C.; Bicci, A.; Dei, D.; Fratini, M.; Pieraccini, M. Remote Survey of the Leaning Tower of Pisa by Interferometric Sensing. *IEEE Geosci. Remote Sens.* **2010**, *7*, 185–189. [CrossRef]

35. Gentile, C. Application of microwave remote sensing to dynamic testing of stay-cables. *Remote Sens.* **2010**, *2*, 36–51. [CrossRef]

36. Montuori, A.; Luzi, G.; Bignami, C.; Gaudiosi, I.; Stramondo, S.; Crosetto, M.; Buongiorno, M.F. The Interferometric Use of Radar Sensors for the Urban Monitoring of Structural Vibrations and Surface Displacements. *IEEE J. Sel. Top. Appl. Earth Obs. Remote Sens.* **2016**, *9*, 3761–3776. [CrossRef]

37. Castagnetti, C.; Bassoli, E.; Vincenzi, L.; Mancini, F. Dynamic Assessment of Masonry Towers Based on Terrestrial Radar Interferometer and Accelerometers. *Sensors* **2019**, *19*, 1319. [CrossRef]

38. Coppi, F.; Gentile, C.; Ricci, P. A Software Tool for Processing the Displacement Time Series Extracted from Raw Radar Data. *AIP Conf. Proc.* **2010**, *1*, 190–201.

39. Iglesias, R.; Fabregas, X.; Aguasca, A.; Mallorqui, J.J.; Carlos, L.M.; Gili, J.A.; Corominas, J. Atmospheric Phase Screen Compensation in Ground-Based SAR with a Multiple-Regression Model over Mountainous Regions. *IEEE Geosci. Remote Sens.* **2014**, *52*, 2436–2449. [CrossRef]

40. Shimizu, N.; Mizuta, Y.; Kondo, H. A new GPS real-time monitoring system for deformation measurements and its application. In Proceedings of the 8th FIG International Symposium on Deformation Measurements, Hong Kong, 25–28 June 1996.

41. Li, J.; Chen, C.; Zheng, C.; Liu, S.; Tang, Z. Extraction of physical parameters from photo acoustic spectroscopy using wavelet transform. *J. Appl. Phys.* **2011**, *6*, 063110–063116. [CrossRef]

42. Sang, Y.F. A review on the applications of wavelet transform in hydrology time series analysis. *Atmos. Res.* **2013**, *122*, 8–15. [CrossRef]

Review

State-of-the-Art Review on Determining Prestress Losses in Prestressed Concrete Girders

Marco Bonopera [1,*], Kuo-Chun Chang [2] and Zheng-Kuan Lee [1]

[1]   Bridge Engineering Division, National Center for Research on Earthquake Engineering (NCREE), Taipei 10668, Taiwan; zklee@ncree.narl.org.tw

[2]   Department of Civil Engineering, National Taiwan University, Taipei 10617, Taiwan; ciekuo@ntu.edu.tw

*   Correspondence: marco.bonopera@unife.it

Received: 25 August 2020; Accepted: 14 October 2020; Published: 16 October 2020

Abstract: Prestressing methods were used to realize long-span bridges in the last few decades. For their predictive maintenance, devices and dynamic nondestructive procedures for identifying prestress losses were mainly developed since serviceability and safety of Prestressed Concrete (PC) girders depend on the effective state of prestressing. In fact, substantial long term prestress losses can induce excessive deflections and cracking in large span PC bridge girders. However, old unsolved problematics as well as new challenges exist since a variation in prestress force does not significantly affect the vibration responses of such PC girders. As a result, this makes uncertain the use of natural frequencies as appropriate parameters for prestress loss determinations. Thus, amongst emerging techniques, static identification based on vertical deflections has preliminary proved to be a reliable method with the goal to become a dominant approach in the near future. In fact, measured vertical deflections take accurately and instantaneously into account the changes of structural geometry of PC girders due to prestressing losses on the equilibrium conditions, in turn caused by the combined effects of tendon relaxation, concrete creep and shrinkage, and parameters of real environment as, e.g., temperature and relative humidity. Given the current state of quantitative and principled methodologies, this paper represents a state-of-the-art review of some important research works on determining prestress losses conducted worldwide. The attention is principally focused on a static nondestructive method, and a comparison with dynamic ones is elaborated. Comments and recommendations are made at proper places, while concluding remarks including future studies and field developments are mentioned at the end of the paper.

Keywords: bending test; bridge; *"compression-softening"* theory; frequency; inverse problem; nondestructive testing (NDT) method; prestressed concrete (PC) girder; prestress force determination; prestress loss; vertical deflection measurement

---

1. Introduction

The first applications of prestressing methods to concrete structures go back to the first half of the 20th century. Today prestressing is widely used for many applications, ranging from small members, such as railway sleepers, to more important structures such as bridges, long and light precast flooring, and roofing elements for constructions. Serviceability and safety of Prestressed Concrete (PC) structures rely on the effective state of prestress force [1]. In fact, prestressing methods are principally utilized to reduce deflections and to partially counterbalance the effect of dead and live loads in the case of bridges [2]. As a result, an extreme loss of prestressing may cause excessive deflections or jeopardize the performance of large span PC girders by indicating cracking phenomena [3]. For this reason, devices and dynamic approaches capable of determining prestress losses largely developed. In particular, dynamic Structural Health Monitoring (SHM) techniques also generated for damage identification purposes based on the vibration responses of the span girders, so preventing maintenance, repair, or replacement of a bridge [4–6]. Damage detection techniques as well developed using different equipment and methods as, e.g., parked vehicles inducing frequency variation [7], long-gauge Fiber Bragg Grating (FBG) [8], or hybrid vibration testing data [9]. Therefore, the operating state of bridges can be controlled. However, prestress losses can be directly, simply, and accurately determined over time if the internal tendons of PC girders are instrumented by load cells, vibrating wire strain gauges, or elasto-magnetic sensors during construction [10–13]. Besides, FBG sensors can be embedded in seven-wire strands along PC girders for long term monitoring of tensile forces [14,15]. Although instrumentation of external tendons is easy during their serviceability, NonDestructive Testing (NDT) methods are required. Nevertheless, as far as the influence of prestressing on the dynamics of PC girders is concerned, the discussion is still ongoing.

Several procedures and equations are available in design codes for predicting prestress losses. According to ACI 318-2019 [16] and PCI DH [17], reasonable estimations can be calculated. For unusual design conditions and special structures, a more detailed procedure established by PCI CPL [18] can be considered. AASHTO LRFD [19] adopted new procedures since the previous prestress loss methods led to unrealistic applications with high strength concrete. However, AASHTO Standard specifications [20] remain in accordance with AASHTO LRFD [21]. PCI BDM [22] includes both AASHTO Standard [20] and LRFD [21] methods.

A series of studies were conducted to measure prestress losses in PC girders, and to compare them versus design code estimations. Among these works, there are laboratory tests of old PC girders removed from existing bridges, and experiments including fabrication, testing, and field monitoring of PC members under service. Table 1 in [23] and Table 6 in [24] summarize an extensive literature review on references, PC member identification (type, old time), testing place, experimental technique used, time of study, and measured losses. As observed in these tables, measured prestress losses exceeded those predicted by design codes in some cases. On the other hand, measured prestress losses which were in line with the values expected by codes were obtained in PC girders, which exceeded the allowable compressive stress limit [25]. To overcome this problem, Caro et al. [26] used the ECADA+ method [27] to measure the effective prestress forces in a number of PC specimens for over 1 year and, consequently, compared the results with prestress losses estimated by several codes. Although design code-based predictions can be considered as quite satisfactory, they are very conservative [28]. Accordingly, there are difficulties in determining prestress losses related to factors including, inter alia, assumptions about the properties of prestressing systems and time-dependent phenomena, such as long term degradation processes, tendon relaxation, creep and shrinkage of concrete, and parameters of the real environment [29,30].

Given the current state of quantitative and principled methodologies, this paper represents a state-of-the-art review of some important research works conducted worldwide on determining prestress losses in PC girders. At first, laboratory, numerical investigations, and testing methods are reviewed. Secondly, the article focuses on a static NDT method, and a comparison with dynamic ones is elaborated, since old unsolved problematics as well as new challenges exist because a variation in prestress force does not significantly influence the vibration responses of PC girders. Consequently, this makes uncertain the use of natural frequencies as appropriate parameters for prestress loss determinations. Thus, amongst emerging techniques, static identification based on vertical deflections has preliminary proved to be a reliable method with the aim to become a dominant testing approach in the near future. In fact, measured vertical deflections take accurately and instantaneously into account the changes of structural geometry of PC girders due to prestressing losses on the equilibrium conditions, in turn caused by the combined effects of relaxation of tendon, concrete creep and shrinkage, and parameters of the real environment as, e.g., temperature and relative humidity. Comments and recommendations are made at proper places, whilst concluding remarks including future investigations and field developments are mentioned at the end of the article.

2. Works Conducted by Researchers Worldwide

Research was devoted to analyzing the influence of prestress force on the dynamic behavior of PC girders. A state-of-the-art review of the primary contributions on this topic is presented herein, including laboratory, numerical investigations, and testing methods.

2.1. Laboratory Investigations

Hop [31] monitored the vibration responses of a number of PC beams. Investigation focused on the influence of prestress forces on frequency and damping. The author found that applying an increase in prestress force, acting unevenly on the member, would increase the vibration frequency. In many cases, it was measured that the application of further levels of prestressing increase would result in drop of vibration frequency.

Similar experimental results were obtained by Saiidi et al. [32] on a PC girder with concentric tendon. The research proved an increase of the first eigenfrequency from 11.41 Hz for the case of null prestressing, to 15.07 Hz for the maximum magnitude of prestressing. Besides, the authors observed that an increase in prestressing seems to influence microcrack closure and, consequently, increment stiffness and natural frequencies of PC beams.

Miyamoto et al. [33] tested the dynamic response of PC girders, strengthened with external tendons. According to their results, prestress forces applied to external tendons influence the frequency vibrations of PC girders.

Lu and Law [34] tested a PC beam with a straight concentric tendon. Two conditions were studied, id est, with and without the prestress force of 66.7 kN. The authors noted that the prestressing induced an increment in the first three eigenfrequencies within a range of 0.4–2.1%.

Xiong and Zhang [35] tested three simply supported PC girders with different configurations of external tendons. The authors observed that the natural frequency initially increased with the increase in prestress force. Vice versa, the natural frequency decreased after cracks induced by prestressing.

Kim et al. [36] tested a PC girder with many damage scenarios of prestress losses. Starting from a state of absence of prestress losses, the prestress force was then gradually reduced to zero. During this unloading process, vibration measurements allowed to estimate reductions of the first four eigenfrequencies up to values of 4.0–4.4% from the initial stage to the final one.

Jang et al. [37] tested a number of six PC beams with a bonded tendon. By applying continuously an increase of prestressing from 0 to 523 kN, the authors observed a progressive increase of the first eigenfrequency from 7.6 to 8.7 Hz.

Noh et al. [38] performed tests on three PC girders with different configurations of tendons. The researchers detected that the natural frequency generally increases as tension force in prestressing steel increases. Additionally, they observed that natural frequencies of PC girders are affected by other parameters, such as beam camber, geometric stiffness of tendon, and stiffness effect of beam-tendon system.

The results of the aforementioned relevant works, id est, of Hop [31], Saiidi et al. [32], Kim et al. [36], and Jang et al. [37], declared that for lower values of prestress force, an increase in prestressing generates an increase in eigenfrequencies, especially for the fundamental frequency. Conversely, for higher levels of prestressing, the rate of increase of the eigenfrequencies tends to decrease. Moreover, changes in vibration frequencies were observed to be higher for smaller eccentricities of prestressing tendon. By considering such investigations, it can generally be claimed that prestress force slightly affects the dynamics of PC girders (Table 1). Nevertheless, relevance of these modifications depends on many characteristics (e.g., cracking and nonlinearity of concrete, bonding and eccentricity of tendon) which generate counterbalancing effects making it difficult to determine an evident relation between dynamic properties and prestress force.

Table 1. Laboratory investigations on the relation between prestress force and dynamics of Prestressed Concrete (PC) girders.

Author	Year	Vibration Test	Result and Finding
Hop [31]	1991	UnCracked PC girders	Increase in prestress force slightly increases the natural frequencies
Saiidi et al. [32]	1994	UnCracked PC girder	(*as above*)
Miyamoto et al. [33]	2000	UnCracked PC girders	(*as above*)
Lu and Law [34]	2006	UnCracked PC girder	(*as above*)
Xiong and Zhang [35]	2009	Cracked PC girders	Increase in prestress force slightly decreases the natural frequencies
Kim et al. [36]	2010	Cracked PC girder	Increase in prestress force slightly increases the natural frequencies
Jang et al. [37]	2010	UnCracked PC girders	(*as above*)
Noh et al. [38]	2015	UnCracked PC girders	(*as above*)

2.2. Numerical Investigations

Numerical studies were conducted by addressing the effect of prestress force on the dynamics of PC girders. During simulation of a moving force identification method which assumed the effects of prestressing, Chan and Yung [39] found that the natural frequencies of a PC bridge decrease with an increase in prestress force. This is well-known as the *"compression-softening"* effect and, in general, occurs in Euler–Bernoulli beams and PC members preserved against crack formation [40–42].

Kim et al. [43] investigated prestress loss identifications in PC girders based on the measurement variations of natural frequencies. Comparison between the experimental results obtained by Saiidi et al. [32] and the previsions of their model validated their method.

Law and Lu [44] analyzed the time-domain response of a PC girder under dynamic excitation. By comparing their results of numerical simulations with theoretical findings, the authors identified the prestress force in the time domain by displacement and strain measurements. According to their findings, the natural frequencies decrease as the prestressing increases.

Hamed and Frostig [45] developed a nonlinear solution for modeling the behavior of PC girders with a bonded or an unbonded tendon. Based on the derived governing equations, the authors proved that prestressing does not influence the natural frequencies of PC girders.

Jaiswal [46] underlined that the increase of a PC girder's stiffness (and frequency) depends on the tendon's eccentricity, thus inducing greater moment and stiffening effect along the member.

Limongelli et al. [47] investigated the prediction of early warning signs of deterioration in a PC beam caused by prestress losses. The investigators pointed out that vibration frequencies of PC girders significantly change only under the effects of crack initiation or crack re-opening.

Gan et al. [48] validated the experiments done by Jang et al. [37] and Noble et al. [49] by a Finite Element (FE) model, where the influence of prestressing on the natural frequencies was simulated by the existence of early-age shrinkage cracks inside the concrete.

Bonopera et al. [50,51] found that the fundamental frequency of uncracked PC girders with a parabolic tendon is unaffected by the prestress force because the course of the *"compression–softening"* theory being cancelled out by the increase of elastic modulus due to the concrete's consolidation/hardening with time [52].

Luna Vera et al. [53] investigated the flexural performance of two PC girders under uncracked and cracked conditions for further applications of SHM. In relation to the uncracked state, the authors affirmed that changes in natural frequency due to prestress losses are negligible.

Looking at the relevant works above mentioned, id est, at Hamed and Frostig [45], Jaiswal [46], Limongelli et al. [47], and Luna Vera et al. [53], no significant agreement between the effect of prestress force and dynamics of uncracked PC girders was observed (Table 2). As a result, natural frequencies are generally considered inappropriate parameters for determining prestress losses, as also indicated by Saiidi et al. [32] and Bonopera et al. [50,51].

Table 2. Numerical investigations on the relation between prestress force and dynamics of PC girders.

Author	Year	Numerical Solution	Dynamic Model	Result and Finding
Chan and Yung [39]	2000	Analytical	UnCracked PC girders	Increase in prestress force slightly decreases the natural frequencies
Kim et al. [43]	2004	Analytical simulation of tests performed by Saiidi et al. [32]	UnCracked PC girders	Increase in prestress force slightly increases the natural frequencies
Law and Lu [44]	2005	Analytical	UnCracked PC girders	Increase in prestress force slightly decreases the natural frequencies
Hamed and Frostig [45]	2006	Analytical	UnCracked PC girders	Increase in prestress force does not affect the natural frequencies
Jaiswal [46]	2008	FE	UnCracked PC girders	(*as above*)
Limongelli et al. [47]	2016	Analytical	UnCracked and Cracked PC girders	(*as above*)
Gan et al. [48]	2019	FE simulation of tests performed by Jang et al. [37] and Noble et al. [49]	Cracked PC girders	Increase in prestress force slightly increases the natural frequencies
Bonopera et al. [50]	2019	Analytical	UnCracked PC girders	Increase in prestress force does not affect the natural frequencies
Luna Vera et al. [53]	2020	Analytical	UnCracked and Cracked PC girders	(*as above*)
Bonopera et al. [51]	2021	Analytical	UnCracked PC girders	(*as above*)

2.3. Testing Methods

Testing methods are in general required for estimating prestress losses, and include five typologies (Table 3): (1) static load testing to determine crack initiation or crack re-opening loads to obtain the available compressive stress in the bottom flange of a PC girder [23,28,54,55]. (2) Severing the prestressing tendon by cutting it into a representative exposed length after placing strain gauges on the tendon [23,56]. (3) Relating the tension in the tendon to a vertical deflection recorded when known weights are suspended from it on a representative exposed length [54,57]. (4) Determining the side pressure to close the induced crack in a small cylindrical hole drilled adjacent to the tendon in the bottom flange of a PC girder [55,58]. (5) Vibration testing to determine the natural frequencies and/or dynamic responses of a PC girder [59–65].

Table 3. Testing methods for determining prestress losses in PC girders.

Method	Reference	Testing on a PC Girder	Result and Finding
(1) Destructive	[23,28,54,55]	Static test to determine crack initiation or crack re-opening loads to obtain the compressive stress in the bottom flange	It causes damages
(2) Destructive	[23,56]	Severing the prestressing tendon by cutting it into an exposed length after placing strain gauges	It causes damages

Table 3. *Cont.*

Method	Reference	Testing on a PC Girder	Result and Finding
(3) Semidestructive	[54,57]	Relating the tension in the tendon to a vertical deflection recorded when weights are suspended from it on an exposed length	It causes partial damages
(4) Nondestructive	[55,58]	Determining the side pressure to close the induced crack in a small cylindrical hole drilled adjacent to the tendon in the bottom flange	The embedded tendon, if is too close to the concrete surface, could be damaged by the deep hole-cut
(5) Nondestructive	[59–68]	Vibration test to determine natural frequencies and/or dynamic responses	It requires an accurate selection of the mode shape

Methods 1 and 2 are destructive approaches which cause damages. In Method 1, a large amount of vertical loads must be applied to involve crack initiation, or to force a main crack to form in the same location as a monitored crack.

Method 3 is a semidestructive test for exposed tendons, and involves accurately determining the exposed length for calculations.

Method 4 is a NDT approach for embedded tendons which involves a negligible local damage along the PC span girder. This is not possible when the tendon is too close to the concrete surface because it could be damaged by the deep hole-cut. Notably, further studies in a controlled environment with monitoring of actual prestress forces were suggested in order to validate its applicability for PC girders with a parabolic tendon [55].

Method 5 represents a series of NDT techniques which require local vibration measurements along the PC girder. With regards, Law et al. [66], Li et al. [67], and Xiang et al. [68] performed numerical simulations using the dynamics of PC girders to moving vehicular loads. In their methods, prestress force identifications were obtained using vibration measures if the prestress force in the tendon is assumed as equivalent to an external compressive force applied to the beam ends [69–82]. Consequently, the natural frequency of the PC girder tends to increase with a decrease in prestress force according to the *"compression–softening"* theory. Notably, whether prestressed beams are subjected to the *"compression–softening"* effect was extensively discussed as, e.g., in the literature review by Noble et al. [49,83]. Most of the dynamic methods, cited therein, are based on the identified modal characteristics of a PC girder, id est, natural frequencies and/or mode shapes. The modal characteristics depend on the PC girder's stiffness and, accordingly, are affected by the prestress force. Particularly, such vibration-based identifications require an accurate selection of the mode shape to be utilized in the procedures. In fact, selecting the optimal frequency a priori is challenging, and different frequencies yield varying degrees of accuracy in prestress force determinations.

3. Static NDT Methods

Amongst vibration-based techniques, static identifications using vertical deflections have proved to be reliable methods for axial force identification in beam elements. Indeed, measured vertical deflections take accurately into account the changes of structural geometry of the members due to axial force variations on the equilibrium conditions [40,41,71,74,77,84] (Figure 1a,b). Experimental simulations were as well conducted on members belonging to space frames and trusses [78,79,85]. Likewise, Bonopera et al. [80] verified the feasibility of estimating prestress force in a PC girder using vertical deflections measured by three-point bending tests. It is also worth noting that this approach only adopts static parameters, thus, in contrast to vibration-based techniques, does not require selecting experimental data for use in the algorithms.

(a)   (b)

Figure 1. Static NDT methods proposed by Tullini et al. [71,74]: (**a**) cast iron disks suspended at the reference beam; (**b**) arrangement of the dial indicators for vertical deflection measurements during an experimental simulation. Copyright © 2012, 2013 Elsevier Ltd. Reprinted with permission.

3.1. Brief on Works Conducted by Bonopera et al. (2018)

The static approach was originally developed for detecting axial force in compressed steel beams [77]. Subsequently, it was employed for identifying prestress forces in PC girders [80]. In this last case, the reference model comprised a simply supported Euler–Bernoulli beam of 250 mm in width, 400 mm in height, and length L of 6.62 m, made with a high strength concrete, and prestressed by a straight unbonded tendon, where the prestress force N was assumed as an external compressive force eccentrically applied to the end constraints $N\,e$ (Figure 2a). The cross sectional second moment of the area of the PC beam's section I was equal to 1.3333×10^9 mm^4. Besides, a bending deflection $v^{(1)}$ along the aforementioned beam's model, of 0.01 mm in accuracy, was properly approximated by multiplying the corresponding first-order deflection by the "*magnification factor*" of the second-order effects, id est, according to the "*compression-softening*" theory [40–42] (Figure 2b).

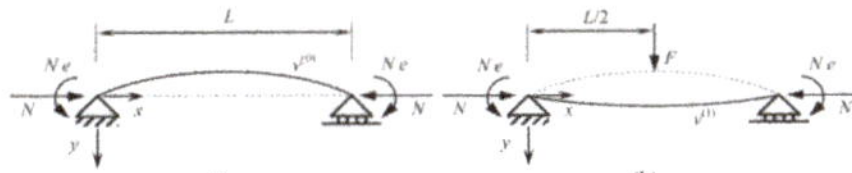

(a)   (b)

Figure 2. Reference model of the PC girder [80]: (**a**) deflection curve $v^{(0)}$ after the application of the eccentric prestress force N; (**b**) deflection curve $v^{(1)}$ after the application of the vertical load F to the deflection curve $v^{(0)}$. The dashed lines represent the initial deflection curves. Copyright © 2018, World Scientific Publishing Co. Pte Ltd. Reprinted with permission.

Experiments on a PC beam specimen, having the test configuration above mentioned (Figure 2a,b), were arranged in the laboratory of the National Center for Research on Earthquake Engineering (NCREE) of Taipei, Taiwan [80], where a research program, based on testing uncracked PC bridge member prototypes, began in 2015 (Figure 3). All geometrical dimensions were checked by laser rangefinder and caliper, of 0.01 mm in tolerance, after the PC beam was fixed on the simple supports. First, deflected-shape measurements $v^{(1)}$ along its span, obtained from 27 three-point bending tests with different applied prestress forces N, measured by a load cell placed at both end constraints [86], were examined to verify the accuracy of the assumptions of the beam's model (Figure 2b). The span of the PC beam specimen was specifically instrumented to short term measure such static deflection shapes, with an accuracy of 0.01 mm, by a set of displacement transducers. Temperature and relative humidity of real environment of the PC beam were not continuously recorded during testing. Second, based on the "*magnification factor*" formula of the "*compression-softening*" theory [40–42], prestress force determinations were obtained using two series of vertical deflections $v^{(1)}$, id est, those recorded at the quarter $v_2^{(1)}$ and at the midspan $v_4^{(1)}$ of the PC beam, respectively. Information regarding the flexural rigidity of the PC beam were in addition required. In detail, average values of the chord elastic modulus E_{aver} of the high strength concrete, with an accuracy of 1 MPa, were estimated by compression tests on a series of cylinders cast at the same time of the PC beam [87] to determine its increment caused by the concrete's consolidation/hardening with time [52].

291

Figure 3. Three-point bending tests on a PCI girder conducted in the laboratory of NCREE, Taiwan [50]: (**a**) indoor test rig; (**b**) application of a FBG-Differential Settlement Measurement (DSM) liquid-level system for vertical deflection measurements along the span. Copyright © 2019 Elsevier Ltd. Adapted with permission.

3.2. Prestress Force Determinations Obtained by Bonopera et al. (2018)

Table 4 lists the prestress force determinations N_a obtained using the vertical deflections $v_2^{(1)}$ and corresponding experimental values $\Psi = F\,L^3/E_{aver}\,I$ in Equation (8a) (Test 1), as well as the vertical deflections $v_4^{(1)}$ and corresponding parameters Ψ in Equation (8b) (Test 2), respectively. Equations (8a) and (8b) are illustrated in Bonopera et al. [80]. In specific, the nine test combinations represent the best prestress force determinations among the three test repetitions performed (Section 3.1). The chord elastic modulus E_{aver} of the high strength concrete, for each day of execution of the experiments (Section 3.1), was utilized as parameter in the identification process. The corresponding first-order vertical deflections were instead estimated by Equations (4a) and (4b), as similarly reported in Bonopera et al. [80]. Table 4 additionally shows the related percentage errors $\Delta = (N_a - N)/N$. In general, poor estimates N_a were obtained when prestressing N equal to 617 and 620 kN were assigned (Figure 2a,b). Vice versa, the test combinations with prestressing that induced second-order effects greater than 6.5%, id est, when $N \geqq 721$ kN, furnished excellent identifications of prestress forces N_a. In fact, in this last case, estimation errors were lower, in absolute value, than 6.1%.

Table 4. Prestress force determinations N_a based on Equation (8a) (Test 1) and Equation (8b) (Test 2), and corresponding measured parameters for each test day [80]. Copyright © 2018, World Scientific Publishing Co. Pte Ltd. Adapted with permission.

Days of Concrete Curing	E_{aver} (MPa)	N (kN)	F (kN)	Test 1-$v_2^{(1)}$ Deflections at a Quarter		Test 2-$v_4^{(1)}$ Deflections at the Midspan	
				N_a (kN)	Δ (%)	N_a (kN)	Δ (%)
426	34,870	620	20.2	777	25.3	789	27.3
		620	22.6	857	38.2	857	38.2
		617	25.0	386	−37.4	550	−10.9
427	37,618	724	20.1	729	0.7	732	1.1
		721	22.6	721	0.0	761	5.5
		721	25.1	715	−0.8	718	−0.4
433	38,791	820	20.2	846	3.2	823	0.4
		820	22.9	825	0.6	825	0.6
		820	25.1	824	0.5	870	6.1

Sensitivity analyses were elaborated for the prestress force determinations based on Equations (8a) and (8b). The vertical deflections $v_2^{(1)}$ and $v_4^{(1)}$, calculated with 0.01 mm in accuracy by Equations (3a) and (3b) reported in Bonopera et al. [80], and parameter Ψ were modified to generate possible experimental errors. In detail, deflections $v_2^{(1)}$, $v_4^{(1)}$, and Ψ were alternatively multiplied by 0.99 and 1.01 to reproduce 14 combinations of simulated values for nine different assumed prestress forces N. The average value of the applied vertical loads, $F_{aver} = 22.6$ kN (Figure 2b), was taken into account in the manipulations. Figure 4 depicts a comparison between the worst determined N_a

and assumed values N conducted using vertical deflections $v_2^{(1)}$ and $v_4^{(1)}$, both of which yielded a constant error of about ± 107 kN. Based on all the results obtained, a favorable correspondence between analytical N and experimental determinations of prestress force N_a was found when midspan deflections $v_4^{(1)}$ were taken into account.

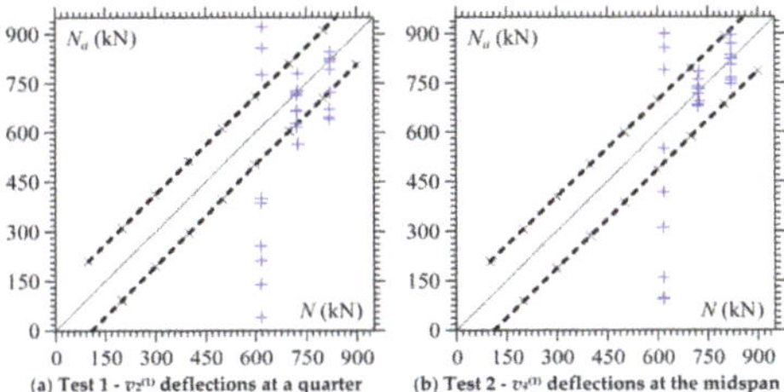

Figure 4. Prestress force determinations N_a based on (**a**) Equation (8a) (Test 1) and (**b**) Equation (8b) (Test 2). Symbols + refer to the comparison between determined N_a and measured values of prestress force N for all the 27 test combinations. The dashed lines with symbol × represent the sensitivity analyses.

4. Concluding Remarks

A state-of-the-art review of some important research works conducted worldwide on determining prestress losses in PC girders allowed to analyze various information and trace future developments. Some references affirm that a variation in prestress force does not significantly affect the vibration responses of PC girders. Accordingly, this makes uncertain the use of natural frequencies as appropriate parameters for prestress loss determinations. Vice versa, most of laboratory works show a slight increment in the eigenfrequencies under the increase in prestress force. This behavior is related to the concrete mechanics, and is a main consequence of the effect of crack and microcrack closure along PC girders. However, vibration-based identification methods require an accurate selection of the mode shape because different natural frequencies provide varying degrees of accuracy in prestress force evaluations.

By considering these characteristics, the manuscript focused then on the static NDT method preliminary proposed by Bonopera et al. [80] through laboratory simulations on an uncracked PC beam specimen. The procedure can accurately and instantaneously determine the effective prestress force using vertical deflection measurements, of 0.01 mm in accuracy, under actual ambient conditions [51,81]. The precision of estimation improved when the PC beam was subjected to a high prestress force and, moreover, when midspan deflections were taken into account. Information regarding the flexural rigidity of a PC girder under investigation are also necessary. With regards, an average value of the chord elastic modulus, of 1 MPa in accuracy, must be estimated by compression tests on a set of concrete cores, drilled along its span, at the time of deflection measurements. Besides, the static NDT method does not require any direct measure of the tension force in the tendon and, mostly, in contrast to dynamic NDT ones, does not request selecting experimental data for use in the algorithms (Table 5).

Table 5. Characteristics of the static NDT method preliminary proposed by Bonopera et al. [80] for determining prestress losses in PC girders.

Advantages	Disadvantages
(1) Precise determinations by vertical deflection measurements of 0.01 mm in accuracy	(1) Determination of the concrete elastic modulus by compression tests on a set of drilled cores
(2) No requirement of direct measure of the tension force in the tendon	(2) Vertical deflections, of 0.01 mm in accuracy, are not always easy to measure in situ
(3) No requirement of any selecting experimental data	
(4) Determinations take into account the combined effects of tendon relaxation, concrete creep and shrinkage, and parameters of the real environment	

In conclusion, to make the NDT method applicable in situ, further studies should focus on the measurement of vertical deflections induced by bending tests with vehicle loading along PC bridge girders [88–90], in which their constraint stiffness should be evaluated with unknown boundary conditions. In fact, static vertical deflections take accurately and instantaneously into account the changes of structural geometry due to prestressing losses on the equilibrium conditions [51,80,81], in turn caused by the combined effects of tendon relaxation, concrete creep and shrinkage, and parameters of real environment of the PC girder as, e.g., temperature and relative humidity [49] (Figure 3a). The FBG-DSM liquid-level system [77,91] is instead an effective measurement device because it can provide bridge deflections over long distances up to 0.01 mm in accuracy (Figure 3b), referred to an absolute point, without any external physical reference and requirements of good environmental conditions, accessibility and visibility in situ.

Author Contributions: Methodology, M.B.; software, Z.-K.L.; validation, M.B., K.-C.C. and Z.-K.L.; formal analysis, M.B.; investigation, M.B. and K.-C.C.; resources, Z.-K.L.; data curation, M.B. and Z.-K.L.; writing—original draft preparation, M.B.; writing—review and editing, M.B.; visualization, M.B. and Z.-K.L.; supervision, K.-C.C.; project administration, M.B.; funding acquisition, M.B. and K.-C.C. All authors have read and agreed to the published version of the manuscript.

Funding: This work was funded by the National Applied Research Laboratories Project of Taiwan (grant number: NCREE–06109A1700), and Ministry of Science and Technology (MOST) of Taiwan in the framework of the project "Recruitment of Visiting Science and Technology Personnel with MOST Funding" (grant number: MOST 108–2811–E–492–001).

Acknowledgments: The authors would like to thank the National Applied Research Laboratories Project of Taiwan (NCREE–06109A1700), and Ministry of Science and Technology (MOST) of Taiwan in the framework of the project "Recruitment of Visiting Science and Technology Personnel with MOST Funding" (MOST 108–2811–E–492–001) for their financial support.

Conflicts of Interest: The authors declare no conflict of interest.

References

1. Breccolotti, M.; Materazzi, A. Prestress Losses and Camber Growth in Wing-Shaped Structural Members. *PCI J.* **2015**, *60*, 98–117. [CrossRef]
2. Robertson, I.N. Prediction of vertical deflections for a long-span prestressed concrete bridge structure. *Eng. Struct.* **2005**, *27*, 1820–1827. [CrossRef]
3. Huang, H.; Huang, S.-S.; Pilakoutas, K. Modeling for assessment of long-term behavior of prestressed concrete box-girder bridges. *J. Bridg. Eng.* **2018**, *23*, 04018002. [CrossRef]
4. Domaneschi, M.; Limongelli, M.; Martinelli, L. Multi-site damage localization in a suspension bridge via aftershock monitoring. *Ing. Sismica* **2013**, *30*, 56–72.
5. Domaneschi, M.; Limongelli, M.; Martinelli, L. Damage detection and localization on a benchmark cable-stayed bridge. *Earthq. Struct.* **2015**, *8*, 1113–1126. [CrossRef]
6. Limongelli, M.P.; Tirone, M.; Surace, C. Non-destructive monitoring of a prestressed bridge with a data-driven method. *Proc. SPIE Int. Soc. Opt. Eng.* **2017**, *10170*, 1017033.
7. He, W.-Y.; Ren, W.-X. Structural damage detection using a parked vehicle induced frequency variation. *Eng. Struct.* **2018**, *170*, 34–41. [CrossRef]
8. Chen, S.Z.; Wu, G.; Feng, D.C. Damage detection of highway bridges based on long-gauge strain response under stochastic traffic flow. *Mech. Syst. Signal Process.* **2019**, *127*, 551–572. [CrossRef]
9. Chen, G.W.; Chen, X.; Omenzetter, P. Modal parameter identification of a multiple-span post-tensioned concrete bridge using hybrid vibration testing data. *Eng. Struct.* **2020**, *219*, 110953. [CrossRef]
10. Saiidi, M.; Shields, J.; O'Connor, D.; Hutchens, E. Variation of prestress forces in a prestressed concrete bridge during the first 30 months. *PCI J.* **1996**, *41*, 66–72. [CrossRef]
11. Saiidi, M.; Hutchens, E.; Gardella, D. Bridge prestress losses in dry climate. *J. Bridge Eng.* **1998**, *3*, 111–116. [CrossRef]
12. Guo, T.; Chen, Z.; Lu, S.; Yao, R. Monitoring and analysis of long-term prestress losses in post-tensioned concrete beams. *Measurement* **2018**, *122*, 573–581. [CrossRef]

13. Kim, J.; Kim, J.W.; Park, S. Investigation of applicability of an embedded EM sensor to measure the tension of a PSC girder. *J. Sens.* **2019**, *2019*, 2469647. [CrossRef]

14. Cho, K.; Kim, S.T.; Cho, J.R.; Park, Y.H. Estimation of tendon force distribution in prestressed concrete girders using smart strand. *Appl. Sci.* **2017**, *7*, 1319. [CrossRef]

15. Sung, H.J.; Do, T.M.; Kim, J.M.; Kim, Y.S. Long-term monitoring of ground anchor tensile forces by FBG sensors embedded tendon. *Smart. Struct. Syst.* **2017**, *19*, 269–277. [CrossRef]

16. ACI (American Concrete Institute). *Building Code Requirements for Structural Concrete and Commentary*; ACI: Farmington Hills, MI, USA, 2019; pp. 318–319.

17. PCI. *Design Handbook*, 7th ed.; Precast/Prestressed Concrete Institute: Chicago, IL, USA, 2010.

18. PCI Commitee on Prestress Losses. Recommendations for estimating prestress losses. *PCI J.* **1975**, *20*, 43–75.

19. AASHTO (American Association of State Highway and Transportation Officials). *AASHTO LRFD Bridge Design Specifications*, 8th ed.; American Association of State Highway and Transportation Officials: Washington, DC, USA, 2017.

20. AASHTO (American Association of State Highway and Transportation Officials). *AASHTO Standard Specifications for Highways Bridges*, 17th ed.; American Association of State Highway and Transportation Officials: Washington, DC, USA, 2002.

21. AASHTO (American Association of State Highway and Transportation Officials). *AASHTO LRFD Bridge Design Specifications*, 3rd ed.; American Association of State Highway and Transportation Officials: Washington, DC, USA, 2004.

22. PCI. *Bridge Design Manual*, 3rd ed.; Precast/Prestressed Concrete Institute: Chicago, IL, USA, 2011.

23. Baran, E.; Shield, C.K.; French, C.E. A comparison of methods for experimentally determining prestress losses in pretensioned prestressed concrete girders. In *Historic Innovations in Prestressed Concrete*; Russell, B.W., Gross, S.P., Eds.; American Concrete Institute: Farmington Hills, MI, USA, 2005; pp. 161–179.

24. Al-Omaishi, N.; Tadros, M.K.; Seguirant, S.J. Estimating prestress loss in pretensioned, high-strength concrete members. *PCI J.* **2009**, *54*, 132–159. [CrossRef]

25. Hale, W.M.; Russell, B.W. Effect of allowable compressive stress at release on prestress losses and on the performance of precast, prestressed concrete bridge girders. *PCI J.* **2006**, *51*, 14–25. [CrossRef]

26. Caro, L.A.; Martí-Vargas, J.R.; Serna, P. Prestress losses evaluation in prestressed concrete prismatic specimens. *Eng. Struct.* **2013**, *48*, 704–715. [CrossRef]

27. Martí-Vargas, J.R.; Caro, L.A.; Serna, P. Experimental technique for measuring the long-term transfer length in prestressed concrete. *Strain* **2013**, *49*, 125–134. [CrossRef]

28. Garber, D.B.; Gallardo, J.M.; Deschenes, D.J.; Bayrak, O. Experimental investigation of prestress losses in full-scale bridge girders. *ACI Struct. J.* **2015**, *112*, 553–564. [CrossRef]

29. Bažant, Z.P.; Jirásek, M.; Hubler, M.H.; Carol, I. RILEM draft recommendation: TC-242-MDC multi-decade creep and shrinkage of concrete: Material model and structural analysis. Model B4 for creep, drying shrinkage and autogenous shrinkage of normal and high-strength concretes with multi-decade applicability. *Mater. Struct.* **2015**, *48*, 753–770.

30. Kolínský, V.; Vítek, J.L. New model for concrete creep and shrinkage prediction and its application. *Solid State Phenom.* **2016**, *249*, 125–130. [CrossRef]

31. Hop, T. The effect of degree of prestressing and age of concrete beams on frequency and damping of their free vibration. *Mater. Struct.* **1991**, *24*, 210–220. [CrossRef]

32. Saiidi, M.; Douglas, B.; Feng, S. Prestress force effect on vibration frequency of concrete bridges. *J. Struct. Eng.* **1994**, *120*, 2233–2241. [CrossRef]

33. Miyamoto, A.; Tei, K.; Nakamura, H.; Bull, J.W. Behavior of prestressed beam strengthened with external tendons. *J. Struct. Eng.* **2000**, *126*, 1033–1044. [CrossRef]

34. Lu, Z.; Law, S. Identification of prestress force from measured structural responses. *Mech. Syst. Signal Pr.* **2006**, *20*, 2186–2199. [CrossRef]

35. Xiong, H.X.; Zhang, Y.T. Theoretical analysis of natural frequency of externally prestressed concrete beam based on rigidity correction. *Acad. J. Xian Jiaotong Univ.* **2009**, *21*, 31–35.

36. Kim, J.T.; Park, J.H.; Hong, D.S.; Park, W.S. Hybrid health monitoring of prestressed concrete girder bridges by sequential vibration-impedance approaches. *Eng. Struct.* **2010**, *32*, 115–128. [CrossRef]

37. Jang, J.; Lee, H.; Hwang, K.; Song, Y. A sensitivity analysis of the key parameters for the prediction of the pre-stress force on bonded tendons. *Nucl. Eng. Technol.* **2010**, *42*, 319–328. [CrossRef]

38. Noh, M.H.; Seong, T.R.; Lee, J.; Park, K.S. Experimental investigation of dynamic behavior of prestressed girders with internal tendons. *Int. J. Steel Struct.* **2015**, *15*, 401–414. [CrossRef]

39. Chan, T.H.T.; Yung, T.H. A theoretical study of force identification using prestressed concrete bridges. *Eng. Struct.* **2000**, *22*, 1529–1537. [CrossRef]

40. Timoshenko, S.P.; Gere, J.M. *Theory of Elastic Stability*; McGraw-Hill: New York, NY, USA, 1961.

41. Bažant, Z.P.; Cedolin, L. *Stability of Structures*; Oxford University Press: New York, NY, USA, 1991.

42. Young, W.C.; Budynas, R.G. Dynamic and Temperature Stresses. In *Roark's Formulas for Stress and Strain*; Young, W.C., Budynas, R.G., Eds.; McGraw–Hill: New York, NY, USA, 2002; pp. 767–768.

43. Kim, J.T.; Yun, C.B.; Ryu, Y.S.; Cho, H.M. Identification of prestress-loss in PSC beams using modal information. *Struct. Eng. Mech.* **2004**, *17*, 467–482. [CrossRef]

44. Law, S.; Lu, Z. Time domain responses of a prestressed beam and prestress identification. *J. Sound Vib.* **2005**, *288*, 1011–1025. [CrossRef]

45. Hamed, E.; Frostig, Y. Natural frequencies of bonded and unbonded prestressed beams-prestress force effects. *J. Sound Vib.* **2006**, *295*, 28–39. [CrossRef]

46. Jaiswal, O.R. Effect of prestressing on the first flexural natural frequency of beams. *Struct. Eng. Mech.* **2008**, *28*, 515–524. [CrossRef]

47. Limongelli, M.P.; Siegert, D.; Merliot, E.; Waeytens, J.; Bourquin, F.; Vidal, R.; Le Corvec, V.; Gueguen, I.; Cottineau, L.M. Damage detection in a post tensioned concrete beam—Experimental investigation. *Eng. Struct.* **2016**, *128*, 15–25. [CrossRef]

48. Gan, B.Z.; Chiew, S.P.; Lu, Y.; Fung, T.C. The effect of prestressing force on natural frequencies of concrete beams—A numerical validation of existing experiments by modelling shrinkage crack closure. *J. Sound Vib.* **2019**, *455*, 20–31. [CrossRef]

49. Noble, D.; Nogal, M.; O'Connor, A.; Pakrashi, V. The effect of prestress force magnitude and eccentricity on the natural bending frequencies of uncracked prestressed concrete beams. *J. Sound Vib.* **2016**, *365*, 22–44. [CrossRef]

50. Bonopera, M.; Chang, K.C.; Chen, C.C.; Sung, Y.C.; Tullini, N. Experimental study on the fundamental frequency of prestressed concrete bridge beams with parabolic unbonded tendons. *J. Sound Vib.* **2019**, *455*, 150–160. [CrossRef]

51. Bonopera, M.; Chang, K.C.; Lin, T.K.; Tullini, N. Influence of prestressing on the behavior of uncracked concrete beams with a parabolic bonded tendon. *Struct. Eng. Mech.* **2021**, *77*. Accepted.

52. Singh, B.P.; Yazdani, N.; Ramirez, G. Effect of a time dependent concrete modulus of elasticity on prestress losses in bridge girders. *Int. J. Concr. Struct. Mater.* **2013**, *7*, 183–191. [CrossRef]

53. Luna Vera, O.S.; Oshima, Y.; Kim, C.W. Flexural performance correlation with natural bending frequency of post-tensioned concrete beam: Experimental investigation. *J. Civ. Struct. Health Monit.* **2020**, *10*, 135–151. [CrossRef]

54. Labia, Y.; Saiidi, M.S.; Douglas, B. Full-scale testing and analysis of 20-year-old pretensioned concrete box girders. *ACI Struct. J.* **1997**, *94*, 471–482.

55. Bagge, N.; Nilimaa, J.; Elfgren, L. In-situ methods to determine residual prestress forces in concrete bridges. *Eng. Struct.* **2017**, *135*, 41–52. [CrossRef]

56. Halsey, J.T.; Miller, R. Destructive testing of two forty-year-old prestressed concrete bridge beams. *PCI J.* **1996**, *41*, 84–93. [CrossRef]

57. Civjan, S.A.; Jirsa, J.O.; Carrasquillo, R.L.; Fowler, D.W. Instrument to evaluate remaining prestress in damaged prestressed concrete bridge girders. *PCI J.* **1998**, *43*, 62–69. [CrossRef]

58. Azizinamini, A.; Keeler, B.J.; Rohde, J.; Mehrabi, A.B. Application of a new nondestructive evaluation technique to a 25-year-old prestressed concrete girder. *PCI J.* **1996**, *41*, 82–95. [CrossRef]

59. Kim, J.T.; Ryu, Y.S.; Yun, C.B. Vibration based method to detect pre–stress loss in beam type bridges. *Smart Struct. Mater.* **2003**, *5057*, 559–569. [CrossRef]

60. Bu, J.Q.; Wang, H.Y. Effective pre–stress identification for a simply supported PRC beam bridge by BP neural network method. *J. Vib. Shock* **2011**, *30*, 155–159.

61. Xu, J.; Sun, Z. Pre–stress force identification for eccentrically prestressed concrete beam from beam vibration response. *Technol. Sci. Press SL* **2011**, *5*, 107–115.

62. Shi, L.; He, H.; Yan, W. Prestress force identification for externally pre–stressed concrete beam based on frequency equation and measured frequencies. *Math. Probl. Eng.* **2014**, *2014*, 840937. [CrossRef]

63. Xiang, Z.; Chan, T.H.T.; Thambiratnam, D.P.; Nguyen, A. Prestress and excitation force identification in a prestressed concrete box-girder bridge. *Comput. Concr.* **2017**, *20*, 617–625.

64. Breccolotti, M. On the evaluation of prestress loss in PRC beams by means of dynamic techniques. *Int. J. Concr. Struct. Mater.* **2018**, *12*, 1. [CrossRef]

65. Breccolotti, M.; Pozza, F. Prestress evaluation in continuous PSC bridges by dynamic identification. *Struct. Monit. Maint.* **2018**, *5*, 463–488.

66. Law, S.S.; Wu, S.Q.; Shi, Z.Y. Moving load and prestress identification using wavelet based method. *J. Appl. Mech. Trans. ASME* **2008**, *75*, 021014. [CrossRef]

67. Li, H.; Lv, Z.; Liu, J. Assessment of prestress force in bridges using structural dynamic responses under moving vehicles. *Math. Probl. Eng.* **2013**, *2013*, 435939. [CrossRef]

68. Xiang, Z.; Chan, T.H.T.; Thambiratnam, D.P.; Nguyen, T. Synergic identification of prestress force and moving load on prestressed concrete beam based on virtual distortion method. *Smart Struct. Syst.* **2016**, *17*, 917–933. [CrossRef]

69. Tullini, N.; Laudiero, F. Dynamic identification of beam axial loads using one flexural mode shape. *J. Sound Vib.* **2008**, *318*, 131–147. [CrossRef]

70. Bahra, A.S.; Greening, P.D. Identifying multiple axial load patterns using measured vibration data. *J. Sound Vib.* **2011**, *330*, 3591–3605. [CrossRef]

71. Tullini, N.; Rebecchi, G.; Laudiero, F. Bending tests to estimate the axial force in tie–rods. *Mech. Res. Commun.* **2012**, *44*, 57–64. [CrossRef]

72. Maes, K.; Peeters, J.; Reynders, E.; Lombaert, G.; De Roeck, G. Identification of axial forces in beam members by local vibration measurements. *J. Sound Vib.* **2013**, *332*, 5417–5432. [CrossRef]

73. Rebecchi, G.; Tullini, N.; Laudiero, F. Estimate of the axial force in slender beams with unknown boundary conditions using one flexural mode shape. *J. Sound Vib.* **2013**, *332*, 4122–4135. [CrossRef]

74. Tullini, N. Bending tests to estimate the axial force in slender beams with unknown boundary conditions. *Mech. Res. Commun.* **2013**, *53*, 15–23. [CrossRef]

75. Luong, H.T.M.; Zabel, V.; Lorenz, W.; Rohrmann, R.G. Vibration-based model updating and identification of multiple axial forces in truss structures. *Procedia Eng.* **2017**, *188*, 385–392. [CrossRef]

76. Luong, H.T.M.; Zabel, V.; Lorenz, W.; Rohrmann, R.G. Non-destructive assessment of the axial stress state in iron and steel truss structures by dynamic measurements. *Procedia Eng.* **2017**, *199*, 3380–3385. [CrossRef]

77. Bonopera, M.; Chang, K.C.; Chen, C.C.; Lee, Z.K.; Tullini, N. Axial load detection in compressed steel beams using FBG–DSM sensors. *Smart. Struct. Syst.* **2018**, *21*, 53–64.

78. Bonopera, M.; Chang, K.C.; Chen, C.C.; Lin, T.K.; Tullini, N. Compressive column load identification in steel space frames using second–order deflection–based methods. *Int. J. Struct. Stab. Dyn.* **2018**, *18*, 1850092. [CrossRef]

79. Bonopera, M.; Chang, K.C.; Chen, C.C.; Lin, T.K.; Tullini, N. Bending tests for the structural safety assessment of space truss members. *Int. J. Space Struct.* **2018**, *33*, 138–149. [CrossRef]

80. Bonopera, M.; Chang, K.C.; Chen, C.C.; Sung, Y.C.; Tullini, N. Feasibility study of prestress force prediction for concrete beams using second–order deflections. *Int. J. Struct. Stab. Dyn.* **2018**, *18*, 1850124. [CrossRef]

81. Bonopera, M.; Chang, K.C.; Chen, C.C.; Sung, Y.C.; Tullini, N. Prestress force effect on fundamental frequency and deflection shape of PCI beams. *Struct. Eng. Mech.* **2018**, *67*, 255–265.

82. Kernicky, T.; Whelan, M.; Al-Shaer, E. Dynamic identification of axial force and boundary restraints in tie rods and cables with uncertainty quantification using Set Inversion Via Interval Analysis. *J. Sound Vib.* **2018**, *423*, 401–420. [CrossRef]

83. Noble, D.; Nogal, M.; O'Connor, A.; Pakrashi, V. Dynamic impact testing on post–tensioned steel rectangular hollow sections; An investigation into the "compression–softening" effect. *J. Sound Vib.* **2015**, *355*, 246–263. [CrossRef]

84. Tullini, N.; Rebecchi, G.; Laudiero, F. Reliability of the tensile force identification in ancient tie-rods using one flexural mode shape. *Int. J. Archit. Herit.* **2019**, *13*, 402–410. [CrossRef]

85. Turco, E. Identification of axial forces on statically indeterminate pin-jointed trusses by a nondestructive mechanical test. *Open Civ. Eng. J.* **2013**, *7*, 50–57. [CrossRef]

86. Hsiao, J.K. Prestress loss distributions along simply supported pretensioned concrete beams. *Electron. J. Struct. Eng.* **2017**, *16*, 18–25.

87. ASTM (American Society for Testing and Materials). *Annual Book of ASTM standards, Section 4: Construction vol. 04.02*; Concrete & Aggregates: West Conshohocken, PA, USA, 2016.

88. Chiu, Y.T.; Lin, T.K.; Hung, H.H.; Sung, Y.C.; Chang, K.C. Integration of in–situ load experiments and numerical modeling in a long–term bridge monitoring system on a newly–constructed widened section of freeway in Taiwan. *Smart Struct. Syst.* **2014**, *13*, 1015–1039. [CrossRef]

89. Sung, Y.C.; Lin, T.K.; Chiu, Y.T.; Chang, K.C.; Chen, K.L.; Chang, C.C. A bridge safety monitoring system for prestressed composite box–girder bridges with corrugated steel webs based on in–situ loading experiments and a long–term monitoring database. *Eng. Struct.* **2016**, *126*, 571–585. [CrossRef]

90. Jeon, J.C.; Lee, H.H. Development of displacement estimation method of girder bridges using measured strain signal induced by vehicular loads. *Eng. Struct.* **2019**, *186*, 203–215. [CrossRef]

91. Bonopera, M.; Chang, K.C.; Chen, C.C.; Lee, Z.K.; Sung, Y.C.; Tullini, N. Fiber Bragg grating–differential settlement measurement system for bridge displacement monitoring: Case study. *J. Bridge Eng.* **2019**, *24*, 05019011. [CrossRef]

Publisher's Note: MDPI stays neutral with regard to jurisdictional claims in published maps and institutional affiliations.

 *applied sciences*

Review

Application of the Subspace-Based Methods in Health Monitoring of Civil Structures: A Systematic Review and Meta-Analysis

Hoofar Shokravi [1,*], Hooman Shokravi [2], Norhisham Bakhary [1,3], Mahshid Heidarrezaei [4], Seyed Saeid Rahimian Koloor [5] and Michal Petrů [5]

1   School of Civil Engineering, Faculty of Engineering, Universiti Teknologi Malaysia, Skudai 81310, Johor, Malaysia; norhisham@utm.my
2   Department of Civil Engineering, Islamic Azad University, Tabriz 5157944533, Iran; hooman.shokravi@gmail.com
3   Institute of Noise and Vibration, Universiti Teknologi Malaysia, City Campus, Jalan Semarak, Kuala Lumpur 54100, Malaysia
4   School of Chemical and Energy Engineering, Faculty of Engineering, Universiti Teknologi Malaysia, Skudai 81310, Johor, Malaysia; mahshid@ibd.utm.my
5   Institute for Nanomaterials, Advanced Technologies and Innovation (CXI), Technical University of Liberec (TUL), Studentska 2, 461 17 Liberec, Czech Republic; s.s.r.koloor@gmail.com (S.S.R.K.); michalpetru@tul.cz (M.P.)
*   Correspondence: hf.shokravi@gmail.com

Received: 21 April 2020; Accepted: 15 May 2020; Published: 22 May 2020

Abstract: A large number of research studies in structural health monitoring (SHM) have presented, extended, and used subspace system identification. However, there is a lack of research on systematic literature reviews and surveys of studies in this field. Therefore, the current study is undertaken to systematically review the literature published on the development and application of subspace system identification methods. In this regard, major databases in SHM, including Scopus, Google Scholar, and Web of Science, have been selected and preferred reporting items for systematic reviews and meta-analyses (PRISMA) has been applied to ensure complete and transparent reporting of systematic reviews. Along this line, the presented review addresses the available studies that employed subspace-based techniques in the vibration-based damage detection (VDD) of civil structures. The selected papers in this review were categorized into authors, publication year, name of journal, applied techniques, research objectives, research gap, proposed solutions and models, and findings. This study can assist practitioners and academicians for better condition assessment of structures and to gain insight into the literature.

Keywords: subspace system identification; data-driven stochastic subspace identification (SSI-DATA); covariance-driven stochastic subspace identification (SSI-COV); combined subspace system identification; PRISMA; damage detection; vibration-based damage detection

---

1. Introduction

Structural health monitoring (SHM) is an emerging multidisciplinary field for damage detection and condition monitoring of structures [1,2]. Due to the complexity of civil structures and the associated ambient-induced uncertainty, the development of a reliable SHM is a challenging task. Vibration-based damage detection (VDD) is a promising field in SHM that deals with assessing the health state of structures using vibration parameters [3–5]. The key factor in VDD is to establish a reliable analytical model of a dynamic structure to estimate vibration parameters. Several researchers have reviewed literature on the vibration testing and damage detection of structures. Fan and Qiao [6] provided a comprehensive review of VDD methods. Reynders [7] reviewed the applicability of damage detection system using vibration behavior of structure. Das et al. [8] conducted a comparative study to evaluate different VDD methods. Moughty and Casas [9] performed a review of VDD techniques for small to medium span bridges.

System identification methods provide a powerful tool to construct an analytical model of a dynamic system [10–13]. Subspace system identification aims to establish a mathematical model for resolving practical problems in various branches of science and technology, such as chemistry [14,15], computer science [16], electrical engineering [17], industrial engineering [18], bioscience [19] and even finance [20]. Using subspace system identification for modal analysis is a well-established field in the dynamics of structures [21,22]. VDD methods rely on observable variations in changes in modal parameters (resonant frequency, damping, and mode shape) or their derivatives as indicators of damage existence. Song et al. [23] and Reynders [7] reviewed subspace system identification for its use in VDD and modal analysis.

Structures in VDD can be broadly divided into two categories of: (1) mechanical engineering structures, such as airplanes [24], vehicle test rig [25], ship [26], and (2) civil engineering structures, such as bridges [27], buildings [28], offshore jackets [29], and dams [30]. It is difficult to sustain any clear distinction between mechanical and civil engineering structures but, as a general idea, they could be differentiated based on their characteristics. In general, mechanical and civil engineering structure are usually subjected to different loading and boundary conditions. Civil structures are stationary, massive, and heavy [31] and they have simple structural and geometrical configuration. Civil engineering structures can be modeled in the form of simple structural elements such as beams (e.g., in bridges and wind turbines) and frames (e.g., in buildings and offshore jackets). Shells and plates are mainly used in liquid retaining and transmitting structures (e.g., in dams, and pipes). However VDD methods are not suited for structures with changing dynamic characteristics such as dams and water reservoirs. Hence, the focus of the studies on VDD of civil engineering structures is to apply their developed algorithms on beam and frame structures. Though the requirement and deployment challenges for each class of VDD structure are different, diverse techniques are essential.

Subspace system identification is one of the popular methods in time-domain that was first proposed by Van Overschee and De Moor [32] to derive modal parameters. Peeters and De Roeck [33] enhanced its computational efficiency by extending the method to handle stochastic input data. Peeters and De Roeck also utilized stabilization diagram for subspace system identification to improve the quality of the identified results [34]. Overschee et al. [32] extended the concept of weight matrices in subspace system identification as a basis for using the column space of the extended observability matrix.

Based on the incorporated input and output data, identification methods can be classified into two categories: the methods that incorporates input-output measurements to identify system parameters; so-called input–output methods, and the approaches that just use unknown output measurements, termed as output-only methods [13,35]. Since output-only methods take all excitation forces as an unknown output, the obtained results are not controllable and repeatable. Moreover, the accuracy of results is greatly affected by variation in noise level [36–38]. Despite the mentioned challenges, output-only methods are preferred over input–output methods due to the technical difficulties associated with artificial exciting of large civil engineering structures that is the main requirement of input-output methods [39,40]. Kim et al. [41,42] conducted a comparison between input–output and output-only subspace system identification methods using a model of a support-excited multi-story frame structure. Modal parameters were extracted from an input–output state-space model and the obtained results were compared to the ones obtained from output-only response data. Higher accuracy was achieved using the input–output method.

The input–output algorithm is still a tempting choice for earthquake induced excitation. Mellinger et al. [43] developed a new scheme for modal identification using output-only and input–output methods. The quality of identified system parameters was evaluated using Monte Carlo analysis in terms of accuracy of estimations and noise robustness. It was inferred that using input information provides more reliable results for modal identification. Xin et al. [44] evaluated the performance of data-driven stochastic subspace identification (SSI-DATA) using test data from offshore jacket-type platform. The efficiency and efficacy of three different excitation signals of impact, step relaxation and ground motion were investigated using both input-output and output-only algorithms. All procedures had excellent agreement with estimated modal frequencies of stronger modes. However, less accurate results were reported for damping ratios.

Stochastic subspace identification has been successfully applied for the modal analysis of several civil engineering structures [45,46]. Different authors have used the identical term of "SSI" to denote two different phenomena of "stochastic subspace identification" and "subspace system identification" [47–49]. In order to avoid confusion with the term "SSI", from now on, SSI is only given to refer "stochastic subspace identification" category and no abbreviation is going to be used for subspace system identification throughout this paper.

Recently a large number of subspace-based methods have been applied for VDD of civil structures. However, the previously conducted surveys have not kept pace with the changing environment and diversity in this field. Therefore, there is a need for a systematic review and meta-analysis focusing on the most important recent studies conducted in the considered area. The presented review systematically addresses the available studies that employed subspace-based techniques in the VDD of civil structures and describes some contributions towards the development and application of a subspace system identification algorithm in recent years. Some new perspectives are considered in the current study including classification of the selected papers.

The outline of this review paper is as follows: Section 2 reviews literature about subspace-based dynamic identification and damage detection. The research framework including the PRISMA methodology is outlined in Section 3. Section 4 describes the results and the relation between key parameters in the selected papers. Finally, Section 5 ends with the concluding remarks and recommendations for future studies.

2. Literature Review

The pioneering works in the field of SHM of civil structures have used forced-vibration as their excitation source [50]. Input–output system models, termed also as combined subspace system identification, could be simply adapted to identify dynamic parameters in forced excitation. Nowadays, a combined subspace system identification method is generally applied in modal analysis and the health monitoring of seismic-excited civil structures. Potenza et al. [51] adapted subspace system identification algorithm for seismic monitoring of historical structure by means of an advanced wireless sensor network. Zhong and Chang [52] proposed a technique that adopted an orthogonal projection to eliminate the effect of earthquake input and noise. The obtained results for combined subspace system identification algorithm are more accurate than the ones extracted from output-only identification techniques [41]. However, forced vibration and seismic motions are not always practical solutions for SHM in civil engineering due to the associated interruption in serviceability and the potential hazard to the safety [53].

2.1. Classification of the Subspace System Identification Methods

Recent researches greatly deals with application of ambient excitation for damage detection and modal analysis of the in-service structures. Output-only subspace system identification also referred as stochastic subspace identification (SSI), could be simply adapted to identify dynamic parameters in ambient excitation. In a pioneering work, Overschee et al. [32] introduced stochastic subspace identification together with combined and deterministic models within a unified framework. The proposed stochastic subspace method used Hankel block matrix of the output data to analyze system and to extract state space model. Due to the direct use of the response data in the identification process, the method is named data-driven stochastic subspace identification or SSI-DATA. The state sequence matrix is calculated before deriving state-space equation. SSI-DATA is a numerically robust algorithm that uses QR decomposition to project future data on the past subspace [54]. The methodology of SSI-DATA is provided in Figure 1.

The Hankel block matrix is directly constructed from measured vibration data as: $H_{0|2i-1} = \frac{1}{\sqrt{j}} \begin{bmatrix} Y_{0|i-1} \\ Y_{i|2i-1} \end{bmatrix}$ where i and j are the number of block rows and columns, respectively. $Y_{0|2i-1}$ is the measured data.

By QR decomposition of the obtained Hankel block matrix it yields: $H_{0|2i-1} = RQ$ where:

$$R = \begin{bmatrix} R_{11} & 0 & 0 \\ R_{21} & R_{22} & 0 \\ R_{31} & R_{32} & R_{33} \end{bmatrix} \text{ and } Q = \begin{bmatrix} Q_1^T \\ Q_2^T \\ Q_3^T \end{bmatrix}$$

Projection matrices are defined based on the obtained decompositions as:

$$P_i = \begin{bmatrix} R_{21} \\ R_{31} \end{bmatrix} Q_1^T, \ P_{i-1} = [R_{31} \ \ R_{32}] \begin{bmatrix} Q_1^T \\ Q_2^T \end{bmatrix} \text{ where: } Y_{i|i} = [R_{21} \ \ R_{22}] \begin{bmatrix} Q_1^T \\ Q_2^T \end{bmatrix}$$

By calculating the SVD of the projection matrix into orthogonal matrices of U and V together with the principle angle of Σ using: $P_i = U\Sigma V^T$

By factorization of the P_i the Kalman filter state sequence $\hat{S}_i = O_i^+ P_i$ and observability matrix $O_i = U\Sigma^{1/2} T^T$ can be obtained by the formula: $P_i = O_i \hat{S}_i$. O_i^+ is the pseudo-inverse of O_i

Using similar factorization of the P_i and $\hat{S}_i$ the one-step ahead projection P_{i-1} and Kalman state sequence $\hat{S}_{i+1}$ can be calculated as: $\hat{S}_{i+1} = [O_i^+]^+ P_i$ where O_i^+ can be obtained by deleting the last l rows.

Solving the least squares for the state space matrix A and output matrix C yields: $\begin{bmatrix} A \\ C \end{bmatrix} = \begin{bmatrix} \hat{S}_{i+1} \\ Y_{i|i} \end{bmatrix} \hat{S}_i^+$

Figure 1. Methodology of the data-driven stochastic subspace identification (SSI-DATA) technique.

The introduced identification method by Overschee et al. [32] has received considerable attention due to its well-defined algorithm and data structure. However, the aforementioned algorithm is not suitable for complex data categories with a large number of sensors, large number of modes of interest, and existing turbulence or no-stationarity. To deal with the shortcomings of the algorithm proposed by Overschee et al. [32], several researchers proposed improving the convergence rates of transfer matrices to deal with large number of sensor data [55–57]. Studies such as those of Peeters and De Roeck [33] or Reynders and De Roeck [58] suggested to reduce the data complexity using subset data, so-called reference sensors. Advance processing of measurement data before the estimation of observability matrix [59–61] and introduction of recursive identification systems [62–64] are among the proposed solutions. In order to deal with complex data, Döhler and Mevel [65] introduced a new SSI-DATA algorithm using multi-order system identification. In this method a fast computation scheme using multiple-order observability matrix is suggested to solve the least squares problem. The computational burden of the proposed algorithm is much lower than the conventional algorithms. In another research, Döhler and Mevel [27] proposed an efficient SSI algorithm by reformulation and computation of uncertainty bounds. The obtained results from application of the method on Z24 Bridge showed that the algorithm is both computationally and memory efficient.

Nowadays, wireless sensor networks (WSNs) are widely used in SHM. However, the computational load is one of the main concerns regarding the application of WSNs. Hence, it is necessary to significantly reduce the computational burden and data processing efforts. Centralized algorithms are not suitable for sensor applications due to impractical computational and communication load, as well as its increased vulnerability. Cho et al. [66] presented a decentralized SSI-DATA algorithm implemented on the Imote2-based WSNs. The results obtained from an experimental test of a five-story shear building shows a similar accuracy for the centralized and decentralized subspace system identification algorithms.

Classical covariance-based subspace algorithms [67–69] took advantage of using output data to calculate covariance. To deal with output-only measurement, Peeters and De Roeck [33] used covariance between outputs and a reference outputs for health monitoring of ambient excited civil structures. The proposed SSI-COV method used correlation functions for modal identification. In this method, the response signal of the applied ambient excitation is considered as Gaussian white noise, equal to the covariance of the response signal. The methodology of SSI-COV is provided in Figure 2.

Using SSI-COV to extract damage features or modal parameters is a common practice in VDD. Basseville et al. [70] proposed using residual of SSI-COV and a local statistical approach for VDD. Sun et al. [71] defined a nonlinear subspace-based distance using covariance of the response signal in the Hankel matrix. The distance index indicates the deviation from the normal state, and reflects structural states. Zarbaf et al. [72] derived a frequency stabilization diagram using SSI-COV method. Then, hierarchical clustering was deployed to the stabilization diagrams to identify natural frequencies of each stay cable.

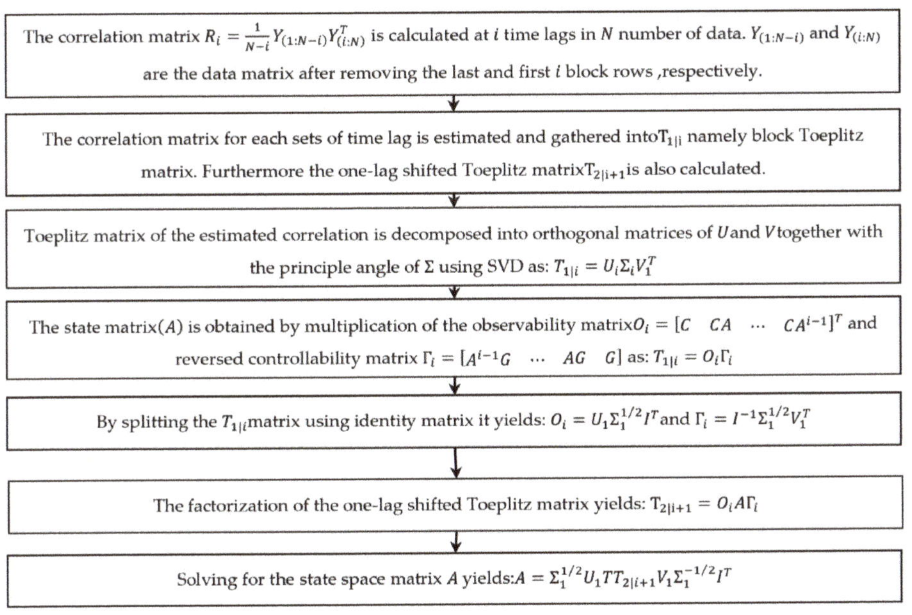

Figure 2. Methodology of the covariance-driven stochastic subspace identification (SSI-COV) technique.

For most VDD methods, it has been of great interest to study the effect of damage on eigenstructure of dynamic systems. Most of the VDD methods use modal parameters as their damage index. The dynamic characteristic of a structure can be extracted using eigensolutions [54].

2.2. Application of Subspace System Identification for Modal Analysis

Subspace-based identification methods are widely used for modal parameter estimation in time-domain [73]. For most VDD methods, it has been of great interest to study the effect of damage on natural frequencies, mode shapes and damping ratios of a dynamic systems [74–76]. Table 1 shows a number of studies that have used the subspace algorithm for modal analysis.

Table 1. Some examples of the schemes that use subspace algorithm for modal analysis.

References	Extraction Method	Test Model	Specification
Saeed et al. [77]	RSSI-COV (SubID)	Composite beam and an CACTUS aluminium plate	Iterative procedure is used to improve identification results. A stabilization histogram is applied to spurious mode elimination
Reynders et al. [78]	SSI-ICOV (CSI-ic)	Simulated model of an industrial process tower	Hybrid vibration testing or OMAX model was adopted in this study.
Li & Chang [49]	Recursive SSI-COV-IV	Numerical models of a SDOF structure and ASCE steel frame structure	Model identification was conducted for a system with time-varying measurement noise
Loendersloot et al. [79]	RD–SSIcov	Numerical model and a small scale wind turbine tower	The random decrement (RD) method was selected in this study for its noise reduction capabilities.
Miguel et al. [80]	SSI-COV	Numerical examples and a laboratory model of cantilever beams	The model is appropriate to handle incomplete measurements data and truncated mode shapes
Reynders & De Roeck [58]	CSI/ref	Z24 bridge benchmark structure	Stabilization diagram is adopted for post processing of modal data combined deterministic–stochastic subspace identification is used for modal analysis using this algorithm
Urgessa [81]	McKelvey frequency domain subspace algorithm	Uncontrolled cantilever plate	Natural frequency was predicted with an average error of 3.2% and damping ratio had average error of 2.8%
Goursat et al. [82]	used crystal clear stochastic subspace identification (CC-SSI)	Ariane 5 launch vehicle	Clear results even in the case of nonstationary data are obtained using this algorithm
Weng & Loh [83]	RSSI	3-story steel frame & 2-story reinforced concrete frame	In this methodSVD algorithm is replaced by an advanced algorithm to update LQ decomposition.
Zhang et al. [84]	Improved SSI	A numerical example of 7 Degrees of freedom (DOF) and experimental model of Chaotianmen bridge	Less computing time due to not having QR decomposition CH matrix is constructed as a replacement for Hankel matrix Spurious modes are removed using model similarity index
Döhler et al. [26]	Fast CC-SSI	Operational data from a ship	Fast multi-order computation
Hong et al. [85]	ECCA-based SSI algorithm	FE model and experimental wind tunnel bridge model	Enhanced results are achieved for weakly excited modes and noisy response signal

The methodology of calculating modal parameters from state-space parameters of subspace system identification algorithm is presented in Figure 3.

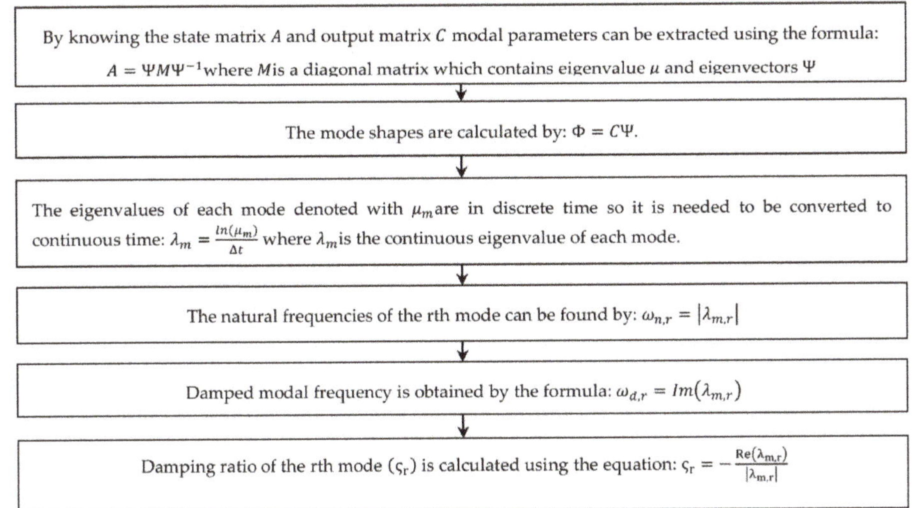

Figure 3. Methodology of the calculating modal parameters from the state-space parameters of subspace system identification algorithm.

Vibration-based SHM is concurrently subject of intensive investigation. Most of the VDD methods use modal parameters to extract dynamic characteristic of structure.

2.3. Comparison with Other Algorithms

In recent years, several studies have been conducted to compare the performance of subspace system identification with other time domain (TD), frequency domain, (FD) and time frequency domain (TFD). This subsection provides a review of the studies with focus on advantages and drawbacks of the subspace system identification. Rainieri et al. [86] assessed the performance of SSI-COV and FDD for the modal identification of ambient excited structures. The results indicated that subspace system identification is a more appropriate choice for modal identification of closely spaced modal frequencies, however coupling effect yielded unreliable result for second pairs of the closely spaced natural frequencies. Furthermore, subspace system identification had the drawback of requiring human judgment to determine system order.

Giraldo et al. [87] presented an analytical comparison among eigensystem realization algorithm (ERA), subspace system identification, and auto-regressive moving average (ARMA) techniques for modal identification of ambient-excited structures. It is indicated that subspace system identification has provided the most accurate results for analytical and experimental tests. Magalhães et al. [88] compared SSI-COV and poly-reference least squares complex frequency (p-LSCF) algorithms using field data obtained from a concrete arch bridge. Both SSI-COV and p-LSCF found to give good results for mode shapes and natural frequency. However, better results were obtained for the daily variation of damping ratio using p-LSCF. Moaveni et al. [28] used SSI-DATA, multiple-reference natural excitation technique combined with eigensystem realization algorithm (MNExT-ERA) [89], enhanced frequency domain decomposition (EFDD) [90], deterministic-stochastic subspace identification (DSI) [91], observer/Kalman filter identification (OKID)-ERA [92] and general realization algorithm (GRA) [93] for modal identification of a full-scale structure on a shaking table. The mode shapes identified by the subspace system identification algorithm were the most accurate. The measured damping ratio for SSI-DATA and MNeXT-ERA was higher than the ones obtained from EFDD.

Wang et al. [94] studied performance of subspace system identification, ERA, ARMA and Ibrahim time-domain (ITD) methods. A more stable result was reported for modal identification in a numerical model using subspace system identification. However, ERA outperforms for field testing. Kim and Lynch [95] studied subspace system identification and FDD methods. Resolution problem was reported for FDD with output-only measurements data. Cunha et al. [96] compared the modal identification results of SSI-COV and FDD. The obtained results for both of the methods were too similar. Liu et al. [97] implemented modal analysis of the Lupu Bridge in Shanghai using subspace system identification, ERA, PolyMAX, polynomial power spectrum method (PPM), power spectrum z-transform method (PZM), EFDD, frequency spatial domain decomposition (FSDD), and wavelet transform (WT) under ambient excitation. The PolyMAX, PPM, PZM, EFDD, and FSDD are in FD. Subspace system identification and ERA are TD methods used in modal identification of structures whereas WT is in time/frequency-domain. Subspace system identification provided the most accurate results for modal parameters, but computational burden of the algorithm was found to be significant.

Ceravolo and Abbiati [98] conducted a comparative study among ERA applied to RDS, AR and SSI-DATA. All of the methods were robust enough to deal with modal identification in ambient condition, but subspace system identification showed superior performance. Generally, the comparison showed that subspace system identification algorithm outperformed for identification of natural frequency, mode shape, and damping ratio. However, the computational burden of the algorithm and determining user-defined parameters are two challenges that were reported as the main downside of using subspace system identification algorithm. In the next subsection, conducted studies to overcome these challenges and improve the performance of the subspace-based algorithms are highlighted.

2.4. Challenges in the Practical Application

Several research studies have been conducted to enhance performance of the subspace system identification method. In this sub-section, the focus is on the problems involved in practical application of subspace-based damage detection. Among them merging sensors data, determining the optimum position for sensors, dealing with nonstationarity in the vibration signal, removing the uncertainties caused by environmental factors, eliminating spurious modes, improving performance of an identification scheme, determining the number of block rows and system order in subspace system identification are of the topics that is widely studied in subspace system identification. Most of these challenges are not specific to subspace system identification but generalize to all system identification methods.

In practical modal analysis of large civil engineering structures, dynamic response cannot be measured from all degrees of freedom (DOFs) in one setup. Merging sensor data, so called data aggregation, is used to reduce the number of transmissions in decentralized networks. Peeters [60] presented a subspace system identification approach to merge sensor data of different measurement setups with overlapping reference sensors. One of the solutions to merge multi-setup sensor data is to identify natural frequencies separately and merge the results in the next step. In this case inconsistency may arise due to mismatch of the identified frequencies. Another multi-setup method to deal with this problem is to merge the successive measurements, and to process them globally, instead of merging the identified natural frequencies. These methods are called post- and pre-identification merging method. Simultaneous measurement is considered as another choice for merging sensor data away from the multi-setup method. Mevel et al. [99] proposed post-identification method using SSI-COV for merging multiple non-simultaneously measured vibration responses through gluing natural frequencies and pole matching. Döhler et al. [100] used three subspace-based approaches of PoGER, PreGER and PreGER for merging non-simultaneously recorded measurement data. In another research, Döhler and Mevel [101] addressed a modular and scalable approach to solve the problem of dimension explosion in merging multi-setups. Furthermore, Döhler et al. [102] evaluated the statistical uncertainty in identified modal parameters using subspace system identification in multi-setup configuration. Orlowitz et al. [103] conducted a comparative study to investigate the relative advantages of multi-setup and simultaneous methods for merging multi-setup configuration. The post-identification method showed a better correlation of mode shapes and natural frequencies, however, for the structures with changing dynamic characteristics such as dams and water reservoirs.

Subspace system identification has shown great potential in identification of dynamic parameters in civil structures. It was shown by Benveniste and Mevel [104] that the subspace algorithm is robust against nonstationarity caused by parameters such as varying operating load. Benveniste and Mevel [104] studied the impact of nonstationarity in the vibration signal on consistency of subspace system identification algorithm. It is reported that subspace algorithm ensures consistency against nonstationarity. Alıcıoğlu and Luş [105] assessed the effect of structural complexity and ambient uncertainty on identified modal parameters using SSI-COV and SSI-DATA techniques. It was demonstrated that the algorithm performed reliably in the identification of natural frequencies and improved efficiency was achieved by adopting a stabilization diagram. Clustering analysis was found to be promising to automate selecting of real modes.

Separating the effect of externally acting agents such as operational and environmental factors is important for successful damage detection. Several researchers have studied the effect of environmental variation in dynamic identification, as shown in Table 2. Hence, some researchers reported measuring externally acting agents along with measurement of the vibration response.

Table 2. Influence of environmental and operational condition on damage detection of structures.

Reference	Test Model	Environmental and Operational Effect
Sohn et al. [106]	Alamosa Canyon Bridge	5% daily change in natural frequency due to temperature variation
Liu and DeWolf [107]	Real-scale bridge	4–5% variation in natural frequencies during spring and winter were observed.
Nayeri et al. [108]	a full-scale 17-story building	Correlation between modal frequency and temperature is reported in a 24-h period.
Cornwell et al. [106]	Alamosa Canyon Bridge.	6% variation in modal frequencies have been recorded
Wood [109]	Bridge beam	Damp air caused decrease in natural frequency of structures
Xia et al. [110]	Reinforced concrete slab	2% increase was recorded when relative humidity was ranged from 15% to 80%.
Farrar et al. [74] and Alampalli [111]	Alamosa Canyon Bridge	Variation in modal parameters is entirely dependent on the targeted structure
Peeters et al. [112]	Z24 bridge	Frequency variation due to ambient, shaker and impact excitations was very small
Peeters and De Roek [113]	Z24-Bridge	Temperature differentials across the bridge deck as the driving forces for natural frequency variations.
Ni et al. [114]	Ting Kau Bridge	Temperature variation changes modal frequencies with variance ranged from 0.20% to 1.52% in the first ten modes.
Kim et al. [115]	Experimental model of a Euler–Bernoulli beam	Natural frequencies variation/ambient temperature from 0 °C to 30 °C was 19%, 10%, 13% and 7% for 1st, 2nd, 3rd and 4th modes, respectively.

Spiridonakos et al. [116] incorporated the variance of the uncertainties caused by humidity and temperature in identification of the modal parameters using subspace system identification. Two polynomial chaos expansion and independent component analysis were conducted to isolate structural variations caused by deviation of acting agents and extraction of structural features, respectively. Loh and Chen [117] addressed covariance-driven recursive stochastic subspace identification (RSSI-COV) for isolating environmental effect from anomaly caused by damage. Huynh et al. [118] analyzed the wind-induced vibration due to typhoons with various wind speeds. Deraemaeker [119] evaluated the robustness of subspace system identification method by introducing uncertainty into the FE model. It was shown that, other than the effect of externally acting agents, the inherent performance of an identification scheme plays an important role in accuracy of the estimation result. Then studying of the detectability of the dynamic parameters is of paramount importance. Magalhães et al. [120] studied the effect of several factors, including the proximity of natural frequencies, non-proportional damping, and accuracy of the identification algorithms, on the quality of the extracted damping ratios. Rainieri and Fabbrocino [121] investigated the influence of the number of block rows and system order on estimation accuracy in subspace system identification algorithm. The most robust identification using a subspace system identification algorithm is obtained when the number of data goes to infinity. Short-length data cause estimation bias in modal identification. The bias error is intensified when dealing with systems having high damping and high frequency. Wang et al. [122] proposed a combined subspace system identification and ARX algorithms for VDD of Hammerstein systems. Li et al. [123] developed a subspace system identification algorithm to eliminate spurious modes caused by non-white noise. Brasiliano et al. [124] investigated the effect of non-structural elements on vibration parameters using SSI-COV and SSI-DATA. Cara et al. [125] discussed the modal contribution in each mode to the recorded vibration signal. In some structural systems ambient excitation is the only practical means to excite civil structure as a result; some of the modes are not influenced. Ashari et al. [35] introduced injecting auxiliary input to the subspace system identification algorithm to extract the unexcited modes. Several methods are used to introduce uncertainty including adding Gaussian perturbation into natural frequency or damping coefficients, adding independent Gaussian noise at each mode-shape measurement location and adding uncorrelated noise on the extracted vibration response.

Some other researchers studied the specific cases that may occur in practice. Pridham and Wilson [126] investigated the use of correlation–driven SSI to estimate damping ratio from short-length data sets. Banfi and Carassale [127] studied the effect of environmental variability and short-length measurement data in determining modal parameters. Marchesiello et al. [128] proposed short-time stochastic subspace identification (ST-SSI) to deal with time-variant identification. Markovsky [129] developed a subspace system identification algorithm for dynamic system with missing data. Brownjohn and Carden [130] compared the degree of uncertainty in black box identification from the author's experiences. Carden and Mita [131] summarized the challenges to extract accurate confidence intervals in the modal identification of civil structures using subspace system identification.

As demonstrated above, the most researched challenges in implementation of subspace system identification algorithm deal with merging multi-setup sensor data and improving the performance of the subspace algorithm for the identification of the modal parameters using short-length measurement data. In the next subsection, the use of subspace system identification in the development of software is presented.

2.5. The Software Packages

The subspace method has been used in many structural monitoring and modal analysis software programs. In this subsection, the software packages that used subspace system identification for modal identification and SHM are further investigated. ARTeMIS is a self-stand tool suite that utilized CC-SSI for operational modal analysis [132]. Reynders and De Roeck [58] developed MACEC for modal analysis in TD and FD. SSI-COV, SSI-DATA, combined deterministic-stochastic subspace identification (CSI), and their reference-based generalization (SSI-data/ref, SSI/ref and CSI/ref) are adopted in the software package. MACEC 3.2 is the latest version of the software [133]. ModalVIEW [134] software was developed under LabVIEW which used subspace system identification algorithm for modal analysis. Hu et al. [135] presented structural modal identification (SMI) and continuous structural modal identification (CSMI) for modal analysis within the LabVIEW environment. Goursat and Mevel [136] proposed COSMAD toolbox in Scilab, for in-operation damage identification that used SSI-COV as the basic identification tool in the software. Chang et al. [137] introduced structural modal identification toolsuite (SMIT) to study the modal parameters of natural frequency, mode shapes, and damping ratio.

Operational modal analysis (OMA) [138] is another software program that uses subspace system identification for the dynamic identification of structure and it has been used for the modal identification of several structures such as Berta Bridge [139] and Berke Arch Dam [140]. LMS Cada-X [141] is another software program employing subspace algorithm. The software is developed by LMS International in Leuven, Belgium. TestLab [142] is another software by LMS that was used extensively for modal analysis. The software also used a subspace algorithm for parameter identification. Automated operational modal analysis (AOMA) [143] utilized a strong identification and stabilization diagram. The algorithm uses one user-defined parameter.

3. Methodology

For the research methodology of the present review paper, the preferred reporting items for systematic reviews and meta-analyses (PRISMA) is proposed by Moher et al. [144]. PRISMA statement consists of two main parts of systematic reviews and meta-analysis. Systematic reviews provide objective summaries of researches carried out on a specific field. An explicit and systematic method is used for identification, selection, appraisal, collecting and analysis of the data to answer clearly formulated questions about the studies included in the review. This is highly useful especially in wide research area to encompass the researches that focus on narrow aspect of the field [145]. The provided explicit framework to conduct the review is to ensure the procedure is objective and replicable by others. Meta-analysis is referred to as the statistical analysis recommended for integrating findings of the included studies. The main goal of using PRISMA statement is to help authors to improve reporting of literature reviews [146–149]. The PRISMA statement has been used in several studies to provide comprehensive literature review in various fields. In order to conduct the present review study, a three step procedure including search in literature, choosing the eligible published papers and data extraction and summarizing is employed.

3.1. Literature Search

Literature search was carried out by consulting three databases of Scopus, Web of Science, and Google Scholar for systematic review of the applications and methodologies on subspace-based SHM. Defining keywords for a systematic review and meta-analysis is more than just important. Selecting keywords from subject heading is of the best tools for efficient retrieval and survey of data from database [150]. Hence, in the first step, the following combinations were used in the keyword search: ("subspace system identification" AND ("structural health monitoring" OR "damage detection" OR "fault detection" OR "modal")). Duplicates and unrelated articles; assessed from title screening; were excluded from the study. Following the database searches and title screening, eligibility of the retrieved records were assessed through abstract screening. The search process was iterative, and the studies that met the inclusion and exclusion criteria were continuously extracted till the end of the study. Moreover, the search terms were refined in the process of becoming familiar with literature. Other search keyword were also added in the course of the review process such as a combination of ("subspace system identification" AND ("output-only" OR "ambient excitation" OR "civil" OR "stochastic")).

It is now about 25 years or more since subspace system identification was linked as an approach to the dynamic identification and SHM of civil structures. The literature search and eligibility assessment study shows that the time period 1995–2019 can be divided into two time intervals. The 1995–2008 can be characterized to development of the theoretical foundation and conceptualization of the framework that is discussed in introduction section. Hence, to deal with application and application-related topics more specifically, the scope of the literature search was limited to the papers published in the time frame of 2008–2019. An evaluation process was conducted to determine whether a publication must be retained in the final list.

The literature search was confined to the English language journal papers and the relevant works in the form of book chapters, non-indexed conference papers, editorial notes, master dissertations, doctoral theses, and textbooks were excluded from the review. Abstract review is the first screening of the papers for inclusion or exclusion that is conducted based on the pass/fail criteria. Using this criteria a total of 90 scholarly papers were identified. The duplicated records with redundant information were removed from the final search results. In this stage, 67 papers remained. All the above identified articles were thoroughly read based on topics and abstracts while unrelated studies were removed. Totally, 69 potentially related studies qualified, as shown in Figure 4.

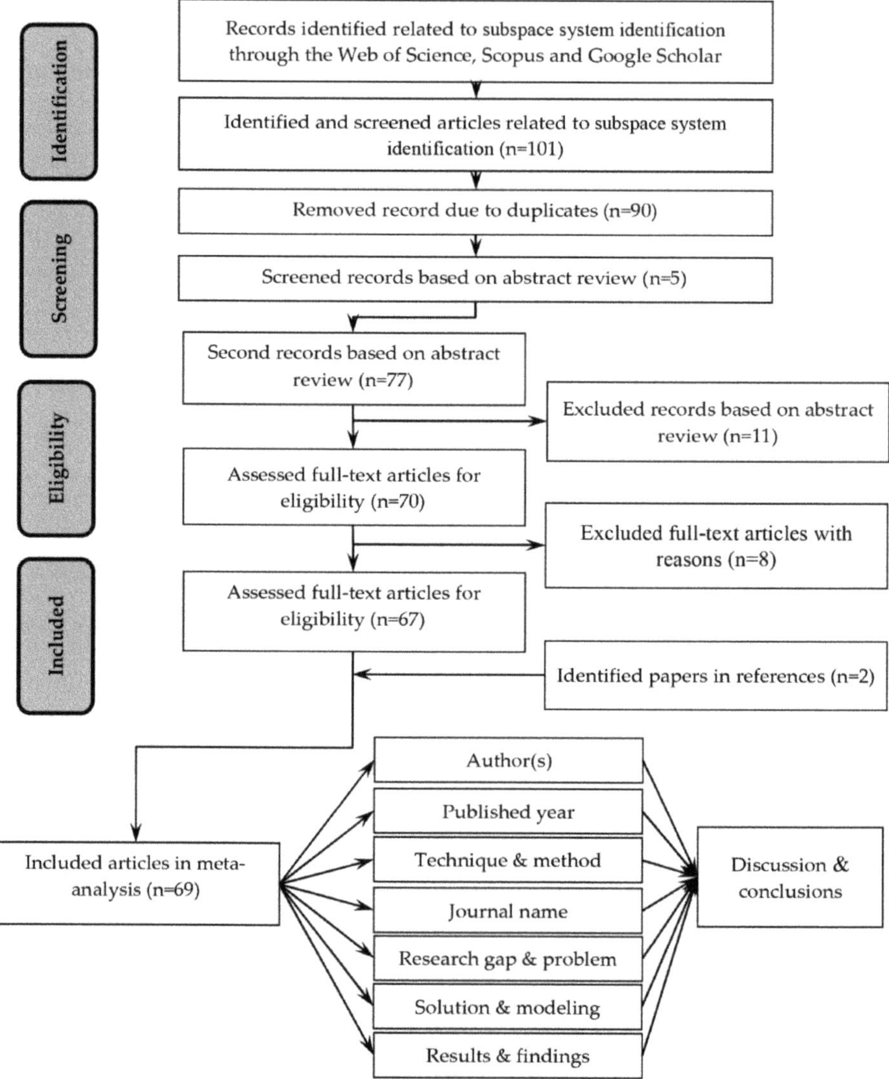

Figure 4. Study flowcharts for the identification, screening, eligibility and inclusion of articles.

3.2. Articles Eligibility

Article eligibility was assessed based on full-text reading of each manuscript obtained from the above process. In the final step all identified articles were carefully read in its entirety to confirm the significance and relevance to the review topic. In several previous studies, the combined subspace method is used for modal identification and SHM of civil engineering structures under the seismic excitation. However, the ambient excitation is the most common procedure for SHM of civil engineering structures; as a result the focus in the literature search is more on SSI-COV and SSI-DATA rather than the combined method. In the end, 69 articles were selected for the application of SSI in SHM of civil structures from 31 scholarly international journals between 2008 and 2019 that satisfied the inclusion criteria.

3.3. Summarizing and Data Extraction

In the final step of our methodology, finally 69 articles were reviewed and summarized. Furthermore, all articles were reviewed based on various criteria such as the used technique and method, research gap and results and findings. We believe that, the reviewing, and classifying of articles can help to extract valuable and important information. Consequently, several recommendations were given for future studies. It is noteworthy that the difficult part during the accomplishment of the PRISMA method was to extract the implicit methodology in abstracts and the context of the selected articles. Hence, in order to provide sufficient information and unbiased decisions regarding the approach applied in the analysis, in most cases, the full manuscript was searched. The authors believe that this review could help the readers to find the most relevant and appropriate published studies regarding subspace system identification.

4. Distribution of the Subspace-Based Damage Detection Techniques

4.1. Distribution of the Papers on SSI-DATA Approach

Table A1 in Appendix A shows those studies which used SSI-DATA technique. A total of 31 studies have used SSI-DATA method alone or combined with other methods in various test structures such as beams and 2D frames, 3D frames structures and buildings, and bridges and other structures.

WSNs are promising future use technology and now are applied for SHM of civil engineering structures. Some of the studies in application of SSI-DATA algorithm are dealt with the limitations of WSNs facilities for data transmission and developing dense networks of low-cost wireless sensors for complex infrastructures [66,151,152]. To deal with the limitations of WSN facilities for data transmission Cho et al. [66] presented a decentralized SSI-DATA algorithm implemented on Imote2-based WSNs. An experimental test of a five-story shear building was used as the verification test. The identification results obtained from decentralized and centralized SSI techniques were close to each other. Kurata et al. [151] developed a novel internet-enabled wireless structural monitoring system for large-scale civil infrastructures. A wireless monitoring system was installed on New Carquinez Bridge to verify the applicability of the proposed framework. The obtained results verified the stable and reliable application of the proposed system on a large number of nodes. Kim and Lynch [152] introduced an indirect SSI-DATA algorithm based on Markov parameters customized for decentralized WSNs. The proposed strategy is verified by dynamic testing of a cantilevered balcony in a historic building. System properties were identified with a high accuracy.

FE model updating is a powerful tool in SHM to ensure that FE analysis reflects the real behavior of structures. Several researches on SSI-DATA were focused on practical limitation of FE updating and to validate a reliable FE model [153–155]. In order to validate FE models by applying identification methods, Nozari et al. [153] implemented an FE model updating framework to identify damage in a ten-story reinforced concrete building. Due to the limitations of experimental responses and measurement errors, the optimization in FE updating problem may reach multiple solutions in the search domain. To deal with this problem, Shabbir and Omenzetter [154] applied a methodology using particle swarm optimization (PSO) with sequential Niche technique (SNT) for FE model updating of a pedestrian cable-stayed bridge. It was shown that the proposed methodology gives more confidence for model updating. In order to know the dynamic behavior of complex buildings subjected to near-fault earthquakes, Foti et al. [155] used output-only EFDD and SSI-DATA to identify modal parameters of two buildings to update an FE model of the damaged structures. Testing was conducted on a complex building which was heavily damaged in an earthquake. After a series of improvements of the model, satisfactory agreement has been reached.

Several researches have been conducted to improve performance of classical SSI-DATA to be applied on continuous time SHM and enhance the efficiency [84,156–158]. In order to track the current structural state from building seismic responses, Chen and Loh [156] developed two recursive SSI-DATA algorithms using BonaFide LQ renewing algorithm and inversion lemma algorithm. Two sets of building seismic response data from a three-story steel structure and a four-story-reinforced concrete elementary school building were used for verification of proposed methods. The results show that subspace system identification inversion with forgetting factor could provide more accurate estimation of the stiffness change. Li et al. [157] developed a reference-based subspace system identification technique to identify structural flexibility using modal scaling factors. A numerical model of an RC bridge and a laboratory-scale simply supported beam were presented to illustrate the robustness of the proposed method. The examples, successfully illustrated the robustness of the proposed method. Dai et al. [158] presented a modified subspace system identification method for modal analysis of structures under harmonic excitation with frequencies close to natural frequencies of the structure. In this method, Hankel matrix was modified by adding harmonic vectors. Application of the algorithm on numerical lumped-mass dynamic system model and an in-service utility-scale wind turbine tower resulted in accurate estimation of the modal parameters. Zhang et al. [84] introduced a CH matrix as a replacement for a Hankel matrix and replaced a projection operator with the classical QR decomposition. A seven-DOF numerical model and experimental test of Chaotianmen Bridge were used to verify the method. An improved computational efficiency without losing the quality and separation of the spurious modes are the advantages achieved using the proposed algorithm. Further details of the selected papers of this section can be found in Table A1.

4.2. Distribution of the Papers on SSI-COV Approach

Table A2 in Appendix A shows the studies with focus on the SSI-COV approach. From the data presented in this table, a total of 25 studies used SSI-COV in various structures including beams and 2D frames, 3D frames structures and buildings, and bridges. Some of these studies integrated SSI-COV approach with preprocessing or postprocessing stages [72,159–163].

In order to smoothen input signal and yield reliable modal parameters, Loh et al. [159] adopted singular spectrum analysis (SSA), for preprocessing of the response signal, and a stabilization diagram for postprocessing of the extracted modal parameters, respectively. The experimental test was carried out for the validation of the proposed algorithm using the long-term monitoring data of Canton Tower high-rise slender structure. It was found that the use of SSA as a pre-processing tool for SSI-COV improved the identifiability of modes using a stabilization diagram. To estimate the tension forces of the cables in cable-stayed bridges, Zarbaf et al. [72] adopted hierarchical clustering algorithm to identify natural frequencies of each stay cable in Veterans' Glass City Skyway Bridge. The agreement between the estimated results and the measured tension forces was good. Due to the need for the removal of bias and variance errors in the modal parameter estimation, Reynders et al. [161] used first-order sensitivity of the modal parameters and stabilization diagram to remove bias errors. A simulation model and measured vibration data of a beam and a mast structure were used for the verification purpose. The practicability of the proposed method was confirmed in a real-world application.

In order to improve the identifiability of the weakly excited modes Zhang et al. [162] introduced component energy index (CEI) and an alternative stabilization diagram to identify spurious and physical modes. A simulation model of a seven-DOF mass-spring-dashpot (MSD) system and the experimental model of a metallic frame structure subject to wind load were used for verification of the proposed scheme. Good performance was observed especially for the measurement data with low SNR. In order to identify structural changes in presence of environmental variation, Carden and Brownjohn [163] proposed a fuzzy clustering algorithm to extract state parameters from real and numerical poles. Data from Z24 Bridge and Republic Plaza Office Building in Singapore were used for experimental verification of the method. The inflicted damage on the Z24 Bridge was successfully identified using the proposed method. The shifts in modes of the Plaza Office Building in Singapore were also clearly captured.

Several studies on SSI-COV were concerned with discrimination environmental and operational effect during the identification process by improving the inherent performance of the SSI-COV algorithm. Döhler et al. [164] presented an efficient and fast SSI-COV damage detection that is robust to changes in the excitation covariance. Three numerical applications were presented. It is reported that the new approach can better detect and separate different levels of damage.

Several researches on SSI-COV dealt with improving the damage detection process by introducing a damage sensitive and noise-insensitive features [71,165–169]. To discriminate changes in modal parameters caused by damage from those occurred due to environmental factors, Basseville et al. [165] designed a damage detection algorithm using null space residual, $\chi 2$ test and a statistical nuisance rejection. A vertical beam made of steel and aluminum was tested under controlled ambient temperature for verification of the presented scheme. The relevance of the presented algorithm was illustrated using a laboratory-scale test structure. Balmès et al. [170] proposed the use of subspace residual as damage feature and $\chi 2$ tests to discriminate the effect of noise from estimated modal parameters. A simulated bridge deck with controlled temperature variations was used for verification of the proposed method. Efficiency of the method on simulation model for various temperature cases was confirmed. Zhou et al. [168] used a residual of the subspace system identification and global $\chi 2$-tests for damage detection. A full-scale bridge benchmark was validated by numerical simulation. It is reported that the damage in tower was detected in the same time. In order to consider nonlinearity of structures for identification of modal characteristics Sun et al. [71] defined a nonlinear subspace distance as damage feature. The proposed index was validated by the data obtained from a viscoelastic sandwich structure (VSS) subjected to an accelerated ageing. It is shown that the designed index is very effective to evaluate the health state in the structure. Ren et al. [169] adopted Mahalanobis and Euclidean distance decision functions for the pattern recognition of a proposed damage index. One numerical signal and two simulated FE dynamic beam models were used for the verification of the proposed procedure. The method was capable of locating damage in FE beam structures. Details of selected papers which adopted the SSI-COV approach in their identification process are presented in Table A2.

4.3. Distribution of the Papers on Combined Subspace System Identification Approaches

Table A3 in Appendix A shows the studies which used combined subspace system identification techniques. Based on results presented in the table, a total of 13 studies have used combined subspace system identification algorithms for analysis of various test structures. Though subspace system identification algorithm is originally a TD identification approach, some researchers have developed the FD version of the combined subspace system identification algorithm for identification of the vibration parameters [81,171]. In order to meet interpretation challenges associated with system identification obtained from measured sensor data, Urgessa [81] presented two FD system identification methods by adopting ERA and the McKelvey subspace system identification approaches. FE model of a plate structure was used for verification of the proposed algorithms. The methods were able to predict natural frequencies and damping ratio with a high accuracy. Akçay [171] proposed a two-step subspace algorithm by calculating minimal realization of the power spectrum samples and a canonical spectral factor. A numerical example is provided to illustrate the performance of the proposed algorithm. Serious drawbacks regarding reliable performance of the algorithm dealing with short data records and corrupted data were reported. Several studies are concerned with improving the performance of the combined subspace system identification algorithm [41,42,172,173] to deal with these problem. Kim and Lynch [41,42] presented a theoretical framework to extract actual physical parameters of structures using a physics-based model and a data-driven mathematical model. Numerical model of a multi-DOF shear building structure and experimental verification test of a six-story steel frame structure under support excitation were tested. The proposed grey-box framework has shown a promising performance for SHM of civil engineering structures exposed to base motions. Gandino et al. [172] developed a novel multivariate input–output SSI-COV formulation for modal parameter identification. A 15-DOFs numerical example and an experimental application consisting of a thin-walled metallic structure were used for verification. The obtained results were similar to those reached by data-driven method. Verhaegen and Hansson [173] introduced data-driven input-output N2SID using convex nuclear norm optimization. Mathematical formulations are furnished to derive the theory of the N2SID algorithm. The sequence for derivation of the system parameters from N2SID was clearly demonstrated. Table A3 provides the information of the selected papers which applied combined subspace system identification approaches.

4.4. Comparison among Identification Methods

Several subspace system identification methods have been applied for modal identification and VDD of civil structures. These methods are in the form of output-only or input–output algorithms. Output-only algorithms are used for vibration analysis of ambient excited structures. SSI-DATA and SSI-COV techniques are the two main output-only subspace system identification algorithms. SSI-COV algorithm uses the covariance of the raw time-history to reduce the dimensionality of the measurement data. Data reduction in SSI-DATA is performed using QR projection of the Hankel matrix. Both subspace system identification algorithms use SVD to determine the order of a dynamic system. The calculation of the covariance matrix is faster compared to calculation of the QR decomposition which is much slower. However, both algorithms are reported to perform well for the estimation of the modal parameters whereas SSI-DATA is expected to be theoretically more robust due to avoiding squaring up of the measurement data. Combined subspace system identification algorithm is used for identification of system parameters with known input data. More reliable results are obtained by using the input-output subspace system identification compared to the output-only scheme. Several algorithms are introduced based on the classical SSI-COV, SSI-DATA and the combined method to improve the performance of the subspace system identification for SHM application. The performance is enhanced either by change in structure of the underlying algorithms or by adding preprocessing or postprocessing steps to the original subspace system identification algorithm. In some cases, the subspace system identification algorithm is integrated with other analytical methods to yield higher performance.

4.5. Test Structure's Classification

Selected articles are categorized into five different test structures including 2D structures, 3D frame structures and buildings, bridge structures, multiple test structures, and others. 2D structures are in the forms of simply supported beam, cantilever beam or 2D shear frames. Most of the applied 3D test structures for verification of subspace system identification algorithms in this study were in the form of 1-span shear building tested on shaking table for progressive damage test. Furthermore, some of the algorithms are applied into structures from two different categories such as "bridge, and 3D frame and buildings" which are classified within the multiple test structure groups. The category "others" include structures such as dam, wind turbine, chimney, tensegrity systems and sandwich structures. The distribution of the selected paper list based on test structures and applied subspace system identification methods is shown in Figure 5.

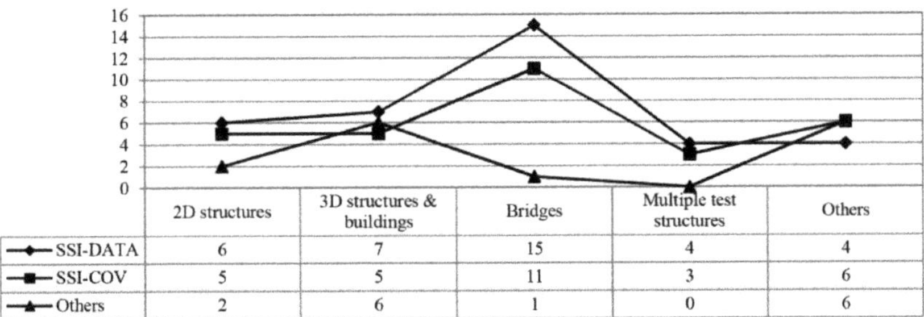

	2D structures	3D structures & buildings	Bridges	Multiple test structures	Others
SSI-DATA	6	7	15	4	4
SSI-COV	5	5	11	3	6
Others	2	6	1	0	6

Figure 5. The distribution of the paper by the test structures and the applied subspace system identification methods.

5. Conclusions

In this review paper, the theory and applications with respect to recent developments of the subspace system identification approach in the modal identification and health monitoring of civil engineering structures are comprehensively reviewed. The applied test structures of these selected papers were classified into five groups. These papers are accessible via three important databases of Scopus, Google Scholar, and Web of Science. To this end, 69 studies were carefully selected about subspace system identification application in health monitoring of the civil engineering structures based on title, abstract, introduction, research method, and conclusion. A number of important issues with respect to subspace system identification application were extracted from the present literature review. The extensive of the selected studies were published in 2016. In total, papers were classified into five test structures including 2D frame structures, 3D frame structures and buildings, models tested on multiple structures and others. In this regard, bridge structures were the most likely candidate structure with 25 papers using SSI-DATA, SSI-COV, and combined subspace system identification approaches. In addition, 31 international journals were considered in the current review paper.

Output-only methods are generally applied for identification of the state-space parameters under ambient excitation where the combined method used seismic or forced vibration excitation. Test structures for input-output subspace system identification are generally 2D or 3D frames or buildings where in output-only subspace system identification, test structures are generally bridges. SSI-DATA is the most researched subspace system identification approach in health monitoring of the civil structures. The obtained results for SSI-DATA and SSI-COV algorithms are overall similar in the case of accuracy, but the computation time SSI-COV is much lower than the SSI-DATA approach. The research works contributed with the SSI-COV are mainly concentrated on improving the quality of the obtained modal parameters using preprocessing or postprocessing techniques. Stabilization diagram is the most applied postprocessing method to select physical modes and distinguish false and spurious modes. Additionally, some studies are conducted to introduce appropriate damage features for SHM. However, the research studies in the SSI-DATA are generally devoted to enhancing the intrinsic structure of the subspace system identification algorithm itself, or integrating with other soft computing approaches to deal with the problem.

This study confirms that subspace based damage detection approaches can help researchers and practitioners to overcome some uncertainties regarding the quality of the condition assessment in various application areas. The present review has some limitations, which are common to these types of studies and can be considered as an object of future studies. First, this review is focused mainly on the application of a subspace system identification algorithm for the health monitoring of civil structures rather than the theory and development of the classical subspace-based techniques. Second, the available papers of the publishers in Web of Science, Scopus, and Google Scholar till the end of November 2019 have been included in the identification process.

This review can be expanded to include future studies. Another limitation is that the collected data were from international journals while non-indexed conferences papers, textbooks, doctoral theses, and masters projects were excluded from the current study. Therefore, in the future studies, the data from the aforementioned resources can be collected and the obtained results can be evaluated with the data reported in this study. However, the authors believe that this paper has comprehensively reviewed the most published papers in international journals focusing on several aspects such as the authors, publication year, technique and methods, research purpose, gap and contribution, solution and modeling, and results and findings. It is recommended that future papers focus on different functions. In this regard, the current review paper presented some opportunities to find gaps that can be addressed for further study directions.

Author Contributions: Resources, H.S. (Hoofar Shokravi), H.S. (Hooman Shokravi), N.B., M.H., S.S.R.K. and M.P., investigation, H.S. (Hoofar Shokravi); writing—original draft preparation, H.S. (Hoofar Shokravi) and H.S. (Hooman Shokravi); writing—review and editing, H.S. (Hoofar Shokravi), H.S. (Hooman Shokravi), N.B. and S.S.R.K., M.H.; visualization, H.S. (Hoofar Shokravi), H.S. (Hooman Shokravi), N.B., S.S.R.K. and M.P., M.H., supervision, N.B.; project administration, H.S. (Hoofar Shokravi), H.S. (Hooman Shokravi), N.B., M.H., S.S.R.K. and M.P.; funding acquisition, N.B., S.S.R.K., and M.P.; All authors have read and agreed to the published version of the manuscript.

Funding: This research was funded by Ministry of Higher Education, Malaysia, and Universiti Teknologi Malaysia (UTM) for their financial support through the Fundamental Research Grant Scheme (grant number: 4F800) and Higher Institution Centre of Excellent grant (grant number: 4J224) The APC was funded by Ministry of Education, Youth, and Sports of the Czech Republic and the European Union (European Structural and Investment Funds Operational Program Research, Development, and Education) in the framework of the project "Modular platform for autonomous chassis of specialized electric vehicles for freight and equipment transportation", Reg. No. CZ.02.1.01/0.0/0.0/16_025/0007293.

Acknowledgments: The authors would like to thank the Ministry of Higher Education, Malaysia, and Universiti Teknologi Malaysia (UTM) for their financial support through the Fundamental Research Grant Scheme (4F800) and Higher Institution Centre of Excellent grant (4J224), Ministry of Education, Youth, and Sports of the Czech Republic and the European Union (European Structural and Investment Funds Operational Program Research, Development, and Education) in the framework of the project "Modular platform for autonomous chassis of specialized electric vehicles for freight and equipment transportation", Reg. No. CZ.02.1.01/0.0/0.0/16_025/0007293.

Conflicts of Interest: The authors declare no conflict of interest.

Nomenclature

ARMA	Auto-regressive moving average
CC-SSI	Crystal clear stochastic subspace identification
CSI	Combined deterministic-stochastic subspace identification
CSMI	Continuous structural modal identification
DOFs	Degrees of freedom
DSI	Deterministic-stochastic subspace identification
EFDD	Enhanced frequency domain decomposition
ERA	Eigensystem realization algorithm
FD	Frequency-domain
GRA	General realization algorithm
ITD	Ibrahim Time-domain
MIMO	Multiple-input multiple-output
MNExT-ERA	Multiple-reference natural excitation technique combined with ERA
MOESP	Multivariable output error state-space
MSD	Mass-spring-dashpot
OKID	Observer/Kalman filter identification
PPM	Polynomial power spectrum method
PRISMA	Preferred reporting items for systematic reviews and meta-analyses
PSO	Particle swarm optimization

PZM	Power spectrum z-transform method
RD	The random decrement
RSSI-COV	Covariance-driven recursive stochastic subspace identification
SHM	Structural health monitoring
SIMO	Single-input multiple-output
SMI	Structural modal identification
SMIT	Structural modal identification toolsuite
SSI	Stochastic ubspace identification
SSI-COV	Covariance-driven stochastic subspace system identification
SSI-DATA	Data-driven stochastic subspace system identification
ST-SSI	Short-time stochastic subspace identification
TARMA	Time-varying analysis method using time-dependent auto-regressive moving average
TD	Time-domain
TFD	Time/frequency domain
VDD	Vibration-based damage detection
VSS	Viscoelastic sandwich structure
WSNs	Wireless sensor networks
WT	Wavelet transform

Appendix A

Table A1. Distribution of the papers based on SSI-DATA techniques.

Author	Method	Research Objective	Research Gap and Problem	Solution and Modeling	Result and Finding
Priori et al. [174]	SSI-DATA	Proposed rules to determine the number of block rows and columns of the Hankel matrix	Need to define optimum value for -defined parameters in SSI	Vibration test on a scaled structure and tests on a real-size RC building.	Rules to determine the lower bound for the user-defined parameters of the SSI algorithm was discussed.
Pioldi and Rizzi [175]	Improved SSI-DATA	Adopted an improved SSI-DATA procedure and a refined FFD algorithm	Need to identify modal parameters from short-duration, non-stationary, earthquake-induced response	A numerical model of a ten-story frame structure under a set of selected earthquakes	Both rFDD and the SSI- methodologies turn out robust results.
Chen and Loh [156]	Improved SSI-DATA	Developed two algorithms of recursive SSI with BonaFide LQ renewing algorithm and matrix inversion lemma algorithm	Need to track structural current state from the building seismic response	A three-story steel structure and a four-story-reinforced concrete an elementary school building	The SSI Inversion with forgetting factor can provide more accurate estimation of the stiffness change.
Li et al. [157]	Reference-based SSI-DATA	Developed a SSI technique to identify structural flexibility using the modal scaling factors	Need to correct estimation of the structural modal scaling factor and flexibility characteristics	A numerical model of a RC bridge and a laboratory-scale simply supported beam	The Examples successfully illustrated the robustness of the proposed method.
Park and Noh Hae [176]	SSI-DATA	Adopted an iterative parameter updating	Need to deal with practical limitation of output-only methods	A numerical model of a 5-story shear building	The modal parameters are estimated with 85–99%. Updating further improves these accuracies.
Nozari et al. [153]	SSI-DATA	Implemented a FE model updating framework to identify damage in a 10-story reinforced concrete building.	Need to validate FE models by applying identification methods	A ten-story reinforced concrete building	The updated model parameters shown considerable variability across different sets.
Dai et al. [158]	SSI-DATA	Presented a modified SSI method for modal identification under harmonic excitation	Need for a SHM system to ensure proper performance and save maintenance costs in wind turbines	A numerical lumped-mass system model and an in-service utility-scale wind turbine tower	The modal parameters of the first two modes were accurately estimated.
Tarinejad and Pourgholi [30]	SSI-DATA	Proposed an algorithms using stochastic realization theory and canonical correlation analysis for operational modal analysis	Need to deal with uncertainties of unknown nature such as ambient noises and measurement errors.	Experimental tests on Shahid-Rajaee arch dam and Pacoima dam	More accurate natural frequencies are obtained compared to those of classic SSI.
Soria et al. [177]	SSI-COV, SSI-DATA & SSI-EM	Studied the influence of the environmental and operational factors using three SSI-based modal analysis techniques	Need to a low-cost vibration-monitoring system	A steel-plated stress-ribbon footbridge was used as the experimental case study	An excellent correlation for the lowest persistent vibration modes was reported.
Loh et al. [178]	SSI-DATA	Used SSI and a technique to remove spurious modes	Need to identification of an earthquake-induced structural response	One 7-story RC building and one mid-isolation building and an isolated bridge	The identified system dynamic parameters were used for seismic assessment of the structures.
Lardies [179]	SSI-DATA	Presented four different algorithms of (i) block Hankel matrix, block observability and block controllability and shifted versions	Need to determine the transition matrix	Numerical model of a two-DOF system and experimental model of a cantilever beam	The same results are obtained using these algorithms.

Table A1. *Cont.*

Author	Method	Research Objective	Research Gap and Problem	Solution and Modeling	Result and Finding
Cho et al. [66]	SSI-DATA	Presented a decentralized SSI-DATA implemented on the Imote2-based WSN	Need to deal with the limitations of WSNs facilities for data transmission	Experimental test of a 5-story shear building model using WSNs	The identification results obtained from the WSNs and the centralized were close to each other.
Shabbir and Omenzetter [154]	SSI-DATA	Proposed a particle swarm optimization with sequential niche technique (SNT) for FE updating	Need to deal with the limitation of FE updating problem	FE model updating of a pedestrian cable-stayed bridge is used to analyze the method	The proposed methodology gives the analyst more confidence for model updating.
Junhee et al. [180]	SSI-DATA	Applied a SSI technique to model guided wave propagation	Need to model complex dynamics behavior of wave propagation.	Welded plates of varying thicknesses	The algorithm was capable to simulate the propagating waves.
Yu et al. [181]	SSI-DATA	Investigated the time-varying system identification in temperature-varying environments.	Need to confirm the applicability of time-varying modal parameter identification algorithm	A steel beam with a removable mass	The effect of the thermal stresses on the natural frequency reduction is revealed
Foti et al. [155]	SSI-DATA	Used output-only EFDD and SSI to identify the modal parameters of two building to update a FE model	Need to know the dynamic behavior of complex buildings subjected to near-fault earthquakes	A complex building which was heavily damaged in an earthquake.	At first low agreement was found but finally satisfactory agreement has been reached.
Kurata et al. [151]	SSI-DATA	Developed a novel internet-enabled wireless structural monitoring system tailored for large-scale civil infrastructures	Need to develop dense networks of low-cost wireless sensors for large and complex infrastructure	Installed wireless monitoring system is on New Carquinez Bridge	The obtained results verified the stable and reliable application of the proposed monitoring system.
Ubertini et al. [182]	SSI-DATA	Proposed an automated SSI-based modal identification procedure, using clustering analysis	Increasing need to diffusion of continuous monitoring systems for structural condition assessment	Two bridges of iron arch bridge and a long-span footbridge	The reliable performance of the automated long term monitoring was verified.
Döhler et al. [102]	SSI-DATA and SSI-COV	Proposed an efficient stochastic SSI algorithm by reformulation and computation of uncertainty bounds	Need to a fast and reliable damage detection algorithm	The field vibrational data of the Z24 Bridge	The algorithm is both computationally and memory efficient.
Döhler and Mevel [27]	SSI-DATA & SSI-COV	Derived a new efficient algorithm for multi-order system identification using SSI method	Need to distinguish the true modes from spurious structural modes	Z24 Bridge data	The presented methods are faster than the conventional algorithms in use.
Kim and Lynch [152]	Indirect SSI-DATA	Introduced a SIMO model of SSI algorithm based on Markov parameters customized for the decentralized WSNs	Need to decentralized data processing due to its advantages consumption.	Dynamic testing of a cantilevered balcony in a historic building	System properties were identified with a high accuracy.
Zhang et al. [84]	Improved SSI-DATA	Introduced a CH matrix as a replacement of Hankel matrix and projection operator for QR decomposition	Need to improve the low computational efficiency of the SSI-DATA	A numerical model of a 7-DOF and an experimental model of Chaotianmen bridge	Computational efficiency and reject of the spurious modes without losing the quality are achieved.
Lardies and Minh-Ngi [183]	SSI-DATA	Applied improved SSI using modal coherence indicator to eliminates spurious modes and Morlet wavelet	Need to overcome the concerns about health state of the tension cables in cable-stayed bridges	Two experiments of stay cables in laboratory scale and Jinma cable-stayed bridge	The robustness and reliability of the subspace and the WT transform methods are demonstrated.

Table A1. *Cont.*

Author	Method	Research Objective	Research Gap and Problem	Solution and Modeling	Result and Finding
Weng and Loh [83]	RSI-DATA & RSSI-DATA	Developed an on-line tracking of the estimated system parameter using response measurements	Need to develop an on-line tracking of modal parameter without human interference	Seismic excitation of a 3-story steel frame and a 2-story reinforced concrete frame	Accurate results were obtained by identifying the model properties.
Carden and Mita [131]	SSI-DATA	Investigated the methods applied to estimate uncertainty and confidence intervals and summarized drawback of each method.	Need to deal with finite lengths of data for modal identification	Numerical models of a MSD system and experimental model of a suspension bridge	The drawbacks for reliable application of residual bootstrapping procedure are reported.
Brownjohn et al. [184]	SSI-DATA	Implemented the SSI procedure in the 'virtual instrument' for SHM of a 183 m reinforced concrete chimney	Need to overcome the concerns about large-amplitude response induced by interference effects	A 183 m reinforced concrete chimney for a coal-fired power station	The damping values show the tune mass damper to have been effective in controlling response.
Hu et al. [135]	SSI-DATA and SSI-COV	Introduced tools for modal identification in LabVIEW named SMI and CSMI	Need to computational tools for modal identification and long term vibration monitoring	Field data collected at Pinhaˉo bridge and Coimbra footbridge	The potential of this software to obtain the natural frequencies and modal damping.
Marchesiello et al. [128]	ST-SSI	Two approaches of continuous wavelet transform and the ST-SSI is proposed and compared.	Need to take into account the effect of system variation in time-variant systems	A pinned–pinned bridge carrying a moving load	CWT was found to suffer from the drawback of edge effects compared to ST-SSI.
Deraemaeker et al. [185]	SSI-DATA	Examined two damage features obtained from SSI and peak indicators	Need to consider the effect of environmental condition in analysis	A numerical bridge model subject to noise and damage	All damages were detected using the proposed procedure.
Alıcıoğlu and Luş [105]	SSI-DATA & SSI-COV	Investigated the performance of output-only SSI-DATA and SSI-COV algorithms	Need to objectively determine the practical benefits of SSI and to find out the potential difficulties	FE model, physical laboratory model of a small scale steel frame and a long span suspension bridge	Both SSI algorithms are found to perform quite satisfactorily for operational modal analysis.
He et al. [186]	SSI-DATA	Simulated the wind-induced vibration response of a Bridge using FE model and stochastic wind excitation model	Need to study systematically the effects of damage scenarios in long-span cable-supported bridges	Simulation of the wind-induced vibration response of Vincent Thomas Bridge,	The framework was validated to study the effects of damage scenarios.

Table A2. Distribution of the papers based on SSI-COV techniques.

Author	Method	Research Objective	Research Gap and Problem	Solution and Modelling	Result and Finding
Zarbaf et al. [72]	SSI-COV	Adopted a hierarchical clustering algorithm to obtain tensions in the stay-cable	Need to estimate the tension forces of cables in cable-stayed bridges	The ambient response of the Veterans' Glass City Skyway Bridge	A good agreement between the estimated results and measured tension forces was observed.
Reynders et al. [187]	SSI-COV	Validated a method for estimating the (co)variance of modal parameters identified using SSI	Need to estimate the variance of modal parameters	A damaged prestressed concrete bridge and a mid-rise building	Good agreement is reported between the predicted uncertainty and the observation data.
Wu et al. [188]	SSI-COV	Developed a new SSI methodology to identify modal parameters of stay cables	Need to extract numerous modes in stay cable	The ambient response of the three stay cables of Chi-Lu Bridge	The feasibility of this new approach is verified successfully.
Zhou et al. [168]	SSI-COV	Used the residual of the SSI and global χ2-tests built on that residual for damage detection.	Need to exploit possible damages in structure using output data	A full-scale bridge benchmark validated by numerical simulation	The damage in tower was detected in the same time.
Karami and Akbarabadi [189]	SSI-COV	Proposed an algorithm in two steps by integrating structural health monitoring with semi-active control strategy	Need to damage detection of large building structures using limited output data	A numerical model of a shear building structure	The algorithm could identify the damage accurately with saving time and cost due.
Attig et al. [160]	SSI-COV	Investigated performance of the combined SSI algorithms and a stabilization diagram for tensegrity systems	Need to identify structural changes in Tensegrity systems	A numerical models of a tripod simplex structure and a Geiger dome	Effectiveness of the proposed methodology was verified using the proposed methodology.
Sun et al. [71]	SSI-COV	Defined a nonlinear subspace distance to detect the deviation from the normal state, and reflects structural states.	Need to consider nonlinearity of the structures for identification of modal characteristics	A VSS subjected to accelerated ageing	The designed index is very effective to evaluate the health state.
Khan et al. [190]	SSI-COV	Employed EDA, outlier analysis and cross correlation to elucidate any defects and anomalies in the data.	Need to distinguish between abnormal data malfunctioning, and anomalies of the sensors	A cable stayed bridge over Sutong Yangtze river	The method was very effective to provide accurate real life results in the continuous SHM of bridges.
Guo et al. [191]	SSI-COV	Proposed a near-real-time hybrid framework for system identification of structures to deal with stationary and transient response	Need to simultaneously deal with stationary and transient responses of the applied excitation loads	Extensive numerical simulations as well as analysis of the internet enabled data of Burj Khalifa	The efficacy of the framework is demonstrated.
Mekki et al. [192]	SSI-COV	Applied a null-space Hankel matrix of correlation estimates	Need to study the dynamic response of structures on composite structures	Numerical and experimental of a one span composite bridge deck, formed by wood and concrete	The first natural frequencies were determined with an uncertainty below 0.15%.
Döhler et al. [164]	SSI-COV	Presented an efficient and fast SSI damage detection that is robust to changes in the excitation covariance	Need to investigate the change in unmeasured ambient excitation properties	Three numerical model	The new approach can detect better and separate different levels of damage.
Tondreau and Deraemaeker [119]	SSI-COV	Studied the effect of noise on the uncertainty of obtained modal parameters using SSI	Need to study the resulting uncertainty for modal analysis using the stochastic SSI method.	A numerical test of a supported beam, and the experimental model of a clamped-free plate	The uncertainty on modal damping and eigenfrequencies may exhibit a non-normal distribution.
Dohler et al. [193]	SSI-COV	SSI-COV together with their confidence interval estimation and a null space-based VDD	Need to consider the intrinsic uncertainty for a robust and automated SHM	A large scale progressive damage test of the S101 Bridge in Austria.	The proposed method is able to clearly indicate the presence of damages.

Table A2. *Cont.*

Author	Method	Research Objective	Research Gap and Problem	Solution and Modelling	Result and Finding
Hong et al. [85]	SSI-COV	Adapted enhanced canonical correlation analysis (ECCA) for state variable estimation	Need to determine model order and prevent failure of identification system	A FE simulation and field measurements of the Carquinez suspension bridge	The reliability of the new algorithm was verified through numerical analyses.
Loh et al. [159]	SSI-COV	Adopted singular spectrum analysis (SSA), for pre-processing and stabilization diagram for post-processing	Need to do some pre-processing to smooth noisy signal,	The experimental test on Canton Tower high-rise slender structure	The use of SSA as a pre-processing tool improved the stabilization diagram identifiablity of modes.
Döhler & Mevel [101]	Modular and scalable SSI-COV	Proposed a modular and scalable SSI approach to improve retrieving the system matrices of a full system	Need to deal with the problem of merging sensor data of non-simultaneously recorded setups	Mathematical formulations	The application of the method for has been verified successfully.
Chauhan [194]	SSI-COV	Developed a unified matrix polynomial approach (UMPA) to explain the SSI algorithm	Need to explain and derive various experimental modal analysis algorithms in an easy way	Mathematical formulations	The sequences for derivation the system parameters from output data are clearly demonstrated.
Ren et al. [169]	SSI-COV	Introduced a new damage feature to reject the environmental effects. Two distance functions adopted for pattern recognition	Need to extract the damage-sensitive but environment-insensitive damage features	One numerical signal and two simulated FE dynamic beam models	The method was capable to locate damage in FE beam structures.
Basseville et al. [165]	SSI-COV	Designed a damage detection algorithm based on null space residual and a $\chi2$ test to exploit the thermal model	Need to discriminate changes in modal parameters caused by damage	A vertical beam made of steel, and aluminium tested under controlled ambient temperature.	Relevance of the presented algorithms was illustrated using the laboratory test case.
Whelan et al. [195]	SSI-COV	Deployed a wireless sensor network with higher sampling rates with reliable large, dense array sensory network	The need to enhance data analysis methods for the data obtained from remote sensor-based SHM	A single-span integral abutment bridge	The feasibility and maturity of the distributed network of wireless sensor was confirmed.
Balmes et al. [170]	SSI-COV	Proposed using subspace residual as damage feature and $\chi2$ tests to discriminate the effect of noise	Need to remove the effect of temperature and other environmental factors for VDD.	A simulated bridge deck with controlled temperature variations	Efficiency of the method on simulation model for various temperature models was confirmed.
Carden and Brownjohn [163]	SSI-COV	Proposed a Fuzzy Clustering Algorithm to extract state parameters from the real and numerical poles	Need to identify structural changes in the presence of environmental variation	The data from Z24 Bridge and the Republic Plaza Office Building (POB) in Singapore	The damage inflicted on the Z24 Bridge and the shifts in modes of the POB were clearly captured.
Reynders et al. [161]	SSI-COV	Used first-order sensitivity of the modal parameter and stabilization to remove bias errors	Need to remove of bias and variance errors in the estimated modal parameters	Simulation model and measured vibration data of a beam and a mast structures	Practicability of the proposed method was confirmed in real-life application.
Balmès et al. [170]	SSI-COV	Investigated damage localisation using clustering in the large-scale FE models.	Need for localization of damage in vibration-based methods.	A FE model of a bridge deck with a large number of elements	The algorithm was able to locate the damage in case of a FE model.
Zhang et al. [162]	SSI-COV	Introduced component energy index together with an alternative stabilization diagram to identify spurious and physical modes	Need to improve the identifiability of weakly excited modes	A 7 DOF MSD system and the experimental model of a metallic frame	Good performance was observed especially for measurements with low SNR.

Table A3. Distribution of the papers based on combined SSI techniques.

Author	Method	Research Objective	Research Gap and Problem	Solution and Modelling	Result and Finding
Marchesiello et al. [196]	Non-linear SSI	Introduced a modal decoupling procedure and the modal mass	Need to deal with variability of the identification results due to nonlinear effects	A multi-storey building model with a local nonlinearity	Significant improvements were highlighted in estimates obtained by the proposed approach.
Shi et al. [197]	MOESP	Used two SSI techniques sequentially and iteratively to extract modal parameters and estimates the ground acceleration.	Need to estimate the structural parameters of a under unknown ground excitation	A numerical and a laboratory test of a 3-story building model	The estimation of structural parameters is satisfactory and fairly robust.
Zhong and Chang [52]	Combined SSI	Adopted an orthogonal projection and IV approach to eliminate the effect of earthquake input and noise	Need for modal identification of time-varying structures under non-stationary earthquake excitation	Numerical model of a four DOF structure and a three DOF experimental building model.	The proposed algorithm can track the modal parameters quite well.
Verhaegen and Hansson [173]	input-output N2SID	Introduced a SSI using convex nuclear norm optimization	Need to an identification scheme for multivariable state space model by improving the classical methods	Mathematical formulations	The sequences for derivation the system parameters from N2SID is clearly demonstrated.
Potenza et al. [51]	SSI-COV & combined SSI	Focused on the seismic monitoring of a historical structure by means of an advanced WSNs	Need to analyse critical issues in the wireless data acquisition	The historical structure of the Basilica S. Maria di Collemaggio.	The monitoring system permitted to update a finite element model in the current damaged conditions.
Al-Gahtani et al. [198]	Deterministic SSI	Performed deterministic SSI on the obtained response signal after applying wavelet de-noising methods	Need to an system identification with low sensitivity to the inflicted noise	A numerically simulated model and experimentally measured rotor	The use of multi-wavelet de-noising result in a more accurate identification.
Gandino et al. [172]	Combined SSI-COV	Developed a novel multivariate SSI-COV-based formulation for modal parameter identification	Need to a reliable SHM systems with no memory limitation and work properly in presence of noise	A 15-DOF numerical example and an experimental application of a thin-walled metallic structure	The obtained results are similar to those reached by data-driven method.
Kim and Lynch [41]	SSI-DATA & combined SSI	Presented a theoretical framework to extract physical parameters using a physics-based and a data-driven models	Need to estimate physical modal parameters of structures	A multi-DOF shear building model and an experimental test of a six-story steel frame.	The proposed grey-box framework has shown a promising performance for SHM of civil structures.
Akçay [171]	Frequency domain subspace	Proposed a subspace algorithm by calculating minimal realization of power spectrum and a canonical spectral factor	Need to deal with the problem of system identification of dynamic systems.	A numerical example	Some drawback regarding reliable performance of the algorithm is highlighted.
Urgessa [81]	McKelvey SSI-FD	Presented two system identification methods based on eigensystem realization and the McKelvey frequency-domain SSI	Need to meet interpretation challenges associated with system identification	FE model of a plate structure	The methods were able to predict natural frequency and damping ratio with high accuracy.
Weng et al. [199]	Input-output SSI	Proposed a damage assessment method by adopting input/output SSI algorithm and a model updating method.	The need to validate FE models by applying input-output identification methods	A1/4-scale six-story steel frame structure and a two-story RC frame	The method was able to detect the damage locations and quantify the damage severity.
Reynders and De Roeck [58]	Combined SSI-DATA	Adopted modal decoupling and a new criterion from model reduction theory for automation of the modal analysis process.	Need to extract frequency content of limited number of modes from the narrow band ambient excitation	Field vibration data obtained from the Z24 Bridge	The most complete set of modes reported so far is obtained.
Kurka and Cambraia [167]	Multivariable combined SSI	Proposed a Multiple-input multiple-output (MIMO) input–output SSI method that uses multi-input and single-output (MISO) realization	A need to provide a robust model order determination using SVD.	Numerical model and a free–free spatial truss	Accurate modal parameters were estimated using this method.

References

1. Ozer, E.; Feng, Q.M. Structural Reliability Estimation with Participatory Sensing and Mobile Cyber-Physical Structural Health Monitoring Systems. *Appl. Sci.* **2019**, *9*, 2840. [CrossRef]
2. Her, S.-C.; Chung, S.-C.; Hou, Q.; Zhu, W.; Yang, Q.; Wang, C. Dynamic Responses Measured by Optical Fiber Sensor for Structural Health Monitoring. *Appl. Sci.* **2019**, *9*, 2956. [CrossRef]
3. Deng, G.; Zhou, Z.; Shao, S.; Chu, X.; Jian, C. A Novel Dense Full-Field Displacement Monitoring Method Based on Image Sequences and Optical Flow Algorithm. *Appl. Sci.* **2020**, *10*, 2118. [CrossRef]
4. Kovačević, S.M.; Bačić, M.; Stipanović, I.; Gavin, K. Categorization of the Condition of Railway Embankments Using a Multi-Attribute Utility Theory. *Appl. Sci.* **2019**, *9*, 5089. [CrossRef]
5. Zhou, L.; Guo, J.; Wen, X.; Ma, J.; Yang, F.; Wang, C.; Zhang, D. Monitoring and Analysis of Dynamic Characteristics of Super High-rise Buildings using GB-RAR: A Case Study of the WGC under Construction, China. *Appl. Sci.* **2020**, *10*, 808. [CrossRef]
6. Fan, W.; Qiao, P. Vibration-based damage identification methods: A review and comparative study. *Struct. Health Monit.* **2011**, *10*, 83–111. [CrossRef]
7. Reynders, E. System identification methods for (operational) modal analysis: Review and comparison. *Arch. Comput. Methods Eng.* **2012**, *19*, 51–124. [CrossRef]
8. Das, S.; Saha, P.; Patro, S.K. Vibration-based damage detection techniques used for health monitoring of structures: A review. *J. Civ. Struct. Health Monit.* **2016**, *6*, 477–507. [CrossRef]
9. Moughty, J.J.; Casas, J.R. Vibration based damage detection techniques for small to medium span bridges: A review and case study. In Proceedings of the 8th European Workshop on Structural Health Monitoring (EWSHM 2016), Bilbao, Spain, 5–8 July 2016.
10. Zhou, C.; Gao, L.; Xiao, H.; Hou, B. Railway Wheel Flat Recognition and Precise Positioning Method Based on Multisensor Arrays. *Appl. Sci.* **2020**, *10*, 1297. [CrossRef]
11. Zou, Y.; Fu, Z.; He, X.; Cai, C.; Zhou, J.; Zhou, S. Wind Load Characteristics of Wind Barriers Induced by High-Speed Trains Based on Field Measurements. *Appl. Sci.* **2019**, *9*, 4865. [CrossRef]
12. Chu, X.; Zhou, Z.; Deng, G.; Duan, X.; Jiang, X. An Overall Deformation Monitoring Method of Structure Based on Tracking Deformation Contour. *Appl. Sci.* **2019**, *9*, 4532. [CrossRef]
13. Shokravi, H.; Shokravi, H.; Bakhary, N.; Koloor, S.S.R.; Petrů, M. Health Monitoring of Civil Infrastructures by Subspace System Identification Method: An Overview. *Appl. Sci.* **2020**, *10*, 2786. [CrossRef]
14. Garg, A.; Mhaskar, P. Subspace identification-based modeling and control of batch particulate processes. *Ind. Eng. Chem. Res.* **2017**, *56*, 7491–7502. [CrossRef]
15. Deng, X. System identification based on particle swarm optimization algorithm. In Proceedings of the 2009 International Conference on Computational Intelligence and Security, Beijing, China, 11–14 December 2009; pp. 259–263.
16. Ramos, J.A.; Mercère, G. Image modeling based on a 2-D stochastic subspace system identification algorithm. *Multidimens. Syst. Signal Process.* **2017**, *28*, 1133–1165. [CrossRef]
17. Yan, Z.; Wang, J.-S.; Wang, S.-Y.; Li, S.-J.; Wang, D.; Sun, W.-Z. Model Predictive Control Method of Simulated Moving Bed Chromatographic Separation Process Based on Subspace System Identification. *Math. Probl. Eng.* **2019**, *2019*, 2391891. [CrossRef]
18. Wahlberg, B.; Jansson, M.; Matsko, T.; Molander, M.A. Experiences from subspace system identification-comments from process industry users and researchers. In *Modeling, Estimation and Control*; Springer: Berlin, Germany, 2007; pp. 315–327.
19. Becker, C.O.; Bassett, D.S.; Preciado, V.M. Large-scale dynamic modeling of task-fMRI signals via subspace system identification. *J. Neural Eng.* **2018**, *15*, 66016. [CrossRef]
20. Romano, R.A.; Pait, F. Matchable-observable linear models and direct filter tuning: An approach to multivariable identification. *IEEE Trans. Autom. Control* **2016**, *62*, 2180–2193. [CrossRef]
21. Pappalardo, C.M.; Guida, D. System Identification and Experimental Modal Analysis of a Frame Structure. *Eng. Lett.* **2018**, *26*, 112–136.
22. Peeters, B.; De Roeck, G. Stochastic system identification for operational modal analysis: A review. *J. Dyn. Syst. Meas. Control* **2001**, *123*, 659. [CrossRef]
23. Song, G.; Wang, C.; Wang, B. Structural Health Monitoring (SHM) of Civil Structures. *Appl. Sci.* **2017**, *7*, 789. [CrossRef]

24. De Cock, K.; Mercere, G.; De Moor, B. Recursive subspace identification for in flight modal analysis of airplanes. In Proceedings of the International Conference on Noise and Vibration Engineering, ISMA 2006, Leuven, Belgium, 20 September 2006.
25. De Cock, K.; Peeters, B.; Vecchio, A.; Van der Auweraer, H.; De Moor, B. Subspace system identification for mechanical engineering. In Proceedings of the International Conference on Noise and Vibration Engineering (ISMA 2002), Leuven, Belgium, 11 June 2002.
26. Döhler, M.; Andersen, P.; Mevel, L. Operational modal analysis using a fast stochastic subspace identification method. In *Topics in Modal Analysis I*; Springer: Berlin, Germany, 2012; Volume 5, pp. 19–24.
27. Döhler, M.; Mevel, L. Efficient multi-order uncertainty computation for stochastic subspace identification. *Mech. Syst. Signal Process.* **2013**, *38*, 346–366. [CrossRef]
28. Moaveni, B.; He, X.; Conte, J.P.; Restrepo, J.I.; Panagiotou, M. System identification study of a 7-story full-scale building slice tested on the UCSD-NEES shake table. *J. Struct. Eng.* **2010**, *137*, 705–717. [CrossRef]
29. Xin, J.F.; Hu, S.-L.J.L.J.; Li, H.J. Experimental modal analysis of jacket-type platforms using data-driven stochastic subspace identification method. In Proceedings of the ASME 31st International Conference on Ocean, Offshore and Arctic Engineering, Rio de Janeiro, Brazil, 1–6 July 2012; American Society of Mechanical Engineers Digital Collection: New York, NY, USA, 2012; Volume 5.
30. Tarinejad, R.; Pourgholi, M. Modal identification of arch dams using balanced stochastic subspace identification. *J. Vib. Control* **2016**, *24*, 2030–2044. [CrossRef]
31. Yang, J.N.; Soong, T.T. Recent advances in active control of civil engineering structures. *Probab. Eng. Mech.* **1988**, *3*, 179–188. [CrossRef]
32. Van Overschee, P.; De Moor, B.L.; Hensher, D.A.; Rose, J.M.; Greene, W.H.; Train, K. *Subspace Identification for the Linear Systems: Theory–Implementation-Application*; Katholieke Universiteit Leuven: Leuven, Belgium; Kluwer Academic Publishers: Boston, MA, USA, 1996.
33. Peeters, B.; De Roeck, G. Reference-based stochastic subspace identification for output-only modal analysis. *Mech. Syst. Signal Process.* **1999**, *13*, 855–878. [CrossRef]
34. Chang, C.-M.; Loh, C.-H. Improved Stochastic Subspace System Identification for Structural Health Monitoring. In *Journal of Physics: Conference Series*; IOP Publishing: Bristol, UK, 2015; Volume 628, p. 12010.
35. Ashari, A.E.; Mevel, L. Auxiliary input design for stochastic subspace-based structural damage detection. *Mech. Syst. Signal Process.* **2013**, *34*, 241–258. [CrossRef]
36. Shokravi, H.; Shokravi, H.; Bakhary, N.; Heidarrezaei, M.; Koloor, S.S.R.; Petru, M. Vehicle-assisted techniques for health monitoring of bridges. *Sensors* **2020**, *20*. Under review.
37. Shokravi, H.; Shokravi, H.; Bakhary, N.; Koloor, S.R.K.; Petru, M. A review on vehicle classification methods and the potential of using smart-vehicle-assisted techniques. *Sensors (Basel)* **2020**, *20*. Under review.
38. Sun, L.; Shang, Z.; Xia, Y.; Bhowmick, S.; Nagarajaiah, S. Review of Bridge Structural Health Monitoring Aided by Big Data and Artificial Intelligence: From Condition Assessment to Damage Detection. *J. Struct. Eng.* **2020**, *146*, 4020073. [CrossRef]
39. Shokravi, H.; Shokravi, H.; Bakhary, N.; Koloor, S.S.R.; Petru, M. A Comparative Study of the Data-driven Stochastic Subspace Methods for Health Monitoring of Structures: A Bridge Case Study. *Appl. Sci.* **2020**, *10*, 132.
40. Abazarsa, F.; Nateghi, F.; Ghahari, S.F.; Taciroglu, E. Extended blind modal identification technique for nonstationary excitations and its verification and validation. *J. Eng. Mech.* **2016**, *142*, 4015078. [CrossRef]
41. Kim, J.; Lynch, J.P. Subspace system identification of support-excited structures—part I: Theory and black-box system identification. *Earthq. Eng. Struct. Dyn.* **2012**, *41*, 2235–2251. [CrossRef]
42. Kim, J.; Lynch, J.P. Subspace system identification of support excited structures—part II: Gray-box interpretations and damage detection. *Earthq. Eng. Struct. Dyn.* **2012**, *41*, 2253–2271. [CrossRef]
43. Mellinger, P.; Döhler, M.; Mevel, L. Variance estimation of modal parameters from output-only and input/output subspace-based system identification. *J. Sound Vib.* **2016**, *379*, 1–27. [CrossRef]
44. Xin, J.; Sheng, J.; Sui, W. Study on the Reason for Difference of Data-Driven and Covariance-driven Stochastic Subspace Identification Method. In Proceedings of the 2012 International Conference on Computer Science and Electronics Engineering, Hangzhou, China, 23–25 March 2012; pp. 356–360.
45. Ren, W.-X.; Zong, Z.-H. Output-only modal parameter identification of civil engineering structures. *Struct. Eng. Mech.* **2004**, *17*, 429–444. [CrossRef]

46. Yu, D.-J.; Ren, W.-X. EMD-based stochastic subspace identification of structures from operational vibration measurements. *Eng. Struct.* **2005**, *27*, 1741–1751. [CrossRef]
47. Huth, O.; Czaderski, C.; Hejll, A.; Feltrin, G.; Motavalli, M. Tendon breakages effect on static and modal parameters of a post-tensioned concrete girder. In *EMPA, Mat Sci & Technol, Struct Engn Res Lab, Dubendorf, Switzerland*; Huth, O., Ed.; Taylor & Francis Ltd.: Abingdon-on-Thames, UK, 2006.
48. Loh, C.H.; Liu, Y.C. Determination of Reliable Control Parameters for Monitoring of Large Flexible Structure Using Recursive Stochastic Subspace Identification. In Proceedings of the 8th International Conference on Structural Dynamics, EURODYN 2011, Leuven, Belgium, 4–6 July 2011; pp. 2189–2196.
49. Li, Z.; Chang, C.C. Tracking of structural dynamic characteristics using recursive stochastic subspace identification and instrumental variable technique. *J. Eng. Mech.* **2011**, *138*, 591–600. [CrossRef]
50. Alwash, M.B. Excitation Sources for Structural Health Monitoring of Bridges. Ph.D. Thesis, University of Saskatchewan, Saskatoon, SK, Canada, 2010.
51. Potenza, F.; Federici, F.; Lepidi, M.; Gattulli, V.; Graziosi, F.; Colarieti, A. Long-term structural monitoring of the damaged Basilica, S. Maria di Collemaggio through a low-cost wireless sensor network. *J. Civ. Struct. Health Monit.* **2015**, *5*, 655–676. [CrossRef]
52. Zhong, K.; Chang, C.C. Recursive Combined Subspace Identification Technique for Tracking Dynamic Characteristics of Structures under Earthquake Excitation. *J. Eng. Mech.* **2016**, *142*, 4016092. [CrossRef]
53. Shokravi, H.; Bakhary, N.H. Comparative analysis of different weight matrices in subspace system identification for structural health monitoring. *IOP Conf. Ser. Mater. Sci. Eng.* **2017**, *271*, 12092. [CrossRef]
54. Rainieri, C.; Fabbrocino, G. *Operational Modal Analysis of Civil Engineering Structures*; Springer: New York, NY, USA, 2014.
55. Bauer, D. Asymptotic properties of subspace estimators. *Automatica* **2005**, *41*, 359–376. [CrossRef]
56. Chiuso, A.; Picci, G. Asymptotic variance of subspace methods by data orthogonalization and model decoupling: A comparative analysis. *Automatica* **2004**, *40*, 1705–1717. [CrossRef]
57. Bauer, D.; Ljung, L. Some facts about the choice of the weighting matrices in Larimore type of subspace algorithms. *Automatica* **2002**, *38*, 763–773. [CrossRef]
58. Reynders, E.; De Roeck, G. Reference-based combined deterministic–stochastic subspace identification for experimental and operational modal analysis. *Mech. Syst. Signal Process.* **2008**, *22*, 617–637. [CrossRef]
59. Cho, Y.M.; Kailath, T. Fast subspace-based system identification: An instrumental variable approach. *Automatica* **1995**, *31*, 903–905. [CrossRef]
60. Peeters, B. System Identification and Damage Detection in Civil Engineering. Ph.D. Thesis, Department of Civil Engineering KU Leuven, Leuven, Belgium, 2000.
61. Mastronardi, N.; Kressner, D.; Sima, V.; Van Dooren, P.; Van Huffel, S. A fast algorithm for subspace state-space system identification via exploitation of the displacement structure. *J. Comput. Appl. Math.* **2001**, *132*, 71–81. [CrossRef]
62. Lovera, M.; Gustafsson, T.; Verhaegen, M. Recursive subspace identification of linear and non-linear Wiener state-space models. *Automatica* **2000**, *36*, 1639–1650. [CrossRef]
63. Mercère, G.; Bako, L.; Lecœuche, S. Propagator-based methods for recursive subspace model identification. *Signal Process.* **2008**, *88*, 468–491. [CrossRef]
64. Oku, H.; Kimura, H. Recursive 4SID algorithms using gradient type subspace tracking. *Automatica* **2002**, *38*, 1035–1043. [CrossRef]
65. Döhler, M.; Mevel, L. Fast multi-order computation of system matrices in subspace-based system identification. *Control Eng. Pract.* **2012**, *20*, 882–894. [CrossRef]
66. Cho, S.; Park, J.-W.; Sim, S.-H. Decentralized system identification using stochastic subspace identification for wireless sensor networks. *Sensors* **2015**, *15*, 8131–8145. [CrossRef] [PubMed]
67. Benveniste, A.; Fuchs, J.J. Single sample modal identification of a nonstationary stochastic process. *Autom. Control IEEE Trans.* **1985**, *30*, 66–74. [CrossRef]
68. Akaike, H. Stochastic theory of minimal realization. *Autom. Control IEEE Trans.* **1974**, *19*, 667–674. [CrossRef]
69. Aoki, M. State Space and ARMA Representation. In *State Space Modeling of Time Series*; Springer: Berlin, Germany, 1987; pp. 30–57.
70. Basseville, M.; Mevel, L.; Goursat, M. Statistical model-based damage detection and localization: Subspace-based residuals and damage-to-noise sensitivity ratios. *J. Sound Vib.* **2004**, *275*, 769–794. [CrossRef]

71. Sun, C.; Zhang, Z.; Luo, X.; Guo, T.; Qu, J.; Li, B. Support vector machine-based Grassmann manifold distance for health monitoring of viscoelastic sandwich structure with material ageing. *J. Sound Vib.* **2016**, *368*, 249–263. [CrossRef]

72. Zarbaf, E.; Norouzi, M.; Allemang Randall, J.; Hunt Victor, J.; Helmicki, A.; Nims Douglas, K. Stay Force Estimation in Cable-Stayed Bridges Using Stochastic Subspace Identification Methods. *J. Bridge Eng.* **2017**, *22*, 4017055. [CrossRef]

73. Yang, H.; Gu, Q.; Aoyama, T.; Takaki, T.; Ishii, I. Dynamics-based stereo visual inspection using multidimensional modal analysis. *IEEE Sens. J.* **2013**, *13*, 4831–4843. [CrossRef]

74. Farrar, C.R.; Doebling, S.W.; Cornwell, P.J.; Straser, E.G. *Variability of modal parameters measured on the Alamosa Canyon Bridge*; Los Alamos National Lab.: New Mexico, NM, USA, 1997; pp. 257–263.

75. Sohn, H.; Farrar, C.R. Damage diagnosis using time series analysis of vibration signals. *Smart Mater. Struct.* **2001**, *10*, 446–451. [CrossRef]

76. Gertler, J.J. Survey of model-based failure detection and isolation in complex plants. *IEEE Control Syst. Mag.* **1988**, *8*, 3–11. [CrossRef]

77. Saeed, K.; Mechbal, N.; Coffignal, G.; Verge, M. Recursive modal parameter estimation using output-only subspace identification for structural health monitoring. In Proceedings of the 2008 16th Mediterranean Conference on Control and Automation, Ajaccio, France, 25–27 June 2008; pp. 77–82.

78. Reynders, E.; Teughels, A.; De Roeck, G. Finite element model updating and structural damage identification using OMAX data. *Mech. Syst. Signal Process.* **2010**, *24*, 1306–1323. [CrossRef]

79. Loendersloot, R.; Schiphorst, F.B.A.; Basten, T.G.H.; Tinga, T. Application of SHM using an autonomous sensor network. In Proceedings of the 9th International Workshop on Structural Health Monitoring, IWSHM, Stanford, CA, USA, 10–12 September 2013; DEStech Publications, Inc.: Lancaster, PA, USA, 2013.

80. Miguel, L.F.F.; Lopez, R.H.; Miguel, L.F.F. A hybrid approach for damage detection of structures under operational conditions. *J. Sound Vib.* **2013**, *332*, 4241–4260. [CrossRef]

81. Urgessa, G.S. Vibration properties of beams using frequency-domain system identification methods. *J. Vib. Control* **2010**, *17*, 1287–1294. [CrossRef]

82. Goursat, M.; Döhler, M.; Mevel, L.; Andersen, P. Crystal clear SSI for operational modal analysis of aerospace vehicles. In *Structural Dynamics*; Springer: Berlin, Germany, 2011; Volume 3, pp. 1421–1430.

83. Weng, J.-H.; Loh, C.-H. Recursive subspace identification for on-line tracking of structural modal parameter. *Mech. Syst. Signal Process.* **2011**, *25*, 2923–2937. [CrossRef]

84. Zhang, G.; Tang, B.; Tang, G. An improved stochastic subspace identification for operational modal analysis. *Measurement* **2012**, *45*, 1246–1256. [CrossRef]

85. Hong, A.L.; Ubertini, F.; Betti, R. New Stochastic Subspace Approach for System Identification and Its Application to Long-Span Bridges. *J. Eng. Mech.* **2013**, *139*, 724–736. [CrossRef]

86. Rainieri, C.; Fabbrocino, G. Automated output-only dynamic identification of civil engineering structures. *Mech. Syst. Signal Process.* **2010**, *24*, 678–695. [CrossRef]

87. Giraldo, D.F.; Song, W.; Dyke, S.J.; Caicedo, J.M. Modal identification through ambient vibration: Comparative study. *J. Eng. Mech.* **2009**, *135*, 759–770. [CrossRef]

88. Magalhães, F.; Reynders, E.; Cunha, Á.; De Roeck, G. Online automatic identification of modal parameters of a bridge using the p-LSCF method. In Proceedings of the IOMAC'09–3rd International Operational Modal Analysis Conference, Ancona, Italy, 4–6 May 2009.

89. Prowell, I.; Elgamal, A.; Luco, J.E.; Conte, J.P. In-situ Ambient Vibration Study of a 900-kw Wind Turbine. *J. Earthq. Eng.* **2019**, 1–22. [CrossRef]

90. Kasımzade, A.A.; Tuhta, S.; Aydın, H.; Günday, F. Investigation of Modal Parameters on Steel Model Bridge Using EFDD Method. In Proceedings of the 2nd International Conference on Technology and Science, Bali, Indonesia, 14–16 November 2019.

91. Li, W.; Vu, V.-H.; Liu, Z.; Thomas, M.; Hazel, B. Extraction of modal parameters for identification of time-varying systems using data-driven stochastic subspace identification. *J. Vib. Control* **2018**, *24*, 4781–4796. [CrossRef]

92. Júnior, J.S.S.; Costa, E.B.M. Fuzzy Modelling Methodologies Based on OKID/ERA Algorithm Applied to Quadrotor Aerial Robots. In *Intelligent Systems: Theory, Research and Innovation in Applications*; Springer: Berlin, Germany, 2020; pp. 295–317.

93. Dautt-Silva, A.; De Callafon, R.A. Optimal Input Shaping with Finite Resolution Computed from Step-Response Experimental Data. In Proceedings of the 2018 Annual American Control Conference (ACC), Milwaukee, WI, USA, 27–29 June 2018; pp. 6703–6708.

94. Wang, S.Q.; Zhang, Y.T.; Feng, Y.X. Comparative study of output-based modal identification methods using measured signals from an offshore platform. In Proceedings of the ASME 2010 29th International Conference on Ocean, Offshore and Arctic Engineering, Shanghai, China, 6–11 June 2010; American Society of Mechanical Engineers: New York, NY, USA, 2010; pp. 561–567.

95. Kim, J.; Lynch, J.P. Comparison study of output-only subspace and frequency-domain methods for system identification of base excited civil engineering structures. In *Civil Engineering Topics*; Springer: Berlin, Germany, 2011; Volume 4, pp. 305–312.

96. Cunha, A.; Caetano, E.; Ribeiro, P.; Müller, G. Vibration-based SHM of a centenary bridge: A comparative study between two different automated OMA techniques. *Preservation* **2011**, *1*, 12.

97. Liu, C.W.; Wu, J.Z.; Zhang, Y.G. Review and prospect on modal parameter identification of spatial lattice structure based on ambient excitation. In *Applied Mechanics and Materials*; Trans Tech Publications Ltd.: Stafa-Zurich, Switzerland, 2011; Volume 94, p. 1271.

98. Ceravolo, R.; Abbiati, G. Time domain identification of structures: Comparative analysis of output-only methods. *J. Eng. Mech.* **2012**, *139*, 537–544. [CrossRef]

99. Mevel, L.; Basseville, M.; Benveniste, A.; Goursat, M. Merging sensor data from multiple measurement set-ups for non-stationary subspace-based modal analysis. *J. Sound Vib.* **2002**, *249*, 719–741. [CrossRef]

100. Döhler, M.; Reynders, E.; Magalhaes, F.; Mevel, L.; De Roeck, G.; Cunha, A. Pre-and post-identification merging for multi-setup OMA with covariance-driven SSI. In *Dynamics of Bridges*; Springer: Berlin, Germna, 2011; Volume 5, pp. 57–70.

101. Döhler, M.; Mevel, L. Modular subspace-based system identification from multi-setup measurements. *IEEE Trans. Autom. Contr.* **2012**, *57*, 2951–2956. [CrossRef]

102. Döhler, M.; Lam, X.-B.; Mevel, L. Uncertainty quantification for modal parameters from stochastic subspace identification on multi-setup measurements. *Mech. Syst. Signal Process.* **2013**, *36*, 562–581. [CrossRef]

103. Orlowitz, E.; Andersen, P.; Brandt, A. Comparison of Simultaneous and Multi-setup Measurement Strategies in Operational Modal Analysis. In Proceedings of the 6th International Operational Modal Analysis Conference (IOMAC'15), Gijon, Spain, 12–14 May 2015.

104. Benveniste, A.; Mevel, L. Nonstationary consistency of subspace methods. *Autom. Control IEEE Trans.* **2007**, *52*, 974–984. [CrossRef]

105. Alıcıoğlu, B.; Luş, H. Ambient vibration analysis with subspace methods and automated mode selection: Case studies. *J. Struct. Eng.* **2008**, *134*, 1016–1029. [CrossRef]

106. Cornwell, P.; Farrar, C.R.; Doebling, S.W.; Sohn, H. Environmental variability of modal properties. *Exp. Tech.* **1999**, *23*, 45–48. [CrossRef]

107. Liu, C.; DeWolf, J.T. Effect of temperature on modal variability of a curved concrete bridge under ambient loads. *J. Struct. Eng.* **2007**, *133*, 1742–1751. [CrossRef]

108. Nayeri, R.D.; Masri, S.F.; Ghanem, R.G.; Nigbor, R.L. A novel approach for the structural identification and monitoring of a full-scale 17-story building based on ambient vibration measurements. *Smart Mater. Struct.* **2008**, *17*, 25006. [CrossRef]

109. Wood, M.G. Damage Analysis of Bridge Structures Using Vibrational Techniques. Ph.D. Thesis, University of Aston in Birmingham, Birmingham, UK, 1992.

110. Xia, Y.; Hao, H.; Zanardo, G.; Deeks, A. Long term vibration monitoring of an RC slab: Temperature and humidity effect. *Eng. Struct.* **2006**, *28*, 441–452. [CrossRef]

111. Alampalli, S. Influence of in-service environment on modal parameters. In Proceedings of the 1998 16th International Modal Analysis Conference IMAC. Part 1 (of 2), Santa Barabara, CA, USA, 2–5 February 1998; New York State Dep of Transportation SEM: New York, NY, USA, 1998; Volume 1, pp. 111–116.

112. Peeters, B.; De Roeck, G. One-year monitoring of the Z 24-Bridge: Environmental effects versus damage events. *Earthq. Eng. Struct. Dyn.* **2001**, *30*, 149–171. [CrossRef]

113. Peeters, B.; Maeck, J.; De Roeck, G. Dynamic monitoring of the Z24-Bridge: Separating temperature effects from damage. In Proceedings of the European COST F3 Conference on System Identification and Structural Health Monitoring, Madrid, Spain, 6–9 June 2000; Volume 3, pp. 377–386.

114. Ni, Y.Q.; Hua, X.G.; Fan, K.Q.; Ko, J.M. Correlating modal properties with temperature using long-term monitoring data and support vector machine technique. *Eng. Struct.* **2005**, *27*, 1762–1773. [CrossRef]
115. Kim, J.T.; Park, J.H.; Lee, B.J. Vibration-based damage monitoring in model plate-girder bridges under uncertain temperature conditions. *Eng. Struct.* **2007**, *29*, 1354–1365. [CrossRef]
116. Spiridonakos, M.D.; Chatzi, E.N.; Sudret, B. Polynomial Chaos Expansion Models for the Monitoring of Structures under Operational Variability. *ASCE-ASME J. Risk Uncertain Eng. Syst. Part A Civ. Eng.* **2016**, *2*, B4016003. [CrossRef]
117. Loh, C.H.; Chen, M.C. Modeling of environmental effects for vibration-based shm using recursive stochastic subspace identification analysis. In *Key Engineering Materials*; Trans Tech Publications Ltd.: Stafa-Zurich, Switzerland, 2013; Volume 558, pp. 52–64.
118. Huynh, T.C.; Park, J.H.; Kim, J.T. Structural identification of cable-stayed bridge under back-to-back typhoons by wireless vibration monitoring. *Measurement* **2016**, *88*, 385–401. [CrossRef]
119. Tondreau, G.; Deraemaeker, A. Numerical and experimental analysis of uncertainty on modal parameters estimated with the stochastic subspace method. *J. Sound Vib.* **2014**, *333*, 4376–4401. [CrossRef]
120. Magalhães, F.; Cunha, Á.; Caetano, E.; Brincker, R. Damping estimation using free decays and ambient vibration tests. *Mech. Syst. Signal Process.* **2010**, *24*, 1274–1290. [CrossRef]
121. Rainieri, C.; Fabbrocino, G. Influence of model order and number of block rows on accuracy and precision of modal parameter estimates in stochastic subspace identification. *Int. J. Lifecycle Perform. Eng.* **2014**, *10*, 317–334. [CrossRef]
122. Wang, J.; Sano, A.; Chen, T.; Huang, B. Identification of Hammerstein systems without explicit parameterisation of non-linearity. *Int. J. Control* **2009**, *82*, 937–952. [CrossRef]
123. Li, D.; Ren, W.-X.; Hu, Y.-D.; Yang, D. Operational modal analysis of structures by stochastic subspace identification with a delay index. *Struct. Eng. Mech.* **2016**, *59*, 187–207. [CrossRef]
124. Brasiliano, A.; Doz, G.; Brito JL, V.; Pimentel, R. Role of non-metallic components on the dynamic behavior of composite footbridges. In Proceedings of the 3rd International Conference Footbridge, London, UK, 2–4 July 2008.
125. Cara, F.J.; Juan, J.; Alarcón, E.; Reynders, E.; De Roeck, G. Modal contribution and state space order selection in operational modal analysis. *Mech. Syst. Signal Process.* **2013**, *38*, 276–298. [CrossRef]
126. Pridham, B.A.; Wilson, J.C. A study of damping errors in correlation-driven stochastic realizations using short data sets. *Probab. Eng. Mech.* **2003**, *18*, 61–77. [CrossRef]
127. Banfi, L.; Carassale, L. Uncertainties in an Application of Operational Modal Analysis. In *Model Validation and Uncertainty Quantification*; Springer: Berlin, Germany, 2016; Volume 3, pp. 107–115.
128. Marchesiello, S.; Bedaoui, S.; Garibaldi, L.; Argoul, P. Time-dependent identification of a bridge-like structure with crossing loads. *Mech. Syst. Signal Process.* **2009**, *23*, 2019–2028. [CrossRef]
129. Markovsky, I. The most powerful unfalsified model for data with missing values. *Syst. Control Lett.* **2016**, *95*, 53–61. [CrossRef]
130. Brownjohn, J.; Carden, P. Reliability of frequency and damping estimates from free vibration response. In Proceedings of the 2nd International Operational Modal Analysis Conference, Copenhagen, Denmark, 30 April–2 May 2007; pp. 23–30.
131. Carden, E.P.; Mita, A. Challenges in developing confidence intervals on modal parameters estimated for large civil infrastructure with stochastic subspace identification. *Struct. Control Health Monit.* **2011**, *18*, 53–78. [CrossRef]
132. ARTeMIS Cho, S.-H. Ambient vibration testing and system identification for tall buildings'. *J. Earthq. Eng. Soc. Korea* **2012**, *16*, 23–33.
133. Reynders, E.; Schevenels, M.; De Roeck, G. *MACEC 3.3: A Matlab Toolbox for Experimental and Operational Modal Analysis-User's Manual*; Kathol Univ Leuven: Leuven, Belgium, 2011.
134. Zhou, Y.; Prader, J.; Weidner, J.; Moon, F.; Aktan, A.E.; Zhang, J.; Yi, W.J. Structural Identification Study of a Steel Multi-Girder Bridge Based on Multiple Reference Impact Test. In Proceedings of the International Symposium on Innovation & Sustainability of Structures in Civil Engineering (ISISS-2013), Harbin, China, 6–7 July 2013.
135. Hu, W.-H.; Cunha, Á.; Caetano, E.; Magalhães, F.; Moutinho, C. LabVIEW toolkits for output-only modal identification and long-term dynamic structural monitoring. *Struct. Infrastruct. Eng.* **2010**, *6*, 557–574. [CrossRef]

136. Goursat, M.; Mevel, L. COSMAD: Identification and diagnosis for mechanical structures with Scilab. In Proceedings of the 2008 IEEE International Conference on Computer-Aided Control Systems, San Antonio, TX, USA, 3–5 September 2008; pp. 353–358.

137. Chang, M.; Pakzad, S.N.; Leonard, R. Modal identification using smit. In *Topics on the Dynamics of Civil Structures*; Springer: Berlin, Germnay, 2012; Volume 1, pp. 221–228.

138. Bayraktar, A.; Altunışık, A.C.; Sevim, B.; Türker, T. Seismic Response of a Historical Masonry Minaret using a Finite Element Model Updated with Operational Modal Testing. *J. Vib. Control.* **2010**, *17*, 129–149. [CrossRef]

139. Kudu, F.N.; Bayraktar, A.; Bakir, P.G.; Türker, T.; Altunişik, A.C. Ambient vibration testing of Berta Highway Bridge with post-tension tendons. *Steel Compos. Struct.* **2014**, *16*, 21–44. [CrossRef]

140. Sevim, B.; Altunisik, A.C.; Bayraktar, A. Structural identification of concrete arch dams by ambient vibration tests. *Adv. Concr. Constr.* **2013**, *1*, 227. [CrossRef]

141. Peeters, B.; Van der Auweraer, H.; Guillaume, P. The integration of operational modal analysis in vibration qualification testing. In Proceedings of the IMAC, Madrid, Spain, 17–20 September 2002; Volume 20, pp. 977–983.

142. Kołakowski, P.; Mroz, A.; Sala, D.; Pawłowski, P.; Sekuła, K.; Świercz, A. Investigation of Dynamic Response of a Railway Bridge Equipped with a Tailored SHM System. In *Key Engineering Materials*; Trans Tech Publications Ltd.: Stafa-Zurich, Switzerland, 2013; Volume 569, pp. 1068–1075.

143. Neu, E.; Janser, F.; Khatibi, A.A.; Orifici, A.C. Fully automated operational modal analysis using multi-stage clustering. *Mech. Syst. Signal Process.* **2017**, *84*, 308–323. [CrossRef]

144. Moher, D.; Liberati, A.; Tetzlaff, J.; Altman, D.G. Preferred reporting items for systematic reviews and meta-analyses: The PRISMA statement. *Ann. Intern. Med.* **2009**, *151*, 264–269. [CrossRef] [PubMed]

145. Budgen, D.; Brereton, P. Performing systematic literature reviews in software engineering. In Proceedings of the 28th International Conference on Software Engineering, Shanghai, China, 20–28 May 2006; pp. 1051–1052.

146. Liberati, A.; Altman, D.G.; Tetzlaff, J.; Mulrow, C.; Gøtzsche, P.C.; Ioannidis, J.P.; Clarke, M.; Devereaux, P.J.; Kleijnen, J.; Moher, D. The PRISMA statement for reporting systematic reviews and meta-analyses of studies that evaluate health care interventions: Explanation and elaboration. *PLoS Med.* **2009**, *6*, e1000100. [CrossRef] [PubMed]

147. Hughes-Morley, A.; Young, B.; Waheed, W.; Small, N.; Bower, P. Factors affecting recruitment into depression trials: Systematic review, meta-synthesis and conceptual framework. *J. Affect. Disord.* **2015**, *172*, 274–290. [CrossRef]

148. Consedine, N.S.; Tuck, N.L.; Ragin, C.R.; Spencer, B.A. Beyond the black box: A systematic review of breast, prostate, colorectal, and cervical screening among native and immigrant African-descent Caribbean populations. *J. Immigr. Minor. Health* **2015**, *17*, 905–924. [CrossRef]

149. Phillips, P.J.; Newton, E.M. Meta-analysis of face recognition algorithms. Autom. Face Gesture Recognition, 2002. In Proceedings of the Fifth IEEE International Conference on Automatic Face Gesture Recognition, Washington, DC, USA, 21 May 2002; pp. 235–241.

150. Higgins, J.P.T.; Green, S. *Cochrane Handbook for Systematic Reviews of Interventions*; The Cochrane Collaboration: London, UK, 2011.

151. Kurata, M.; Kim, J.; Lynch, J.P.; Van Der Linden, G.W.; Sedarat, H.; Thometz, E.; Hipley, P.; Sheng, L.H. Internet-Enabled Wireless Structural Monitoring Systems: Development and Permanent Deployment at the New Carquinez Suspension Bridge. *J. Struct. Eng.* **2013**, *139*, 1688–1702. [CrossRef]

152. Kim, J.; Lynch, J.P. Autonomous Decentralized System Identification by Markov Parameter Estimation Using Distributed Smart Wireless Sensor Networks. *J. Eng. Mech.* **2012**, *138*, 478–490. [CrossRef]

153. Nozari, A.; Behmanesh, I.; Yousefianmoghadam, S.; Moaveni, B.; Stavridis, A. Effects of variability in ambient vibration data on model updating and damage identification of a 10-story building. *Eng. Struct.* **2017**, *151*, 540–553. [CrossRef]

154. Shabbir, F.; Omenzetter, P. Particle Swarm Optimization with Sequential Niche Technique for Dynamic Finite Element Model Updating. *Comput. Civ. Infrastruct. Eng.* **2015**, *30*, 359–375. [CrossRef]

155. Foti, D.; Gattulli, V.; Potenza, F. Output-Only Identification and Model Updating by Dynamic Testing in Unfavorable Conditions of a Seismically Damaged Building. *Comput. Civ. Infrastruct. Eng.* **2014**, *29*, 659–675. [CrossRef]

156. Chen, J.-D.; Loh, C.-H. Tracking modal parameters of building structures from experimental studies and earthquake response measurements. *Struct. Health Monit.* **2017**, *16*, 551–567. [CrossRef]
157. Li, P.J.; Xia, Q.; Zhang, J. Subspace Flexibility Identification Adaptive to Different Types of Input Forces. *Int. J. Struct. Stab. Dyn.* **2017**, *18*, 1850067. [CrossRef]
158. Dai, K.; Wang, Y.; Huang, Y.; Zhu, W.; Xu, Y. Development of a modified stochastic subspace identification method for rapid structural assessment of in-service utility-scale wind turbine towers. *Wind Energy* **2017**, *20*, 1687–1710. [CrossRef]
159. Loh, C.H.; Liu, Y.C.; Ni, Y.Q. SSA-based stochastic subspace identification of structures from output-only vibration measurements. *Smart Struct. Syst.* **2012**, *10*, 331–351. [CrossRef]
160. Attig, M.; Abdelghani, M.; Kahla, N.B. Output-only modal identification of tensegrity structures. *Eng. Struct. Technol.* **2016**, *8*, 52–64. [CrossRef]
161. Reynders, E.; Pintelon, R.; De Roeck, G. Uncertainty bounds on modal parameters obtained from stochastic subspace identification. *Mech. Syst. Signal Process.* **2008**, *22*, 948–969. [CrossRef]
162. Zhang, Z.; Fan, J.; Hua, H. Simulation and experiment of a blind subspace identification method. *J. Sound Vib.* **2008**, *311*, 941–952. [CrossRef]
163. Carden, E.P.; Brownjohn, J.M.W. Fuzzy Clustering of Stability Diagrams for Vibration-Based Structural Health Monitoring. *Comput. Civ. Infrastruct. Eng.* **2008**, *23*, 360–372. [CrossRef]
164. Döhler, M.; Mevel, L.; Hille, F. Subspace-based damage detection under changes in the ambient excitation statistics. *Mech. Syst. Signal Process.* **2014**, *45*, 207–224. [CrossRef]
165. Basseville, M.; Bourquin, F.; Mevel, L.; Nasser, H.; Treyssède, F. Handling the temperature effect in vibration monitoring: Two subspace-based analytical approaches. *J. Eng. Mech.* **2010**, *136*, 367. [CrossRef]
166. Balmès, É.; Basseville, M.; Mevel, L.; Nasser, H. Handling the temperature effect in vibration monitoring of civil structures: A combined subspace-based and nuisance rejection approach. *Control Eng. Pract.* **2009**, *17*, 80–87. [CrossRef]
167. Kurka, P.R.G.; Cambraia, H.N. Application of a multivariable input–output subspace identification technique in structural analysis. *J. Sound Vib.* **2008**, *312*, 461–475. [CrossRef]
168. Zhou, W.; Li, S.; Li, H. Damage Detection for SMC Benchmark Problem: A Subspace-Based Approach. *Int. J. Struct. Stab. Dyn.* **2016**, *16*, 1640025. [CrossRef]
169. Ren, W.X.; Lin, Y.Q.; Fang, S.E. Structural damage detection based on stochastic subspace identification and statistical pattern recognition: I. Theory. *Smart Mater. Struct.* **2011**, *20*, 115009. [CrossRef]
170. Balmes, E.; Basseville, M.; Mevel, L.; Nasser, H.; Zhou, W. Statistical model-based damage localization: A combined subspace-based and substructuring approach. *Struct. Control Health Monit.* **2008**, *15*, 857–875. [CrossRef]
171. Akçay, H. Frequency domain subspace-based identification of discrete-time singular power spectra. *Signal Process.* **2012**, *92*, 2075–2081. [CrossRef]
172. Gandino, E.; Garibaldi, L.; Marchesiello, S. Covariance-driven subspace identification: A complete input–output approach. *J. Sound Vib.* **2013**, *332*, 7000–7017. [CrossRef]
173. Verhaegen, M.; Hansson, A. N2SID: Nuclear norm subspace identification of innovation models. *Automatica* **2016**, *72*, 57–63. [CrossRef]
174. Priori, C.; De Angelis, M.; Betti, R. On the selection of user-defined parameters in data-driven stochastic subspace identification. *Mech. Syst. Signal Process.* **2018**, *100*, 501–523. [CrossRef]
175. Pioldi, F.; Rizzi, E. Earthquake-induced structural response output-only identification by two different Operational Modal Analysis techniques. *Earthq. Eng. Struct. Dyn.* **2017**, *47*, 257–264. [CrossRef]
176. Park, S.-K.; Noh Hae, Y. Updating Structural Parameters with Spatially Incomplete Measurements Using Subspace System Identification. *J. Eng. Mech.* **2017**, *143*, 4017040. [CrossRef]
177. Soria, J.; Díaz, I.M.; García-Palacios, J.H.; Ibán, N. Vibration Monitoring of a Steel-Plated Stress-Ribbon Footbridge: Uncertainties in the Modal Estimation. *J. Bridge Eng.* **2016**, *21*, C5015002. [CrossRef]
178. Loh, C.-H.; Chao, S.-H.; Weng, J.-H.; Wu, T.-H. Application of subspace identification technique to long-term seismic response monitoring of structures. *Earthq. Eng. Struct. Dyn.* **2015**, *44*, 385–402. [CrossRef]
179. Lardies, J. Modal Parameter Identification from Output Data Only: Equivalent Approaches. *Shock Vib.* **2015**, *2015*, 10. [CrossRef]
180. Junhee, K.; Kiyoung, K.; Hoon, S. Subspace model identification of guided wave propagation in metallic plates. *Smart Mater. Struct.* **2014**, *23*, 35006.

181. Yu, K.; Yang, K.; Bai, Y. Experimental investigation on the time-varying modal parameters of a trapezoidal plate in temperature-varying environments by subspace tracking-based method. *J. Vib. Control* **2014**, *21*, 3305–3319. [CrossRef]
182. Ubertini, F.; Gentile, C.; Materazzi, A.L. Automated modal identification in operational conditions and its application to bridges. *Eng. Struct.* **2013**, *46*, 264–278. [CrossRef]
183. Lardies, J.; Minh-Ngi, T. Modal parameter identification of stay cables from output-only measurements. *Mech. Syst. Signal Process.* **2011**, *25*, 133–150. [CrossRef]
184. Brownjohn, J.M.W.; Carden, E.P.; Goddard, C.R.; Oudin, G. Real-time performance monitoring of tuned mass damper system for a 183m reinforced concrete chimney. *J. Wind Eng. Ind. Aerodyn.* **2010**, *98*, 169–179. [CrossRef]
185. Deraemaeker, A.; Reynders, E.; De Roeck, G.; Kullaa, J. Vibration-based structural health monitoring using output-only measurements under changing environment. *Mech. Syst. Signal Process.* **2008**, *22*, 34–56. [CrossRef]
186. He, X.; Moaveni, B.; Conte, J.P.; Elgamal, A.; Masri, S.F. Modal Identification Study of Vincent Thomas Bridge Using Simulated Wind-Induced Ambient Vibration Data. *Comput. Civ. Infrastruct. Eng.* **2008**, *23*, 373–388. [CrossRef]
187. Reynders, E.; Maes, K.; Lombaert, G.; De Roeck, G. Uncertainty quantification in operational modal analysis with stochastic subspace identification: Validation and applications. *Mech. Syst. Signal Process.* **2016**, *66*, 13–30. [CrossRef]
188. Wu, W.H.; Wang, S.W.; Chen, C.C.; Lai, G. Application of stochastic subspace identification for stay cables with an alternative stabilization diagram and hierarchical sifting process. *Struct. Control Health Monit.* **2016**, *23*, 1194–1213. [CrossRef]
189. Karami, K.; Akbarabadi, S. Developing a Smart Structure Using Integrated Subspace-Based Damage Detection and Semi-Active Control. *Comput. Civ. Infrastruct. Eng.* **2016**, *31*, 887–903. [CrossRef]
190. Khan, I.U.; Shan, D.S.; Li, Q.; He, J.; Nan, F.L. Data interpretation and continuous modal parameter identification of cable stayed bridge. *Open Civ. Eng. J.* **2015**, *9*, 577–591. [CrossRef]
191. Guo, Y.; Kwon, D.K.; Kareem, A. Near-Real-Time Hybrid System Identification Framework for Civil Structures with Application to Burj Khalifa. *J. Struct. Eng.* **2015**, *142*, 4015132. [CrossRef]
192. Ben Mekki, O.; Siegert, D.; Toutlemonde, F.; Mevel, L.; Goursat, M. A New Composite Bridge: Feasibility Validation and Vibration Monitoring. *Mech. Adv. Mater. Struct.* **2015**, *22*, 850–863. [CrossRef]
193. Dohler, M.; Hille, F.; Mevel, L.; Rucker, W. Structural health monitoring with statistical methods during progressive damage test of S101 Bridge. *Eng. Struct.* **2014**, *69*, 183–193. [CrossRef]
194. Chauhan, S. Using the unified matrix polynomial approach (UMPA) for the development of the stochastic subspace identification (SSI) algorithm. *J. Vib. Control* **2012**, *19*, 1950–1961. [CrossRef]
195. Whelan, M.J.; Gangone, M.V.; Janoyan, K.D.; Jha, R. Real-time wireless vibration monitoring for operational modal analysis of an integral abutment highway bridge. *Eng. Struct.* **2009**, *31*, 2224–2235. [CrossRef]
196. Marchesiello, S.; Fasana, A.; Garibaldi, L. Modal contributions and effects of spurious poles in nonlinear subspace identification. *Mech. Syst. Signal Process.* **2016**, *74*, 111–132. [CrossRef]
197. Shi, Y.; Li, Z.; Chang, C.C. Output-only subspace identification of structural properties and unknown ground excitation for shear-beam buildings. *Adv. Mech. Eng.* **2016**, *8*, 1687814016679908. [CrossRef]
198. Al-Gahtani, O.; El-Gebeily, M.; Khulief, Y. Output-Only Identification of System Parameters from Noisy Measurements by Multiwavelet Denoising. *Adv. Mech. Eng.* **2014**, *6*, 218328. [CrossRef]
199. Weng, J.H.; Loh, C.H.; Yang, J.N. Experimental Study of Damage Detection by Data-Driven Subspace Identification and Finite-Element Model Updating. *J. Struct. Eng.* **2009**, *135*, 1533–1544. [CrossRef]

MDPI

St. Alban-Anlage 66

4052 Basel

Switzerland

Tel. +41 61 683 77 34

Fax +41 61 302 89 18

www.mdpi.com

Applied Sciences Editorial Office

E-mail: applsci@mdpi.com

www.mdpi.com/journal/applsci

9 783036 524047